Linear Algebra: Concepts and Methods

Any student of linear algebra will welcome this textbook, which provides a thorough treatment of this key topic. Blending practice and theory, the book enables the reader to learn and comprehend the standard methods, with an emphasis on understanding how they actually work. At every stage, the authors are careful to ensure that the discussion is no more complicated or abstract than it needs to be, and focuses on the most fundamental topics.

- Hundreds of examples and exercises, including solutions, give students plenty of hands-on practice
- End-of-chapter sections summarise material to help students consolidate their learning
- Ideal as a course text and for self-study
- Instructors can use the many examples and exercises to supplement their own assignments
- Both authors have extensive experience of undergraduate teaching and of preparation of distance learning materials.

Martin Anthony is Professor of Mathematics at the London School of Economics (LSE), and Academic Coordinator for Mathematics on the University of London International Programmes for which LSE has academic oversight. He has over 20 years' experience of teaching students at all levels of university, and is the author of four books, including (with N. L. Biggs) the textbook *Mathematics for Economics and Finance: Methods and Modelling* (Cambridge University Press, 1996). He also has extensive experience of preparing distance learning materials.

Michele Harvey lectures at the London School of Economics, where she has taught and developed the linear algebra part of the core foundation course in mathematics for over 20 years. Her dedication to helping students learn mathematics has been widely recognised. She is also Chief Examiner for the Advanced Linear Algebra course on the University of London International Programmes and has co-authored with Martin Anthony the study guides for Advanced Linear Algebra and Linear Algebra on this programme.

Linear Algebra: Concepts and Methods

MARTIN ANTHONY and MICHELE HARVEY

Department of Mathematics
The London School of Economics and Political Science

CAMBRIDGE
UNIVERSITY PRESS

CAMBRIDGE
UNIVERSITY PRESS

Shaftesbury Road, Cambridge CB2 8EA, United Kingdom

One Liberty Plaza, 20th Floor, New York, NY 10006, USA

477 Williamstown Road, Port Melbourne, VIC 3207, Australia

314–321, 3rd Floor, Plot 3, Splendor Forum, Jasola District Centre, New Delhi – 110025, India

103 Penang Road, #05–06/07, Visioncrest Commercial, Singapore 238467

Cambridge University Press is part of Cambridge University Press & Assessment, a department of the University of Cambridge.

We share the University's mission to contribute to society through the pursuit of education, learning and research at the highest international levels of excellence.

www.cambridge.org
Information on this title: www.cambridge.org/9780521279482

© Cambridge University Press & Assessment 2012

First published 2012
10th printing 2021

A catalogue record for this publication is available from the British Library

ISBN   978-0-521-27948-2   Paperback

To Colleen, Alistair, and my parents. And, just for Alistair, here's one of those sideways moustaches: }

(MA)

To Bill, for his support throughout, and to my father, for his encouragement to study mathematics.

(MH)

Contents

Preface

Linear algebra is one of the core topics studied at university level by students on many different types of degree programme. Alongside calculus, it provides the framework for mathematical modelling in many diverse areas. This text sets out to introduce and *explain* linear algebra to students from any discipline. It covers all the material that would be expected to be in most first-year university courses in the subject, together with some more advanced material that would normally be taught later.

The book has drawn on our extensive experience over a number of years in teaching first- and second-year linear algebra to LSE undergraduates and in providing self-study material for students studying at a distance. This text represents our best effort at distilling from our experience what it is that we think works best in helping students not only to *do* linear algebra, but to *understand* it. We regard *understanding* as essential. 'Understanding' is not some fanciful intangible, to be dismissed because it does not constitute a 'demonstrable learning outcome': it is at the heart of what higher education (rather than merely more education) is about. Linear algebra is a coherent, and beautiful, part of mathematics: manipulation of matrices and vectors leads, with a dash of abstraction, to the underlying concepts of vector spaces and linear transformations, in which contexts the more mechanical, manipulative, aspects of the subject *make sense*. It is worth striving for understanding, not only because of the inherent intellectual satisfaction, but because it pays off in other ways: it helps a student to work with the methods and techniques because he or she knows *why* these work and *what they mean.*

Large parts of the material in this book have been adapted and developed from lecture notes prepared by MH for the Mathematical Methods course at the LSE, a long-established course which has a large audience, and which has evolved over many years. Other parts have been influenced by MA's teaching of non-specialist first-year courses and second-year linear algebra. Both of us have written self-study materials for students; some of the book is based on material originally produced by us for the programmes in economics, management, finance and the social sciences by distance and flexible learning offered by the University of London International Programmes (www.londoninternational. ac.uk).

We have attempted to write a user-friendly, fairly interactive and helpful text, and we intend that it could be useful not only as a course text, but for self-study. To this end, we have written in what we hope is an open and accessible – sometimes even conversational – style, and have included 'learning outcomes' and many 'activities' and 'exercises'. We have also provided a very short introduction just to indicate some of the background which a reader should, ideally, possess (though if some of that is lacking, it can easily be acquired in passing).

Reading a mathematics book properly cannot be a passive activity: the reader should interrogate the text and have pen and paper at the ready to check things. To help in this, the chapters contain many activities – prompts to a reader to be an 'active' reader, to pause for thought and really make sure they understand what has just been written, or to think ahead and anticipate what is to come next. At the end of chapters, there are comments on most of the activities, which a reader can consult to confirm his or her understanding.

The main text of each chapter ends with a brief list of 'learning outcomes'. These are intended to highlight the main aspects of the chapter, to help a reader review and consolidate what has been read.

There are carefully designed exercises towards the end of each chapter, with full solutions (not just brief answers) provided at the end of the book. These exercises vary in difficulty from the routine to the more challenging, and they are one of the key ingredients in helping a reader check his or her understanding of the material. Of course, these are best made use of by attempting them seriously before consulting the solution. (It's all very easy to read and agree with a solution, but unless you have truly grappled with the exercise, the benefits of doing so will be limited.)

We also provide sets of additional exercises at the end of each chapter, which we call Problems as the solutions are not given. We hope they will be useful for assignments by teachers using this book, who will be able to obtain solutions from the book's webpage. Students will gain confidence by tackling, and solving, these problems, and will be able to check many of their answers using the techniques given in the chapter.

Over the years, many people – students and colleagues – have influenced and informed the way we approach the teaching of linear algebra, and we thank them all.

Preliminaries: before we begin

This short introductory chapter discusses some very basic aspects of mathematics and mathematical notation that it would be useful to be comfortable with before proceeding. We imagine that you have studied most (if not all) of these topics in previous mathematics courses and that nearly all of the material is revision, but don't worry if a topic is new to you. We will mention the main results which you will need to know. If you are unfamiliar with a topic, or if you find any of the topics difficult, then you should look up that topic in any basic mathematics text.

Sets and set notation

A set may be thought of as a collection of objects. A set is usually described by listing or describing its *members* inside curly brackets. For example, when we write $A = \{1, 2, 3\}$, we mean that the objects belonging to the set A are the numbers $1, 2, 3$ (or, equivalently, the set A consists of the numbers $1, 2$ and 3). Equally (and this is what we mean by 'describing' its members), this set could have been written as

$$A = \{n \mid n \text{ is a whole number and } 1 \leq n \leq 3\}.$$

Here, the symbol $\mid$ stands for 'such that'. (Sometimes, the symbol ':' is used instead.) As another example, the set

$$B = \{x \mid x \text{ is a reader of this book}\}$$

has as its members all of you (and nothing else). When x is an object in a set A, we write $x \in A$ and say 'x belongs to A' or 'x is a member of A'.

The set which has no members is called the *empty set* and is denoted by Ø. The empty set may seem like a strange concept, but it has its uses.

We say that the set S is a *subset* of the set T, and we write $S \subseteq T$, or $S \subset T$, if every member of S is a member of T. For example, $\{1, 2, 5\} \subseteq \{1, 2, 4, 5, 6, 40\}$. The difference between the two symbols is that $S \subset T$ means that S is a *proper subset* of T, meaning not all of T, and $S \subseteq T$ means that S is a subset of T and possibly (but not necessarily) all of T. So in the example just given we could have also written $\{1, 2, 5\} \subset \{1, 2, 4, 5, 6, 40\}$.

Given two sets A and B, the *union* $A \cup B$ is the set whose members belong to A or B (or both A and B); that is,

$$A \cup B = \{x \mid x \in A \text{ or } x \in B\}.$$

For example, if $A = \{1, 2, 3, 5\}$ and $B = \{2, 4, 5, 7\}$, then $A \cup B = \{1, 2, 3, 4, 5, 7\}$.

Similarly, we define the *intersection* $A \cap B$ to be the set whose members belong to both A and B:

$$A \cap B = \{x \mid x \in A \text{ and } x \in B\}.$$

So, if $A = \{1, 2, 3, 5\}$ and $B = \{2, 4, 5, 7\}$, then $A \cap B = \{2, 5\}$.

Numbers

There are some standard notations for important sets of numbers. The set $\mathbb{R}$ of *real numbers*, the 'normal' numbers you are familiar with, may be thought of as the points on a line. Each such number can be described by a decimal representation.

The set of real numbers $\mathbb{R}$ includes the following subsets: $\mathbb{N}$, the set of natural numbers, $\mathbb{N} = \{1, 2, 3, \dots\}$, also referred to as the positive integers; $\mathbb{Z}$, the set of all integers, $\{\dots, -3, -2, -1, 0, 1, 2, 3, \dots\}$; and $\mathbb{Q}$, the set of rational numbers, which are numbers that can be written as fractions, p/q, with $p, q \in \mathbb{Z}$, $q \neq 0$. In addition to the real numbers, there is the set $\mathbb{C}$ of *complex* numbers. You may have seen these before, but don't worry if you have not; we cover the basics at the start of Chapter 13, when we need them.

The *absolute value* of a real number a is defined by

$$|a| = \begin{cases} a & \text{if } a \geq 0 \\ -a & \text{if } a \leq 0 \end{cases}.$$

So the absolute value of a equals a if a is non-negative (that is, if $a \geq 0$), and equals $-a$ otherwise. For instance, $|6| = 6$ and $|-2.5| = 2.5$. Note that

$$\sqrt{a^2} = |a|,$$

since by $\sqrt{x}$ we always mean the non-negative square root to avoid ambiguity. So the two solutions of the equation $x^2 = 4$ are $x = \pm 2$ (meaning $x = 2$ or $x = -2$), but $\sqrt{4} = 2$.

The absolute value of real numbers satisfies the following inequality:

$$|a + b| \le |a| + |b|, \qquad a, b \in \mathbb{R}.$$

Having defined $\mathbb{R}$, we can define the set $\mathbb{R}^2$ of *ordered pairs* (x, y) of real numbers. Thus, $\mathbb{R}^2$ is the set usually depicted as the set of points in a plane, x and y being the coordinates of a point with respect to a pair of axes. For instance, $(-1, 3/2)$ is an element of $\mathbb{R}^2$ lying to the left of and above $(0, 0)$, which is known as the *origin*.

Mathematical terminology

In this book, as in most mathematics texts, we use the words 'definition', 'theorem' and 'proof', and it is important not to be daunted by this language if it is unusual to you. A definition is simply a precise statement of what a particular idea or concept means. Definitions are hugely important in mathematics, because it is a precise subject. A theorem is just a statement or result. A proof is an explanation as to why a theorem is true. As a fairly trivial example, consider the following:

Definition: An integer n is *even* if it is a multiple of 2; that is, if $n = 2k$ for some integer k.

Note that this is a precise statement *telling us* what the word 'even' means. It is *not* to be taken as a 'result': it's defining what the word 'even' means.

Theorem: *The sum of two even integers is even. That is, if m, n are even, so is $m + n$.*

Proof: Suppose m, n are even. Then, by the definition, there are integers k, l such that $m = 2k$ and $n = 2l$. Then

$$m + n = 2k + 2l = 2(k + l).$$

Since $k + l$ is an integer, it follows that $m + n$ is even. $\square$

Note that, as here, we often use the symbol $\square$ to denote the end of a proof. This is just to make it clear where the proof ends and the following text begins.

Occasionally, we use the term 'corollary'. A corollary is simply a result that is a consequence of a theorem and perhaps isn't 'big' enough to be called a theorem in its own right.

Don't worry about this terminology if you haven't met it before. It will become familiar as you work through the book.

Basic algebra

Algebraic manipulation

You should be capable of manipulating simple algebraic expressions and equations.

You should be proficient in:

- collecting up terms; for example, $2a + 3b - a + 5b = a + 8b$
- multiplication of variables; for example,

$$a(-b) - 3ab + (-2a)(-4b) = -ab - 3ab + 8ab = 4ab$$

- expansion of bracketed terms; for example,

$$-(a - 2b) = -a + 2b,$$
$$(2x - 3y)(x + 4y) = 2x^2 - 3xy + 8xy - 12y^2$$
$$= 2x^2 + 5xy - 12y^2.$$

Powers

When n is a positive integer, the nth *power* of the number a, denoted a^n, is simply the product of n copies of a; that is,

$$a^n = \underbrace{a \times a \times a \times \cdots \times a}_{n \text{ times}}.$$

The number n is called the *power*, *exponent* or *index*. We have the *power rules* (or *rules of exponents*),

$$a^r a^s = a^{r+s}, \quad (a^r)^s = a^{rs},$$

whenever r and s are positive integers.

The power a^0 is defined to be 1.

The definition is extended to negative integers as follows. When n is a positive integer, a^{-n} means $1/a^n$. For example, 3^{-2} is $1/3^2 = 1/9$. The power rules hold when r and s are any integers, positive, negative or zero.

When n is a positive integer, $a^{1/n}$ is the positive nth root of a; this is the positive number x such that $x^n = a$. For example, $a^{1/2}$ is usually denoted by $\sqrt{a}$, and is the positive *square root* of a, so that $4^{1/2} = 2$.

When m and n are integers and n is positive, $a^{m/n}$ is $(a^{1/n})^m$. This extends the definition of powers to the rational numbers (numbers which can be written as fractions). The definition is extended to real numbers by 'filling in the gaps' between the rational numbers, and it can be shown that the rules of exponents still apply.

Quadratic equations

It is straightforward to find the *solution* of a linear equation, one of the form $ax + b = 0$ where $a, b \in \mathbb{R}$. By a solution, we mean a real number x for which the equation is true.

A common problem is to find the set of solutions of a *quadratic* equation

$$ax^2 + bx + c = 0,$$

where we may as well assume that $a \neq 0$, because if $a = 0$ the equation reduces to a linear one. In some cases, the quadratic expression can be factorised, which means that it can be written as the product of two linear terms. For example,

$$x^2 - 6x + 5 = (x - 1)(x - 5),$$

so the equation $x^2 - 6x + 5 = 0$ becomes $(x - 1)(x - 5) = 0$. Now, the only way that two numbers can multiply to give 0 is if at least one of the numbers is 0, so we can conclude that $x - 1 = 0$ or $x - 5 = 0$; that is, the equation has two solutions, 1 and 5.

Although factorisation may be difficult, there is a general method for determining the solutions to a quadratic equation using the *quadratic formula*, as follows. Suppose we have the quadratic equation $ax^2 + bx + c = 0$, where $a \neq 0$. Then the solutions of this equation are

$$x_1 = \frac{-b - \sqrt{b^2 - 4ac}}{2a} \qquad x_2 = \frac{-b + \sqrt{b^2 - 4ac}}{2a}.$$

The term $b^2 - 4ac$ is called the *discriminant*.

- If $b^2 - 4ac > 0$, the equation has *two* real solutions as given above.
- If $b^2 - 4ac = 0$, the equation has *exactly one* solution, $x = -b/(2a)$. (In this case, we say that this is a solution of multiplicity two.)

- If $b^2 - 4ac < 0$, the equation has *no* real solutions. (It will have complex solutions, but we explain this in Chapter 13.)

For example, consider the equation $2x^2 - 7x + 3 = 0$. Using the quadratic formula, we have

$$x = \frac{-b \pm \sqrt{b^2 - 4ac}}{2a} = \frac{7 \pm \sqrt{49 - 4(2)(3)}}{2(2)} = \frac{7 \pm 5}{4}.$$

So the solutions are $x = 3$ and $x = \frac{1}{2}$.

The equation $x^2 + 6x + 9 = 0$ has one solution of multiplicity 2; its discriminant is $b^2 - 4ac = 36 - 9(4) = 0$. This equation is most easily solved by recognising that $x^2 + 6x + 9 = (x + 3)^2$, so the solution is $x = -3$.

On the other hand, consider the quadratic equation

$$x^2 - 2x + 3 = 0;$$

here we have $a = 1, b = -2, c = 3$. The quantity $b^2 - 4ac$ is negative, so this equation has no real solutions. This is less mysterious than it may seem. We can write the equation as $(x - 1)^2 + 2 = 0$. Rewriting the left-hand side of the equation in this form is known as *completing the square*. Now, the square of a number is always greater than or equal to 0, so the quantity on the left of this equation is always at least 2 and is therefore never equal to 0. The quadratic formula for the solutions to a quadratic equation is obtained using the technique of completing the square. Quadratic polynomials which cannot be written as a product of linear terms (so ones for which the discriminant is negative) are said to be *irreducible*.

Polynomial equations

A polynomial of degree n in x is an expression of the form

$$P_n(x) = a_0 + a_1 x + a_2 x^2 + \cdots + a_n x^n,$$

where the a_i are real constants, $a_n \neq 0$, and x is a real variable. For example, a quadratic expression such as those discussed above is a polynomial of degree 2.

A polynomial equation of degree n has at most n solutions. For example, since

$$x^3 - 7x + 6 = (x - 1)(x - 2)(x + 3),$$

the equation $x^3 - 7x + 6 = 0$ has three solutions; namely, $1, 2, -3$. The solutions of the equation $P_n(x) = 0$ are called the *roots* or *zeros* of the polynomial. Unfortunately, there is no general straightforward formula (as there is for quadratics) for the solutions to $P_n(x) = 0$ for polynomials P_n of degree larger than 2.

To find the solutions to $P(x) = 0$, where P is a polynomial of degree n, we use the fact that if α is such that $P(\alpha) = 0$, then $(x - \alpha)$ must be a factor of $P(x)$. We find such an a by trial and error and then write $P(x)$ in the form $(x - \alpha)Q(x)$, where $Q(x)$ is a polynomial of degree $n - 1$.

As an example, we'll use this method to factorise the cubic polynomial $x^3 - 7x + 6$. Note that if this polynomial can be expressed as a product of linear factors, then it will be of the form

$$x^3 - 7x + 6 = (x - r_1)(x - r_2)(x - r_3),$$

where its constant term is the product of the roots: $6 = -r_1 r_2 r_3$. (To see this, just substitute $x = 0$ into both sides of the above equation.) So if there is an integer root, it will be a factor of 6. We will try $x = 1$. Substituting this value for x, we do indeed get $1 - 7 + 6 = 0$, so $(x - 1)$ is a factor. Then we can deduce that

$$x^3 - 7x + 6 = (x - 1)(x^2 + \lambda x - 6)$$

for some number λ, as the coefficient of x^2 must be 1 for the product to give x^3, and the constant term must be -6 so that $(-1)(-6) = 6$, the constant term in the cubic. It only remains to find λ. This is accomplished by comparing the coefficients of either x^2 or x in the cubic polynomial and the product. The coefficient of x^2 in the cubic is 0, and in the product the coefficient of x^2 is obtained from the terms $(-1)(x^2) + (x)(\lambda x)$, so that we must have $\lambda - 1 = 0$ or $\lambda = 1$. Then

$$x^3 - 7x + 6 = (x - 1)(x^2 + x - 6),$$

and the quadratic term is easily factorised into $(x - 2)(x + 3)$; that is,

$$x^3 - 7x + 6 = (x - 1)(x - 2)(x + 3).$$

Trigonometry

The trigonometrical functions, $\sin \theta$ and $\cos \theta$ (the *sine function* and *cosine function*), are very important in mathematics. You should know their geometrical meaning. (In a right-angled triangle, $\sin \theta$ is the ratio of the length of the side opposite the angle θ to the length of the hypotenuse, the longest side of the triangle; and $\cos \theta$ is the ratio of the length of the side adjacent to the angle to the length of the hypotenuse.)

It is important to realise that throughout this book angles are measured in *radians* rather than *degrees*. The conversion is as follows: 180 degrees equals π radians, where π is the number $3.141\ldots$ It is good practice *not* to expand π or multiples of π as decimals, but to leave them in terms of the symbol π. For example, since 60 degrees is one-third of 180 degrees, it follows that in radians 60 degrees is $\pi/3$.

The sine and cosine functions are related by the fact that $\cos x = \sin(x + \frac{\pi}{2})$, and they always take a value between 1 and -1. Table 1 gives some important values of the trigonometrical functions.

There are some useful results about the trigonometrical functions, which we use now and again. In particular, for any angles θ and ϕ, we have

$$\sin^2\theta + \cos^2\theta = 1,$$
$$\sin(\theta + \phi) = \sin\theta\cos\phi + \cos\theta\sin\phi$$

and

$$\cos(\theta + \phi) = \cos\theta\cos\phi - \sin\theta\sin\phi.$$

Table 1

θ	$\sin\theta$	$\cos\theta$
0	0	1
$\pi/6$	1/2	$\sqrt{3}/2$
$\pi/4$	$1/\sqrt{2}$	$1/\sqrt{2}$
$\pi/3$	$\sqrt{3}/2$	1/2
$\pi/2$	1	0

A little bit of logic

It is very important to understand the formal meaning of the word 'if' in mathematics. The word is often used rather sloppily in everyday life, but has a very precise mathematical meaning. Let's give an example. Suppose someone tells you 'If it rains, then I wear a raincoat', and suppose that this is a true statement. Well, then suppose it rains. You can certainly conclude the person will wear a raincoat. But what if it does not rain? Well, you can't conclude anything. The statement only tells you about what happens *if* it rains. If it does not, then the person might, or might not, wear a raincoat. You have to be clear about this: an 'if–then' statement only tells you about what follows *if* something particular happens.

More formally, suppose P and Q are mathematical statements (each of which can therefore be either true or false). Then we can form the statement denoted $P \implies Q$ ('P implies Q' or, equivalently, 'if P, then Q'), which means 'if P is true, then Q is true'. For instance, consider the theorem we used as an example earlier. This says that if m, n are even integers, then so is $m + n$. We can write this as

$$m, n \text{ even integers} \implies m + n \text{ is even.}$$

The *converse* of a statement $P \implies Q$ is $Q \implies P$ and whether that is true or not is a separate matter. For instance, the converse of the statement just made is

$$m + n \text{ is even} \implies m, n \text{ even integers.}$$

This is *false*. For instance, $1 + 3$ is even, but 1 and 3 are not.

If, however, both statements $P \implies Q$ and $Q \implies P$ are true, then we say that Q is true *if and only if* P is. Alternatively, we say that P and Q are *equivalent*. We use the single piece of notation $P \iff Q$ instead of the two separate $P \implies Q$ and $Q \implies P$.

1

Matrices and vectors

Matrices and vectors will be the central objects in our study of linear algebra. In this chapter, we introduce matrices, study their properties and learn how to manipulate them. This will lead us to a study of vectors, which can be thought of as a certain type of matrix, but which can more usefully be viewed geometrically and applied with great effect to the study of lines and planes.

1.1   What is a matrix?

Definition 1.1 (Matrix) A matrix is a rectangular array of numbers or symbols. It can be written as

$$A = \begin{pmatrix} a_{11} & a_{12} & \cdots & a_{1n} \\ a_{21} & a_{22} & \cdots & a_{2n} \\ \vdots & \vdots & \ddots & \vdots \\ a_{m1} & a_{m2} & \cdots & a_{mn} \end{pmatrix}.$$

We denote this array by the single letter A or by (a_{ij}), and we say that A has m rows and n columns, or that it is an $m \times n$ matrix. We also say that A is a matrix of *size* $m \times n$.

The number a_{ij} in the ith row and jth column is called the (i, j) entry. Note that the first subscript on a_{ij} always refers to the row and the second subscript to the column.

Example 1.2 The matrix

$$A = \begin{pmatrix} 2 & 1 & 7 & 8 \\ 0 & -2 & 5 & -1 \\ 4 & 9 & 3 & 0 \end{pmatrix}$$

is a 3×4 matrix whose entries are integers. For this matrix, $a_{23} = 5$, since this is the entry in the second row and third column.

Activity 1.3 In Example 1.2 above, what is a_{32}?

A *square* matrix is an $n \times n$ matrix; that is, a matrix with the same number of rows as columns. The *diagonal* of a square matrix is the list of entries $a_{11}, a_{22}, \ldots, a_{nn}$.

A *diagonal* matrix is a square matrix with all the entries which are not on the diagonal equal to 0. So A is diagonal if it is $n \times n$ and $a_{ij} = 0$ if $i \neq j$. Then A looks as follows:

$$\begin{pmatrix} a_{11} & 0 & \cdots & 0 \\ 0 & a_{22} & \cdots & 0 \\ \vdots & \vdots & \ddots & \vdots \\ 0 & 0 & \cdots & a_{nn} \end{pmatrix}.$$

Activity 1.4 Which of these matrices are diagonal?

$$\begin{pmatrix} -3 & 0 & 0 \\ 0 & 2 & 1 \\ 0 & 0 & 1 \end{pmatrix}, \quad \begin{pmatrix} 0 & 0 & 0 \\ 0 & -1 & 0 \\ 0 & 0 & 2 \end{pmatrix}, \quad \begin{pmatrix} 2 & 0 & 0 \\ 0 & 1 & 0 \end{pmatrix}.$$

Definition 1.5 (Equality) Two matrices are *equal* if they are the same size and if corresponding entries are equal. That is, if $A = (a_{ij})$ and $B = (b_{ij})$ are both $m \times n$ matrices, then

$$A = B \iff a_{ij} = b_{ij} \qquad 1 \le i \le m, \ 1 \le j \le n.$$

1.2 Matrix addition and scalar multiplication

If A and B are two matrices, then provided they are the same size we can add them together to form a new matrix $A + B$. We define $A + B$ to be the matrix whose entries are the sums of the corresponding entries in A and B.

Definition 1.6 (Addition) If $A = (a_{ij})$ and $B = (b_{ij})$ are both $m \times n$ matrices, then

$$A + B = (a_{ij} + b_{ij}) \qquad 1 \le i \le m, \ 1 \le j \le n.$$

We can also multiply any matrix by a real number, referred to as a *scalar* in this context. If λ is a scalar and A is a matrix, then λA is the matrix whose entries are λ times each of the entries of A.

Definition 1.7 (Scalar multiplication) If $A = (a_{ij})$ is an $m \times n$ matrix and $\lambda \in \mathbb{R}$, then

$$\lambda A = (\lambda a_{ij}) \qquad 1 \le i \le m, \ 1 \le j \le n.$$

Example 1.8

$$A + B = \begin{pmatrix} 3 & 1 & 2 \\ 0 & 5 & -2 \end{pmatrix} + \begin{pmatrix} -1 & 1 & 4 \\ 2 & -3 & 1 \end{pmatrix} = \begin{pmatrix} 2 & 2 & 6 \\ 2 & 2 & -1 \end{pmatrix}$$

$$-2A = -2 \begin{pmatrix} 3 & 1 & 2 \\ 0 & 5 & -2 \end{pmatrix} = \begin{pmatrix} -6 & -2 & -4 \\ 0 & -10 & 4 \end{pmatrix}.$$

1.3   Matrix multiplication

Is there a way to multiply two matrices together? The answer is sometimes, depending on the sizes of the matrices. If A and B are matrices such that the number of columns of A is equal to the number of rows of B, then we can define a matrix C which is the product of A and B. We do this by saying what the entry c_{ij} of the product matrix AB should be.

Definition 1.9 (Matrix multiplication) If A is an $m \times n$ matrix and B is an $n \times p$ matrix, then the product is the matrix $AB = C = (c_{ij})$ with

$$c_{ij} = a_{i1}b_{1j} + a_{i2}b_{2j} + \cdots + a_{in}b_{nj}.$$

Although this formula looks daunting, it is quite easy to use in practice. What it says is that the element in row i and column j of the product is obtained by taking each entry of row i of A and multiplying it by the corresponding entry of column j of B, then adding these n products together.

$$\text{row } i \text{ of } A \longrightarrow \begin{pmatrix} & & & \\ a_{i1} & a_{i2} & \cdots & a_{in} \\ & & & \end{pmatrix} \begin{pmatrix} b_{1j} \\ b_{2j} \\ \vdots \\ b_{nj} \end{pmatrix}.$$

$$\underset{\uparrow}{\text{column } j \text{ of } B}$$

What size is $C = AB$? The matrix C must be $m \times p$ since it will have one entry for each of the m rows of A and each of the p columns of B.

Example 1.10 In the following product, the element in row 2 and column 1 of the product matrix (indicated in bold type) is found, as described above, by using the row and column printed in bold type.

$$AB = \begin{pmatrix} 1 & 1 & 1 \\ \mathbf{2} & \mathbf{0} & \mathbf{1} \\ 1 & 2 & 4 \\ 2 & 2 & -1 \end{pmatrix} \begin{pmatrix} \mathbf{3} & 0 \\ \mathbf{1} & 1 \\ \mathbf{-1} & 3 \end{pmatrix} = \begin{pmatrix} 3 & 4 \\ \mathbf{5} & 3 \\ 1 & 14 \\ 9 & -1 \end{pmatrix}.$$

This entry is 5 because

$$(2)(3) + (0)(1) + (1)(-1) = 5.$$

Notice the sizes of the three matrices. A is 4×3, B is 3×2, and the product AB is 4×2.

We shall see in later chapters that this definition of matrix multiplication is exactly what is needed for applying matrices in our study of linear algebra.

It is an important consequence of this definition that:

- $AB \neq BA$ in general. That is, matrix multiplication is not 'commutative'.

To see just how non-commutative matrix multiplication is, let's look at some examples, starting with the two matrices A and B in the example above. The product AB is defined, but the product BA is not even defined. Since A is 4×3 and B is 3×2, it is not possible to multiply the matrices in the order BA.

Now consider the matrices

$$A = \begin{pmatrix} 2 & 1 & 3 \\ 1 & 2 & 1 \end{pmatrix} \quad \text{and} \quad B = \begin{pmatrix} 3 & 1 \\ 1 & 0 \\ 1 & 1 \end{pmatrix}.$$

Both products AB and BA are defined, but they are different sizes, so they cannot be equal. What sizes are they?

Activity 1.11 Answer the question just posed concerning the sizes of AB and BA. Multiply the matrices to find the two product matrices, AB and BA.

Even if both products are defined and the same size, it is still generally true that $AB \neq BA$.

Activity 1.12 Investigate this last claim. Write down two different 2×2 matrices A and B and find the products AB and BA. For example, you could use

$$A = \begin{pmatrix} 1 & 2 \\ 3 & 4 \end{pmatrix} \quad B = \begin{pmatrix} 1 & 1 \\ 0 & 1 \end{pmatrix}.$$

1.4 Matrix algebra

Matrices are useful because they provide a compact notation and we can perform algebra with them.

For example, given a matrix equation such as

$$3A + 2B = 2(B - A + C),$$

we can solve this for the matrix C using the rules of algebra. You must always bear in mind that to perform the operations they must be defined. In this equation, it is understood that all the matrices A, B and C are the same size, say $m \times n$.

We list the rules of algebra satisfied by the operations of addition, scalar multiplication and matrix multiplication. The sizes of the matrices are dictated by the operations being defined. The first rule is that addition is 'commutative':

- $A + B = B + A$.

This is easily shown to be true. The matrices A and B must be of the same size, say $m \times n$, for the operation to be defined, so both $A + B$ and $B + A$ are $m \times n$ matrices for some m and n. They also have the same entries. The (i, j) entry of $A + B$ is $a_{ij} + b_{ij}$ and the (i, j) entry of $B + A$ is $b_{ij} + a_{ij}$, but $a_{ij} + b_{ij} = b_{ij} + a_{ij}$ by the properties of real numbers. So the matrices $A + B$ and $B + A$ are equal.

On the other hand, as we have seen, matrix multiplication is not commutative: $AB \neq BA$ in general.

We have the following 'associative' laws:

- $(A + B) + C = A + (B + C)$,
- $\lambda(AB) = (\lambda A)B = A(\lambda B)$,
- $(AB)C = A(BC)$.

These rules allow us to remove brackets. For example, the last rule says that we will get the same result if we first multiply AB and then multiply by C on the right as we will if we first multiply BC and then multiply by A on the left, so the choice is ours.

We can show that all these rules follow from the definitions of the operations, just as we showed the commutativity of addition. We need to know that the matrices on the left and on the right of the equals sign have the same size and that corresponding entries are equal. Only the associativity of multiplication presents any complications, but you just need to carefully write down the (i, j) entry of each side and show that, by rearranging terms, they are equal.

Activity 1.13 Think about these rules. What sizes are each of the matrices? Write down the (i, j) entry for each of the matrices $\lambda(AB)$ and $(\lambda A)(B)$ and prove that the matrices are equal.

Similarly, we have three 'distributive' laws:

- $A(B + C) = AB + AC$,
- $(B + C)A = BA + CA$,
- $\lambda(A + B) = \lambda A + \lambda B$.

Why do we need both of the first two rules (which state that matrix multiplication distributes through addition)? Well, since matrix multiplication is not commutative, we cannot conclude the second distributive rule from the first; we have to prove it is true separately. These statements can be proved from the definitions of the operations, as above, but we will not take the time to do this here.

If A is an $m \times n$ matrix, what is the result of $A - A$? We obtain an $m \times n$ matrix all of whose entries are 0. This is an 'additive identity'; that is, it plays the same role for matrices as the number 0 does for numbers, in the sense that $A + 0 = 0 + A = A$. There is a zero matrix of any size $m \times n$.

Definition 1.14 (Zero matrix) A *zero matrix*, denoted 0, is an $m \times n$ matrix with all entries zero:

$$\begin{pmatrix} 0 & 0 & \cdots & 0 & 0 \\ 0 & 0 & \cdots & 0 & 0 \\ \vdots & \vdots & \ddots & \vdots & \vdots \\ 0 & 0 & \cdots & 0 & 0 \end{pmatrix}.$$

Then:

- $A + 0 = A$,
- $A - A = 0$,
- $0A = 0$, $A0 = 0$,

where the sizes of the zero matrices above must be compatible with the size of the matrix A.

We also have a 'multiplicative identity', which acts like the number 1 does for multiplication of numbers.

Definition 1.15 (Identity matrix) The $n \times n$ *identity matrix*, denoted I_n or simply I, is the diagonal matrix with $a_{ii} = 1$,

$$I = \begin{pmatrix} 1 & 0 & \cdots & 0 \\ 0 & 1 & \cdots & 0 \\ \vdots & \vdots & \ddots & \vdots \\ 0 & 0 & \cdots & 1 \end{pmatrix}.$$

If A is any $m \times n$ matrix, then:

- $AI = A$ and $IA = A$,

where it is understood that the identity matrices are the appropriate size for the products to be defined.

Activity 1.16 What size is the identity matrix if A is $m \times n$ and $IA = A$?

Example 1.17 We can apply these rules to solve the equation, $3A + 2B = 2(B - A + C)$ for C. We will pedantically apply each rule so that you can see how it is being used. In practice, you don't need to put in all these steps, just implicitly use the rules of algebra. We begin by removing the brackets using the distributive rule.

$$
\begin{aligned}
&3A + 2B = 2B - 2A + 2C && \text{(distributive rule)} \\
&3A + 2B - 2B && \text{(add } -2B \text{ to both sides)} \\
&\quad = 2B - 2A + 2C - 2B \\
&3A + (2B - 2B) && \text{(commutativity, associativity} \\
&\quad = -2A + 2C + (2B - 2B) && \text{of addition)} \\
&3A + 0 = -2A + 2C + 0 && \text{(additive inverse)} \\
&3A = -2A + 2C && \text{(additive identity)} \\
&3A + 2A = -2A + 2C + 2A && \text{(add } 2A \text{ to both sides)} \\
&5A = 2C && \text{(commutativity, associativity of} \\
& && \quad \text{addition, additive identity)} \\
&C = \tfrac{5}{2}A && \text{(scalar multiplication).}
\end{aligned}
$$

1.5 Matrix inverses

1.5.1 The inverse of a matrix

If $AB = AC$, can we conclude that $B = C$? The answer is 'no', as the following example shows.

Example 1.18 If

$$A = \begin{pmatrix} 0 & 0 \\ 1 & 1 \end{pmatrix}, \quad B = \begin{pmatrix} 1 & -1 \\ 3 & 5 \end{pmatrix}, \quad C = \begin{pmatrix} 8 & 0 \\ -4 & 4 \end{pmatrix},$$

then the matrices B and C are *not* equal, but

$$AB = AC = \begin{pmatrix} 0 & 0 \\ 4 & 4 \end{pmatrix}.$$

Activity 1.19 Check this by multiplying out the matrices.

On the other hand, if $A + 5B = A + 5C$, then we can conclude that $B = C$ because the operations of addition and scalar multiplication have inverses. If we have a matrix A, then the matrix $-A = (-1)A$ is an additive inverse because it satisfies $A + (-A) = 0$. If we multiply a matrix A by a non-zero scalar c, we can 'undo' this by multiplying cA by $1/c$.

What about matrix multiplication? Is there a multiplicative inverse? The answer is 'sometimes'.

Definition 1.20 (Inverse matrix) The $n \times n$ matrix A is *invertible* if there is a matrix B such that

$$AB = BA = I,$$

where I is the $n \times n$ identity matrix. The matrix B is called the inverse of A and is denoted by A^{-1}.

Notice that the matrix A must be square, and that both I and $B = A^{-1}$ must also be square $n \times n$ matrices, for the products to be defined.

Example 1.21 Let $A = \begin{pmatrix} 1 & 2 \\ 3 & 4 \end{pmatrix}$. Then with

$$B = \begin{pmatrix} -2 & 1 \\ \frac{3}{2} & -\frac{1}{2} \end{pmatrix},$$

we have $AB = BA = I$, therefore $B = A^{-1}$.

Activity 1.22 Check this. Multiply the matrices to show that $AB = I$ and $BA = I$, where I is the 2×2 identity matrix.

You might have noticed that we have said that B is *the* inverse of A. This is because an invertible matrix has only one inverse. We will prove this.

Theorem 1.23 *If A is an $n \times n$ invertible matrix, then the matrix A^{-1} is unique.*

Proof: Assume the matrix A has two inverses, B and C, so that $AB = BA = I$ and $AC = CA = I$. We will show that B and C must actually be the same matrix; that is, that they are equal. Consider the product CAB. Since matrix multiplication is associative and $AB = I$, we have

$$CAB = C(AB) = CI = C.$$

On the other hand, again by associativity,

$$CAB = (CA)B = IB = B$$

since $CA = I$. We conclude that $C = B$, so there is only one inverse matrix of A.   □

Not all square matrices will have an inverse. We say that A is *invertible* or *non-singular* if it has an inverse. We say that A is *non-invertible* or *singular* if it has no inverse.

For example, the matrix

$$\begin{pmatrix} 0 & 0 \\ 1 & 1 \end{pmatrix}$$

(used in Example 1.18 of this section) is *not* invertible. It is not possible for a matrix to satisfy

$$\begin{pmatrix} 0 & 0 \\ 1 & 1 \end{pmatrix} \begin{pmatrix} a & b \\ c & d \end{pmatrix} = \begin{pmatrix} 1 & 0 \\ 0 & 1 \end{pmatrix}$$

since the $(1,1)$ entry of the product is 0 and $0 \neq 1$.

On the other hand, if

$$A = \begin{pmatrix} a & b \\ c & d \end{pmatrix}, \quad \text{where} \quad ad - bc \neq 0,$$

then A has the inverse

$$A^{-1} = \frac{1}{ad - bc} \begin{pmatrix} d & -b \\ -c & a \end{pmatrix}.$$

Activity 1.24 Check that this is indeed the inverse of A, by showing that if you multiply A on the left or on the right by this matrix, then you obtain the identity matrix I.

This tells us how to find the inverse of any 2×2 invertible matrix. If

$$A = \begin{pmatrix} a & b \\ c & d \end{pmatrix},$$

the scalar $ad - bc$ is called the *determinant* of the matrix A, denoted $|A|$. We shall see more about the determinant in Chapter 3. So if

$|A| = ad - bc \neq 0$, then to construct A^{-1} we take the matrix A, switch the main diagonal entries and put minus signs in front of the other two entries, then multiply by the scalar $1/|A|$.

Activity 1.25 Use this to find the inverse of the matrix $\begin{pmatrix} 1 & 2 \\ 3 & 4 \end{pmatrix}$, and check your answer by looking at Example 1.21.

If $AB = AC$, and A is invertible, can we conclude that $B = C$? This time the answer is 'yes', because we can multiply each side of the equation on the left by A^{-1}:

$$A^{-1}AB = A^{-1}AC \quad \Longrightarrow \quad IB = IC \quad \Longrightarrow \quad B = C.$$

But be careful! If $AB = CA$, then we cannot conclude that $B = C$, only that $B = A^{-1}CA$.

It is not possible to '*divide*' by a matrix. We can only multiply on the right or left by the inverse matrix.

1.5.2 Properties of the inverse

If A is an invertible matrix, then, by definition, A^{-1} exists and $AA^{-1} = A^{-1}A = I$. This statement also says that the matrix A is the inverse of A^{-1}; that is,

- $(A^{-1})^{-1} = A$.

It is important to understand the definition of an inverse matrix and be able to use it. Essentially, if we can find a matrix that satisfies the definition, then that matrix is the inverse, and the matrix is invertible. For example, if A is an invertible $n \times n$ matrix, then:

- $(\lambda A)^{-1} = \dfrac{1}{\lambda} A^{-1}$.

This statement says that the matrix λA is invertible, and its inverse is given by the matrix $C = (1/\lambda)A^{-1}$. To prove this is true, we just need to show that the matrix C satisfies $(\lambda A)C = C(\lambda A) = I$. This is straightforward using matrix algebra:

$$(\lambda A)\left(\frac{1}{\lambda}A^{-1}\right) = \lambda\frac{1}{\lambda}AA^{-1} = I \quad \text{and} \quad \left(\frac{1}{\lambda}A^{-1}\right)(\lambda A) = \frac{1}{\lambda}\lambda A^{-1}A = I.$$

If A and B are invertible $n \times n$ matrices, then using the definition of the inverse you can show the following important fact:

- $(AB)^{-1} = B^{-1}A^{-1}$.

This last statement says that if A and B are invertible matrices of the same size, then the product AB is invertible and its inverse is the product of the inverses *in the reverse order*. The proof of this statement is left as an exercise. (See Exercise 1.3.)

1.6   Powers of a matrix

If A is a square matrix, what do we mean by A^2? We naturally mean the product of A with itself, $A^2 = AA$. In the same way, if A is an $n \times n$ matrix and $r \in \mathbb{N}$, then:

$$A^r = \underbrace{A A \ldots A}_{r \text{ times}}.$$

Powers of matrices obey a number of rules, similar to powers of numbers. First, if A is an $n \times n$ matrix and $r \in \mathbb{N}$, then:

- $(A^r)^{-1} = (A^{-1})^r$.

This follows immediately from the definition of an inverse matrix and the associativity of matrix multiplication. Think about what it says: that the inverse of the product of A times itself r times is the product of A^{-1} times itself r times.

The usual rules of exponents hold: for integers r, s,

- $A^r A^s = A^{r+s}$,
- $(A^r)^s = A^{rs}$.

As r and s are positive integers and matrix multiplication is associative, these properties are easily verified in the same way as they are with real numbers.

Activity 1.26   Verify the above three properties.

1.7   The transpose and symmetric matrices

1.7.1   The transpose of a matrix

If we interchange the rows and columns of a matrix, we obtain another matrix, known as its *transpose*.

Definition 1.27 (Transpose) The *transpose* of an $m \times n$ matrix

$$A = (a_{ij}) = \begin{pmatrix} a_{11} & a_{12} & \cdots & a_{1n} \\ a_{21} & a_{22} & \cdots & a_{2n} \\ \vdots & \vdots & \ddots & \vdots \\ a_{m1} & a_{m2} & \cdots & a_{mn} \end{pmatrix}$$

is the $n \times m$ matrix

$$A^{\mathrm{T}} = (a_{ji}) = \begin{pmatrix} a_{11} & a_{21} & \cdots & a_{m1} \\ a_{12} & a_{22} & \cdots & a_{m2} \\ \vdots & \vdots & \ddots & \vdots \\ a_{1n} & a_{2n} & \cdots & a_{mn} \end{pmatrix}.$$

So, on forming the transpose of a matrix, row i of A becomes column i of A^{T}.

Example 1.28 If $A = \begin{pmatrix} 1 & 2 \\ 3 & 4 \end{pmatrix}$ and $B = (1 \quad 5 \quad 3)$, then

$$A^{\mathrm{T}} = \begin{pmatrix} 1 & 3 \\ 2 & 4 \end{pmatrix}, \qquad B^{\mathrm{T}} = \begin{pmatrix} 1 \\ 5 \\ 3 \end{pmatrix}.$$

Notice that the diagonal entries of a square matrix do not move under the operation of taking the transpose, as a_{ii} remains a_{ii}. So if D is a diagonal matrix, then $D^{\mathrm{T}} = D$.

1.7.2 Properties of the transpose

If we take the transpose of a matrix A by switching the rows and columns, and then take the transpose of the resulting matrix, then we get back to the original matrix A. This is summarised in the following equation:

- $(A^{\mathrm{T}})^{\mathrm{T}} = A$.

Two further properties relate to scalar multiplication and addition:

- $(\lambda A)^{\mathrm{T}} = \lambda A^{\mathrm{T}}$ and
- $(A + B)^{\mathrm{T}} = A^{\mathrm{T}} + B^{\mathrm{T}}$.

These follow immediately from the definition. In particular, the (i, j) entry of $(\lambda A)^{\mathrm{T}}$ is λa_{ji}, which is also the (i, j) entry of λA^{T}.

The next property tells you what happens when you take the transpose of a product of matrices:

- $(AB)^{\mathrm{T}} = B^{\mathrm{T}} A^{\mathrm{T}}$.

This can be stated as: *The transpose of the product of two matrices is the product of the transposes in the reverse order.*

Showing that this is true is slightly more complicated, since it involves matrix multiplication. It is more important to understand why the product of the transposes must be in the reverse order: the following activity explores this.

Activity 1.29 If A is an $m \times n$ matrix and B is $n \times p$, look at the sizes of the matrices $(AB)^T$, A^T, B^T. Show that only the product $B^T A^T$ is always defined. Show also that its size is equal to the size of $(AB)^T$.

If A is an $m \times n$ matrix and B is $n \times p$, then, from Activity 1.29, you know that $(AB)^T$ and $B^T A^T$ are the same size. To prove that $(AB)^T = B^T A^T$, you need to show that the (i, j) entries are equal. You can try this as follows.

Activity 1.30 The (i, j) entry of $(AB)^T$ is the (j, i) entry of AB, which is obtained by taking row j of A and multiplying each term by the corresponding entry of column i of B. We can write this as

$$\left((AB)^T\right)_{ij} = a_{j1}b_{1i} + a_{j2}b_{2i} + \cdots + a_{jn}b_{ni}.$$

Do the same for the (i, j) entry of $B^T A^T$ and show that you obtain the same number.

The final property in this section states that the inverse of the transpose of an invertible matrix is the transpose of the inverse; that is, if A is invertible, then:

- $(A^T)^{-1} = (A^{-1})^T.$

This follows from the previous property and the definition of inverse. We have

$$A^T(A^{-1})^T = (A^{-1}A)^T = I^T = I$$

and, in the same way, $(A^{-1})^T A^T = I$. Therefore, by the definition of the inverse of a matrix, $(A^{-1})^T$ must be the inverse of A^T.

1.7.3   Symmetric matrices

Definition 1.31 (Symmetric matrix) A matrix A is *symmetric* if it is equal to its transpose, $A = A^T$.

Only square matrices can be symmetric. If A is symmetric, then $a_{ij} = a_{ji}$. That is, entries diagonally opposite to each other must be equal, or, in other words, the matrix is symmetric about its diagonal.

Activity 1.32 Fill in the missing numbers if the matrix A is symmetric:

$$A = \begin{pmatrix} 1 & 4 \\ & 2 \\ 5 & & 3 \end{pmatrix} = \begin{pmatrix} 1 & \\ & -7 \\ & & \end{pmatrix} = A^{\mathrm{T}}.$$

If D is a diagonal matrix, then $d_{ij} = 0 = d_{ji}$ for all $i \neq j$. So all diagonal matrices are symmetric.

1.8 Vectors in $\mathbb{R}^n$

1.8.1 Vectors

An $n \times 1$ matrix is a *column vector*, or simply a *vector*

$$\mathbf{v} = \begin{pmatrix} v_1 \\ v_2 \\ \vdots \\ v_n \end{pmatrix},$$

where each v_i is a real number. The numbers $v_1, v_2, \ldots, v_n$, are known as the *components* (or *entries*) of the vector $\mathbf{v}$.

We can also define a *row vector* to be a $1 \times n$ matrix.

In this text, when we simply use the term *vector*, we shall mean a column vector.

In order to distinguish vectors from scalars, and to emphasise that they are vectors and not general matrices, we will write vectors in lowercase boldface type. (When writing by hand, vectors should be underlined to avoid confusion with scalars.)

Addition and scalar multiplication are defined for vectors as for $n \times 1$ matrices:

$$\mathbf{v} + \mathbf{w} = \begin{pmatrix} v_1 + w_1 \\ v_2 + w_2 \\ \vdots \\ v_n + w_n \end{pmatrix}, \qquad \lambda\mathbf{v} = \begin{pmatrix} \lambda v_1 \\ \lambda v_2 \\ \vdots \\ \lambda v_n \end{pmatrix}.$$

For a fixed positive integer n, the set of vectors (together with the operations of addition and scalar multiplication) form the set $\mathbb{R}^n$, usually called *Euclidean n-space*.

We will often write a column vector as the transpose of a row vector. Although

$$\mathbf{x} = \begin{pmatrix} x_1 \\ x_2 \\ \vdots \\ x_n \end{pmatrix} = (x_1 \quad x_2 \quad \cdots \quad x_n)^{\mathrm{T}},$$

we will usually write $\mathbf{x} = (x_1, x_2, \cdots, x_n)^{\mathrm{T}}$, with commas separating the entries. A matrix does not have commas; however, we will use them in order to clearly distinguish the separate components of the vector.

For vectors $\mathbf{v}_1, \mathbf{v}_2, \ldots, \mathbf{v}_k$ in $\mathbb{R}^n$ and scalars $\alpha_1, \alpha_2, \ldots, \alpha_k$ in $\mathbb{R}$, the vector

$$\mathbf{v} = \alpha_1 \mathbf{v}_1 + \cdots + \alpha_k \mathbf{v}_k \in \mathbb{R}^n$$

is known as a *linear combination* of the vectors $\mathbf{v}_1, \ldots, \mathbf{v}_k$.

A *zero vector*, denoted $\mathbf{0}$, is a vector with all of its entries equal to 0. There is one zero vector in each space $\mathbb{R}^n$. As with matrices, this vector is an additive identity, meaning that for any vector $\mathbf{v} \in \mathbb{R}^n$, $\mathbf{0} + \mathbf{v} = \mathbf{v} + \mathbf{0} = \mathbf{v}$. Further, multiplying any vector $\mathbf{v}$ by the scalar zero results in the zero vector: $0\mathbf{v} = \mathbf{0}$.

Although the matrix product of two vectors $\mathbf{v}$ and $\mathbf{w}$ in $\mathbb{R}^n$ cannot be calculated, it is possible to form the matrix products $\mathbf{v}^{\mathrm{T}}\mathbf{w}$ and $\mathbf{v}\mathbf{w}^{\mathrm{T}}$. The first is a 1×1 matrix, and the latter is an $n \times n$ matrix.

Activity 1.33 Calculate $\mathbf{a}^{\mathrm{T}}\mathbf{b}$ and $\mathbf{a}\mathbf{b}^{\mathrm{T}}$ for $\mathbf{a} = \begin{pmatrix} 1 \\ 2 \\ 3 \end{pmatrix}$, $\mathbf{b} = \begin{pmatrix} 4 \\ -2 \\ 1 \end{pmatrix}$.

1.8.2   The inner product of two vectors

For $\mathbf{v}, \mathbf{w} \in \mathbb{R}^n$, the 1×1 matrix $\mathbf{v}^{\mathrm{T}}\mathbf{w}$ can be identified with the real number, or scalar, which is its unique entry. This turns out to be particularly useful, and is known as the *inner product* of $\mathbf{v}$ and $\mathbf{w}$.

Definition 1.34 (inner product) Given two vectors

$$\mathbf{v} = \begin{pmatrix} v_1 \\ v_2 \\ \vdots \\ v_n \end{pmatrix}, \quad \mathbf{w} = \begin{pmatrix} w_1 \\ w_2 \\ \vdots \\ w_n \end{pmatrix},$$

the *inner product,* denoted $\langle \mathbf{v}, \mathbf{w} \rangle$, is the real number given by

$$\langle \mathbf{v}, \mathbf{w} \rangle = \left\langle \begin{pmatrix} v_1 \\ v_2 \\ \vdots \\ v_n \end{pmatrix}, \begin{pmatrix} w_1 \\ w_2 \\ \vdots \\ w_n \end{pmatrix} \right\rangle = v_1 w_1 + v_2 w_2 + \cdots + v_n w_n.$$

The inner product, $\langle \mathbf{v}, \mathbf{w} \rangle$, is also known as the *scalar product* of $\mathbf{v}$ and $\mathbf{w}$, or as the *dot product*. In the latter case, it is denoted by $\mathbf{v} \cdot \mathbf{w}$.

The inner product of $\mathbf{v}$ and $\mathbf{w}$ is precisely the scalar quantity (that is, the number) given by

$$\mathbf{v}^\mathrm{T} \mathbf{w} = \begin{pmatrix} v_1 & v_2 & \cdots & v_n \end{pmatrix} \begin{pmatrix} w_1 \\ w_2 \\ \vdots \\ w_n \end{pmatrix} = v_1 w_1 + v_2 w_2 + \cdots + v_n w_n,$$

so that we can write

$$\langle \mathbf{v}, \mathbf{w} \rangle = \mathbf{v}^\mathrm{T} \mathbf{w}.$$

Example 1.35 If $\mathbf{x} = (1, 2, 3)^\mathrm{T}$ and $\mathbf{y} = (2, -1, 1)^\mathrm{T}$, then

$$\langle \mathbf{x}, \mathbf{y} \rangle = 1(2) + 2(-1) + 3(1) = 3.$$

It is important to realise that the inner product is just a number, a *scalar*, not another vector or a matrix.

The inner product on $\mathbb{R}^n$ satisfies certain basic properties as shown in the next theorem.

Theorem 1.36 *The inner product*

$$\langle \mathbf{x}, \mathbf{y} \rangle = x_1 y_1 + x_2 y_2 + \cdots + x_n y_n, \qquad \mathbf{x}, \mathbf{y} \in \mathbb{R}^n$$

satisfies the following properties for all $\mathbf{x}, \mathbf{y}, \mathbf{z} \in \mathbb{R}^n$ *and for all* $\alpha \in \mathbb{R}$:

(i) $\langle \mathbf{x}, \mathbf{y} \rangle = \langle \mathbf{y}, \mathbf{x} \rangle$,
(ii) $\alpha \langle \mathbf{x}, \mathbf{y} \rangle = \langle \alpha \mathbf{x}, \mathbf{y} \rangle = \langle \mathbf{x}, \alpha \mathbf{y} \rangle$,
(iii) $\langle \mathbf{x} + \mathbf{y}, \mathbf{z} \rangle = \langle \mathbf{x}, \mathbf{z} \rangle + \langle \mathbf{y}, \mathbf{z} \rangle$,
(iv) $\langle \mathbf{x}, \mathbf{x} \rangle \geq 0$, *and* $\langle \mathbf{x}, \mathbf{x} \rangle = 0$ *if and only if* $\mathbf{x} = \mathbf{0}$.

Proof: We have, by properties of real numbers

$$\langle \mathbf{x}, \mathbf{y} \rangle = x_1 y_1 + x_2 y_2 + \cdots + x_n y_n$$
$$= y_1 x_1 + y_2 x_2 + \cdots + y_n x_n = \langle \mathbf{y}, \mathbf{x} \rangle,$$

which proves (i). We leave the proofs of (ii) and (iii) as an exercise. For (iv), note that

$$\langle \mathbf{x}, \mathbf{x} \rangle = x_1^2 + x_2^2 + \cdots + x_n^2$$

is a sum of squares, so $\langle \mathbf{x}, \mathbf{x} \rangle \geq 0$, and $\langle \mathbf{x}, \mathbf{x} \rangle = 0$ if and only if each term x_i^2 is equal to 0; that is, if and only if each $x_i = 0$, so $\mathbf{x}$ is the zero vector, $\mathbf{x} = \mathbf{0}$. $\square$

Activity 1.37 Prove properties (ii) and (iii). Show, also, that these two properties are equivalent to the single property

$$\langle \alpha \mathbf{x} + \beta \mathbf{y}, \mathbf{z} \rangle = \alpha \langle \mathbf{x}, \mathbf{z} \rangle + \beta \langle \mathbf{y}, \mathbf{z} \rangle.$$

From the definitions, it is clear that it is not possible to combine vectors in different Euclidean spaces, either by addition or by taking the inner product. If $\mathbf{v} \in \mathbb{R}^n$ and $\mathbf{w} \in \mathbb{R}^m$, with $m \neq n$, then these vectors live in different 'worlds', or, more precisely, in different 'vector spaces'.

1.8.3   Vectors and matrices

If A is an $m \times n$ matrix, then the columns of A are vectors in $\mathbb{R}^m$. If $\mathbf{x} \in \mathbb{R}^n$, then the product $A\mathbf{x}$ is an $m \times 1$ matrix, so is also a vector in $\mathbb{R}^m$. There is a fundamental relationship between these vectors, which we present here as an example of matrix manipulation. We list it as a theorem so that we can refer back to it later.

Theorem 1.38   *Let A be an $m \times n$ matrix*

$$A = \begin{pmatrix} a_{11} & a_{12} & \cdots & a_{1n} \\ a_{21} & a_{22} & \cdots & a_{2n} \\ \vdots & \vdots & \ddots & \vdots \\ a_{m1} & a_{m2} & \cdots & a_{mn} \end{pmatrix},$$

and denote the columns of A by the column vectors $\mathbf{c}_1, \mathbf{c}_2, \ldots, \mathbf{c}_n$, so that

$$\mathbf{c}_i = \begin{pmatrix} a_{1i} \\ a_{2i} \\ \vdots \\ a_{mi} \end{pmatrix}, \quad i = 1, \ldots, n.$$

Then if $\mathbf{x} = (x_1, x_2, \ldots, x_n)^{\mathrm{T}}$ is any vector in $\mathbb{R}^n$,

$$A\mathbf{x} = x_1 \mathbf{c}_1 + x_2 \mathbf{c}_2 + \cdots + x_n \mathbf{c}_n.$$

The theorem states that the matrix product $A\mathbf{x}$, which is a vector in $\mathbb{R}^m$, can be expressed as a linear combination of the column vectors of A. Before you look at the proof, try to carry out the calculation yourself, to see how it works. Just write both the left-hand side and the right-hand side of the equality as a single $m \times 1$ vector.

Proof: We have

$$
A\mathbf{x} = \begin{pmatrix} a_{11} & a_{12} & \cdots & a_{1n} \\ a_{21} & a_{22} & \cdots & a_{2n} \\ \vdots & \vdots & \ddots & \vdots \\ a_{m1} & a_{m2} & \cdots & a_{mn} \end{pmatrix} \begin{pmatrix} x_1 \\ x_2 \\ \vdots \\ x_n \end{pmatrix}
$$

$$
= \begin{pmatrix} a_{11}x_1 + a_{12}x_2 + \cdots + a_{1n}x_n \\ a_{21}x_1 + a_{22}x_2 + \cdots + a_{2n}x_n \\ \vdots \\ a_{m1}x_1 + a_{m2}x_2 + \cdots + a_{mn}x_n \end{pmatrix}
$$

$$
= x_1 \begin{pmatrix} a_{11} \\ a_{21} \\ \vdots \\ a_{m1} \end{pmatrix} + x_2 \begin{pmatrix} a_{12} \\ a_{22} \\ \vdots \\ a_{m2} \end{pmatrix} + \cdots + x_n \begin{pmatrix} a_{1n} \\ a_{2n} \\ \vdots \\ a_{mn} \end{pmatrix}
$$

$$
= x_1\mathbf{c}_1 + x_2\mathbf{c}_2 + \cdots x_n\mathbf{c}_n.
$$

$\square$

There are many useful ways to view this relationship, as we shall see in later chapters.

1.9 Developing geometric insight

Vectors have a broader use beyond that of being special types of matrices. It is possible that you have some previous knowledge of vectors; for example, in describing the displacement of an object from one point to another in $\mathbb{R}^2$ or in $\mathbb{R}^3$. Before we continue our study of linear algebra, it is important to consolidate this background, for it provides valuable geometric insight into the definitions and uses of vectors in higher dimensions. Parts of the next section may be a review for you.

1.9.1 Vectors in $\mathbb{R}^2$

The set $\mathbb{R}$ can be represented as points along a horizontal line, called a *real-number line*. In order to represent pairs of real numbers, (a_1, a_2), we use a *Cartesian plane*, a plane with both a horizontal axis and a vertical axis, each axis being a copy of the real-number line, and we mark $A = (a_1, a_2)$ as a *point* in this plane. We associate this point with the vector $\mathbf{a} = (a_1, a_2)^{\mathrm{T}}$, as representing a *displacement* from the origin

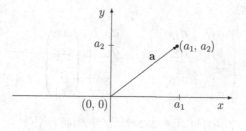

Figure 1.1 A
position vector, **a**

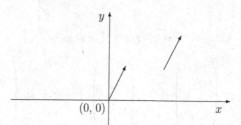

Figure 1.2
Displacement
vectors, **v**

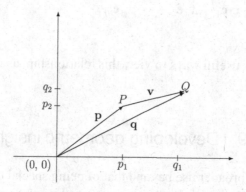

Figure 1.3 If
$\mathbf{v} = (v_1, v_2)^{\mathsf{T}}$, then
$q_1 = p_1 + v_1$ and
$q_2 = v_2 + p_2$

(the point $(0, 0)$) to the point A. In this context, **a** is the *position vector* of the point A. This displacement is illustrated by an arrow, or directed line segment, with the initial point at the origin and the terminal point at A, as shown in Figure 1.1.

Even if a displacement does not begin at the origin, two displacements of the same length and the same direction are considered to be equal. So, for example, the two arrows in Figure 1.2 represent the same vector, $\mathbf{v} = (1, 2)^{\mathsf{T}}$.

If an object is displaced from a point, say $(0, 0)$, the origin, to a point P by the displacement **p**, and then displaced from P to Q by the displacement **v**, then the total displacement is given by the vector from 0 to Q, which is the position vector **q**. So we would expect vectors to satisfy $\mathbf{q} = \mathbf{p} + \mathbf{v}$, both geometrically (in the sense of a displacement) and algebraically (by the definition of vector addition). This is certainly true in general, as illustrated in Figure 1.3.

Figure 1.4
$\mathbf{p} + \mathbf{v} = \mathbf{v} + \mathbf{p}$

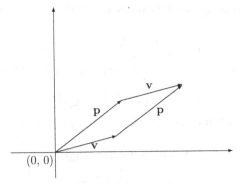

Figure 1.5 A
right-angled triangle
to determine the
length of a vector

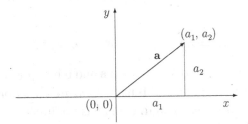

The order of displacements does not matter (nor does the order of vector addition), so $\mathbf{q} = \mathbf{v} + \mathbf{p}$. For this reason, the addition of vectors is said to follow the *parallelogram law*. This is illustrated in Figure 1.4.

From the equation $\mathbf{q} = \mathbf{p} + \mathbf{v}$, we have $\mathbf{v} = \mathbf{q} - \mathbf{p}$. This is the displacement from P to Q. To help you determine in which direction the vector $\mathbf{v}$ points, think of $\mathbf{v} = \mathbf{q} - \mathbf{p}$ as the vector which is added to the vector $\mathbf{p}$ in order to obtain the vector $\mathbf{q}$.

If $\mathbf{v}$ represents a displacement, then $2\mathbf{v}$ must represent a displacement in the same direction, but twice as far, and $-\mathbf{v}$ represents an equal displacement in the opposite direction. This interpretation is compatible with the definition of scalar multiplication.

Activity 1.39 Sketch the vector $\mathbf{v} = (1, 2)^{\mathrm{T}}$ in a coordinate system. Then sketch $2\mathbf{v}$ and $-\mathbf{v}$. Looking at the coordinates on your sketch, what are the components of $2\mathbf{v}$ and $-\mathbf{v}$?

We have stated that a vector has both a *length* and a *direction*. Given a vector $\mathbf{a} = (a_1, a_2)^{\mathrm{T}}$, its length, denoted by $\|\mathbf{a}\|$, can be calculated using Pythagoras' theorem applied to the right triangle shown in Figure 1.5.

So the *length* of $\mathbf{a}$ is the scalar quantity

$$\|\mathbf{a}\| = \sqrt{a_1^2 + a_2^2}.$$

The length of a vector can be expressed in terms of the inner product,

$$\|\mathbf{a}\| = \sqrt{\langle \mathbf{a}, \mathbf{a} \rangle},$$

simply because $\langle \mathbf{a}, \mathbf{a} \rangle = a_1^2 + a_2^2$. A *unit* vector is a vector of length 1.

Example 1.40 If $\mathbf{v} = (1, 2)^{\mathrm{T}}$, then $\|\mathbf{v}\| = \sqrt{1^2 + 2^2} = \sqrt{5}$. The vector $\mathbf{u} = \left(\frac{1}{\sqrt{5}}, \frac{2}{\sqrt{5}}\right)^{\mathrm{T}}$ is a unit vector in the same direction as $\mathbf{v}$.

Activity 1.41 Check this. Calculate the length of $\mathbf{u}$.

The *direction* of a vector is essentially given by the components of the vector. If we have two vectors $\mathbf{a}$ and $\mathbf{b}$ which are (non-zero) scalar multiples, say

$$\mathbf{a} = \lambda\mathbf{b}, \qquad \lambda \in \mathbb{R}, \quad (\lambda \neq 0),$$

then $\mathbf{a}$ and $\mathbf{b}$ are *parallel*. If $\lambda > 0$, then $\mathbf{a}$ and $\mathbf{b}$ have the *same direction*. If $\lambda < 0$, then we say that $\mathbf{a}$ and $\mathbf{b}$ have *opposite directions*.

The zero vector, $\mathbf{0}$, has length 0 and has no direction. For any other vector, $\mathbf{v} \neq \mathbf{0}$, there is one unit vector in the same direction as $\mathbf{v}$, namely

$$\mathbf{u} = \frac{1}{\|\mathbf{v}\|}\mathbf{v}.$$

Activity 1.42 Write down a unit vector, $\mathbf{u}$, which is parallel to the vector $\mathbf{a} = (4, 3)^{\mathrm{T}}$. Then write down a vector, $\mathbf{w}$, of length 2 which is in the opposite direction to $\mathbf{a}$.

1.9.2 Inner product

The inner product in $\mathbb{R}^2$ is closely linked with the geometrical concepts of *length* and *angle*. If $\mathbf{a} = (a_1, a_2)^{\mathrm{T}}$, we have already seen that

$$\|\mathbf{a}\|^2 = \langle \mathbf{a}, \mathbf{a} \rangle = a_1^2 + a_2^2.$$

Let $\mathbf{a}, \mathbf{b}$ be two vectors in $\mathbb{R}^2$, and let θ denote the angle between them.[1] By this we shall always mean the angle θ such that $0 \leq \theta \leq \pi$. If $\theta < \pi$, the vectors $\mathbf{a}$, $\mathbf{b}$ and $\mathbf{c} = \mathbf{b} - \mathbf{a}$ form a triangle, where $\mathbf{c}$ is the side opposite the angle θ, as, for example, in Figure 1.6.

The *law of cosines* (which is a generalisation of Pythagoras' theorem) applied to this triangle gives us the important relationship stated in the following theorem.

---

[1] Angles are always measured in radians, not degrees, here. So, for example 45 degrees is $\pi/4$ radians,

Figure 1.6 Two vectors and the angle between them

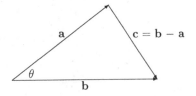

Theorem 1.43 *Let* **a, b** $\in \mathbb{R}^2$ *and let* θ *denote the angle between them. Then*

$$\langle \mathbf{a}, \mathbf{b} \rangle = \|\mathbf{a}\| \, \|\mathbf{b}\| \cos \theta.$$

Proof: The law of cosines states that $c^2 = a^2 + b^2 - 2ab \cos \theta$, where $c = \|\mathbf{b} - \mathbf{a}\|$, $a = \|\mathbf{a}\|$, $b = \|\mathbf{b}\|$. That is,

$$\|\mathbf{b} - \mathbf{a}\|^2 = \|\mathbf{a}\|^2 + \|\mathbf{b}\|^2 - 2\|\mathbf{a}\| \, \|\mathbf{b}\| \cos \theta. \tag{1}$$

Expanding the inner product and using its properties, we have

$$\|\mathbf{b} - \mathbf{a}\|^2 = \langle \mathbf{b} - \mathbf{a}, \mathbf{b} - \mathbf{a} \rangle = \langle \mathbf{b}, \mathbf{b} \rangle + \langle \mathbf{a}, \mathbf{a} \rangle - 2\langle \mathbf{a}, \mathbf{b} \rangle,$$

so that

$$\|\mathbf{b} - \mathbf{a}\|^2 = \|\mathbf{a}\|^2 + \|\mathbf{b}\|^2 - 2\langle \mathbf{a}, \mathbf{b} \rangle. \tag{2}$$

Comparing equations (1) and (2) above, we conclude that

$$\langle \mathbf{a}, \mathbf{b} \rangle = \|\mathbf{a}\| \, \|\mathbf{b}\| \cos \theta.$$

$\square$

Theorem 1.43 has many geometrical consequences. For example, we can use it to find the angle between two vectors by using

$$\cos \theta = \frac{\langle \mathbf{a}, \mathbf{b} \rangle}{\|\mathbf{a}\| \, \|\mathbf{b}\|}.$$

Example 1.44 Let $\mathbf{v} = \begin{pmatrix} 1 \\ 2 \end{pmatrix}$, $\mathbf{w} = \begin{pmatrix} 3 \\ 1 \end{pmatrix}$, and let θ be the angle between them. Then

$$\cos \theta = \frac{5}{\sqrt{5}\sqrt{10}} = \frac{1}{\sqrt{2}},$$

so that $\theta = \dfrac{\pi}{4}$.

Since

$$\langle \mathbf{a}, \mathbf{b} \rangle = \|\mathbf{a}\| \, \|\mathbf{b}\| \cos \theta,$$

and since $-1 \leq \cos \theta \leq 1$ for any real number θ, the maximum value of the inner product is $\langle \mathbf{a}, \mathbf{b} \rangle = \|\mathbf{a}\| \, \|\mathbf{b}\|$. This occurs precisely when

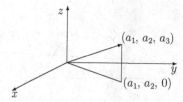

Figure 1.7 Diagram
for Activity 1.45

$\cos \theta = 1$; that is, when $\theta = 0$. In this case, the vectors **a** and **b** are *parallel* and in the *same direction*. If they point in opposite directions, then $\theta = \pi$ and we have $\langle \mathbf{a}, \mathbf{b} \rangle = -\|\mathbf{a}\| \, \|\mathbf{b}\|$. The inner product will be positive if and only if the angle between the vectors is acute, meaning that $0 \le \theta < \frac{\pi}{2}$. It will be negative if the angle is obtuse, meaning that $\frac{\pi}{2} < \theta \le \pi$.

The non-zero vectors **a** and **b** are *orthogonal* (or *perpendicular* or, sometimes, *normal*) when the angle between them is $\theta = \frac{\pi}{2}$. Since $\cos(\frac{\pi}{2}) = 0$, this is precisely when their inner product is zero. We restate this important fact:

- The vectors **a** and **b** are orthogonal if and only if $\langle \mathbf{a}, \mathbf{b} \rangle = 0$.

1.9.3   Vectors in $\mathbb{R}^3$

Everything we have said so far about the geometrical interpretation of vectors and the inner product in $\mathbb{R}^2$ extends to $\mathbb{R}^3$. In particular, if

$$\mathbf{a} = \begin{pmatrix} a_1 \\ a_2 \\ a_3 \end{pmatrix},$$

then

$$\|\mathbf{a}\| = \sqrt{a_1^2 + a_2^2 + a_3^2}.$$

Activity 1.45 Show this. Sketch a position vector $\mathbf{a} = (a_1, a_2, a_3)^{\mathrm{T}}$ in $\mathbb{R}^3$. Drop a perpendicular to the xy-plane as in Figure 1.7, and apply Pythagoras' theorem twice to obtain the result.

The vectors **a**, **b** and $\mathbf{c} = \mathbf{b} - \mathbf{a}$ in $\mathbb{R}^3$ lie in a plane and the law of cosines can still be applied to establish the result that

$$\langle \mathbf{a}, \mathbf{b} \rangle = \|\mathbf{a}\| \, \|\mathbf{b}\| \cos \theta,$$

where θ is the angle between the vectors.

Figure 1.8 The line $y = 2x$. The vector shown is $\mathbf{v} = (1, 2)^{\mathrm{T}}$

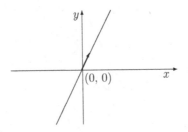

Activity 1.46 Calculate the angles of the triangle with sides $\mathbf{a}, \mathbf{b}, \mathbf{c}$ and show it is an isosceles right-angled triangle, where

$$\mathbf{a} = \begin{pmatrix} 1 \\ 2 \\ 2 \end{pmatrix}, \qquad \mathbf{b} = \begin{pmatrix} -1 \\ 1 \\ 4 \end{pmatrix}, \qquad \mathbf{c} = \mathbf{b} - \mathbf{a}.$$

1.10 Lines

1.10.1 Lines in $\mathbb{R}^2$

In $\mathbb{R}^2$, a line is given by a single Cartesian equation, such as $y = ax + b$, and, as such, we can draw a graph of the line in the xy-plane. This line can also be expressed as a single vector equation with one parameter. To see this, look at the following examples.

Example 1.47 Consider the line $y = 2x$. Any point (x, y) on this line must satisfy this equation, and all points that satisfy the equation are on this line (Figure 1.8).

Another way to describe the points on the line is by giving their position vectors. We can let $x = t$, where t is any real number. Then y is determined by $y = 2x = 2t$. So if $\mathbf{x} = (x, y)^{\mathrm{T}}$ is the position vector of a point on the line, then

$$\mathbf{x} = \begin{pmatrix} t \\ 2t \end{pmatrix} = t \begin{pmatrix} 1 \\ 2 \end{pmatrix} = t\mathbf{v}, \quad t \in \mathbb{R}.$$

For example, if $t = 2$, we get the position vector of the point $(2, 4)$ on the line, and if $t = -1$, we obtain the point $(-1, -2)$. As the *parameter* t runs through all real numbers, this vector equation gives the position vectors of all the points on the line.

Starting with the vector equation

$$\mathbf{x} = \begin{pmatrix} x \\ y \end{pmatrix} = t\mathbf{v} = t \begin{pmatrix} 1 \\ 2 \end{pmatrix}, \quad t \in \mathbb{R},$$

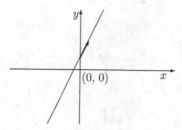

Figure 1.9 The line $y = 2x + 1$. The vector shown is $\mathbf{v} = (1, 2)^{\text{T}}$

we can retrieve the Cartesian equation using the fact that the two vectors are equal if and only if their components are equal. This gives us the two equations $x = t$ and $y = 2t$. Eliminating the parameter t between these two equations yields $y = 2x$.

The line in the above example is a line through the origin. What about a line which does not contain $(0, 0)$?

Example 1.48 Consider the line $y = 2x + 1$. Proceeding as above, we set $x = t$, $t \in \mathbb{R}$. Then $y = 2x + 1 = 2t + 1$, so the position vector of a point on this line is given by

$$\mathbf{x} = \begin{pmatrix} t \\ 2t + 1 \end{pmatrix} = \begin{pmatrix} 0 \\ 1 \end{pmatrix} + \begin{pmatrix} t \\ 2t \end{pmatrix} = \begin{pmatrix} 0 \\ 1 \end{pmatrix} + t \begin{pmatrix} 1 \\ 2 \end{pmatrix}, \quad t \in \mathbb{R}.$$

We can interpret this as follows. To locate any point on the line, first locate *one* particular point which is on the line, for example the y intercept, $(0, 1)$. Then the position vector of *any* point on the line is a sum of two displacements, first going to the point $(0, 1)$ and then going along the line, in a direction parallel to the vector $\mathbf{v} = (1, 2)^{\text{T}}$. It is important to notice that in this case the actual position vector of a point on the line does not lie along the line. Only if the line goes through the origin will that happen.

Activity 1.49 Sketch the line $y = 2x + 1$ and the position vector $\mathbf{q}$ of the point $(3, 7)$ which is on this line. Express $\mathbf{q}$ as the sum of two vectors, $\mathbf{q} = \mathbf{p} + t\mathbf{v}$ where $\mathbf{p} = (0, 1)^{\text{T}}$ and $\mathbf{v} = (1, 2)^{\text{T}}$ for some $t \in \mathbb{R}$, and add these vectors to your sketch.

In the vector equation, any point on the line can be used to locate the line, and any vector parallel to the direction vector, $\mathbf{v}$, can be used to give the direction. So, for example,

$$\begin{pmatrix} x \\ y \end{pmatrix} = \begin{pmatrix} 1 \\ 3 \end{pmatrix} + s \begin{pmatrix} -2 \\ -4 \end{pmatrix}, \quad s \in \mathbb{R}$$

is also a vector equation of this line.

Activity 1.50 If $\mathbf{q} = (3, 7)^T$, what is s in this expression of the line?

As before, we can retrieve the Cartesian equation of the line by equating components of the vector and eliminating the parameter.

Activity 1.51 Do this for each of the vector equations given above for the line $y = 2x + 1$.

In general, any line in $\mathbb{R}^2$ is given by a vector equation with one parameter of the form

$$\mathbf{x} = \mathbf{p} + t\mathbf{v},$$

where $\mathbf{x}$ is the position vector of a point on the line, $\mathbf{p}$ is any particular point on the line and $\mathbf{v}$ is the *direction* of the line.

Activity 1.52 Write down a vector equation of the line through the points $P = (-1, 1)$ and $Q = (3, 2)$. What is the direction of this line? Find a value for c such that the point $(7, c)$ is on the line.

In $\mathbb{R}^2$, two lines are either parallel or intersect in a unique point.

Example 1.53 The lines ℓ_1 and ℓ_2 are given by the following equations (for $t \in \mathbb{R}$)

$$\ell_1 : \begin{pmatrix} x \\ y \end{pmatrix} = \begin{pmatrix} 1 \\ 3 \end{pmatrix} + t \begin{pmatrix} 1 \\ 2 \end{pmatrix},$$

$$\ell_2 : \begin{pmatrix} x \\ y \end{pmatrix} = \begin{pmatrix} 5 \\ 6 \end{pmatrix} + t \begin{pmatrix} -2 \\ 1 \end{pmatrix}.$$

These lines are not parallel, since their direction vectors are not scalar multiples of one another. Therefore, they intersect in a unique point. We can find this point either by finding the Cartesian equation of each line and solving the equations simultaneously, or using the vector equations. We will do the latter. We are looking for a point (x, y) on both lines, so its position vector will satisfy

$$\begin{pmatrix} x \\ y \end{pmatrix} = \begin{pmatrix} 1 \\ 3 \end{pmatrix} + t \begin{pmatrix} 1 \\ 2 \end{pmatrix} = \begin{pmatrix} 5 \\ 6 \end{pmatrix} + s \begin{pmatrix} -2 \\ 1 \end{pmatrix}$$

for some $t \in \mathbb{R}$ and for some $s \in \mathbb{R}$. We need to use different symbols (s and t) in the equations because they are unlikely to be the same number for each line. We are looking for values of s and t which will give us the same point. Equating components of the position vectors of points on the lines, we have

$$\left. \begin{array}{l} 1 + t = 5 - 2s \\ 3 + 2t = 6 + s \end{array} \right\} \Rightarrow \begin{array}{l} 2s + t = 4 \\ -s + 2t = 3 \end{array} \Rightarrow \begin{array}{l} 2s + t = 4 \\ -2s + 4t = 6 \end{array}.$$

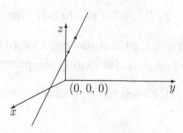

Figure 1.10 A line
in $\mathbb{R}^3$

Adding these last two equations, we obtain $t = 2$, and therefore $s = 1$.
Therefore, the point of intersection is $(3, 7)$:

$$\begin{pmatrix} 1 \\ 3 \end{pmatrix} + 2 \begin{pmatrix} 1 \\ 2 \end{pmatrix} = \begin{pmatrix} 3 \\ 7 \end{pmatrix} = \begin{pmatrix} 5 \\ 6 \end{pmatrix} + 1 \begin{pmatrix} -2 \\ 1 \end{pmatrix}.$$

What is the angle of intersection of these two lines? Since

$$\left\langle \begin{pmatrix} 1 \\ 2 \end{pmatrix}, \begin{pmatrix} -2 \\ 1 \end{pmatrix} \right\rangle = 0,$$

the lines are perpendicular.

1.10.2   Lines in $\mathbb{R}^3$

How can you describe a line in $\mathbb{R}^3$? Because there are three variables
involved, the natural way is to use a vector equation. To describe a line,
you locate one point on the line by its position vector, and then travel
along from that point in a given direction, or in the opposite direction
(Figure 1.10).

Therefore, a line in $\mathbb{R}^3$ is given by a vector equation with one
parameter,

$$\mathbf{x} = \mathbf{p} + t\mathbf{v}, \qquad t \in \mathbb{R},$$

where $\mathbf{x}$ is the position vector of any point on the line, $\mathbf{p}$ is the position
vector of one particular point on the line and $\mathbf{v}$ is the direction of the line,

$$\mathbf{x} = \begin{pmatrix} x \\ y \\ z \end{pmatrix} = \begin{pmatrix} p_1 \\ p_2 \\ p_3 \end{pmatrix} + t \begin{pmatrix} v_1 \\ v_2 \\ v_3 \end{pmatrix}, \qquad t \in \mathbb{R}. \tag{1.10.2}$$

The equation, $\mathbf{x} = t\mathbf{v}$ represents a parallel line through the origin.

Example 1.54 The equations

$$\mathbf{x} = \begin{pmatrix} 1 \\ 3 \\ 4 \end{pmatrix} + t \begin{pmatrix} 1 \\ 2 \\ -1 \end{pmatrix} \quad \text{and} \quad \mathbf{x} = \begin{pmatrix} 3 \\ 7 \\ 2 \end{pmatrix} + s \begin{pmatrix} -3 \\ -6 \\ 3 \end{pmatrix}, \qquad s, t \in \mathbb{R}$$

describe the same line. This is not obvious, so how do we show it?

The lines represented by these equations are parallel since their direction vectors are parallel

$$\begin{pmatrix} -3 \\ -6 \\ 3 \end{pmatrix} = -3 \begin{pmatrix} 1 \\ 2 \\ -1 \end{pmatrix},$$

so they either have no points in common and are parallel, or they have all points in common and are really the same line. Since

$$\begin{pmatrix} 3 \\ 7 \\ 2 \end{pmatrix} = \begin{pmatrix} 1 \\ 3 \\ 4 \end{pmatrix} + 2 \begin{pmatrix} 1 \\ 2 \\ -1 \end{pmatrix},$$

the point $(3, 7, 2)$ is on both lines, so they must have all points in common. We say that the lines are *coincident*.

On the other hand, the lines represented by the equations

$$\mathbf{x} = \begin{pmatrix} 1 \\ 3 \\ 4 \end{pmatrix} + t \begin{pmatrix} 1 \\ 2 \\ -1 \end{pmatrix} \quad \text{and} \quad \mathbf{x} = \begin{pmatrix} 3 \\ 7 \\ 1 \end{pmatrix} + t \begin{pmatrix} -3 \\ -6 \\ 3 \end{pmatrix}, \quad t \in \mathbb{R}$$

are parallel, with no points in common, since there is no value of t for which

$$\begin{pmatrix} 3 \\ 7 \\ 1 \end{pmatrix} = \begin{pmatrix} 1 \\ 3 \\ 4 \end{pmatrix} + t \begin{pmatrix} 1 \\ 2 \\ -1 \end{pmatrix}.$$

Activity 1.55 Verify this last statement.

Now try the following:

Activity 1.56 Write down a vector equation of the line through the points $P = (-1, 1, 2)$ and $Q = (3, 2, 1)$. What is the direction of this line? Is the point $(7, 1, 3)$ on this line? Suppose you want a point on this line of the form $(c, d, 3)$. Find one such point. How many choices do you actually have for the values of c and d?

We can also describe a line in $\mathbb{R}^3$ by Cartesian equations, but this time we need two such equations because there are three variables. Equating components in the vector equation 1.10.2 above, we have

$$x = p_1 + t v_1, \qquad y = p_2 + t v_2, \qquad z = p_3 + t v_3.$$

Solving each of these equations for the parameter t and equating the results, we have the two equations

$$\frac{x - p_1}{v_1} = \frac{y - p_2}{v_2} = \frac{z - p_3}{v_3}, \quad \text{provided } v_i \neq 0, \ i = 1, 2, 3.$$

Example 1.57 To find Cartesian equations of the line

$$\mathbf{x} = \begin{pmatrix} 1 \\ 2 \\ 3 \end{pmatrix} + t \begin{pmatrix} -1 \\ 0 \\ 5 \end{pmatrix}, \quad t \in \mathbb{R},$$

we equate components

$$x = 1 - t, \quad y = 2, \quad z = 3 + 5t,$$

and then solve for t in the first and third equation. The Cartesian equations are

$$1 - x = \frac{z - 3}{5} \quad \text{and} \quad y = 2.$$

This is a line parallel to the xz-plane in $\mathbb{R}^3$. The direction vector has a 0 in the second component, so there is no change in the y direction: the y coordinate has the constant value $y = 2$.

In $\mathbb{R}^2$, two lines are either parallel or intersect in a unique point. In $\mathbb{R}^3$, more can happen. Two lines in $\mathbb{R}^3$ either intersect in a unique point, are parallel or are *skew*, which means that they lie in parallel planes and are not parallel.

  Try to imagine what skew lines look like. If you are in a room with a ceiling parallel to the floor, imagine a line drawn in the ceiling. It is possible for you to draw a parallel line in the floor, but instead it is easier to draw a line in the floor which is not parallel to the one in the ceiling. These lines will be skew; they lie in parallel planes (the ceiling and the floor). If you could move the skew line in the floor onto the ceiling, then the lines would intersect in a unique point.

  Two lines are said to be *coplanar* if they lie in the same plane, in which case they are either parallel or intersecting.

Example 1.58 Are the following lines L_1 and L_2 intersecting, parallel or skew?

$$L_1 : \begin{pmatrix} x \\ y \\ z \end{pmatrix} = \begin{pmatrix} 1 \\ 3 \\ 4 \end{pmatrix} + t \begin{pmatrix} 1 \\ 2 \\ -1 \end{pmatrix}, \quad t \in \mathbb{R}$$

$$L_2 : \begin{pmatrix} x \\ y \\ z \end{pmatrix} = \begin{pmatrix} 5 \\ 6 \\ 1 \end{pmatrix} + t \begin{pmatrix} -2 \\ 1 \\ 7 \end{pmatrix}, \quad t \in \mathbb{R}.$$

Activity 1.59 Clearly, the lines are not parallel. Why?

The lines intersect if there is a point (x, y, z) on both lines; that is, if there exist values of the parameters, s, t such that

$$\begin{pmatrix} x \\ y \\ z \end{pmatrix} = \begin{pmatrix} 1 \\ 3 \\ 4 \end{pmatrix} + t \begin{pmatrix} 1 \\ 2 \\ -1 \end{pmatrix} = \begin{pmatrix} 5 \\ 6 \\ 1 \end{pmatrix} + s \begin{pmatrix} -2 \\ 1 \\ 7 \end{pmatrix}.$$

Equating components, we need to solve the three simultaneous equations in two unknowns,

$$\left. \begin{matrix} 1 + t = 5 - 2s \\ 3 + 2t = 6 + s \\ 4 - t = 1 + 7s \end{matrix} \right\} \Rightarrow \begin{matrix} 2s + t = 4 \\ -s + 2t = 3 \\ 7s + t = 3. \end{matrix}$$

We have already seen in Example 1.53 that the first two equations have the unique solution, $s = 1$, $t = 2$. Substituting these values into the third equation,

$$7s + t = 7(1) + 2 \neq 3,$$

we see that the system has no solution. Therefore, the lines do not intersect and must be skew.

Example 1.60 On the other hand, if we take a new line L_3, which is parallel to L_2 but which passes through the point $(5, 6, -5)$, then the lines

$$L_1 : \begin{pmatrix} x \\ y \\ z \end{pmatrix} = \begin{pmatrix} 1 \\ 3 \\ 4 \end{pmatrix} + t \begin{pmatrix} 1 \\ 2 \\ -1 \end{pmatrix},$$

$$L_3 : \begin{pmatrix} x \\ y \\ z \end{pmatrix} = \begin{pmatrix} 5 \\ 6 \\ -5 \end{pmatrix} + t \begin{pmatrix} -2 \\ 1 \\ 7 \end{pmatrix}, \quad t \in \mathbb{R}$$

do intersect in the unique point $(3, 7, 2)$.

Activity 1.61 Check this. Find the point of intersection of the two lines L_1 and L_3.

1.11   Planes in $\mathbb{R}^3$

On a line, there is essentially one direction in which a point can move, given as all possible scalar multiples of a given direction, but on a plane there are more possibilities. A point can move in two different directions, and in any linear combination of these two directions. So how do we describe a plane in $\mathbb{R}^3$?

The vector parametric equation

$$\mathbf{x} = \mathbf{p} + s\mathbf{v} + t\mathbf{w}, \qquad s, t, \in \mathbb{R}$$

describes the position vectors of points on a plane in $\mathbb{R}^3$ provided that
the vectors $\mathbf{v}$ and $\mathbf{w}$ are non-zero and are not parallel. The vector $\mathbf{p}$ is
the position vector of any particular point on the plane and the vectors
$\mathbf{v}$ and $\mathbf{w}$ are displacement vectors which lie in the plane. By taking all
possible linear combinations $\mathbf{x} = \mathbf{p} + s\mathbf{v} + t\mathbf{w}$, for $s, t \in \mathbb{R}$, we obtain
all the points on the plane.

The equation

$$\mathbf{x} = s\mathbf{v} + t\mathbf{w}, \qquad s, t, \in \mathbb{R}$$

describes a plane through the origin. In this case, the position vector, $\mathbf{x}$,
of any point on the plane lies in the plane.

Activity 1.62 If $\mathbf{v}$ and $\mathbf{w}$ are parallel, what does the equation $\mathbf{x} =$
$\mathbf{p} + s\mathbf{v} + t\mathbf{w}$, $s, t \in \mathbb{R}$, actually represent?

Example 1.63 You have shown that the lines L_1 and L_3 given in example 1.60 intersect in the point $(3, 7, 2)$. Two intersecting lines determine
a plane. A vector equation of the plane containing the two lines is given
by

$$\begin{pmatrix} x \\ y \\ z \end{pmatrix} = \begin{pmatrix} 3 \\ 7 \\ 2 \end{pmatrix} + s \begin{pmatrix} 1 \\ 2 \\ -1 \end{pmatrix} + t \begin{pmatrix} -2 \\ 1 \\ 7 \end{pmatrix}, \qquad s, t \in \mathbb{R}.$$

Why? We know that $(3, 7, 2)$ is a point on the plane, and the directions
of each of the lines must lie in the plane. As s and t run through all
real numbers, this equation gives the position vector of all points on
the plane. Since the point $(3, 7, 2)$ is on both lines, if $t = 0$ we have the
equation of L_1, and if $s = 0$, we get L_3.

Any point which is on the plane can take the place of the vector
$(3, 7, 2)^{\mathrm{T}}$, and any non-parallel vectors which are linear combinations
of $\mathbf{v}$ and $\mathbf{w}$ can replace these in the equation. So, for example,

$$\begin{pmatrix} x \\ y \\ z \end{pmatrix} = \begin{pmatrix} 1 \\ 3 \\ 4 \end{pmatrix} + t \begin{pmatrix} 1 \\ 2 \\ -1 \end{pmatrix} + s \begin{pmatrix} -3 \\ -1 \\ 8 \end{pmatrix}, \qquad s, t \in \mathbb{R}$$

is also an equation of this plane.

Activity 1.64 Verify this. Show that $(1, 3, 4)$ is a point on the plane
given by each equation, and show that $(-3, -1, 8)^{\mathrm{T}}$ is a linear combination of $(1, 2, -1)^{\mathrm{T}}$ and $(-2, 1, 7)^{\mathrm{T}}$.

There is another way to describe a plane in $\mathbb{R}^3$ geometrically, which
is often easier to use. We begin with planes through the origin. Let $\mathbf{n}$
be a given vector in $\mathbb{R}^3$ and consider all position vectors $\mathbf{x}$ which are

orthogonal to **n**. Geometrically, the set of all such vectors describes a plane through the origin in $\mathbb{R}^3$.

Try to imagine this by placing a pencil perpendicular to a table top. The pencil represents a normal vector, the table top a plane, and the point where the pencil is touching the table is the origin of your coordinate system. Then any vector which you can draw on the table top is orthogonal to the pencil, and conversely any point on the table top can be reached by a directed line segment (from the point where the pencil touches the table) which is orthogonal to the pencil.

A vector **x** is orthogonal to **n** if and only if

$$\langle \mathbf{n}, \mathbf{x} \rangle = 0,$$

so this equation gives the position vectors, **x**, of points on the plane. If $\mathbf{n} = (a, b, c)^{\mathrm{T}}$ and $\mathbf{x} = (x, y, z)^{\mathrm{T}}$, then this equation can be written as

$$\langle \mathbf{n}, \mathbf{x} \rangle = \left\langle \begin{pmatrix} a \\ b \\ c \end{pmatrix}, \begin{pmatrix} x \\ y \\ z \end{pmatrix} \right\rangle = 0$$

or

$$ax + by + cz = 0.$$

This is a Cartesian equation of a plane through the origin in $\mathbb{R}^3$. The vector **n** is called a *normal* vector to the plane. Any vector which is parallel to **n** will also be a normal vector and will lead to the same Cartesian equation.

On the other hand, given any Cartesian equation of the form

$$ax + by + cz = 0,$$

then this equation represents a plane through the origin in $\mathbb{R}^3$ with normal vector $\mathbf{n} = (a, b, c)^{\mathrm{T}}$.

To describe a plane which does not go through the origin, we choose a normal vector **n** and one point P on the plane with position vector **p**. We then consider all displacement vectors which lie in the plane with initial point at P. If **x** is the position vector of any point on the plane, then the displacement vector $\mathbf{x} - \mathbf{p}$ lies in the plane, and $\mathbf{x} - \mathbf{p}$ is orthogonal to **n**. Conversely, if the position vector **x** of a point satisfies $\langle \mathbf{n}, \mathbf{x} - \mathbf{p} \rangle = 0$, then the vector $\mathbf{x} - \mathbf{p}$ lies in the plane, so the point (with position vector **x**) is on the plane.

(Again, think about the pencil perpendicular to the table top, only this time the point where the pencil is touching the table is a point, P, on the plane, and the origin of your coordinate system is somewhere else; say, in the corner on the floor.)

The orthogonality condition means that the position vector of any point on the plane is given by the equation

$$\langle \mathbf{n}, \mathbf{x} - \mathbf{p} \rangle = 0.$$

Using properties of the inner product, we can rewrite this as

$$\langle \mathbf{n}, \mathbf{x} \rangle = \langle \mathbf{n}, \mathbf{p} \rangle,$$

where $\langle \mathbf{n}, \mathbf{p} \rangle = d$ is a constant.

If $\mathbf{n} = (a, b, c)^T$ and $\mathbf{x} = (x, y, z)^T$, then

$$ax + by + cz = d$$

is a Cartesian equation of a plane in $\mathbb{R}^3$. The plane goes through the origin if and only if $d = 0$.

Example 1.65 The equation

$$2x - 3y - 5z = 2$$

represents a plane which does not go through the origin, since $(x, y, z) = (0, 0, 0)$ does not satisfy the equation. To find a point on the plane, we can choose any two of the coordinates, say $y = 0$ and $z = 0$, and then the equation tells us that $x = 1$. So the point $(1, 0, 0)$ is on this plane. The components of a normal to the plane can be read from this equation as the coefficients of x, y, z: $\mathbf{n} = (2, -3, -5)^T$.

How does the Cartesian equation of a plane relate to the vector parametric equation of a plane? A Cartesian equation can be obtained from the vector equation algebraically, by eliminating the parameters in the vector equation, and vice versa, as the following example shows.

Example 1.66 Consider the plane

$$\begin{pmatrix} x \\ y \\ z \end{pmatrix} = s \begin{pmatrix} 1 \\ 2 \\ -1 \end{pmatrix} + t \begin{pmatrix} -2 \\ 1 \\ 7 \end{pmatrix} = s\mathbf{v} + t\mathbf{w}, \quad s, t \in \mathbb{R},$$

which is a plane through the origin parallel to the plane in Example 1.63. The direction vectors $\mathbf{v} = (1, 2, -1)^T$ and $\mathbf{w} = (-2, 1, 7)$ lie in the plane.

To obtain a Cartesian equation in x, y and z, we equate the components in this vector equation.

$$x = s - 2t$$
$$y = 2s + t$$
$$z = -s + 7t$$

and eliminate the parameters s and t. We begin by solving the first equation for s, and then substitute this into the second equation to solve for t in terms of x and y,

$$s = x + 2t$$
$$\Rightarrow y = 2(x + 2t) + t = 2x + 5t$$
$$\Rightarrow 5t = y - 2x$$
$$\Rightarrow t = \frac{y - 2x}{5}.$$

We then substitute back into the first equation to obtain s in terms of x and y,

$$s = x + 2\left(\frac{y - 2x}{5}\right) \Rightarrow 5s = 5x + 2y - 4x \Rightarrow s = \frac{x + 2y}{5}.$$

Finally, we substitute for s and t in the third equation, $z = -s + 7t$, and simplify to obtain a Cartesian equation of the plane

$$3x - y + z = 0.$$

Activity 1.67 Carry out this last step to obtain the Cartesian equation of the plane.

This Cartesian equation can be expressed as $\langle \mathbf{n}, \mathbf{x} \rangle = 0$, where

$$\mathbf{n} = \begin{pmatrix} 3 \\ -1 \\ 1 \end{pmatrix}, \quad \mathbf{x} = \begin{pmatrix} x \\ y \\ z \end{pmatrix}.$$

The vector $\mathbf{n}$ is a normal vector to the plane. We can check that $\mathbf{n}$ is, indeed, orthogonal to the plane by taking the inner product with the vectors $\mathbf{v}$ and $\mathbf{w}$, which lie *in* the plane.

Activity 1.68 Do this. Calculate $\langle \mathbf{n}, \mathbf{v} \rangle$ and $\langle \mathbf{n}, \mathbf{w} \rangle$, and verify that both inner products are equal to 0.

Since $\mathbf{n}$ is orthogonal to both $\mathbf{v}$ and $\mathbf{w}$, it is orthogonal to all linear combinations of these vectors, and hence to any vector in the plane. So this plane can equally be described as the set of all position vectors which are orthogonal to $\mathbf{n}$.

Activity 1.69 Using the properties of inner product, show that this last statement is true. That is, if $\langle \mathbf{n}, \mathbf{v} \rangle = 0$ and $\langle \mathbf{n}, \mathbf{w} \rangle = 0$, then $\langle \mathbf{n}, s\mathbf{v} + t\mathbf{w} \rangle = 0$, for any $s, t \in \mathbb{R}$.

Can we do the same for a plane which does not pass through the origin? Consider the following example.

Example 1.70 The plane we just considered in Example 1.66 is parallel to the plane with vector equation

$$\begin{pmatrix} x \\ y \\ z \end{pmatrix} = \begin{pmatrix} 3 \\ 7 \\ 2 \end{pmatrix} + s \begin{pmatrix} 1 \\ 2 \\ -1 \end{pmatrix} + t \begin{pmatrix} -2 \\ 1 \\ 7 \end{pmatrix} = \mathbf{p} + s\mathbf{v} + t\mathbf{w}, \quad s, t \in \mathbb{R},$$

which passes through the point $(3, 7, 2)$. Since the planes are parallel, they will have the same normal vectors. So the Cartesian equation of this plane is of the form

$$3x - y + z = d.$$

Since $(3, 7, 2)$ is a point on the plane, it must satisfy the equation for the plane. Substituting into the equation we find $d = 3(3) - (7) + (2) = 4$ (which is equivalent to finding d by using $d = \langle \mathbf{n}, \mathbf{p} \rangle$). So the Cartesian equation we obtain is

$$3x - y + z = 4.$$

Conversely, starting with a Cartesian equation of a plane, we can obtain a vector equation. Consider the plane just discussed. We are looking for the position vector of a point on the plane whose components satisfy $3x - y + z = 4$, or, equivalently, $z = 4 - 3x + y$. (We can solve for any one of the variables x, y or z, but we chose z for simplicity.) So we are looking for all vectors $\mathbf{x}$ such that

$$\begin{pmatrix} x \\ y \\ z \end{pmatrix} = \begin{pmatrix} x \\ y \\ 4 - 3x + y \end{pmatrix} = \begin{pmatrix} 0 \\ 0 \\ 4 \end{pmatrix} + x \begin{pmatrix} 1 \\ 0 \\ -3 \end{pmatrix} + y \begin{pmatrix} 0 \\ 1 \\ 1 \end{pmatrix}$$

for any $x, y \in \mathbb{R}$. Therefore,

$$\begin{pmatrix} x \\ y \\ z \end{pmatrix} = \begin{pmatrix} 0 \\ 0 \\ 4 \end{pmatrix} + s \begin{pmatrix} 1 \\ 0 \\ -3 \end{pmatrix} + t \begin{pmatrix} 0 \\ 1 \\ 1 \end{pmatrix}, \quad s, t \in \mathbb{R}$$

is a vector equation of the same plane as that given by the original vector equation,

$$\begin{pmatrix} x \\ y \\ z \end{pmatrix} = \begin{pmatrix} 3 \\ 7 \\ 2 \end{pmatrix} + s \begin{pmatrix} 1 \\ 2 \\ -1 \end{pmatrix} + t \begin{pmatrix} -2 \\ 1 \\ 7 \end{pmatrix}, \quad s, t \in \mathbb{R}.$$

It is difficult to spot at a glance that these two different vector equations in fact describe the same plane. There are many ways to show this, but we can use what we know about planes to find the easiest. The planes represented by the two vector equations have the same normal vector $\mathbf{n}$, since the vectors $(1, 0, -3)^T$ and $(0, 1, 1)^T$ are also orthogonal to $\mathbf{n}$. So we know that the two vector equations represent parallel planes. They are the same plane if they have a point in common. It is far easier to find values of s and t for which $\mathbf{p} = (3, 7, 2)^{\mathrm{T}}$ satisfies the new vector equation

$$\begin{pmatrix} 3 \\ 7 \\ 2 \end{pmatrix} = \begin{pmatrix} 0 \\ 0 \\ 4 \end{pmatrix} + s \begin{pmatrix} 1 \\ 0 \\ -3 \end{pmatrix} + t \begin{pmatrix} 0 \\ 1 \\ 1 \end{pmatrix}, \quad s, t \in \mathbb{R}$$

than the other way around (which is by showing that $(0, 0, 4)$ satisfies the original equation) because of the positions of the zeros and ones in these direction vectors.

Activity 1.71 Do this. You should be able to immediately spot the values of s and t which work.

Using the examples we have just done, you should now be able to tackle the following activity:

Activity 1.72 The two lines, L_1 and L_2,

$$L_1: \begin{pmatrix} x \\ y \\ z \end{pmatrix} = \begin{pmatrix} 1 \\ 3 \\ 4 \end{pmatrix} + t \begin{pmatrix} 1 \\ 2 \\ -1 \end{pmatrix},$$

$$L_2: \begin{pmatrix} x \\ y \\ z \end{pmatrix} = \begin{pmatrix} 5 \\ 6 \\ 1 \end{pmatrix} + t \begin{pmatrix} -2 \\ 1 \\ 7 \end{pmatrix}, \quad t \in \mathbb{R}$$

in Example 1.58 are skew, and therefore are contained in parallel planes. Find vector equations and Cartesian equations for these two planes.

Two planes in $\mathbb{R}^3$ are either parallel or intersect in a line. Considering such questions, it is usually easier to use the Cartesian equations of the planes. If the planes are parallel, then this will be obvious from looking at their normal vectors. If they are not parallel, then the line of intersection can be found by solving the two Cartesian equations simultaneously.

Example 1.73 The planes

$$x + 2y - 3z = 0 \quad \text{and} \quad -2x - 4y + 6z = 4$$

are parallel, since their normal vectors are related by

$$(-2, -4, 6)^{\mathrm{T}} = -2(1, 2, -3)^{\mathrm{T}}.$$

The equations do not represent the same plane, since they have no points in common; that is, there are no values of x, y, z which can satisfy both equations. The first plane goes through the origin and the second plane does not.

On the other hand, the planes

$$x + 2y - 3z = 0 \quad \text{and} \quad x - 2y + 5z = 4$$

intersect in a line. The points of intersection are the points (x, y, z) which satisfy *both* equations, so we solve the equations simultaneously. We begin by eliminating the variable x from the second equation, by subtracting the first equation from the second. This will naturally lead us to a vector equation of the line of intersection:

$$\left. \begin{array}{c} x + 2y - 3z = 0 \\ x - 2y + 5z = 4 \end{array} \right\} \quad \Rightarrow \quad \begin{array}{c} x + 2y - 3z = 0 \\ -4y + 8z = 4 \,. \end{array}$$

This last equations tells us that if $z = t$ is any real number, then $y = -1 + 2t$. Substituting these expressions into the first equation, we find $x = 2 - t$. Then a vector equation of the line of intersection is

$$\begin{pmatrix} x \\ y \\ z \end{pmatrix} = \begin{pmatrix} 2 - t \\ -1 + 2t \\ t \end{pmatrix} = \begin{pmatrix} 2 \\ -1 \\ 0 \end{pmatrix} + t \begin{pmatrix} -1 \\ 2 \\ 1 \end{pmatrix}.$$

This can be verified by showing that the point $(2, -1, 0)$ satisfies both Cartesian equations, and that the vector $\mathbf{v} = (-1, 2, 1)^\mathrm{T}$ is orthogonal to the normal vectors of each of the planes (and therefore lies in both planes).

Activity 1.74 Carry out the calculations in the above example and verify that the line is in both planes.

1.12 Lines and hyperplanes in $\mathbb{R}^n$

1.12.1 Vectors and lines in $\mathbb{R}^n$

We can apply similar geometric language to vectors in $\mathbb{R}^n$. We can think of the vector $\mathbf{a} = (a_1, a_2, \ldots, a_n)^\mathrm{T}$ as defining a point in $\mathbb{R}^n$. Using the inner product (defined in Section 1.8.2), we define the *length* of a vector $\mathbf{x} = (x_1, x_2, \ldots, x_n)^\mathrm{T}$ by

$$\|\mathbf{x}\| = \sqrt{x_1^2 + x_2^2 + \cdots + x_n^2} \quad \text{or} \quad \|\mathbf{x}\|^2 = \langle \mathbf{x}, \mathbf{x} \rangle.$$

We say that two vectors, $\mathbf{v}, \mathbf{w} \in \mathbb{R}^n$ are *orthogonal* if and only if

$$\langle \mathbf{v}, \mathbf{w} \rangle = 0.$$

A *line* in $\mathbb{R}^n$ is the set of all points $(x_1, x_2, \ldots, x_n)$ whose position vectors $\mathbf{x}$ satisfy a vector equation of the form

$$\mathbf{x} = \mathbf{p} + t\mathbf{v}, \quad t \in \mathbb{R},$$

where $\mathbf{p}$ is the position vector of one particular point on the line and $\mathbf{v}$ is the direction of the line. If we can write $\mathbf{x} = t\mathbf{v}$, $t \in \mathbb{R}$, then the line goes through the origin.

1.12.2 Hyperplanes

The set of all points $(x_1, x_2, \ldots, x_n)$ which satisfy one Cartesian equation,

$$a_1 x_1 + a_2 x_2 + \cdots + a_n x_n = d,$$

is called a *hyperplane* in $\mathbb{R}^n$.

In $\mathbb{R}^2$, a hyperplane is a line, and in $\mathbb{R}^3$ it is a plane, but for $n > 3$ we simply use the term hyperplane. The vector

$$\mathbf{n} = \begin{pmatrix} a_1 \\ a_2 \\ \vdots \\ a_n \end{pmatrix}$$

is a normal vector to the hyperplane. Writing the Cartesian equation in vector form, a hyperplane is the set of all vectors, $\mathbf{x} \in \mathbb{R}^n$ such that

$$\langle \mathbf{n}, \mathbf{x} - \mathbf{p} \rangle = 0,$$

where the normal vector $\mathbf{n}$ and the position vector $\mathbf{p}$ of a point on the hyperplane are given.

Activity 1.75 How many Cartesian equations would you need to describe a line in $\mathbb{R}^n$? How many parameters would there be in a vector equation of a hyperplane?

1.13 Learning outcomes

You should now be able to:

- explain what is meant by a matrix
- use matrix addition, scalar multiplication and matrix multiplication appropriately (and know when and how these operations are defined)
- manipulate matrices algebraically

- state what is meant by the inverse of a square matrix, a power of a square matrix and the transpose of a matrix, and know the properties of these in order to manipulate them
- explain what is meant by a vector and by Euclidean n-space
- state what is meant by the inner product of two vectors and what properties it satisfies
- state what is meant by the length and direction of a vector, and what is meant by a unit vector
- state the relationship between the inner product and the length of a vector and angle between two vectors
- explain what is meant by two vectors being orthogonal and how to determine this
- find the equations, vector and Cartesian, of lines in $\mathbb{R}^2$, lines and planes in $\mathbb{R}^3$, and work problems involving lines and planes
- state what is meant by a line and by a hyperplane in $\mathbb{R}^n$.

1.14　Comments on activities

Activity 1.3 For this matrix, $a_{32} = 9$.

Activity 1.4 Only the second matrix is diagonal.

Activity 1.11 AB is 2×2 and BA is 3×3,

$$AB = \begin{pmatrix} 10 & 5 \\ 6 & 2 \end{pmatrix} \qquad BA = \begin{pmatrix} 7 & 5 & 10 \\ 2 & 1 & 3 \\ 3 & 3 & 4 \end{pmatrix}.$$

Activity 1.12 $AB = \begin{pmatrix} 1 & 3 \\ 3 & 7 \end{pmatrix} \qquad BA = \begin{pmatrix} 4 & 6 \\ 3 & 4 \end{pmatrix}.$

Activity 1.13 If A is $m \times n$ and B is $n \times p$, then AB is an $m \times p$ matrix. The size of a matrix is not changed by scalar multiplication, so both $\lambda(AB)$ and $(\lambda A)B$ are $m \times p$. Looking at the (i, j) entries of each,

$$(\lambda(AB))_{ij} = \lambda \left(a_{i1}b_{1j} + a_{i2}b_{2j} + \ldots + a_{in}b_{nj} \right)$$
$$= \lambda a_{i1}b_{1j} + \lambda a_{i2}b_{2j} + \ldots + \lambda a_{in}b_{nj}$$
$$= ((\lambda A)B)_{ij},$$

so these two matrices are equal.

Activity 1.16 In this case, I is $m \times m$.

Activity 1.22 $AB = \begin{pmatrix} 1 & 2 \\ 3 & 4 \end{pmatrix} \begin{pmatrix} -2 & 1 \\ \frac{3}{2} & -\frac{1}{2} \end{pmatrix} = \begin{pmatrix} 1 & 0 \\ 0 & 1 \end{pmatrix}$

and $\quad BA = \begin{pmatrix} -2 & 1 \\ \frac{3}{2} & -\frac{1}{2} \end{pmatrix} \begin{pmatrix} 1 & 2 \\ 3 & 4 \end{pmatrix} = \begin{pmatrix} 1 & 0 \\ 0 & 1 \end{pmatrix}.$

Therefore, $A^{-1} = \begin{pmatrix} -2 & 1 \\ \frac{3}{2} & -\frac{1}{2} \end{pmatrix}.$

Activity 1.24 We will show one way (namely, that $AA^{-1} = I$), but you should also show that $A^{-1}A = I$.

$$AA^{-1} = \begin{pmatrix} a & b \\ c & d \end{pmatrix} \frac{1}{ad-bc} \begin{pmatrix} d & -b \\ -c & a \end{pmatrix}$$

$$= \frac{1}{ad-bc} \begin{pmatrix} ad-bc & -ab+ba \\ cd-dc & -bc+ad \end{pmatrix}$$

$$= \begin{pmatrix} 1 & 0 \\ 0 & 1 \end{pmatrix}.$$

Activity 1.26 We will do the first, and leave the others to you. The inverse of A^r is a matrix B such that $A^r B = B A^r = I$. So show that the matrix $B = (A^{-1})^r$ works:

$$A^r (A^{-1})^r = (\underbrace{A A \dots A}_{r \text{ times}})(\underbrace{A^{-1} A^{-1} \dots A^{-1}}_{r \text{ times}}).$$

Removing the brackets (matrix multiplication is associative) and replacing each central $AA^{-1} = I$, the resultant will eventually be $AIA^{-1} = AA^{-1} = I$. To complete the proof, show also that $(A^{-1})^r A^r = I$. Therefore, $(A^r)^{-1} = (A^{-1})^r$.

Activity 1.29 Given the sizes of A and B, the matrix AB is $m \times p$, so $(AB)^T$ is $p \times m$. Also, A^T is $n \times m$ and B^T is $p \times n$, so the only way these matrices can be multiplied is as $B^T A^T$ (unless $m = p$).

Activity 1.30 The (i, j) entry of $B^T A^T$ is obtained by taking row i of B^T, which is column i of B and multiplying each term by the corresponding entry of column j of A^T, which is row j of A, and then summing the products:

$$\left(B^T A^T\right)_{ij} = b_{1i}a_{j1} + b_{2i}a_{j2} + \dots + b_{1n}a_{jn}.$$

This produces the same scalar as the (i, j) entry of $(AB)^T$.

Activity 1.32 The matrix is

$$A = \begin{pmatrix} 1 & 4 & 5 \\ 4 & 2 & -7 \\ 5 & -7 & 3 \end{pmatrix} = A^{\mathrm{T}}.$$

Activity 1.33

$$\mathbf{a}^{\mathrm{T}}\mathbf{b} = (1 \quad 2 \quad 3) \begin{pmatrix} 4 \\ -2 \\ 1 \end{pmatrix} = (3)$$

$$\mathbf{a}\mathbf{b}^{\mathrm{T}} = \begin{pmatrix} 1 \\ 2 \\ 3 \end{pmatrix} (4 \quad -2 \quad 1) = \begin{pmatrix} 4 & -2 & 1 \\ 8 & -4 & 2 \\ 12 & -6 & 3 \end{pmatrix}.$$

Activity 1.37 To prove properties (ii) and (iii), apply the definition to the LHS (left-hand side) of the equation and rearrange the terms to obtain the RHS (right-hand side). For example, for $\mathbf{x}$, $\mathbf{y} \in \mathbb{R}^n$, using the properties of real numbers:

$$\alpha \langle \mathbf{x}, \mathbf{y} \rangle = \alpha(x_1 y_1 + x_2 y_2 + \cdots + x_n y_n)$$
$$= \alpha x_1 y_1 + \alpha x_2 y_2 + \cdots + \alpha x_n y_n$$
$$= (\alpha x_1) y_1 + (\alpha x_2) y_2 + \cdots + (\alpha x_n) y_n = \langle \alpha \mathbf{x}, \mathbf{y} \rangle.$$

Do the same for property (iii).

The single property $\langle \alpha \mathbf{x} + \beta \mathbf{y}, \mathbf{z} \rangle = \alpha \langle \mathbf{x}, \mathbf{z} \rangle + \beta \langle \mathbf{y}, \mathbf{z} \rangle$ implies property (ii) by letting $\beta = 0$ for the first equality and then letting $\alpha = 0$ for the second, and property (iii) by letting $\alpha = \beta = 1$. On the other hand, if properties (ii) and (iii) hold, then

$$\langle \alpha \mathbf{x} + \beta \mathbf{y}, \mathbf{z} \rangle = \langle \alpha \mathbf{x}, \mathbf{z} \rangle + \langle \beta \mathbf{y}, \mathbf{z} \rangle \qquad \text{by property (iii)}$$
$$= \alpha \langle \mathbf{x}, \mathbf{z} \rangle + \beta \langle \mathbf{y}, \mathbf{z} \rangle \qquad \text{by property (ii)} .$$

Activity 1.42 $\|\mathbf{a}\| = 5$, so

$$\mathbf{u} = \frac{1}{5} \begin{pmatrix} 4 \\ 3 \end{pmatrix} \quad \text{and} \quad \mathbf{w} = -\frac{2}{5} \begin{pmatrix} 4 \\ 3 \end{pmatrix}.$$

Activity 1.45 In the figure below

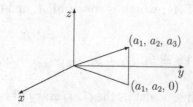

the line from the origin to the point $(a_1, a_2, 0)$ lies in the xy-plane, and by Pythagoras' theorem, it has length $\sqrt{a_1^2 + a_2^2}$. Applying Pythagoras' theorem again to the right triangle shown, we have

$$\|\mathbf{a}\| = \sqrt{\left(\sqrt{a_1^2 + a_2^2}\right)^2 + (a_3)^2} = \sqrt{a_1^2 + a_2^2 + a_3^2}.$$

Activity 1.46 We have

$$\mathbf{a} = \begin{pmatrix} 1 \\ 2 \\ 2 \end{pmatrix}, \qquad \mathbf{b} = \begin{pmatrix} -1 \\ 1 \\ 4 \end{pmatrix}, \qquad \mathbf{c} = \mathbf{b} - \mathbf{a} = \begin{pmatrix} -2 \\ -1 \\ 2 \end{pmatrix}.$$

The cosines of the three angles are given by

$$\frac{\langle \mathbf{a}, \mathbf{b} \rangle}{\|\mathbf{a}\| \|\mathbf{b}\|} = \frac{-1 + 2 + 8}{\sqrt{9}\sqrt{18}} = \frac{1}{\sqrt{2}}$$

$$\frac{\langle \mathbf{a}, \mathbf{c} \rangle}{\|\mathbf{a}\| \|\mathbf{c}\|} = \frac{-2 - 2 + 4}{\sqrt{9}\sqrt{9}} = 0;$$

$$\frac{\langle \mathbf{b}, \mathbf{c} \rangle}{\|\mathbf{b}\| \|\mathbf{c}\|} = \frac{2 - 1 + 8}{\sqrt{18}\sqrt{9}} = \frac{1}{\sqrt{2}}.$$

Thus, the triangle has a right-angle, and two angles of $\pi/4$.

Alternatively, as the vectors $\mathbf{a}$ and $\mathbf{c}$ are orthogonal, and have the same length, it follows immediately that the triangle is right-angled and isosceles.

Activity 1.49 If $t = 3$, then $\mathbf{q} = (3, 7)^{\mathsf{T}}$. You are asked to sketch the position vector $\mathbf{q}$ as this sum to illustrate that the vector $\mathbf{q}$ does locate a point on the line, but the vector $\mathbf{q}$ does *not* lie on the line.

Activity 1.50 Here $s = -1$.

Activity 1.51 We will work through this for the second equation and leave the first for you. We have, for $s \in \mathbb{R}$,

$$\begin{pmatrix} x \\ y \end{pmatrix} = \begin{pmatrix} 1 \\ 3 \end{pmatrix} + s \begin{pmatrix} -2 \\ -4 \end{pmatrix} \Rightarrow \begin{cases} x = 1 - 2s \\ y = 3 - 4s \end{cases} \Rightarrow \frac{1 - x}{2} = s = \frac{3 - y}{4},$$

which yields $2(1 - x) = 3 - y$ or $y = 2x + 1$.

Activity 1.52 A vector equation of the line is

$$\mathbf{x} = \begin{pmatrix} -1 \\ 1 \end{pmatrix} + t \begin{pmatrix} 4 \\ 1 \end{pmatrix} = \mathbf{p} + t\mathbf{v}, \qquad t \in \mathbb{R},$$

where we have used **p** to locate a point on the line, and the direction vector, **v** = **q** − **p**. The point $(7, 3)$ is on the line $(t = 2)$, and this is the only point of this form on the line, since once 7 is chosen for the x coordinate, the y coordinate is determined.

Activity 1.55 Once given, for example, that the x coordinate is $x = 3$, the parameter t of the vector equation is determined. Therefore, so too are the other two coordinates. We saw in the example that $t = 2$ satisfies the first two equations and it certainly does not satisfy the third equation, $1 = 0 - t$.

Activity 1.56 This is similar to the earlier activity in $\mathbb{R}^2$. A vector equation of the line is

$$\mathbf{x} = \begin{pmatrix} -1 \\ 1 \\ 2 \end{pmatrix} + t \begin{pmatrix} 4 \\ 1 \\ -1 \end{pmatrix} = \mathbf{p} + t\mathbf{v}, \quad t \in \mathbb{R}.$$

The point $(7, 1, 3)$ is not on this line, but the point $(-5, 0, 3)$ is on the line. The value $t = -1$ will then satisfy all three component equations. There is, of course, only one possible choice for the values of c and d.

Activity 1.59 The lines are not parallel because their direction vectors are not parallel.

Activity 1.62 If **v** and **w** are parallel, then this equation represents a line in the direction **v**. If **w** = λ**v**, then this line can be written as

$$\mathbf{x} = \mathbf{p} + (s + \lambda t)\mathbf{v}, \quad \text{where } r = s + \lambda t \in \mathbb{R}.$$

Activity 1.69 Using the properties of the inner product, we have for any $s, t \in \mathbb{R}$,

$$\langle \mathbf{n}, s\mathbf{v} + t\mathbf{w} \rangle = s \langle \mathbf{n}, \mathbf{v} \rangle + t \langle \mathbf{n}, \mathbf{w} \rangle = s \cdot 0 + t \cdot 0 = 0.$$

Activity 1.71 Equating components in the vector equation, we have $3 = s$ and $7 = t$ from the first two equations, and these values do satisfy the third equation, $2 = 4 - 3s + t$.

Activity 1.72 The parallel planes must each contain the direction vectors of each of the lines as displacement vectors, so the vector equations of the planes are, respectively

$$\begin{pmatrix} x \\ y \\ z \end{pmatrix} = \begin{pmatrix} 1 \\ 3 \\ 4 \end{pmatrix} + s \begin{pmatrix} 1 \\ 2 \\ -1 \end{pmatrix} + t \begin{pmatrix} -2 \\ 1 \\ 7 \end{pmatrix}$$

and

$$\begin{pmatrix} x \\ y \\ z \end{pmatrix} = \begin{pmatrix} 5 \\ 6 \\ 1 \end{pmatrix} + s \begin{pmatrix} 1 \\ 2 \\ -1 \end{pmatrix} + t \begin{pmatrix} -2 \\ 1 \\ 7 \end{pmatrix},$$

where $s, t \in \mathbb{R}$.

The parallel planes have the same normal vector, which we need for the Cartesian equations. Recall that in Example 1.70, we found a Cartesian equation and a normal vector to the first plane, the plane which contains L_1:

$$3x - y + z = 4, \quad \text{with} \quad \mathbf{n} = \begin{pmatrix} 3 \\ -1 \\ 1 \end{pmatrix}.$$

Note that the point $(1, 3, 4)$ is on this plane because it satisfies the equation, but the point $(5, 6, 1)$ does not. Substituting $(5, 6, 1)$ into the equation $3x - y + z = d$, we find the Cartesian equation of the parallel plane which contains L_2 is

$$3x - y + z = 10.$$

Activity 1.74 As stated, to verify that the line is in both planes, show that its direction vector is perpendicular to the normal vector of each plane, and that the point $(2, -1, 0)$ satisfies both equations.

Activity 1.75 To describe a line in $\mathbb{R}^n$, you need $n - 1$ Cartesian equations. A vector parametric equation of a hyperplane in $\mathbb{R}^n$ would require $n - 1$ parameters.

1.15   Exercises

Exercise 1.1 Given the matrices:

$$A = \begin{pmatrix} 1 & 3 & 5 \\ -1 & 1 & 0 \end{pmatrix}, \quad B = \begin{pmatrix} 1 & 0 & 1 \\ 2 & 1 & 1 \\ 1 & 1 & -1 \end{pmatrix},$$

$$C = \begin{pmatrix} 1 & 1 \\ 3 & 2 \\ -1 & 4 \end{pmatrix}, \quad \mathbf{d} = \begin{pmatrix} 2 \\ -1 \\ 1 \end{pmatrix},$$

which of the following matrix expressions are defined? Compute those which are defined.

(a) $A\mathbf{d}$ (b) $AB + C$ (c) $A + C^{\mathsf{T}}$ (d) $C^{\mathsf{T}}C$ (e) BC

(f) $\mathbf{d}^{\mathsf{T}}B$ (g) $C\mathbf{d}$ (h) $\mathbf{d}^{\mathsf{T}}\mathbf{d}$ (i) $\mathbf{d}\mathbf{d}^{\mathsf{T}}$.

Exercise 1.2 Find, if possible, a matrix A and a constant x such that

$$\begin{pmatrix} 1 & 7 \\ 5 & 0 \\ 9 & 3 \end{pmatrix} A = \begin{pmatrix} -4 & 14 \\ 15 & 0 \\ 24 & x \end{pmatrix}.$$

Exercise 1.3 If A and B are invertible $n \times n$ matrices, then using the definition of the inverse, prove that

$$(AB)^{-1} = B^{-1}A^{-1}.$$

Exercise 1.4 Solve for the matrix A:

$$\left(5A^{\mathsf{T}} + \begin{pmatrix} 1 & 0 \\ 2 & 5 \end{pmatrix}\right)^{\mathsf{T}} = 3A + \begin{pmatrix} 1 & -2 \\ -1 & 3 \end{pmatrix}^{-1}.$$

Exercise 1.5 Suppose A and B are matrices such that A and AB are invertible. Suppose, furthermore, that

$$(AB)^{-1} = 2A^{-1}.$$

Find B.

Exercise 1.6 If B is an $m \times k$ matrix, show that the matrix $B^{\mathsf{T}}B$ is a $k \times k$ symmetric matrix.

Exercise 1.7 Let A be an $m \times n$ matrix and B an $n \times n$ matrix. Simplify, as much as possible, the expression

$$(A^{\mathsf{T}}A)^{-1}A^{\mathsf{T}}(B^{-1}A^{\mathsf{T}})^{\mathsf{T}}B^{\mathsf{T}}B^2B^{-1}$$

assuming that any matrix inverse in the expression is defined.

Exercise 1.8 Write down a vector equation for each of the following lines.

(a) In $\mathbb{R}^2$, the line through the points $(3, 1)$ and $(-2, 4)$.
(b) In $\mathbb{R}^5$, the line through the points $(3, 1, -1, 2, 5)$ and $(-2, 4, 0, 1, 1)$. Is the point $(4, 3, 2, 1, 4)$ on this line?

Exercise 1.9 Find the vector equation of the line in $\mathbb{R}^3$ with Cartesian equations

$$\frac{x-1}{3} = y + 2 = \frac{5-z}{4}.$$

Exercise 1.10 Let

L_1 be the line with equation $\begin{pmatrix} x \\ y \\ z \end{pmatrix} = \begin{pmatrix} 1 \\ 3 \\ 2 \end{pmatrix} + t \begin{pmatrix} -1 \\ 5 \\ 4 \end{pmatrix}$,

L_2 the line through $(8, 0, -3)$ parallel to the vector $\begin{pmatrix} 6 \\ 2 \\ -1 \end{pmatrix}$,

L_3 the line through $(9, 3, 1)$ and $(7, 13, 9)$.

Show that two of the lines intersect, two are parallel and two are skew. Find the angle of intersection of the two intersecting lines.

Exercise 1.11 Referring to the previous exercise, find the vector equation and the Cartesian equation of the plane containing L_1 and L_3.

Exercise 1.12 Show that the line

$$\mathbf{x} = \begin{pmatrix} 2 \\ 3 \\ 1 \end{pmatrix} + t \begin{pmatrix} -1 \\ 4 \\ 2 \end{pmatrix}$$

does not intersect the plane $2x + z = 9$.

Find the equation of the line through the point $(2, 3, 1)$ which is parallel to the normal vector of the plane, and determine at what point it intersects the plane. Hence, or otherwise, find the distance of the line to the plane.

1.16 Problems

Problem 1.1 Given the matrices

$$A = \begin{pmatrix} 2 & 1 \\ 1 & 1 \\ 0 & 3 \end{pmatrix}, \quad \mathbf{b} = \begin{pmatrix} 1 \\ 1 \\ -1 \end{pmatrix}, \quad C = \begin{pmatrix} 1 & 2 & 1 \\ 3 & 0 & -1 \\ 4 & 1 & 1 \end{pmatrix},$$

$$D = \begin{pmatrix} 0 & 1 \\ 2 & 5 \\ 6 & 3 \end{pmatrix},$$

which of the following matrix expressions are defined? Compute those which are defined.

(a) $A\mathbf{b}$ (b) CA (c) $A + C\mathbf{b}$ (d) $A + D$ (e) $\mathbf{b}^T D$

(f) $DA^T + C$ (g) $\mathbf{b}^T \mathbf{b}$ (h) $\mathbf{b}\mathbf{b}^T$ (i) $C\mathbf{b}$.

Problem 1.2 If $\mathbf{a}$ and $\mathbf{b}$ are both column matrices of the same size, $n \times 1$, what is the size of the matrix product $\mathbf{a}^T \mathbf{b}$?

What is the size of the matrix expression $\mathbf{b}^T \mathbf{a}$?

What is the relationship between $\mathbf{a}^T \mathbf{b}$ and $\mathbf{b}^T \mathbf{a}$?

Problem 1.3 Let $B = \begin{pmatrix} 3 & 7 \\ 0 & -1 \end{pmatrix}$ and suppose $B^{-1} = \begin{pmatrix} x & y \\ z & w \end{pmatrix}$.

Solve the system of four equations given by the matrix equation $BB^{-1} = I$,

$$\begin{pmatrix} 3 & 7 \\ 0 & -1 \end{pmatrix} \begin{pmatrix} x & y \\ z & w \end{pmatrix} = \begin{pmatrix} 1 & 0 \\ 0 & 1 \end{pmatrix},$$

to obtain the matrix B^{-1}.

Check your solution by finding B^{-1} using the result in Activity 1.24. Then make absolutely sure B^{-1} is correct by checking that $BB^{-1} = B^{-1}B = I$.

Problem 1.4 Find the matrix A if

$$(A^{-1})^{\mathrm{T}} = \begin{pmatrix} 3 & 5 \\ 1 & 2 \end{pmatrix}.$$

Problem 1.5 A square matrix M is said to be *skew symmetric* if $M = -M^{\mathrm{T}}$.

Given that the 3×3 matrix $A = (a_{ij})$ is symmetric and the 3×3 matrix $B = (b_{ij})$ is skew symmetric, find the missing entries in the following matrices:

$$A = \begin{pmatrix} 1 & & 7 \\ -4 & 6 & \\ & 0 & 2 \end{pmatrix} \qquad B = \begin{pmatrix} & 3 & 0 & 2 \\ -5 & & \end{pmatrix}.$$

Problem 1.6 If A is an $n \times n$ matrix, show that the matrix $A + A^{\mathrm{T}}$ is symmetric and the matrix $A - A^{\mathrm{T}}$ is skew symmetric. (See Problem 1.5.)

Show that any matrix A can be written as the sum of a symmetric matrix and a skew symmetric matrix.

Problem 1.7 Suppose that

$$A = \begin{pmatrix} a & b \\ c & d \end{pmatrix}$$

is a 2×2 matrix such that $AB = BA$ for all 2×2 matrices B. Show that $a = d$, $b = 0$, $c = 0$. Deduce that the only such matrices are scalar multiples of the identity matrix.

Hint: If something is true for *all* 2×2 matrices, then it is true for *any* such matrix. Try some simple choices for B, such as

$$B = \begin{pmatrix} 1 & 0 \\ 0 & 0 \end{pmatrix},$$

and calculate $AB = BA$. Use the fact that two matrices are equal if and only if corresponding entries are equal to derive the conditions on a, b, c, d.

Can you generalise the result to 3×3 matrices? To $n \times n$ matrices?

Problem 1.8 Find a vector equation of the line through the points $A = (4, 5, 1)$ and $B = (1, 3, -2)$.

Find values of c and d such that the points A, B and $C = (c, d, -5)$ are collinear; that is, are points on the same line.

Problem 1.9 Show that the line

$$\mathbf{x} = \begin{pmatrix} 1 \\ 1 \\ 1 \end{pmatrix} + t \begin{pmatrix} 2 \\ 1 \\ -1 \end{pmatrix}, \quad t \in \mathbb{R}$$

intersects the line with Cartesian equations,

$$x = 5, \quad y - 4 = \frac{z - 1}{2},$$

and find the point of intersection.

Problem 1.10 What is the relationship between the lines with equations

$$\mathbf{x} = \begin{pmatrix} 1 \\ 2 \\ 1 \end{pmatrix} + t \begin{pmatrix} 3 \\ -2 \\ 1 \end{pmatrix}, \quad t \in \mathbb{R}; \quad \mathbf{x} = \begin{pmatrix} 8 \\ 4 \\ -3 \end{pmatrix} + t \begin{pmatrix} 1 \\ -2 \\ 2 \end{pmatrix}, \quad t \in \mathbb{R}?$$

Problem 1.11 Find the Cartesian equation of the plane which contains the point $(5, 1, 3)$ and has normal vector $\mathbf{n} = (1, -4, 2)^{\mathrm{T}}$.

Find also a vector (parametric) equation of this plane.

Problem 1.12 Find a Cartesian equation of the plane given by

$$\mathbf{x} = \begin{pmatrix} 1 \\ 1 \\ 1 \end{pmatrix} + s \begin{pmatrix} 2 \\ 1 \\ -1 \end{pmatrix} + t \begin{pmatrix} 0 \\ 1 \\ 2 \end{pmatrix}, \quad s, t \in \mathbb{R}.$$

Show that the equation

$$\mathbf{x} = \begin{pmatrix} 1 \\ 2 \\ 3 \end{pmatrix} + s \begin{pmatrix} 6 \\ 2 \\ -5 \end{pmatrix} + t \begin{pmatrix} 2 \\ -2 \\ -7 \end{pmatrix}, \quad s, t \in \mathbb{R}$$

represents the same plane.

Problem 1.13 Find the equation of the plane containing the two intersecting lines of Problem 1.9.

Problem 1.14 Find the point of intersection of the line

$$\mathbf{x} = \begin{pmatrix} 2 \\ 1 \\ 4 \end{pmatrix} + t \begin{pmatrix} 6 \\ -1 \\ 2 \end{pmatrix}$$

with the plane

$$x + y - 3z = 4.$$

2

Systems of linear equations

Being able to solve systems of many linear equations in many unknowns is a vital part of linear algebra. We use matrices and vectors as essential elements in obtaining and expressing the solutions.

We begin by expressing a system in matrix form and defining elementary row operations on a related matrix, known as the augmented matrix. These operations mimic the standard operations we would use to solve systems of equations by eliminating variables. We then learn a precise algorithm to apply these operations in order to put the matrix in a special form known as reduced echelon form, from which the general solution to the system is readily obtained. The method of manipulating matrices in this way to obtain the solution is known as Gaussian elimination.

We then examine the forms of solutions to systems of linear equations and look at their properties, defining what is meant by a homogeneous system and the null space of a matrix.

2.1 Systems of linear equations

A *system of m linear equations in n unknowns* $x_1, x_2, \ldots, x_n$ is a set of m equations of the form

$$a_{11}x_1 + a_{12}x_2 + \cdots + a_{1n}x_n = b_1$$
$$a_{21}x_1 + a_{22}x_2 + \cdots + a_{2n}x_n = b_2$$
$$\vdots \qquad\qquad \vdots$$
$$a_{m1}x_1 + a_{m2}x_2 + \cdots + a_{mn}x_n = b_m.$$

The numbers a_{ij} are known as the *coefficients* of the system.

Example 2.1 The set of equations

$$x_1 + x_2 + x_3 = 3$$
$$2x_1 + x_2 + x_3 = 4$$
$$x_1 - x_2 + 2x_3 = 5$$

is a system of three linear equations in the three unknowns x_1, x_2, x_3.

Systems of linear equations occur naturally in a number of applications.
    We say that $s_1, s_2, \ldots, s_n$ is a *solution* of the system if *all m* equations hold true when

$$x_1 = s_1, \quad x_2 = s_2, \quad \ldots, \quad x_n = s_n.$$

Sometimes a system of linear equations is known as a set of *simultaneous* equations; such terminology emphasises that a solution is an assignment of values to each of the n unknowns such that *each and every* equation holds with this assignment. It is also referred to simply as a *linear system*.

Example 2.2 The linear system

$$x_1 + x_2 + x_3 + x_4 + x_5 = 3$$
$$2x_1 + x_2 + x_3 + x_4 + 2x_5 = 4$$
$$x_1 - x_2 - x_3 + x_4 + x_5 = 5$$
$$x_1 + x_4 + x_5 = 4.$$

is an example of a system of four equations in five unknowns, x_1, x_2, x_3, x_4, x_5. One solution of this system is

$$x_1 = -1, \quad x_2 = -2, \quad x_3 = 1, \quad x_4 = 3, \quad x_5 = 2,$$

as you can easily verify by substituting these values into the equations. Every equation is satisfied for these values of x_1, x_2, x_3, x_4, x_5. However, this is not the only solution to this system of equations. There are many more.
    On the other hand, the system of linear equations

$$x_1 + x_2 + x_3 + x_4 + x_5 = 3$$
$$2x_1 + x_2 + x_3 + x_4 + 2x_5 = 4$$
$$x_1 - x_2 - x_3 + x_4 + x_5 = 5$$
$$x_1 + x_4 + x_5 = 6.$$

has no solutions. There are no numbers we can assign to the unknowns x_1, x_2, x_3, x_4, x_5 so that all four equations are satisfied.

How do we know this? How do we find all the solutions to a system of linear equations?

We begin by writing a system of linear equations in matrix form.

Definition 2.3 (Coefficient matrix) The matrix $A = (a_{ij})$, whose (i, j) entry is the coefficient a_{ij} of the system of linear equations is called the *coefficient matrix*

$$A = \begin{pmatrix} a_{11} & a_{12} & \cdots & a_{1n} \\ a_{21} & a_{22} & \cdots & a_{2n} \\ \vdots & \vdots & \ddots & \vdots \\ a_{m1} & a_{m2} & \cdots & a_{mn} \end{pmatrix}.$$

Let $\mathbf{x} = (x_1, x_2, \ldots, x_n)^\mathrm{T}$ be the vector of unknowns. Then the product $A\mathbf{x}$ of the $m \times n$ coefficient matrix A and the $n \times 1$ column vector $\mathbf{x}$ is an $m \times 1$ matrix,

$$\begin{pmatrix} a_{11} & a_{12} & \cdots & a_{1n} \\ a_{21} & a_{22} & \cdots & a_{2n} \\ \vdots & \vdots & \ddots & \vdots \\ a_{m1} & a_{m2} & \cdots & a_{mn} \end{pmatrix} \begin{pmatrix} x_1 \\ x_2 \\ \vdots \\ x_n \end{pmatrix} = \begin{pmatrix} a_{11}x_1 + a_{12}x_2 + \cdots + a_{1n}x_n \\ a_{21}x_1 + a_{22}x_2 + \cdots + a_{2n}x_n \\ \vdots & \vdots \\ a_{m1}x_1 + a_{n2}x_2 + \cdots + a_{mn}x_n \end{pmatrix},$$

whose entries are the left-hand sides of our system of linear equations.

If we define another column vector $\mathbf{b}$, whose m components are the right-hand sides b_i, the system is equivalent to the matrix equation

$$A\mathbf{x} = \mathbf{b}.$$

Example 2.4 Consider the following system of three linear equations in the three unknowns, x_1, x_2, x_3:

$$\begin{aligned} x_1 + x_2 + x_3 &= 3 \\ 2x_1 + x_2 + x_3 &= 4 \\ x_1 - x_2 + 2x_3 &= 5. \end{aligned}$$

This system can be written in matrix notation as $A\mathbf{x} = \mathbf{b}$ with

$$A = \begin{pmatrix} 1 & 1 & 1 \\ 2 & 1 & 1 \\ 1 & -1 & 2 \end{pmatrix}, \qquad \mathbf{x} = \begin{pmatrix} x_1 \\ x_2 \\ x_3 \end{pmatrix}, \qquad \mathbf{b} = \begin{pmatrix} 3 \\ 4 \\ 5 \end{pmatrix}.$$

The entries of the matrix A are the coefficients of the x_i. If we perform the matrix multiplication of $A\mathbf{x}$,

$$\begin{pmatrix} 1 & 1 & 1 \\ 2 & 1 & 1 \\ 1 & -1 & 2 \end{pmatrix} \begin{pmatrix} x_1 \\ x_2 \\ x_3 \end{pmatrix} = \begin{pmatrix} x_1 + x_2 + x_3 \\ 2x_1 + x_2 + x_3 \\ x_1 - x_2 + 2x_3 \end{pmatrix},$$

the matrix product is a 3×1 matrix, a column vector. If $A\mathbf{x} = \mathbf{b}$, then

$$\begin{pmatrix} x_1 + x_2 + x_3 \\ 2x_1 + x_2 + x_3 \\ x_1 - x_2 + 2x_3 \end{pmatrix} = \begin{pmatrix} 3 \\ 4 \\ 5 \end{pmatrix},$$

and these two 3×1 matrices are equal if and only if their components are equal. This gives precisely the three linear equations.

2.2   Row operations

Our purpose is to find an efficient means of finding the solutions of systems of linear equations.

Consider Example 2.4. An elementary way of solving a system of linear equations such as

$$\begin{aligned} x_1 + x_2 + x_3 &= 3 \\ 2x_1 + x_2 + x_3 &= 4 \\ x_1 - x_2 + 2x_3 &= 5 \end{aligned}$$

is to begin by eliminating one of the variables from two of the equations. For example, we can eliminate x_1 from the second equation by multiplying the first equation by 2 and then subtracting it from the second equation.

Let's do this. Twice the first equation gives $2x_1 + 2x_2 + 2x_3 = 6$. Subtracting this from the second equation, $2x_1 + x_2 + x_3 = 4$, yields the equation $-x_2 - x_3 = -2$. We can now replace the second equation in the original system by this new equation,

$$\begin{aligned} x_1 + x_2 + x_3 &= 3 \\ -x_2 - x_3 &= -2 \\ x_1 - x_2 + 2x_3 &= 5 \end{aligned}$$

and the new system will have the same set of solutions as the original system.

We can continue in this manner to obtain a simpler set of equations with the same solution set as the original system. Our next step might be to subtract the first equation from the last equation and replace the last equation, to obtain the system

$$\begin{aligned} x_1 + x_2 + x_3 &= 3 \\ x_2 + x_3 &= 2 \\ -2x_2 + x_3 &= 2 \end{aligned}$$

so that the last two equations now only contain the two variables x_2 and x_3. We can then eliminate one of these variables to eventually obtain the solution.

So exactly what operations can we perform on the equations of a linear system without altering the set of solutions? There are three main such types of operation, as follows:

O1   multiply both sides of an equation by a non-zero constant.
O2   interchange two equations.
O3   add a multiple of one equation to another.

These operations do not alter the set of solutions since the restrictions on the variables $x_1, x_2, \ldots, x_n$ given by the new equations imply the restrictions given by the old ones (that is, we can undo the manipulations made to retrieve the old system).

At the same time, we observe that these operations really only involve the coefficients of the variables and the right-hand sides of the equations.

For example, using the same system as above expressed in matrix form, $A\mathbf{x} = \mathbf{b}$, then the matrix

$$(A|\mathbf{b}) = \begin{pmatrix} 1 & 1 & 1 & 3 \\ 2 & 1 & 1 & 4 \\ 1 & -1 & 2 & 5 \end{pmatrix},$$

which is the coefficient matrix A together with the constants $\mathbf{b}$ as the last column, contains all the information we need to use, and rather than manipulating the equations, we can instead manipulate the rows of this matrix. For example, subtracting twice equation 1 from equation 2 is executed by taking twice row 1 from row 2.

These observations form the motivation behind a method to solve systems of linear equations, known as Gaussian elimination. To solve a linear system $A\mathbf{x} = \mathbf{b}$, we first form the *augmented matrix*, denoted $(A|\mathbf{b})$, which is A with column $\mathbf{b}$ tagged on.

Definition 2.5 (Augmented matrix) If $A\mathbf{x} = \mathbf{b}$ is a system of linear equations,

$$A = \begin{pmatrix} a_{11} & a_{12} & \cdots & a_{1n} \\ a_{21} & a_{22} & \cdots & a_{2n} \\ \vdots & \vdots & \ddots & \vdots \\ a_{m1} & a_{m2} & \cdots & a_{mn} \end{pmatrix} \quad \mathbf{x} = \begin{pmatrix} x_1 \\ x_2 \\ \vdots \\ x_n \end{pmatrix} \quad \mathbf{b} = \begin{pmatrix} b_1 \\ b_2 \\ \vdots \\ b_m \end{pmatrix},$$

then the matrix

$$(A|\mathbf{b}) = \begin{pmatrix} a_{11} & a_{12} & \cdots & a_{1n} & b_1 \\ a_{21} & a_{22} & \cdots & a_{2n} & b_2 \\ \vdots & \vdots & \ddots & \vdots & \vdots \\ a_{m1} & a_{m2} & \cdots & a_{mn} & b_m \end{pmatrix}$$

is called the *augmented matrix* of the linear system.

From the operations listed above for manipulating the equations of the linear system, we define corresponding operations on the rows of the augmented matrix.

Definition 2.6 (Elementary row operations) These are:

RO1   multiply a row by a non-zero constant.
RO2   interchange two rows.
RO3   add a multiple of one row to another.

2.3   Gaussian elimination

We will describe a systematic method for solving systems of linear equations by an *algorithm* which uses row operations to put the augmented matrix into a form from which the solution of the linear system can be easily read. This method is known as **Gaussian elimination** or **Gauss–Jordan elimination**. To illustrate the algorithm, we will use two examples: the augmented matrix $(A|\mathbf{b})$ of the example in the previous section and the augmented matrix $(B|\mathbf{b})$ of a second system of linear equations,

$$(A|\mathbf{b}) = \begin{pmatrix} 1 & 1 & 1 & 3 \\ 2 & 1 & 1 & 4 \\ 1 & -1 & 2 & 5 \end{pmatrix}, \qquad (B|\mathbf{b}) = \begin{pmatrix} 0 & 0 & 2 & 3 \\ 0 & 2 & 3 & 4 \\ 0 & 0 & 1 & 5 \end{pmatrix}.$$

2.3.1   The algorithm: reduced row echelon form

Using the above two examples, we will carry out the algorithm in detail.

(1) Find the leftmost column that is not all zeros.
     The augmented matrices are

$$\begin{pmatrix} 1 & 1 & 1 & 3 \\ 2 & 1 & 1 & 4 \\ 1 & -1 & 2 & 5 \end{pmatrix} \qquad \begin{pmatrix} 0 & 0 & 2 & 3 \\ 0 & 2 & 3 & 4 \\ 0 & 0 & 1 & 5 \end{pmatrix}.$$

So this is column 1 of $(A|\mathbf{b})$ and column 2 of $(B|\mathbf{b})$.

(2) Get a non-zero entry at the top of this column.

The matrix on the left already has a non-zero entry at the top. For the matrix on the right, we interchange row 1 and row 2:

$$\begin{pmatrix} 1 & 1 & 1 & 3 \\ 2 & 1 & 1 & 4 \\ 1 & -1 & 2 & 5 \end{pmatrix} \qquad \begin{pmatrix} 0 & 2 & 3 & 4 \\ 0 & 0 & 2 & 3 \\ 0 & 0 & 1 & 5 \end{pmatrix}.$$

(3) Make this entry 1; multiply the first row by a suitable number or interchange two rows. This 1 entry is called a *leading one.*

The left-hand matrix already had a 1 in this position. For the second matrix, we multiply row 1 by one-half:

$$\begin{pmatrix} 1 & 1 & 1 & 3 \\ 2 & 1 & 1 & 4 \\ 1 & -1 & 2 & 5 \end{pmatrix} \qquad \begin{pmatrix} 0 & 1 & \frac{3}{2} & 2 \\ 0 & 0 & 2 & 3 \\ 0 & 0 & 1 & 5 \end{pmatrix}.$$

(4) Add suitable multiples of the top row to rows below so that all entries below *the leading one become zero.*

For the matrix on the left, we add -2 times row 1 to row 2, then we add -1 times row 1 to row 3. These are the same operations as the ones we performed earlier on the example using the equations. The matrix on the right already has zeros under the leading one:

$$\begin{pmatrix} 1 & 1 & 1 & 3 \\ 0 & -1 & -1 & -2 \\ 0 & -2 & 1 & 2 \end{pmatrix} \qquad \begin{pmatrix} 0 & 1 & \frac{3}{2} & 2 \\ 0 & 0 & 2 & 3 \\ 0 & 0 & 1 & 5 \end{pmatrix}.$$

At any stage, we can read the modified system of equations from the new augmented matrix, remembering that column 1 gives the coefficients of x_1, column 2 the coefficients of x_2 and so on, and that the last column represents the right-hand side of the equations. For example, the matrix on the left is now the augmented matrix of the system

$$\begin{aligned} x_1 + x_2 + x_3 &= 3 \\ -x_2 - x_3 &= -2 \\ -2x_2 + x_3 &= 2. \end{aligned}$$

The next step in the algorithm is

(5) Cover up the top row and apply steps (1) to (4) again.

This time we will work on one matrix at a time. After the first four steps, we have altered the augmented matrix $(A|\mathbf{b})$ to

$$(A|\mathbf{b}) \longrightarrow \begin{pmatrix} 1 & 1 & 1 & 3 \\ 0 & -1 & -1 & -2 \\ 0 & -2 & 1 & 2 \end{pmatrix}.$$

We now ignore the top row. Then the leftmost column which is not all zeros is column 2. This column already has a non-zero entry at the top. We make it into a leading one by multiplying row 2 by -1:

$$\longrightarrow \begin{pmatrix} 1 & 1 & 1 & 3 \\ 0 & 1 & 1 & 2 \\ 0 & -2 & 1 & 2 \end{pmatrix}.$$

This is now a leading one, and we use it to obtain zeros below. We add 2 times row 2 to row 3:

$$\longrightarrow \begin{pmatrix} 1 & 1 & 1 & 3 \\ 0 & 1 & 1 & 2 \\ 0 & 0 & 3 & 6 \end{pmatrix}.$$

Now we cover up the top two rows and start again with steps (1) to (4). The leftmost column which is not all zeros is column 3. We multiply row 3 by one-third to obtain the final leading one:

$$\longrightarrow \begin{pmatrix} 1 & 1 & 1 & 3 \\ 0 & 1 & 1 & 2 \\ 0 & 0 & 1 & 2 \end{pmatrix}.$$

This last matrix is in *row echelon form*, or simply, *echelon form*.

Definition 2.7 (Row echelon form) A matrix is said to be in *row echelon form* (or *echelon form*) if it has the following three properties:

(1) Every non-zero row begins with a leading one.
(2) A leading one in a lower row is further to the right.
(3) Zero rows are at the bottom of the matrix.

Activity 2.8 Check that the above matrix satisfies these three properties.

The term echelon form takes its name from the form of the equations at this stage. Reading from the matrix, these equations are

$$\begin{aligned} x_1 + x_2 + x_3 &= 3 \\ x_2 + x_3 &= 2 \\ x_3 &= 2. \end{aligned}$$

We could now use a method called *back substitution* to find the solution of the system. The last equation tells us that $x_3 = 2$. We can then substitute this into the second equation to obtain x_2, and then use these two values to obtain x_1. This is an acceptable approach, but we can effectively do the same calculations by continuing with row operations. So we continue with one final step of our algorithm.

(6) Begin with the last row and add suitable multiples to each row above to get zeros above *the leading ones.*

Continuing from the row echelon form and using row 3, we replace row 2 with row 2−row 3, and at the same time we replace row 1 with row 1−row 3:

$$(A|\mathbf{b}) \longrightarrow \begin{pmatrix} 1 & 1 & 1 & 3 \\ 0 & 1 & 1 & 2 \\ 0 & 0 & 1 & 2 \end{pmatrix} \longrightarrow \begin{pmatrix} 1 & 1 & 0 & 1 \\ 0 & 1 & 0 & 0 \\ 0 & 0 & 1 & 2 \end{pmatrix}.$$

We now have zeros above the leading one in column 3. There is only one more step to do, and that is to get a zero above the leading one in column 2. So the final step is row 1−row 2:

$$\longrightarrow \begin{pmatrix} 1 & 0 & 0 & 1 \\ 0 & 1 & 0 & 0 \\ 0 & 0 & 1 & 2 \end{pmatrix}.$$

This final matrix is now in *reduced row echelon form*. It has the additional property that every column with a leading one has zeros elsewhere.

Definition 2.9 (Reduced row echelon form) A matrix is said to be in *reduced row echelon form* (or *reduced echelon form*) if it has the following four properties:

(1) Every non-zero row begins with a leading one.
(2) A leading one in a lower row is further to the right.
(3) Zero rows are at the bottom of the matrix.
(4) Every column with a leading one has zeros elsewhere.

If R is the reduced row echelon form of a matrix M, we will sometimes write $R = \mathrm{RREF}(M)$.

The solution can now be read from the matrix. The top row says $x_1 = 1$, the second row says $x_2 = 0$ and the third row says $x_3 = 2$. The original system has been reduced to the matrix equation

$$\begin{pmatrix} 1 & 0 & 0 \\ 0 & 1 & 0 \\ 0 & 0 & 1 \end{pmatrix} \begin{pmatrix} x_1 \\ x_2 \\ x_3 \end{pmatrix} = \begin{pmatrix} 1 \\ 0 \\ 2 \end{pmatrix},$$

giving the solution

$$\begin{pmatrix} x_1 \\ x_2 \\ x_3 \end{pmatrix} = \begin{pmatrix} 1 \\ 0 \\ 2 \end{pmatrix}.$$

This system of equations has a *unique* solution.

We can check that this solution is the correct solution of the original system by substituting it into the equations, or equivalently, by multiplying out the matrices $A\mathbf{x}$ to show that $A\mathbf{x} = \mathbf{b}$.

Activity 2.10 Do this: check that $\begin{pmatrix} 1 & 1 & 1 \\ 2 & 1 & 1 \\ 1 & -1 & 2 \end{pmatrix} \begin{pmatrix} 1 \\ 0 \\ 2 \end{pmatrix} = \begin{pmatrix} 3 \\ 4 \\ 5 \end{pmatrix}$.

We now return to the example $(B|\mathbf{b})$, which we left after the first round of steps (1) to (4), and we apply step (5). We cover up the top row and apply steps (1) to (4) again. We need to have a leading one in the second row, which we achieve by switching row 2 and row 3:

$$(B|\mathbf{b}) \longrightarrow \begin{pmatrix} 0 & 1 & \frac{3}{2} & 2 \\ 0 & 0 & 2 & 3 \\ 0 & 0 & 1 & 5 \end{pmatrix} \longrightarrow \begin{pmatrix} 0 & 1 & \frac{3}{2} & 2 \\ 0 & 0 & 1 & 5 \\ 0 & 0 & 2 & 3 \end{pmatrix}.$$

We obtain a zero under this leading one by replacing row 3 with row $3 + (-2)$ times row 2:

$$\longrightarrow \begin{pmatrix} 0 & 1 & \frac{3}{2} & 2 \\ 0 & 0 & 1 & 5 \\ 0 & 0 & 0 & -7 \end{pmatrix}$$

and then, finally, multiply row 3 by $-\frac{1}{7}$:

$$\longrightarrow \begin{pmatrix} 0 & 1 & \frac{3}{2} & 2 \\ 0 & 0 & 1 & 5 \\ 0 & 0 & 0 & 1 \end{pmatrix}.$$

This matrix is now in row echelon form, but we shall see that there is no point in going on to reduced row echelon form. This last matrix is equivalent to the system

$$\begin{pmatrix} 0 & 1 & \frac{3}{2} \\ 0 & 0 & 1 \\ 0 & 0 & 0 \end{pmatrix} \begin{pmatrix} x_1 \\ x_2 \\ x_3 \end{pmatrix} = \begin{pmatrix} 2 \\ 5 \\ 1 \end{pmatrix}.$$

What is the bottom equation of this system? Row 3 says $0x_1 + 0x_2 + 0x_3 = 1$, that is $0 = 1$, which is impossible! This system has *no solution*.

Putting an augmented matrix into reduced row echelon form using this algorithm is usually the most efficient way to solve a system of linear equations. In a variation of the algorithm, when the leading one is obtained in the second row, it can be used to obtain zeros both below it (as in the algorithm) and also above it. Although this may look attractive, it actually uses more calculations on the remaining columns than the method given here, and this number becomes significant for large n.

2.3.2 Consistent and inconsistent systems

Definition 2.11 (Consistent) A system of linear equations is said to be *consistent* if it has at least one solution. It is *inconsistent* if there are no solutions.

The example above demonstrates the following important fact:

- If the row echelon form (REF) of the augmented matrix $(A|\mathbf{b})$ contains a row $(0\,0\,\cdots\,0\,1)$, then the system is inconsistent.

It is instructive to look at the original systems represented by the augmented matrices above:

$$(A|\mathbf{b}) = \begin{pmatrix} 1 & 1 & 1 & 3 \\ 2 & 1 & 1 & 4 \\ 1 & -1 & 2 & 5 \end{pmatrix} \qquad (B|\mathbf{b}) = \begin{pmatrix} 0 & 0 & 2 & 3 \\ 0 & 2 & 3 & 4 \\ 0 & 0 & 1 & 5 \end{pmatrix}.$$

These are

$$\begin{cases} x_1 + x_2 + x_3 & = & 3 \\ 2x_1 + x_2 + x_3 & = & 4 \\ x_1 - x_2 + 2x_3 & = & 5 \end{cases} \qquad \begin{cases} 2x_3 & = & 3 \\ 2x_2 + 3x_3 & = & 4 \\ x_3 & = & 5. \end{cases}$$

We see immediately that the system $B\mathbf{x} = \mathbf{b}$ is inconsistent since it is not possible for both the top and the bottom equation to hold.

Since these are systems of three equations in three variables, we can interpret these results geometrically. Each of the equations above represents a plane in $\mathbb{R}^3$. The system $A\mathbf{x} = \mathbf{b}$ represents three planes which intersect in the point $(1, 0, 2)$. This is the only point which lies on all three planes. The system $B\mathbf{x} = \mathbf{b}$ represents three planes, two of which are parallel (the horizontal planes $2x_3 = 3$ and $x_3 = 5$), so there is no point that lies on all three planes.

We have been very careful when illustrating the Gaussian elimination method to explain what the row operations were for each step of the algorithm, but it is not necessary to include all this detail. The aim is to use row operations to put the augmented matrix into reduced row echelon form, and then read off the solutions from this form. Where it is useful to indicate the operations, you can do so by writing, for example, $R_2 - 2R_1$, where we always write down the row we are replacing first, so that $R_2 - 2R_1$ indicates 'replace row 2 (R_2) with row 2 plus -2 times row 1 ($R_2 - 2R_1$)'. Otherwise, you can just write down the sequence of matrices linked by *arrows*. It is important to realise that once you have performed a row operation on a matrix, the new matrix obtained is not equal to the previous one; this is why you must use arrows between the steps and not equal signs.

Example 2.12 We repeat the reduction of $(A|\mathbf{b})$ to illustrate this for the system

$$x_1 + x_2 + x_3 = 3$$
$$2x_1 + x_2 + x_3 = 4$$
$$x_1 - x_2 + 2x_3 = 5.$$

Begin by writing down the augmented matrix, then apply the row operations to carry out the algorithm. Here we will indicate the row operations:

$$(A|\mathbf{b}) = \begin{pmatrix} 1 & 1 & 1 & 3 \\ 2 & 1 & 1 & 4 \\ 1 & -1 & 2 & 5 \end{pmatrix} \rightarrow$$

$$\begin{matrix} \\ R_2 - 2R_1 \\ R_3 - R_1 \end{matrix} \begin{pmatrix} 1 & 1 & 1 & 3 \\ 0 & -1 & -1 & -2 \\ 0 & -2 & 1 & 2 \end{pmatrix} \rightarrow$$

$$(-1)R_2 \begin{pmatrix} 1 & 1 & 1 & 3 \\ 0 & 1 & 1 & 2 \\ 0 & -2 & 1 & 2 \end{pmatrix} \rightarrow$$

$$\begin{matrix} \\ \\ R_3 + 2R_2 \end{matrix} \begin{pmatrix} 1 & 1 & 1 & 3 \\ 0 & 1 & 1 & 2 \\ 0 & 0 & 3 & 6 \end{pmatrix} \rightarrow$$

$$\begin{matrix} \\ \\ (\tfrac{1}{3})R_3 \end{matrix} \begin{pmatrix} 1 & 1 & 1 & 3 \\ 0 & 1 & 1 & 2 \\ 0 & 0 & 1 & 2 \end{pmatrix}.$$

The matrix is now in row echelon form. We continue to reduced row echelon form:

$$\begin{matrix} R_1 - R_3 \\ R_2 - R_3 \\ \\ \end{matrix} \begin{pmatrix} 1 & 1 & 0 & 1 \\ 0 & 1 & 0 & 0 \\ 0 & 0 & 1 & 2 \end{pmatrix} \rightarrow$$

$$\begin{matrix} R_1 - R_2 \\ \\ \\ \end{matrix} \begin{pmatrix} 1 & 0 & 0 & 1 \\ 0 & 1 & 0 & 0 \\ 0 & 0 & 1 & 2 \end{pmatrix}.$$

The augmented matrix is now in reduced row echelon form.

Activity 2.13 Use Gaussian elimination to solve the following system of equations:

$$x_1 + x_2 + x_3 = 6$$
$$2x_1 + 4x_2 + x_3 = 5$$
$$2x_1 + 3x_2 + x_3 = 6.$$

Be sure to follow the algorithm to put the augmented matrix into reduced row echelon form using row operations.

2.3.3 Linear systems with free variables

Gaussian elimination can be used to solve systems of linear equations with any number of equations and unknowns. We will now look at an example of a linear system with four equations in five unknowns:

$$x_1 + x_2 + x_3 + x_4 + x_5 = 3$$
$$2x_1 + x_2 + x_3 + x_4 + 2x_5 = 4$$
$$x_1 - x_2 - x_3 + x_4 + x_5 = 5$$
$$x_1 + x_4 + x_5 = 4.$$

The augmented matrix is

$$(A|\mathbf{b}) = \begin{pmatrix} 1 & 1 & 1 & 1 & 1 & 3 \\ 2 & 1 & 1 & 1 & 2 & 4 \\ 1 & -1 & -1 & 1 & 1 & 5 \\ 1 & 0 & 0 & 1 & 1 & 4 \end{pmatrix}.$$

Check that your augmented matrix is correct before you proceed, or you could be solving the wrong system! A good method is to first write down the coefficients by rows, reading across the equations, and then to check the columns do correspond to the coefficients of that variable. Now follow the algorithm to put $(A|\mathbf{b})$ into reduced row echelon form:

$$\begin{array}{c} \longrightarrow \\ R_2 - 2R_1 \\ R_3 - R_1 \\ R_4 - R_1 \end{array} \begin{pmatrix} 1 & 1 & 1 & 1 & 1 & 3 \\ 0 & -1 & -1 & -1 & 0 & -2 \\ 0 & -2 & -2 & 0 & 0 & 2 \\ 0 & -1 & -1 & 0 & 0 & 1 \end{pmatrix}$$

$$\begin{array}{c} (-1)R_2 \\ \longrightarrow \end{array} \begin{pmatrix} 1 & 1 & 1 & 1 & 1 & 3 \\ 0 & 1 & 1 & 1 & 0 & 2 \\ 0 & -2 & -2 & 0 & 0 & 2 \\ 0 & -1 & -1 & 0 & 0 & 1 \end{pmatrix}$$

$$\underset{\substack{R_3 + 2R_2 \\ R_4 + R_2}}{\longrightarrow} \begin{pmatrix} 1 & 1 & 1 & 1 & 1 & 3 \\ 0 & 1 & 1 & 1 & 0 & 2 \\ 0 & 0 & 0 & 2 & 0 & 6 \\ 0 & 0 & 0 & 1 & 0 & 3 \end{pmatrix}$$

$$\underset{(\frac{1}{2})R_3}{\longrightarrow} \begin{pmatrix} 1 & 1 & 1 & 1 & 1 & 3 \\ 0 & 1 & 1 & 1 & 0 & 2 \\ 0 & 0 & 0 & 1 & 0 & 3 \\ 0 & 0 & 0 & 1 & 0 & 3 \end{pmatrix}$$

$$\underset{R_4 - R_3}{\longrightarrow} \begin{pmatrix} 1 & 1 & 1 & 1 & 1 & 3 \\ 0 & 1 & 1 & 1 & 0 & 2 \\ 0 & 0 & 0 & 1 & 0 & 3 \\ 0 & 0 & 0 & 0 & 0 & 0 \end{pmatrix}.$$

This matrix is in row echelon form. We continue to reduced row echelon form, starting with the third row:

$$\underset{\substack{R_1 - R_3 \\ R_2 - R_3 \\ \longrightarrow}}{} \begin{pmatrix} 1 & 1 & 1 & 0 & 1 & 0 \\ 0 & 1 & 1 & 0 & 0 & -1 \\ 0 & 0 & 0 & 1 & 0 & 3 \\ 0 & 0 & 0 & 0 & 0 & 0 \end{pmatrix}$$

$$\underset{\substack{R_1 - R_2 \\ \longrightarrow}}{} \begin{pmatrix} 1 & 0 & 0 & 0 & 1 & 1 \\ 0 & 1 & 1 & 0 & 0 & -1 \\ 0 & 0 & 0 & 1 & 0 & 3 \\ 0 & 0 & 0 & 0 & 0 & 0 \end{pmatrix}.$$

There are only three leading ones in the reduced row echelon form of this matrix. These appear in columns 1, 2 and 4. Since the last row gives no information, but merely states that $0 = 0$, the matrix is equivalent to the system of equations:

$$x_1 + 0 + 0 + 0 + x_5 = 1$$
$$x_2 + x_3 + 0 + 0 = -1$$
$$x_4 + 0 = 3.$$

The form of these equations tells us that we can assign any values to x_3 and x_5, and then the values of x_1, x_2 and x_4 will be determined.

Definition 2.14 (Leading variables) The variables corresponding to the columns with leading ones in the reduced row echelon form of an augmented matrix are called *leading variables*. The other variables are called *non-leading variables*.

In this example, the variables x_1, x_2 and x_4 are leading variables, x_3 and x_5 are non-leading variables. We assign x_3, x_5 the arbitrary values s, t, where s, t represent any real numbers, and then solve for the leading variables in terms of these. We get

$$x_4 = 3, \qquad x_2 = -1 - s, \qquad x_1 = 1 - t.$$

Then we express this solution in vector form:

$$\mathbf{x} = \begin{pmatrix} x_1 \\ x_2 \\ x_3 \\ x_4 \\ x_5 \end{pmatrix} = \begin{pmatrix} 1 - t \\ -1 - s \\ s \\ 3 \\ t \end{pmatrix} = \begin{pmatrix} 1 \\ -1 \\ 0 \\ 3 \\ 0 \end{pmatrix} + s \begin{pmatrix} 0 \\ -1 \\ 1 \\ 0 \\ 0 \end{pmatrix} + t \begin{pmatrix} -1 \\ 0 \\ 0 \\ 0 \\ 1 \end{pmatrix}.$$

Observe that there are *infinitely many solutions*, because any values of $s \in \mathbb{R}$ and $t \in \mathbb{R}$ will give a solution.

The solution given above is called a *general solution* of the system, because it gives a solution for any values of s and t, and any solution of the equation is of this form for some $s, t \in \mathbb{R}$. For any particular assignment of values to s and t, such as $s = 0$, $t = 1$, we obtain a *particular solution* of the system.

Activity 2.15 Let $s = 0$ and $t = 0$ and show (by substituting it into the equation or multiplying $A\mathbf{x}_0$) that $\mathbf{x}_0 = (1, -1, 0, 3, 0)^{\mathrm{T}}$ is a solution of $A\mathbf{x} = \mathbf{b}$. Then let $s = 1$ and $t = 2$ and show that the new vector $\mathbf{x}_1$ you obtain is also a solution.

With practice, you will be able to read the general solution directly from the reduced row echelon form of the augmented matrix. We have

$$(A|\mathbf{b}) \longrightarrow \begin{pmatrix} 1 & 0 & 0 & 0 & 1 & 1 \\ 0 & 1 & 1 & 0 & 0 & -1 \\ 0 & 0 & 0 & 1 & 0 & 3 \\ 0 & 0 & 0 & 0 & 0 & 0 \end{pmatrix}.$$

Locate the leading ones, and note which are the leading variables. Then locate the non-leading variables and assign each an arbitrary parameter. So, as above, we note that the leading ones are in the first, second and fourth column, and so correspond to x_1, x_2 and x_4. Then we assign arbitrary parameters to the non-leading variables; that is, values such as $x_3 = s$ and $x_5 = t$, where s and t represent any real numbers. Then write down the vector $\mathbf{x} = (x_1, x_2, x_3, x_4, x_5)^{\mathrm{T}}$ (as a column) and fill in the values starting with x_5 and working up. We have $x_5 = t$. Then the third row tells us that $x_4 = 3$. We have $x_3 = s$. Now look at the second row, which says $x_2 + x_3 = -1$, or $x_2 = -1 - s$. Then the top row tells us that $x_1 = 1 - t$. In this way, we obtain the solution in vector form.

Activity 2.16 Write down the system of three linear equations in three unknowns represented by the matrix equation $A\mathbf{x} = \mathbf{b}$, where

$$A = \begin{pmatrix} 1 & 2 & 1 \\ 2 & 2 & 0 \\ 3 & 4 & 1 \end{pmatrix}, \qquad \mathbf{x} = \begin{pmatrix} x \\ y \\ z \end{pmatrix}, \qquad \mathbf{b} = \begin{pmatrix} 3 \\ 2 \\ 5 \end{pmatrix}.$$

Use Gaussian elimination to solve the system. Express your solution in vector form. If each equation represents the Cartesian equation of a plane in $\mathbb{R}^3$, describe the intersection of these three planes.

2.3.4 Solution sets

We have seen systems of linear equations which have a unique solution, no solution and infinitely many solutions. It turns out that these are the only possibilities.

Theorem 2.17 *A system of linear equations either has no solutions, a unique solution or infinitely many solutions.*

Proof: To see this, suppose we have a linear system $A\mathbf{x} = \mathbf{b}$ which has two distinct solutions, $\mathbf{p}$ and $\mathbf{q}$. So the system has a solution and it is not unique. Thinking of these vector solutions as determining points in $\mathbb{R}^n$, then we will show that every point on the line through $\mathbf{p}$ and $\mathbf{q}$ is also a solution. Therefore, as soon as there is more than one solution, there must be infinitely many.

If $\mathbf{p}$ and $\mathbf{q}$ are vectors such that $A\mathbf{p} = \mathbf{b}$ and $A\mathbf{q} = \mathbf{b}$, $\mathbf{p} \neq \mathbf{q}$, then the equation of the line through $\mathbf{p}$ and $\mathbf{q}$ is

$$\mathbf{v} = \mathbf{p} + t(\mathbf{q} - \mathbf{p}) \qquad t \in \mathbb{R}.$$

Then for any vector $\mathbf{v}$ on the line, we have $A\mathbf{v} = A(\mathbf{p} + t(\mathbf{q} - \mathbf{p}))$.
Using the distributive laws,

$$A\mathbf{v} = A\mathbf{p} + tA(\mathbf{q} - \mathbf{p}) = A\mathbf{p} + t(A\mathbf{q} - A\mathbf{p}) = \mathbf{b} + t(\mathbf{b} - \mathbf{b}) = \mathbf{b}.$$

Therefore, $\mathbf{v}$ is also a solution for any $t \in \mathbb{R}$, so there are infinitely many of them. $\qquad \square$

Notice that in this proof the vector $\mathbf{w} = \mathbf{q} - \mathbf{p}$ satisfies the equation $A\mathbf{x} = \mathbf{0}$. This leads us to our next topic.

2.4 Homogeneous systems and null space

2.4.1 Homogeneous systems

Definition 2.18 A *homogeneous system of linear equations* is a linear system of the form $A\mathbf{x} = \mathbf{0}$.

There is one easy, but important, fact about homogeneous systems:

- A homogeneous system $A\mathbf{x} = \mathbf{0}$ is always consistent.

Why? Because $A\mathbf{0} = \mathbf{0}$, so the system always has the solution $\mathbf{x} = \mathbf{0}$. For this reason, $\mathbf{x} = \mathbf{0}$ is called the *trivial solution*.

The following fact can now be seen:

- If $A\mathbf{x} = \mathbf{0}$ has a *unique* solution, then it must be the trivial solution, $\mathbf{x} = \mathbf{0}$.

If we form the augmented matrix, $(A \mid \mathbf{0})$, of a homogeneous system, then the last column will consist entirely of zeros. This column will remain a column of zeros throughout the entire row reduction, so there is no point in writing it. Instead, we use Gaussian elimination on the coefficient matrix A, *remembering* that we are solving $A\mathbf{x} = \mathbf{0}$.

Example 2.19 Find the solution of the homogeneous linear system,

$$
\begin{aligned}
x + y + 3z + w &= 0 \\
x - y + z + w &= 0 \\
y + 2z + 2w &= 0.
\end{aligned}
$$

We reduce the coefficient matrix A to reduced row echelon form,

$$
A = \begin{pmatrix} 1 & 1 & 3 & 1 \\ 1 & -1 & 1 & 1 \\ 0 & 1 & 2 & 2 \end{pmatrix} \longrightarrow \begin{pmatrix} 1 & 1 & 3 & 1 \\ 0 & -2 & -2 & 0 \\ 0 & 1 & 2 & 2 \end{pmatrix}
$$

$$
\longrightarrow \begin{pmatrix} 1 & 1 & 3 & 1 \\ 0 & 1 & 1 & 0 \\ 0 & 1 & 2 & 2 \end{pmatrix} \longrightarrow \begin{pmatrix} 1 & 1 & 3 & 1 \\ 0 & 1 & 1 & 0 \\ 0 & 0 & 1 & 2 \end{pmatrix}
$$

$$
\longrightarrow \begin{pmatrix} 1 & 1 & 0 & -5 \\ 0 & 1 & 0 & -2 \\ 0 & 0 & 1 & 2 \end{pmatrix} \longrightarrow \begin{pmatrix} 1 & 0 & 0 & -3 \\ 0 & 1 & 0 & -2 \\ 0 & 0 & 1 & 2 \end{pmatrix}.
$$

Activity 2.20 Work through the above calculation and state what row operation is being done at each stage. For example, the first operation is $R_2 - R_1$.

Now we write down the solution from the reduced row echelon form of the matrix. (Remember that this is the reduced row echelon of the coefficient matrix A, representing the homogeneous system $A\mathbf{x} = \mathbf{0}$.)

The solution is

$$\mathbf{x} = \begin{pmatrix} x \\ y \\ z \\ w \end{pmatrix} = t \begin{pmatrix} 3 \\ 2 \\ -2 \\ 1 \end{pmatrix}, \quad t \in \mathbb{R},$$

which is a line through the origin, $\mathbf{x} = t\mathbf{v}$, with $\mathbf{v} = (3, 2, -2, 1)^{\mathrm{T}}$. There are infinitely many solutions, one for every $t \in \mathbb{R}$.

This example illustrates the following fact.

Theorem 2.21 *If A is an $m \times n$ matrix with $m < n$, then $A\mathbf{x} = \mathbf{0}$ has infinitely many solutions.*

Proof: The system is always consistent (because it is homogeneous) and the solutions are found by reducing the coefficient matrix A. If A is $m \times n$, then the reduced row echelon form of A contains at most m leading ones, so there are at most m leading variables. Therefore, there must be $n - m$ non-leading variables. Since $m < n$, $n - m > 0$, which means $n - m \geq 1$. This says that there is at least one non-leading variable. So the solution involves at least one arbitrary parameter which can take on any real value. Hence, there are infinitely many solutions. □

What about a linear system $A\mathbf{x} = \mathbf{b}$? If A is $m \times n$ with $m < n$, does $A\mathbf{x} = \mathbf{b}$ have infinitely many solutions? The answer is: provided the system is consistent, then there are infinitely many solutions. So the system either has no solutions, or infinitely many. The following examples demonstrate both possibilities.

Example 2.22 The linear system

$$x + y + z = 6$$
$$x + y + z = 1$$

is inconsistent, since there are no values of x, y, z which can satisfy both equations. These equations represent parallel planes in $\mathbb{R}^3$.

Example 2.23 On the other hand, consider the system

$$x + y + 3z + w = 2$$
$$x - y + z + w = 4$$
$$y + 2z + 2w = 0.$$

We will show that this is consistent and has infinitely many solutions. Notice that the coefficient matrix of this linear system is the same matrix A as that used in Example 2.19.

The augmented matrix is

$$(A|\mathbf{b}) = \begin{pmatrix} 1 & 1 & 3 & 1 & 2 \\ 1 & -1 & 1 & 1 & 4 \\ 0 & 1 & 2 & 2 & 0 \end{pmatrix}.$$

Activity 2.24 Show that the reduced row echelon form of the augmented matrix is

$$\begin{pmatrix} 1 & 0 & 0 & -3 & 1 \\ 0 & 1 & 0 & -2 & -2 \\ 0 & 0 & 1 & 2 & 1 \end{pmatrix}.$$

Then write down the solution.

The general solution of this system,

$$\mathbf{x} = \begin{pmatrix} x \\ y \\ z \\ w \end{pmatrix} = \begin{pmatrix} 1 \\ -2 \\ 1 \\ 0 \end{pmatrix} + t \begin{pmatrix} 3 \\ 2 \\ -2 \\ 1 \end{pmatrix} = \mathbf{p} + t\mathbf{v} \qquad t \in \mathbb{R},$$

is a line which does *not* go through the origin. It is parallel to the line of solutions of the homogeneous system, $A\mathbf{x} = \mathbf{0}$, and goes through the point determined by $\mathbf{p}$. This should come as no surprise, since the coefficient matrix forms the first four columns of the augmented matrix. Compare the solution sets:

$$A\mathbf{x} = \mathbf{0} : \qquad\qquad\qquad A\mathbf{x} = \mathbf{b} :$$
$$\text{RREF}(A) \qquad\qquad\qquad \text{RREF}(A|\mathbf{b})$$

$$\begin{pmatrix} 1 & 0 & 0 & -3 \\ 0 & 1 & 0 & -2 \\ 0 & 0 & 1 & 2 \end{pmatrix} \qquad \begin{pmatrix} 1 & 0 & 0 & -3 & 1 \\ 0 & 1 & 0 & -2 & -2 \\ 0 & 0 & 1 & 2 & 1 \end{pmatrix}$$

$$\mathbf{x} = t \begin{pmatrix} 3 \\ 2 \\ -2 \\ 1 \end{pmatrix} \qquad\qquad \mathbf{x} = \begin{pmatrix} 1 \\ -2 \\ 1 \\ 0 \end{pmatrix} + t \begin{pmatrix} 3 \\ 2 \\ -2 \\ 1 \end{pmatrix}.$$

The reduced row echelon form of the augmented matrix of a system $A\mathbf{x} = \mathbf{b}$ will always contain the information needed to solve $A\mathbf{x} = \mathbf{0}$, since the matrix A is the first part of $(A|\mathbf{b})$. We therefore have the following definition.

Definition 2.25 (Associated homogeneous system) Given a system of linear equations, $A\mathbf{x} = \mathbf{b}$, the linear system $A\mathbf{x} = \mathbf{0}$ is called the *associated homogeneous system*.

The solutions of the associated homogeneous system form an important part of the solution of the system $A\mathbf{x} = \mathbf{b}$, as we shall see in the next section.

Activity 2.26 Look at the reduced row echelon form of A in Example 2.19,

$$\begin{pmatrix} 1 & 0 & 0 & -3 \\ 0 & 1 & 0 & -2 \\ 0 & 0 & 1 & 2 \end{pmatrix}.$$

Explain why you can tell from this matrix that for all $\mathbf{b} \in \mathbb{R}^3$, the linear system $A\mathbf{x} = \mathbf{b}$ is consistent with infinitely many solutions.

Activity 2.27 Solve the system of equations $A\mathbf{x} = \mathbf{b}$ given by

$$x_1 + 2x_2 + x_3 = 1$$
$$2x_1 + 2x_2 = 2$$
$$3x_1 + 4x_2 + x_3 = 2.$$

Find also the general solution of the associated homogeneous system, $A\mathbf{x} = \mathbf{0}$. Describe the configuration of intersecting planes for each system of equations ($A\mathbf{x} = \mathbf{b}$ and $A\mathbf{x} = \mathbf{0}$).

2.4.2 Null space

It is clear from what we have just seen that the general solution to a consistent linear system $A\mathbf{x} = \mathbf{b}$ involves solutions to the system $A\mathbf{x} = \mathbf{0}$. This set of solutions is given a special name: the *null space* or *kernel* of the matrix A. This null space, denoted $N(A)$, is the set of all solutions $\mathbf{x}$ to $A\mathbf{x} = \mathbf{0}$, where $\mathbf{0}$ is the zero vector. That is:

Definition 2.28 (Null space) For an $m \times n$ matrix A, the null space of A is the subset of $\mathbb{R}^n$ given by

$$N(A) = \{\mathbf{x} \in \mathbb{R}^n \mid A\mathbf{x} = \mathbf{0}\},$$

where $\mathbf{0} = (0, 0, \ldots, 0)^{\mathrm{T}}$ is the zero vector of $\mathbb{R}^m$.

We now formalise the connection between the solution set of a consistent linear system, and the null space of the coefficient matrix of the system.

Theorem 2.29 *Suppose that A is an $m \times n$ matrix, that $\mathbf{b} \in \mathbb{R}^m$ and that the system $A\mathbf{x} = \mathbf{b}$ is consistent. Suppose that $\mathbf{p}$ is any solution of $A\mathbf{x} = \mathbf{b}$. Then the set of all solutions of $A\mathbf{x} = \mathbf{b}$ consists precisely of the vectors $\mathbf{p} + \mathbf{z}$ for $\mathbf{z} \in N(A)$; that is,*

$$\{\mathbf{x} \mid A\mathbf{x} = \mathbf{b}\} = \{\mathbf{p} + \mathbf{z} \mid \mathbf{z} \in N(A)\}.$$

Proof: To show the two sets are equal, we show that each is a subset of the other. This means showing that $\mathbf{p} + \mathbf{z}$ is a solution for any $\mathbf{z}$ in the null space of A, and that all solutions, $\mathbf{x}$, of $A\mathbf{x} = \mathbf{b}$ are of the form $\mathbf{p} + \mathbf{z}$ for some $\mathbf{z} \in N(A)$.

We start with $\mathbf{p} + \mathbf{z}$. If $\mathbf{z} \in N(A)$, then

$$A(\mathbf{p} + \mathbf{z}) = A\mathbf{p} + A\mathbf{z} = \mathbf{b} + \mathbf{0} = \mathbf{b},$$

so $\mathbf{p} + \mathbf{z}$ is a solution of $A\mathbf{x} = \mathbf{b}$; that is, $\mathbf{p} + \mathbf{z} \in \{\mathbf{x} \mid A\mathbf{x} = \mathbf{b}\}$. This shows that

$$\{\mathbf{p} + \mathbf{z} \mid \mathbf{z} \in N(A)\} \subseteq \{\mathbf{x} \mid A\mathbf{x} = \mathbf{b}\}.$$

Conversely, suppose that $\mathbf{x}$ is any solution of $A\mathbf{x} = \mathbf{b}$. Because $\mathbf{p}$ is also a solution, we have $A\mathbf{p} = \mathbf{b}$ and

$$A(\mathbf{x} - \mathbf{p}) = A\mathbf{x} - A\mathbf{p} = \mathbf{b} - \mathbf{b} = \mathbf{0},$$

so the vector $\mathbf{z} = \mathbf{x} - \mathbf{p}$ is a solution of the system $A\mathbf{z} = \mathbf{0}$; in other words, $\mathbf{z} \in N(A)$. But then $\mathbf{x} = \mathbf{p} + \mathbf{z}$, where $\mathbf{z} \in N(A)$. This shows that all solutions are of the form, $\mathbf{p} + \mathbf{z}$ for some $\mathbf{z} \in N(A)$; that is,

$$\{\mathbf{x} \mid A\mathbf{x} = \mathbf{b}\} \subseteq \{\mathbf{p} + \mathbf{z} \mid \mathbf{z} \in N(A)\}.$$

So the two sets are equal, as required. $\square$

We call the above result the 'Principle of Linearity'. It says that the general solution of a consistent linear system $A\mathbf{x} = \mathbf{b}$ is equal to any one particular solution $\mathbf{p}$ (where $A\mathbf{p} = \mathbf{b}$) plus the general solution of the associated homogeneous system.

$$\{\text{solutions of } A\mathbf{x} = \mathbf{b}\} = \mathbf{p} + \{\text{solutions of } A\mathbf{x} = \mathbf{0}\}.$$

In light of this result, let's have another look at some of the examples we worked earlier. In Example 2.23, we observed that the solutions of

$$\begin{aligned} x + y + 3z + w &= 2 \\ x - y + z + w &= 4 \\ y + 2z + 2w &= 0. \end{aligned}$$

are of the form

$$\mathbf{x} = \begin{pmatrix} x \\ y \\ z \\ w \end{pmatrix} = \begin{pmatrix} 1 \\ -2 \\ 1 \\ 0 \end{pmatrix} + t \begin{pmatrix} 3 \\ 2 \\ -2 \\ 1 \end{pmatrix} = \mathbf{p} + t\mathbf{v}, \quad t \in \mathbb{R},$$

where $\mathbf{x} = t\mathbf{v}$ is the general solution we had found of the associated homogeneous system (in Example 2.19). It is clear that $\mathbf{p}$ is a particular solution of the linear system (take $t = 0$), so this solution is of the form described in the theorem.

Now refer back to the first two examples $A\mathbf{x} = \mathbf{b}$ and $B\mathbf{x} = \mathbf{b}$, which we worked through in Section 2.3.1. (For convenience, we'll call the variables x, y, z rather than x_1, x_2, x_3.)

$$\begin{cases} x + y + z = 3 \\ 2x + y + z = 4 \\ x - y + 2z = 5 \end{cases} \qquad \begin{cases} 2z = 3 \\ 2y + 3z = 4 \\ z = 5. \end{cases}$$

The echelon forms of the augmented matrices we found were

$$(A|\mathbf{b}) \longrightarrow \begin{pmatrix} 1 & 0 & 0 & 1 \\ 0 & 1 & 0 & 0 \\ 0 & 0 & 1 & 2 \end{pmatrix}, \qquad (B|\mathbf{b}) \longrightarrow \begin{pmatrix} 0 & 1 & \frac{3}{2} & 2 \\ 0 & 0 & 1 & 5 \\ 0 & 0 & 0 & 1 \end{pmatrix}.$$

The first system, $A\mathbf{x} = \mathbf{b}$, has a unique solution, $\mathbf{p} = (1, 0, 2)^{\mathrm{T}}$, and the second system, $B\mathbf{x} = \mathbf{b}$, is inconsistent.

The reduced row echelon form of the matrix A is the identity matrix (formed from the first three columns of the reduced augmented matrix). Therefore, the homogeneous system $A\mathbf{x} = \mathbf{0}$ will only have the trivial solution. The unique solution of $A\mathbf{x} = \mathbf{b}$ is of the form $\mathbf{x} = \mathbf{p} + \mathbf{0}$, which conforms with the Principle of Linearity.

This principle does not apply to the inconsistent system $B\mathbf{x} = \mathbf{b}$. However, the associated homogeneous system is consistent. Notice that the homogeneous system is

$$\begin{cases} 2z = 0 \\ 2y + 3z = 0 \\ z = 0. \end{cases}$$

which represents the intersection of two planes, since the equations $2z = 0$ and $z = 0$ each represent the xy-plane. To find the solution, we continue to reduce the matrix B to reduced row echelon form.

$$B \longrightarrow \begin{pmatrix} 0 & 1 & \frac{3}{2} \\ 0 & 0 & 1 \\ 0 & 0 & 0 \end{pmatrix} \longrightarrow \begin{pmatrix} 0 & 1 & 0 \\ 0 & 0 & 1 \\ 0 & 0 & 0 \end{pmatrix}.$$

The non-leading variable is x, so we set $x = t$, and the solution is

$$\mathbf{x} = \begin{pmatrix} t \\ 0 \\ 0 \end{pmatrix} = t \begin{pmatrix} 1 \\ 0 \\ 0 \end{pmatrix}, \quad t \in \mathbb{R},$$

which is a line through the origin; namely, the x axis. So the plane $2y + 3z = 0$ intersects the xy-plane along the x axis.

We summarise what we have noticed so far:

- If $A\mathbf{x} = \mathbf{b}$ is consistent, the solutions are of the form $\mathbf{x} = \mathbf{p} + \mathbf{z}$, where $\mathbf{p}$ is any one particular solution and $\mathbf{z} \in N(A)$ the null space of A.

    If $A\mathbf{x} = \mathbf{b}$ has a unique solution, then $A\mathbf{x} = \mathbf{0}$ has only the trivial solution.

    If $A\mathbf{x} = \mathbf{b}$ has infinitely many solutions, then $A\mathbf{x} = \mathbf{0}$ has infinitely many solutions.

- $A\mathbf{x} = \mathbf{b}$ may be inconsistent, but $A\mathbf{x} = \mathbf{0}$ is always consistent.

Activity 2.30 Look at the example we solved in Section 2.3.3 on page 71.

$$x_1 + x_2 + x_3 + x_4 + x_5 = 3$$
$$2x_1 + x_2 + x_3 + x_4 + 2x_5 = 4$$
$$x_1 - x_2 - x_3 + x_4 + x_5 = 5$$
$$x_1 + x_4 + x_5 = 4.$$

Show that the solution we found is of the form

$$\mathbf{x} = \mathbf{p} + s\mathbf{v} + t\mathbf{w}, \, s, t \in \mathbb{R},$$

where $\mathbf{p}$ is a particular solution of $A\mathbf{x} = \mathbf{b}$ and $s\mathbf{v} + t\mathbf{w}$ is a general solution of the associated homogeneous system $A\mathbf{x} = \mathbf{0}$.

2.5 Learning outcomes

You should now be able to:

- express a system of linear equations in matrix form as $A\mathbf{x} = \mathbf{b}$ and know what is meant by the coefficient matrix and the augmented matrix
- put a matrix into reduced row echelon form using row operations and following the algorithm
- recognise consistent and inconsistent systems of equations by the row echelon form of the augmented matrix

- solve a system of m linear equations in n unknowns using Gaussian elimination
- express the solution in vector form
- interpret systems with three unknowns as intersections of planes in $\mathbb{R}^3$
- say what is meant by a homogeneous system of equations and what is meant by the associated homogeneous system of any linear system of equations
- explain why the solution of a consistent system of linear equations $A\mathbf{x} = \mathbf{b}$ is the sum of a particular solution and a general solution of the associated homogeneous system
- say what is meant by the null space of a matrix.

2.6   Comments on activities

Activity 2.13 Put the augmented matrix into reduced row echelon form. It should take five steps:

$$\begin{pmatrix} 1 & 1 & 1 & 6 \\ 2 & 4 & 1 & 5 \\ 2 & 3 & 1 & 6 \end{pmatrix} \longrightarrow (1) \longrightarrow (2) \longrightarrow (3) \longrightarrow (4)$$

$$\longrightarrow \begin{pmatrix} 1 & 0 & 0 & 2 \\ 0 & 1 & 0 & -1 \\ 0 & 0 & 1 & 5 \end{pmatrix},$$

from which you can read the solution, $\mathbf{x} = (2, -1, 5)^{\mathrm{T}}$. We will state the row operations at each stage. To obtain (1), do $R_2 - 2R_1$ and $R_3 - 2R_1$; for (2) switch R_2 and R_3; for (3) do $R_3 - 2R_2$. The augmented matrix is now in row echelon form, so starting with the bottom row, for (4), do $R_2 + R_3$ and $R_1 - R_3$. The final operation, $R_1 - R_2$, will yield the matrix in reduced row echelon form.

Activity 2.15 Multiply the matrices below as instructed to obtain $\mathbf{b}$:

$$A\mathbf{x}_0 = \begin{pmatrix} 1 & 1 & 1 & 1 & 1 \\ 2 & 1 & 1 & 1 & 2 \\ 1 & -1 & -1 & 1 & 1 \\ 1 & 0 & 0 & 1 & 1 \end{pmatrix} \begin{pmatrix} 1 \\ -1 \\ 0 \\ 3 \\ 0 \end{pmatrix} = \begin{pmatrix} 3 \\ 4 \\ 5 \\ 4 \end{pmatrix},$$

$$A\mathbf{x}_1 = \begin{pmatrix} 1 & 1 & 1 & 1 & 1 \\ 2 & 1 & 1 & 1 & 2 \\ 1 & -1 & -1 & 1 & 1 \\ 1 & 0 & 0 & 1 & 1 \end{pmatrix} \begin{pmatrix} -1 \\ -2 \\ 1 \\ 3 \\ 2 \end{pmatrix} = \begin{pmatrix} 3 \\ 4 \\ 5 \\ 4 \end{pmatrix}.$$

Activity 2.16 The equations are:

$$x_1 + 2x_2 + x_3 = 3$$
$$2x_1 + 2x_2 = 2$$
$$3x_1 + 4x_2 + x_3 = 5.$$

Put the augmented matrix into reduced row echelon form:

$$\begin{pmatrix} 1 & 2 & 1 & 3 \\ 2 & 2 & 0 & 2 \\ 3 & 4 & 1 & 5 \end{pmatrix} \longrightarrow \begin{pmatrix} 1 & 2 & 1 & 3 \\ 0 & -2 & -2 & -4 \\ 0 & -2 & -2 & -4 \end{pmatrix}$$

$$\longrightarrow \cdots \longrightarrow \begin{pmatrix} 1 & 2 & 1 & 3 \\ 0 & 1 & 1 & 2 \\ 0 & 0 & 0 & 0 \end{pmatrix} \longrightarrow \begin{pmatrix} 1 & 0 & -1 & -1 \\ 0 & 1 & 1 & 2 \\ 0 & 0 & 0 & 0 \end{pmatrix}.$$

So we have solution

$$\begin{pmatrix} x \\ y \\ z \end{pmatrix} = \begin{pmatrix} -1 + t \\ 2 - t \\ t \end{pmatrix} = \begin{pmatrix} -1 \\ 2 \\ 0 \end{pmatrix} + t \begin{pmatrix} 1 \\ -1 \\ 1 \end{pmatrix},$$

for $t \in \mathbb{R}$. This is the equation of a line in $\mathbb{R}^3$. So the three planes intersect in a line.

Activity 2.26 This is the reduced row echelon form of the coefficient matrix, A. The reduced row echelon form of any augmented matrix, $(A|\mathbf{b})$, will have as its first four columns the same four columns. As there is a leading one in every row, it is impossible to have a row of the form $(0\ 0\ \ldots\ 0\ 1)$, so the system will be consistent. There will be one free (non-leading) variable, (fourth column, say $x_4 = t$), so there will be infinitely many solutions.

Activity 2.27 Using row operations to reduce the augmented matrix to echelon form, we obtain

$$\begin{pmatrix} 1 & 2 & 1 & 1 \\ 2 & 2 & 0 & 2 \\ 3 & 4 & 1 & 2 \end{pmatrix} \rightarrow \begin{pmatrix} 1 & 2 & 1 & 1 \\ 0 & -2 & -2 & 0 \\ 0 & -2 & -2 & -1 \end{pmatrix}$$

$$\rightarrow \begin{pmatrix} 1 & 2 & 1 & 1 \\ 0 & 1 & 1 & 0 \\ 0 & -2 & -2 & -1 \end{pmatrix} \rightarrow \begin{pmatrix} 1 & 2 & 1 & 1 \\ 0 & 1 & 1 & 0 \\ 0 & 0 & 0 & -1 \end{pmatrix}.$$

There is no reason to reduce the matrix further, for we can now conclude that the original system of equations is inconsistent: there is no solution. For the homogeneous system, $A\mathbf{x} = \mathbf{0}$, the row echelon form of A consists of the first three columns of the echelon form of the augmented matrix. So starting from these and continuing to reduced row echelon form, we obtain

$$A = \begin{pmatrix} 1 & 2 & 1 \\ 2 & 2 & 0 \\ 3 & 4 & 1 \end{pmatrix} \rightarrow \cdots \rightarrow \begin{pmatrix} 1 & 2 & 1 \\ 0 & 1 & 1 \\ 0 & 0 & 0 \end{pmatrix} \rightarrow \begin{pmatrix} 1 & 0 & -1 \\ 0 & 1 & 1 \\ 0 & 0 & 0 \end{pmatrix}.$$

Setting the non-leading variable x_3 to $x_3 = t$, we find that the null space of A consists of all vectors, $\mathbf{x}$, of the following form:

$$\mathbf{x} = t \begin{pmatrix} 1 \\ -1 \\ 1 \end{pmatrix}, \quad t \in \mathbb{R}.$$

The system of equations $A\mathbf{x} = \mathbf{0}$ has infinitely many solutions.

Geometrically, the associated homogeneous system represents the equations of three planes, all of which pass through the origin. These planes intersect in a line through the origin. The equation of this line is given by the solution we found.

The original system represents three planes with no common points of intersection. No two of the planes in either system are parallel. Why? Look at the normals to the planes: no two of these are parallel, so no two planes are parallel. These planes intersect to form a kind of triangular prism; any two planes intersect in a line, and the three lines of intersection are parallel, but there are no points which lie on all three planes. (If you have trouble visualising this, take three cards, place one flat on the table, and then get the other two to balance on top, forming a triangle when viewed from the side.)

2.7 Exercises

Exercise 2.1 Write down the augmented matrix for each of the following systems of equations, and use it to solve the system by reducing the augmented matrix to reduced row echelon form.

(a)
$$\begin{cases} x - y + z = -3 \\ -3x + 4y - z = 2 \\ x - 3y - 2z = 7 \end{cases}$$

(b)
$$\begin{cases} 2x - y + 3z = 4 \\ x + y - z = 1 \\ 5x + 2y = 7. \end{cases}$$

Interpret the solutions to each of the above systems as intersections of planes, describing them geometrically.

Exercise 2.2 Solve each of the following systems of equations.

(a) $\begin{cases} -x + y - 3z = 0 \\ 3x - 2y + 10z = 0 \\ -2x + 3y - 5z = 0 \end{cases}$

(b) $\begin{cases} -x + y - 3z = 6 \\ 3x - 2y + 10z = -10 \\ -2x + 3y - 5z = 9. \end{cases}$

Exercise 2.3 Find the vector equation of the line of intersection of the planes

$$3x_1 + x_2 + x_3 = 3 \quad \text{and} \quad x_1 - x_2 - x_3 = 1.$$

What is the intersection of these two planes and the plane

$$x_1 + 2x_2 + 2x_3 = 1?$$

Exercise 2.4 Solve the system of equations $A\mathbf{x} = \mathbf{b}$, where

$$A = \begin{pmatrix} 2 & 3 & 1 & 1 \\ 1 & 2 & 0 & -1 \\ 3 & 4 & 2 & 4 \end{pmatrix}, \qquad \mathbf{b} = \begin{pmatrix} 4 \\ 1 \\ 9 \end{pmatrix}$$

using Gaussian elimination. (Put the augmented matrix into reduced row echelon form.) Express your solution in vector form as $x = \mathbf{p} + t\mathbf{v}$, where t is a real number. Check your solution by calculating $A\mathbf{p}$ and $A\mathbf{v}$.

Write down the reduced row echelon form of the matrix A. Referring to this reduced row echelon form, answer the following two questions and justify each answer.

(i)   Is there a vector $\mathbf{d} \in \mathbb{R}^3$ for which $A\mathbf{x} = \mathbf{d}$ is inconsistent?
(ii) Is there a vector $\mathbf{d} \in \mathbb{R}^3$ for which $A\mathbf{x} = \mathbf{d}$ has a unique solution?

Exercise 2.5 Find the reduced row echelon form of the matrix

$$C = \begin{pmatrix} 1 & 2 & -1 & 3 & 8 \\ -3 & -1 & 8 & 6 & 1 \\ -1 & 0 & 3 & 1 & -2 \end{pmatrix}.$$

(a) If C is the *augmented matrix* of a system of equations $A\mathbf{x} = \mathbf{b}$, $C = (A|\mathbf{b})$, what are the solutions? What Euclidean space are they in?

(b) If C is the *coefficient matrix* of a homogeneous system of equations, $C\mathbf{x} = \mathbf{0}$, what are the solutions? What Euclidean space are they in?

(c) Let $\mathbf{w} = (1, 0, 1, 1, 1)^\mathrm{T}$. Find $\mathbf{d}$ such that $C\mathbf{w} = \mathbf{d}$. Then write down a general solution of $C\mathbf{x} = \mathbf{d}$.

Exercise 2.6 Find the null space of the matrix

$$B = \begin{pmatrix} 1 & 1 & 3 \\ 4 & 1 & 2 \\ -2 & 1 & 5 \\ 3 & -5 & 1 \end{pmatrix}.$$

Let c_1, c_2, c_3 denote the columns of B. Find $\mathbf{d} = c_1 + 2c_2 - c_3$. Then write down all solutions of $B\mathbf{x} = \mathbf{d}$.

2.8 Problems

Problem 2.1 Write down the augmented matrix for each of the following systems of equations, and use it to solve the system by reducing the augmented matrix to reduced row echelon form.

(a) $\begin{cases} x + y + z = 2 \\ \quad\;\; 2y + z = 0 \\ -x + y - z = -4. \end{cases}$

(b) $\begin{cases} x + y + 2z = 2 \\ \quad\;\; 2y + z = 0 \\ -x + y - z = 0. \end{cases}$

(c) $\begin{cases} x + y + 2z = 2 \\ \quad\;\; 2y + z = 0 \\ -x + y - z = -2. \end{cases}$

(d) $\begin{cases} -3x - y + z = 0 \\ -2x + 3y + 2z = 0 \\ \quad x + 2y + 3z = 0. \end{cases}$

Interpret the solutions to each of the above systems as intersections of planes, describing them geometrically.

Problem 2.2 Find the general solution of the following system of linear equations using Gaussian elimination.

$$\begin{cases} x_1 + x_2 + x_3 + x_5 = 1 \\ 3x_1 + 3x_2 + 6x_3 + 3x_4 + 9x_5 = 6 \\ 2x_1 + 2x_2 + 4x_3 + x_4 + 6x_5 = 5. \end{cases}$$

The general solution should be expressed in vector form.

Problem 2.3 Express the following system of equations in matrix form:

$$-5x + y - 3z - 8w = 3$$
$$3x + 2y + 2z + 5w = 3$$
$$x + z + 2w = -1.$$

Show that the system is consistent and find the general solution.
   Then write down the general solution of the associated homogeneous system of equations.

Problem 2.4 Given that the matrix below is in reduced row echelon form, find the missing entries (as indicated by *). Replace every * which has to be a 0 with a 0. Replace every * which has to be a 1 with a 1. Replace every * which does *not* have to be either a 0 or a 1 with a 2.

$$C = \begin{pmatrix} * & * & * & * & * \\ * & 1 & * & * & * \\ * & * & * & -4 & 3 \end{pmatrix}.$$

If C is the reduced row echelon form of the augmented matrix of a system of linear equations, $A\mathbf{x} = \mathbf{b}$, then write down the solution of the system in vector form.
   If C is the reduced row echelon form of a matrix B, write down the general solution of $B\mathbf{x} = \mathbf{0}$.

Problem 2.5 Consider the following matrices and vector:

$$A = \begin{pmatrix} 3 & 1 & -1 \\ 1 & 1 & 0 \\ 2 & 1 & 2 \end{pmatrix}, \quad B = \begin{pmatrix} 1 & 0 & 5 & 3 & 2 \\ 0 & 2 & 4 & 2 & 2 \\ -1 & 5 & 5 & 0 & 1 \end{pmatrix}, \quad \mathbf{b} = \begin{pmatrix} 11 \\ 2 \\ -6 \end{pmatrix}.$$

Solve each of the systems $A\mathbf{x} = \mathbf{b}$ and $B\mathbf{x} = \mathbf{b}$ using Gaussian elimination, and express your solution in vector form.

Problem 2.6 Put the matrix

$$B = \begin{pmatrix} 1 & 1 & 4 & 5 \\ 2 & 1 & 5 & 5 \\ 0 & 1 & 3 & 2 \\ -1 & 1 & 2 & 2 \end{pmatrix}$$

into reduced row echelon form.

(a) The homogeneous system of equations $B\mathbf{x} = \mathbf{0}$ represents how many equations in how many unknowns? Is there a non-trivial solution? If so, find the general solution of $B\mathbf{x} = \mathbf{0}$.
(b) Is there a vector $\mathbf{b} \in \mathbb{R}^4$ for which $B\mathbf{x} = \mathbf{b}$ is inconsistent? Write down such a vector $\mathbf{b}$ if one exists and verify that $B\mathbf{x} = \mathbf{b}$ is inconsistent.

(c) Write down a vector $\mathbf{d} \in \mathbb{R}^4$ for which $B\mathbf{x} = \mathbf{d}$ is consistent. Then write down the general solution of $B\mathbf{x} = \mathbf{d}$.

Problem 2.7 Let

$$A = \begin{pmatrix} 4 & -1 & 1 \\ -1 & 4 & -1 \\ 1 & -1 & 4 \end{pmatrix}, \qquad \mathbf{x} = \begin{pmatrix} x \\ y \\ z \end{pmatrix}.$$

Write out the system of linear equations $A\mathbf{x} = 6\mathbf{x}$ and find all its solutions.

Problem 2.8 Let

$$B = \begin{pmatrix} 1 & 0 & 1 \\ 0 & 1 & 1 \\ -1 & 0 & 3 \\ 3 & 1 & 2 \end{pmatrix}, \qquad \mathbf{b} = \begin{pmatrix} a \\ b \\ c \\ d \end{pmatrix}.$$

Find an equation which the components a, b, c, d of the vector $\mathbf{b}$ must satisfy for the system of equations $B\mathbf{x} = \mathbf{b}$ to be consistent.

If $B\mathbf{x} = \mathbf{b}$ is consistent for a given vector $\mathbf{b}$, will the solution be unique? Justify your answer.

Problem 2.9 (Portfolio theory) A *portfolio* is a row vector

$$Y = (y_1 \quad \cdots \quad y_m)$$

in which y_i is the number of units of asset i held by an investor. After a year, say, the value of the assets will increase (or decrease) by a certain percentage. The change in each asset depends on *states* the economy will assume, predicted as a *returns matrix*, $R = (r_{ij})$, where r_{ij} is the factor by which investment i changes in one year if state j occurs.

Suppose an investor has assets in $y_1 = $ land, $y_2 = $ bonds and $y_3 = $ stocks, and that the returns matrix is

$$R = \begin{pmatrix} 1.05 & 0.95 & 1.0 \\ 1.05 & 1.05 & 1.05 \\ 1.20 & 1.26 & 1.23 \end{pmatrix}.$$

Then the *total* values of the portfolio in one year's time are given by YR, where $(YR)_j$ is the total value of the portfolio if state j occurs.

(a) Find the total values of the portfolio $W = (5000 \quad 2000 \quad 0)$ in one year for each of the possible states.
(b) Show that $U = (600 \quad 8000 \quad 1000)$ is a *riskless portfolio*; that is, it has the same value in all states j.

(c) An *arbitrage portfolio,* $Y = (y_1 \quad \cdots \quad y_m)$ is one which costs
nothing $(y_1 + \cdots + y_m = 0)$, cannot lose $((YR)_j \geq 0$ for all $j)$,
and in at least one state makes a profit $((YR)_j > 0$ for some $j)$.
Show that $Z = (1000 \quad -2000 \quad 1000)$ is an arbitrage portfolio.
(The bond asset of -2000 indicates that this sum was borrowed
from the bank.)
Can you find a more profitable arbitrage vector than this one?

(d) Let R be an $m \times n$ returns matrix, and let $\mathbf{u} = (1, 1, \ldots, 1)^{\mathrm{T}} \in \mathbb{R}^n$.
If the system $R\mathbf{x} = \mathbf{u}$ has a solution $\mathbf{p} = (p_1, \ldots, p_m)^{\mathrm{T}}$ with
$p_i > 0$, then the components p_i of $\mathbf{p}$ are called *state prices* and
the investor is said to be taking part in a 'fair game'. If state prices
exist, then there are no arbitrage vectors for R, and if state prices
do not exist, then arbitrage vectors do exist. Show that state prices
for the given matrix, R, do not exist.

Problem 2.10 (Conditioning of matrices) Some systems of linear
equations lead to 'ill-conditioned' matrices. These occur if a small
difference in the coefficients or constants yields a large difference in the
solution, in particular when the numbers involved are decimal approx-
imations of varying degree of accuracy.

The following two systems of equations represent the same problem,
the first to two decimal places accuracy, the second to three decimal
places accuracy. Solve them using Gaussian elimination and note the
significant difference in the solutions you obtain.

$$
\begin{cases} x + y = 51.11 \\ x + 1.02y = 2.22 \end{cases}
\qquad
\begin{cases} x + y = 51.106 \\ x + 1.016y = 2.218. \end{cases}
$$

3

Matrix inversion and determinants

In this chapter, all matrices will be square $n \times n$ matrices, unless explicitly stated otherwise. Only a square matrix can have an inverse, and the determinant is only defined for a square matrix.

We want to answer the following two questions: When is a matrix A invertible? How can we find the inverse matrix?

3.1 Matrix inverse using row operations

3.1.1 Elementary matrices

Recall the three elementary row operations:

RO1 multiply a row by a non-zero constant.
RO2 interchange two rows.
RO3 add a multiple of one row to another.

These operations change a matrix into a new matrix. We want to examine this process more closely. Let A be an $n \times n$ matrix and let A_i denote the ith row of A. Then we can write A as a column of n rows,

$$A = \begin{pmatrix} a_{11} & a_{12} & \cdots & a_{1n} \\ a_{21} & a_{22} & \cdots & a_{2n} \\ \vdots & \vdots & \ddots & \vdots \\ a_{n1} & a_{n2} & \cdots & a_{nn} \end{pmatrix} = \begin{pmatrix} A_1 \\ A_2 \\ \vdots \\ A_n \end{pmatrix}.$$

We use this row notation to indicate row operations. For example, what row operations are indicated below?

$$\begin{pmatrix} A_1 \\ 3A_2 \\ \vdots \\ A_n \end{pmatrix}, \quad \begin{pmatrix} A_2 \\ A_1 \\ \vdots \\ A_n \end{pmatrix}, \quad \begin{pmatrix} A_1 \\ A_2 + 4A_1 \\ \vdots \\ A_n \end{pmatrix}.$$

The first is multiply row 2 by 3, the second is interchange row 1 and row 2, and the third is add 4 times row 1 to row 2. Each of these represents new matrices after the row operation has been executed.

Now look at a product of two $n \times n$ matrices A and B. The $(1, 1)$ entry in the product is the inner product of row 1 of A and column 1 of B. The $(1, 2)$ entry is the inner product of row 1 of A and column 2 of B, and so on. In fact, row 1 of the product matrix AB is obtained by taking the product of the row A_1 with the matrix B; that is, $A_1 B$. This is true of each row of the product; that is, each row i of the product AB is obtained by calculating $A_i B$. So we can express the product AB as

$$\begin{pmatrix} a_{11} & a_{12} & \cdots & a_{1n} \\ a_{21} & a_{22} & \cdots & a_{2n} \\ \vdots & \vdots & \ddots & \vdots \\ a_{n1} & a_{n2} & \cdots & a_{nn} \end{pmatrix} \begin{pmatrix} b_{11} & b_{12} & \cdots & b_{1n} \\ b_{21} & b_{22} & \cdots & b_{2n} \\ \vdots & \vdots & \ddots & \vdots \\ b_{n1} & b_{n2} & \cdots & b_{nn} \end{pmatrix} = \begin{pmatrix} A_1 B \\ A_2 B \\ \vdots \\ A_n B \end{pmatrix}.$$

Now consider the effect of a row operation on a product AB. The first matrix below is the product AB after the row operation 'add 4 times row 1 of AB to row 2 of AB'

$$\begin{pmatrix} A_1 B \\ A_2 B + 4A_1 B \\ \vdots \\ A_n B \end{pmatrix} = \begin{pmatrix} A_1 B \\ (A_2 + 4A_1)B \\ \vdots \\ A_n B \end{pmatrix} = \begin{pmatrix} A_1 \\ A_2 + 4A_1 \\ \vdots \\ A_n \end{pmatrix} B.$$

In the second matrix, we have used the distributive rule to write

$$A_2 B + 4A_1 B = (A_2 + 4A_1)B.$$

But compare this matrix to the row form of a product of two matrices given above. You can see that it is what would result if we took the matrix obtained from A after the same row operation, and multiplied that by B.

We have shown that the matrix obtained by the row operation, 'add 4 times row 1 to row 2' on the product AB is equal to the product of the matrix obtained by the same row operation on A, with the matrix B. The same argument works for any row operation in general, so that

  (matrix obtained by a row operation on AB)
     = (matrix obtained by a row operation on A)B.

This is true for any $n \times n$ matrices A and B.

  Now take $A = I$, the identity matrix. Since $IB = B$, the previous statement now says that:

• The matrix obtained by a row operation on $B =$ (the matrix obtained by a row operation on I)B.

This leads us to the following definition:

Definition 3.1 (Elementary matrix) An *elementary matrix*, E, is an $n \times n$ matrix obtained by doing exactly *one* row operation on the $n \times n$ identity matrix, I.

For example,

$$\begin{pmatrix} 1 & 0 & 0 \\ 0 & 3 & 0 \\ 0 & 0 & 1 \end{pmatrix}, \qquad \begin{pmatrix} 0 & 1 & 0 \\ 1 & 0 & 0 \\ 0 & 0 & 1 \end{pmatrix}, \qquad \begin{pmatrix} 1 & 0 & 0 \\ 4 & 1 & 0 \\ 0 & 0 & 1 \end{pmatrix}$$

are elementary matrices. The first has had row 2 multiplied by 3, the second has had row 1 and row 2 interchanged, and the last matrix has had 4 times row 1 added to row 2.

Activity 3.2 Which of the matrices below are elementary matrices?

$$\begin{pmatrix} 2 & 1 & 0 \\ 0 & 1 & 0 \\ 0 & 0 & 1 \end{pmatrix}, \qquad \begin{pmatrix} 0 & 1 & 0 \\ 1 & 0 & 0 \\ -1 & 0 & 1 \end{pmatrix}, \qquad \begin{pmatrix} 1 & 0 & 0 \\ 0 & 1 & 0 \\ -1 & 0 & 1 \end{pmatrix}.$$

Write the first matrix as the product of two elementary matrices.

Elementary matrices provide a useful tool to relate a matrix to its reduced row echelon form. We have shown above that the matrix obtained from a matrix B after performing one row operation is equal to a product EB, where E is the elementary matrix obtained from I by that same row operation.

  For example, suppose we want to put the matrix

$$B = \begin{pmatrix} 1 & 2 & 4 \\ 1 & 3 & 6 \\ -1 & 0 & 1 \end{pmatrix}$$

into reduced row echelon form. Our first step is

$$B = \begin{pmatrix} 1 & 2 & 4 \\ 1 & 3 & 6 \\ -1 & 0 & 1 \end{pmatrix} \xrightarrow{R_2 - R_1} \begin{pmatrix} 1 & 2 & 4 \\ 0 & 1 & 2 \\ -1 & 0 & 1 \end{pmatrix}.$$

We perform the same operation on the identity matrix to obtain an elementary matrix, which we will denote by E_1

$$I = \begin{pmatrix} 1 & 0 & 0 \\ 0 & 1 & 0 \\ 0 & 0 & 1 \end{pmatrix} \xrightarrow{R_2 - R_1} \begin{pmatrix} 1 & 0 & 0 \\ -1 & 1 & 0 \\ 0 & 0 & 1 \end{pmatrix} = E_1.$$

Then the matrix $E_1 B$ is

$$E_1 B = \begin{pmatrix} 1 & 0 & 0 \\ -1 & 1 & 0 \\ 0 & 0 & 1 \end{pmatrix} \begin{pmatrix} 1 & 2 & 4 \\ 1 & 3 & 6 \\ -1 & 0 & 1 \end{pmatrix} = \begin{pmatrix} 1 & 2 & 4 \\ 0 & 1 & 2 \\ -1 & 0 & 1 \end{pmatrix},$$

which is the matrix obtained from B after the row operation.

We now want to look at the invertibility of elementary matrices and row operations. First, note that any elementary row operation can be undone by an elementary row operation.

RO1 is multiply a row by a non-zero constant.
To undo **RO1**, multiply the row by 1/(constant).

RO2 is interchange two rows.
To undo **RO2**, interchange the rows again.

RO3 is add a multiple of one row to another.
To undo **RO3**, subtract the multiple of one row from the other.

If we obtain an elementary matrix by performing one row operation on the identity, and another elementary matrix from the row operation which 'undoes' it, then multiplying these matrices together will return the identity matrix. That is, they are inverses of one another. This argument establishes the following theorem:

Theorem 3.3 *Any elementary matrix is invertible, and the inverse is also an elementary matrix.*

Activity 3.4 Let

$$E = \begin{pmatrix} 1 & 0 & 0 \\ -4 & 1 & 0 \\ 0 & 0 & 1 \end{pmatrix}.$$

Write down E^{-1}. Then show that $E E^{-1} = I$ and $E^{-1} E = I$.

We saw earlier in our example that multiplying $E_1 B$ we obtain

$$E_1 B = \begin{pmatrix} 1 & 0 & 0 \\ -1 & 1 & 0 \\ 0 & 0 & 1 \end{pmatrix} \begin{pmatrix} 1 & 2 & 4 \\ 1 & 3 & 6 \\ -1 & 0 & 1 \end{pmatrix} = \begin{pmatrix} 1 & 2 & 4 \\ 0 & 1 & 2 \\ -1 & 0 & 1 \end{pmatrix}.$$

We can undo this row operation and return the matrix B by multiplying on the left by E_1^{-1}:

$$\begin{pmatrix} 1 & 0 & 0 \\ 1 & 1 & 0 \\ 0 & 0 & 1 \end{pmatrix} \begin{pmatrix} 1 & 2 & 4 \\ 0 & 1 & 2 \\ -1 & 0 & 1 \end{pmatrix} = \begin{pmatrix} 1 & 2 & 4 \\ 1 & 3 & 6 \\ -1 & 0 & 1 \end{pmatrix}.$$

3.1.2 Row equivalence

Definition 3.5 If A and B are $m \times n$ matrices, we say that A is *row equivalent* to B if and only if there is a sequence of elementary row operations to transform A into B.

This is an example of what is known as an *equivalence relation*. This means it satisfies three important conditions; it is:

- reflexive: $A \sim A$,
- symmetric: $A \sim B \Rightarrow B \sim A$,
- transitive: $A \sim B$ and $B \sim C \Rightarrow A \sim C$.

Activity 3.6 Argue why it is true that row equivalence is an equivalence relation; that is, explain why *row equivalence* as defined above satisfies these three conditions.

The existence of an algorithm for putting a matrix A into reduced row echelon form by a sequence of row operations has the consequence that every matrix is row equivalent to a matrix in reduced row echelon form. This fact is stated in the following theorem.

Theorem 3.7 *Every matrix is row equivalent to a matrix in reduced row echelon form.*

3.1.3 The main theorem

We are now ready to answer the first of our questions: 'When is a matrix invertible?' We collect our results in the following theorem.

Theorem 3.8 *If A is an $n \times n$ matrix, then the following statements are equivalent (meaning if any one of these statements is true for A, then all the statements are true):*

(1) A^{-1} *exists.*
(2) $A\mathbf{x} = \mathbf{b}$ *has a unique solution for any* $\mathbf{b} \in \mathbb{R}^n$.
(3) $A\mathbf{x} = \mathbf{0}$ *only has the trivial solution,* $\mathbf{x} = \mathbf{0}$.
(4) *The reduced row echelon form of* A *is* I.

Proof: If we show that $(1) \Rightarrow (2) \Rightarrow (3) \Rightarrow (4) \Rightarrow (1)$, then any one statement will imply all the others, so the statements are equivalent.

$(1) \implies (2)$. We assume that A^{-1} exists, and consider the system of linear equations $A\mathbf{x} = \mathbf{b}$ where $\mathbf{x}$ is the vector of unknowns and $\mathbf{b}$ is any vector in $\mathbb{R}^n$. We use the matrix A^{-1} to solve for $\mathbf{x}$ by multiplying the equation on the left by A^{-1}. We have

$$A^{-1}A\mathbf{x} = A^{-1}\mathbf{b} \implies I\mathbf{x} = A^{-1}\mathbf{b} \implies \mathbf{x} = A^{-1}\mathbf{b}.$$

This shows that $\mathbf{x} = A^{-1}\mathbf{b}$ is the only possible solution; and it *is* a solution, since $A(A^{-1}\mathbf{b}) = (AA^{-1})\mathbf{b} = I\mathbf{b} = \mathbf{b}$. So $A\mathbf{x} = \mathbf{b}$ has a unique solution for any $\mathbf{b} \in \mathbb{R}^n$.

$(2) \implies (3)$. If $A\mathbf{x} = \mathbf{b}$ has a unique solution for all $\mathbf{b} \in \mathbb{R}^n$, then this is true for $\mathbf{b} = \mathbf{0}$. The unique solution of $A\mathbf{x} = \mathbf{0}$ must be the trivial solution, $\mathbf{x} = \mathbf{0}$.

$(3) \implies (4)$. If the only solution of $A\mathbf{x} = \mathbf{0}$ is $\mathbf{x} = \mathbf{0}$, then there are no free (non-leading) variables and the reduced row echelon form of A must have a leading one in every column. Since the matrix is square and a leading one in a lower row is further to the right, A must have a leading one in every row. Since every column with a leading one has zeros elsewhere, this can only be the $n \times n$ identity matrix.

$(4) \implies (1)$. We now make use of elementary matrices. If A is row equivalent to I, then there is a sequence or row operations which reduce A to I, so there must exist elementary matrices $E_1, \ldots, E_r$ such that

$$E_r E_{r-1} \cdots E_1 A = I.$$

Each elementary matrix has an inverse. We use these to solve the above equation for A, by first multiplying the equation on the left by E_r^{-1}, then by E_{r-1}^{-1}, and so on, to obtain

$$A = E_1^{-1} \cdots E_{r-1}^{-1} E_r^{-1} I.$$

This says that A is a product of invertible matrices, hence invertible. (Recall from Chapter 1 that if A and B are invertible matrices of the same size, then the product AB is invertible and its inverse is the product of the inverses in the reverse order, $(AB)^{-1} = B^{-1}A^{-1}$.)
    This proves the theorem. $\qquad\square$

3.1.4   Using row operations to find the inverse matrix

From the proof of Theorem 3.8, we have

$$A = E_1^{-1} \cdots E_r^{-1},$$

where the matrices E_i are the elementary matrices corresponding to the row operations used to reduce A to the identity matrix, I. Then, taking the inverse of both sides,

$$A^{-1} = (E_1^{-1} \cdots E_r^{-1})^{-1} = E_r \cdots E_1 = E_r \cdots E_1 I.$$

This tells us that if we apply the *same* row operations to the matrix I that we use to reduce A to I, then we will obtain the matrix A^{-1}. That is, if

$$E_r E_{r-1} \cdots E_1 A = I,$$

then

$$A^{-1} = E_r \cdots E_1 I.$$

This gives us a method to find the inverse of a matrix A. We start with the matrix A and we form a new, larger matrix by placing the identity matrix to the right of A, obtaining the matrix denoted $(A|I)$. We then use row operations to reduce this to $(I|B)$. If this is not possible (which will become apparent) then the matrix is not invertible. If it can be done, then A is invertible and $B = A^{-1}$.

Example 3.9 We use this method to find the inverse of the matrix

$$A = \begin{pmatrix} 1 & 2 & 4 \\ 1 & 3 & 6 \\ -1 & 0 & 1 \end{pmatrix}.$$

In order to determine if the matrix is invertible and, if so, to determine the inverse, we form the matrix

$$(A \mid I) = \left(\begin{array}{ccc|ccc} 1 & 2 & 4 & 1 & 0 & 0 \\ 1 & 3 & 6 & 0 & 1 & 0 \\ -1 & 0 & 1 & 0 & 0 & 1 \end{array} \right).$$

(We have separated A from I by a vertical line just to emphasise how this matrix is formed. It is also helpful in the calculations.) Then we carry out elementary row operations.

$$
\begin{array}{c} R_2 - R_1 \\ R_3 + R_1 \end{array}
\left(\begin{array}{ccc|ccc}
1 & 2 & 4 & 1 & 0 & 0 \\
0 & 1 & 2 & -1 & 1 & 0 \\
0 & 2 & 5 & 1 & 0 & 1
\end{array} \right)
$$

$$
R_3 - 2R_2
\left(\begin{array}{ccc|ccc}
1 & 2 & 4 & 1 & 0 & 0 \\
0 & 1 & 2 & -1 & 1 & 0 \\
0 & 0 & 1 & 3 & -2 & 1
\end{array} \right)
$$

$$
\begin{array}{c} R_1 - 4R_3 \\ R_2 - 2R_3 \end{array}
\left(\begin{array}{ccc|ccc}
1 & 2 & 0 & -11 & 8 & -4 \\
0 & 1 & 0 & -7 & 5 & -2 \\
0 & 0 & 1 & 3 & -2 & 1
\end{array} \right)
$$

$$
R_1 - 2R_2
\left(\begin{array}{ccc|ccc}
1 & 0 & 0 & 3 & -2 & 0 \\
0 & 1 & 0 & -7 & 5 & -2 \\
0 & 0 & 1 & 3 & -2 & 1
\end{array} \right).
$$

This is now in the form $(I|B)$, so we deduce that A is invertible and that

$$
A^{-1} = \left(\begin{array}{ccc}
3 & -2 & 0 \\
-7 & 5 & -2 \\
3 & -2 & 1
\end{array} \right).
$$

It is very easy to make mistakes when row reducing a matrix, so the next thing you should do is *check* that $AA^{-1} = I$.

Activity 3.10 Do this. Check that when you multiply AA^{-1}, you get the identity matrix I. (In order to establish that this is the inverse matrix, you should also show $A^{-1}A = I$, but we will forgo that here. We'll come back to this issue shortly.)

If the matrix A is not invertible, what will happen? By Theorem 3.8, if A is not invertible, then the reduced row echelon form of A cannot be I, so there will be a row of zeros in the row echelon form of A.

Activity 3.11 Find the inverse, if it exists, of each of the following matrices

$$
A = \left(\begin{array}{ccc}
-2 & 1 & 3 \\
0 & -1 & 1 \\
1 & 2 & 0
\end{array} \right), \qquad
B = \left(\begin{array}{ccc}
2 & 1 & 3 \\
0 & -1 & 1 \\
1 & 2 & 0
\end{array} \right).
$$

3.1.5 Verifying an inverse

At this stage, in order to show that a square matrix B is the inverse of the $n \times n$ matrix A, it seems we have to show that *both* statements, $AB = I$ and $BA = I$, are true. However, after we have proved the following theorem (which follows from Theorem 3.8), we will be able to deduce from the single statement $AB = I$ that A and B must be inverses of one another.

Theorem 3.12 *If A and B are $n \times n$ matrices and $AB = I$, then A and B are each invertible matrices, and $A = B^{-1}$ and $B = A^{-1}$.*

Proof: If we show that the homogeneous system of equations $B\mathbf{x} = \mathbf{0}$ has only the trivial solution, $\mathbf{x} = \mathbf{0}$, then by Theorem 3.8 this will prove that B is invertible. So we consider the matrix equation $B\mathbf{x} = \mathbf{0}$ and multiply both sides of this equation on the left by the matrix A. We have

$$B\mathbf{x} = \mathbf{0} \Longrightarrow A(B\mathbf{x}) = A\mathbf{0} \Longrightarrow (AB)\mathbf{x} = \mathbf{0}.$$

But we are given that $AB = I$, so that

$$(AB)\mathbf{x} = \mathbf{0} \Longrightarrow I\mathbf{x} = \mathbf{0} \Longrightarrow \mathbf{x} = \mathbf{0},$$

which shows that the only solution of $B\mathbf{x} = \mathbf{0}$ is the trivial solution. We therefore conclude that B is invertible, so the matrix B^{-1} exists.

We now multiply both sides of the equation $AB = I$ on the right by the matrix B^{-1}. We have

$$AB = I \Longrightarrow (AB)B^{-1} = IB^{-1} \Longrightarrow A(BB^{-1}) = B^{-1} \Longrightarrow A = B^{-1}.$$

So A is the inverse of B, and therefore A is also an invertible matrix. Then taking inverses of both sides of the last equation, we conclude that $A^{-1} = (B^{-1})^{-1} = B$. □

3.2　Determinants

3.2.1　Determinant using cofactors

The *determinant* of a square matrix A is a particular number associated with A, written $|A|$ or det A. This number will provide a quick way to determine whether or not a matrix A is invertible. In view of this, suppose A is a 2×2 matrix, and that we wish to determine A^{-1} using row operations. Then we form the matrix $(A \mid I)$ and attempt to row reduce A to I. We assume $a \neq 0$, otherwise we would begin by switching rows:

$$(A \mid I) = \begin{pmatrix} a & b & \big| & 1 & 0 \\ c & d & \big| & 0 & 1 \end{pmatrix} \xrightarrow{(1/a)R_1} \begin{pmatrix} 1 & b/a & \big| & 1/a & 0 \\ c & d & \big| & 0 & 1 \end{pmatrix}$$

$$\xrightarrow{R_2 - cR_1} \begin{pmatrix} 1 & b/a & \big| & 1/a & 0 \\ 0 & d - cb/a & \big| & -c/a & 1 \end{pmatrix} \xrightarrow{aR_2} \begin{pmatrix} 1 & b/a & \big| & 1/a & 0 \\ 0 & (ad - bc) & \big| & -c & a \end{pmatrix},$$

which shows that A^{-1} exists if and only if $ad - bc \neq 0$.

For a 2×2 matrix, the *determinant* is given by the formula

$$\left| \begin{pmatrix} a & b \\ c & d \end{pmatrix} \right| = \begin{vmatrix} a & b \\ c & d \end{vmatrix} = ad - bc.$$

Note the vertical bars used in the notation for the determinant of a matrix (and note also that we usually dispense with the large parentheses around the matrix when we write its determinant).

For example,

$$\begin{vmatrix} 1 & 2 \\ 3 & 4 \end{vmatrix} = (1)(4) - (2)(3) = -2.$$

To extend this definition to $n \times n$ matrices, we define the determinant of an $n \times n$ matrix recursively in terms of $(n-1) \times (n-1)$ determinants. So the determinant of a 3×3 matrix is given in terms of 2×2 matrices, and so on. To do this, we will need the following two definitions.

Definition 3.13 Suppose A is an $n \times n$ matrix. The (i, j) *minor* of A, denoted by M_{ij}, is the determinant of the $(n-1) \times (n-1)$ matrix obtained by removing the ith row and jth column of A.

Definition 3.14 The (i, j) *cofactor* of a matrix A is

$$C_{ij} = (-1)^{i+j} M_{ij}.$$

So the cofactor is equal to the minor if $i + j$ is even, and it is equal to the negative of the minor if $i + j$ is odd.

Example 3.15 Let

$$A = \begin{pmatrix} 1 & 2 & 3 \\ 4 & 1 & 1 \\ -1 & 3 & 0 \end{pmatrix}.$$

Then the minor M_{23} and the cofactor C_{23} are

$$M_{23} = \begin{vmatrix} 1 & 2 \\ -1 & 3 \end{vmatrix} = 5, \qquad C_{23} = (-1)^{(2+3)} M_{23} = -5.$$

There is a simple way to associate the cofactor C_{ij} with the entry a_{ij} of the matrix. Locate the entry a_{ij} and cross out the row and the column containing a_{ij}. Then evaluate the determinant of the $(n-1) \times (n-1)$ matrix which remains. This is the minor, M_{ij}. Then give it a '+' or '−' sign according to the position of a_{ij} on the following pattern:

$$\begin{pmatrix} + & - & + & - & \cdots \\ - & + & - & + & \cdots \\ + & - & + & - & \cdots \\ \vdots & \vdots & \vdots & \vdots & \ddots \end{pmatrix}.$$

Activity 3.16 Write down the cofactor C_{13} for the matrix A above using this method.

If A is an $n \times n$ matrix, the *determinant* of A is given by

$$|A| = \begin{vmatrix} a_{11} & a_{12} & \cdots & a_{1n} \\ a_{21} & a_{22} & \cdots & a_{2n} \\ \vdots & \vdots & \ddots & \vdots \\ a_{n1} & a_{n2} & \cdots & a_{nn} \end{vmatrix} = a_{11}C_{11} + a_{12}C_{12} + \cdots + a_{1n}C_{1n}.$$

This is called the *cofactor expansion of* $|A|$ *by row one*. It is a recursive definition, meaning that the determinant of an $n \times n$ matrix is given in terms of some $(n-1) \times (n-1)$ determinants.

Example 3.17 We calculate the determinant of the matrix A in Example 3.15:

$$|A| = 1C_{11} + 2C_{12} + 3C_{13}$$
$$= 1\begin{vmatrix} 1 & 1 \\ 3 & 0 \end{vmatrix} - 2\begin{vmatrix} 4 & 1 \\ -1 & 0 \end{vmatrix} + 3\begin{vmatrix} 4 & 1 \\ -1 & 3 \end{vmatrix}$$
$$= 1(-3) - 2(1) + 3(13) = 34.$$

Activity 3.18 Calculate the determinant of the matrix

$$M = \begin{pmatrix} -1 & 2 & 1 \\ 0 & 2 & 3 \\ 1 & 1 & 4 \end{pmatrix}.$$

You might ask: 'Why is the cofactor expansion given by row 1, rather than any other row?' In fact, it turns out that using a cofactor expansion by any row or column of A will give the same number $|A|$, as the following theorem states.

Theorem 3.19 *If A is an $n \times n$ matrix, then the determinant of A can be computed by multiplying the entries of any row (or column) by their cofactors and summing the resulting products:*

$$|A| = a_{i1}C_{i1} + a_{i2}C_{i2} + \ldots + a_{in}C_{in}$$
(cofactor expansion by row i)
$$|A| = a_{1j}C_{1j} + a_{2j}C_{2j} + \ldots + a_{nj}C_{nj}$$
(cofactor expansion by column j).

We will look into the proof of this result later, but first note that this allows you to choose any row or any column of a matrix to find its determinant using a cofactor expansion. So we should choose a row or column which gives the simplest calculations.

Obtaining the correct value for $|A|$ is important, so it is a good idea to check your result by calculating the determinant by another row or column.

Example 3.20 In the matrix of Example 3.15, instead of using the cofactor expansion by row 1 as shown above, we can choose to evaluate the determinant of the matrix A by row 3 or column 3, which will involve fewer calculations since $a_{33} = 0$. To check the result $|A| = 34$, we will evaluate the determinant again; this time using column 3. Remember the correct cofactor signs:

$$|A| = \begin{vmatrix} 1 & 2 & 3 \\ 4 & 1 & 1 \\ -1 & 3 & 0 \end{vmatrix} = 3 \begin{vmatrix} 4 & 1 \\ -1 & 3 \end{vmatrix} - 1 \begin{vmatrix} 1 & 2 \\ -1 & 3 \end{vmatrix} + 0 = 3(13) - (5) = 34.$$

Activity 3.21 Check your calculation of the determinant of the matrix

$$M = \begin{pmatrix} -1 & 2 & 1 \\ 0 & 2 & 3 \\ 1 & 1 & 4 \end{pmatrix}$$

in the previous activity by expanding by a different row or column. Choose one with fewer calculations.

3.2.2 Determinant as a sum of elementary signed products

We will give an informal proof of Theorem 3.19, because it is useful to understand how the definition of determinant works. This section can be safely omitted (meaning you can simply accept the theorem without proof and move on), but you might find it worth your while to read through it.

For a 2×2 matrix, the cofactor expansion by row 1 is equivalent to the definition given on page 99:

$$\begin{vmatrix} a_{11} & a_{12} \\ a_{21} & a_{22} \end{vmatrix} = a_{11}a_{22} - a_{12}a_{21}.$$

Notice that each term of the sum is a product of entries, one from each row and one from each column. Indeed, a_{11} is the entry from row 1 and column 1, and a_{22} is not in either: it comes from row 2 and column 2. Similarly, the second term, $a_{12}a_{21}$, is the only different way of taking one entry from each row and each column of the matrix.

For a 3×3 matrix, the cofactor expansion by row 1 yields,

$$\begin{vmatrix} a_{11} & a_{12} & a_{13} \\ a_{21} & a_{22} & a_{23} \\ a_{31} & a_{32} & a_{33} \end{vmatrix} = a_{11} \begin{vmatrix} a_{22} & a_{23} \\ a_{32} & a_{33} \end{vmatrix} - a_{12} \begin{vmatrix} a_{21} & a_{23} \\ a_{31} & a_{33} \end{vmatrix} + a_{13} \begin{vmatrix} a_{21} & a_{22} \\ a_{31} & a_{32} \end{vmatrix}$$

$$= a_{11}(a_{22}a_{33} - a_{23}a_{32}) - a_{12}(a_{21}a_{33} - a_{23}a_{31}) + a_{13}(a_{21}a_{32} - a_{22}a_{31}).$$

Then $|A|$ is the sum of the products:

$$\begin{array}{llll} & a_{11}a_{22}a_{33} & +a_{12}a_{23}a_{31} & +a_{13}a_{21}a_{32} \\ - & a_{11}a_{23}a_{32} & -a_{12}a_{21}a_{33} & -a_{13}a_{22}a_{31}. \end{array} \qquad (*)$$

The row indices of each product are in ascending order, 123, and the column indices are:

$$\begin{array}{ccc} 123 & 231 & 312 \\ 132 & 213 & 321. \end{array}$$

These are the six permutations of the numbers 1,2,3.

Definition 3.22 A *permutation* of a set of integers $\{1, 2, 3, \ldots, n\}$ is an arrangement of these integers in some order with no omissions and no repetitions.

To find all permutations of a set of numbers, we can use a permutation tree:

```
 1    2    3        ←  3 choices  ⎫
/\   /\   /\                      ⎪
2 3  1 3  1 2       ←  2 choices  ⎬   3 · 2 · 1 = 3!
| |  | |  | |                     ⎪
3 2  3 1  2 1       ←  1 choice   ⎭
```

In the above expansion of $|A|$, each term has the row indices arranged in ascending order and the column indices form a different permutation of the numbers 1,2,3. We know, therefore, that each term of the sum is a different product of entries, one from each row and one from each column of A, and the set of six products contains all ways in which this can happen.

But what about the minus signs? An *inversion* is said to occur in a permutation whenever a larger integer precedes a smaller one. For example,

$$1\,2\,3 \leftarrow \text{no inversions} \qquad 1\,3\,2 \leftarrow \text{one inversion.}$$

A permutation is said to be *even* if the total number of inversions is even. It is *odd* if the total number of inversions is odd.

To find the *total number of inversions* of a permutation, we can start at the left and find the total number of integers to the right of the first integer which are smaller than the first integer. Then go to the next integer to the right and do the same. Continue until the end, and then add up all these numbers.

Example 3.23 Consider the permutation 5 2 3 4 1. We apply the method just described. This tells us that total number of inversions is

$$4 + 1 + 1 + 1 = 7,$$

so this permutation is odd. The total number of inversions gives the minimum number of steps that it takes to put these numbers into ascending order, where in each step you are only allowed to switch the positions of two adjacent numbers. For the permutation 5 2 3 4 1, this can be done in seven steps as follows:

$$52341 \rightarrow 25341 \rightarrow 23541 \rightarrow 23451$$
$$\rightarrow 23415 \rightarrow 23145 \rightarrow 21345 \rightarrow 12345.$$

If we look again at the list of products ($*$), we find that the permutations of the column indices corresponding to the products with a plus sign are all even, and those corresponding to the products with a minus sign are all odd.

Definition 3.24 An *elementary product* from an $n \times n$ matrix A is a product of n entries, no two of which come from the same row or column. A *signed elementary product* has the row indices arranged in *ascending* order, multiplied by -1 if the column indices are an *odd* permutation of the numbers 1 to n.

We are now ready to give an intrinsic (but completely impractical) definition of determinant.

Definition 3.25 (Determinant) The *determinant* of an $n \times n$ matrix A is the sum of all signed elementary products of A.

A cofactor expansion is a clever way to obtain this sum of signed elementary products. You choose the entries from one row, say, and then cross out that row and the column containing the entry to obtain the cofactor, and each stage of calculating the cofactor repeats the process. At the heart of a proof of Theorem 3.19 is the fact that each possible cofactor expansion is the sum of all signed elementary products, and so all the cofactor expansions are equal to each other.

Activity 3.26 Expand the determinant

$$|A| = \begin{vmatrix} a_{11} & a_{12} & a_{13} \\ a_{21} & a_{22} & a_{23} \\ a_{31} & a_{32} & a_{33} \end{vmatrix}$$

using the cofactor expansion by column 2, and show that you get the same list of signed elementary products as we obtained in ($*$) on page 102.

For very large matrices, using a cofactor expansion is impractical. For example,

$$|A| = \begin{vmatrix} 1 & -4 & 3 & 2 \\ 2 & -7 & 5 & 1 \\ 1 & 2 & 6 & 0 \\ 2 & -10 & 14 & 4 \end{vmatrix} = 1C_{11} + (-4)C_{12} + 3C_{13} + 2C_{14}$$

would require calculating four 3×3 determinants. Fortunately, there is a better way. To simplify the calculations, we will turn once again to row operations. But first we need to establish some useful results on determinants, which follow directly from Theorem 3.19.

3.3 Results on determinants

We now look at some standard and useful properties of determinants.

Theorem 3.27 *If A is an $n \times n$ matrix, then*

$$|A^{\mathrm{T}}| = |A|.$$

Proof: This theorem follows immediately from Theorem 3.19. The cofactor expansion by row i of $|A^{\mathrm{T}}|$ is precisely the same, number for number, as the cofactor expansion by column i of $|A|$. $\qquad\square$

Each of the following three statements follows from Theorem 3.19. (They are 'corollaries', meaning consequences, of the theorem.) As a result of Theorem 3.27, it follows that each is true if the word *row* is replaced by *column*. We will need these results in the next section. In all of them, we assume that A is an $n \times n$ matrix.

Corollary 3.28 *If a row of A consists entirely of zeros, then $|A| = 0$.*

Proof: If we evaluate the determinant by the cofactor expansion using the row of zeros, then each cofactor is multiplied by 0 and the sum will be zero. To visualise this, expand the determinant below using row 1:

$$\begin{vmatrix} 0 & 0 & \cdots & 0 \\ a_{21} & a_{22} & \cdots & a_{2n} \\ \vdots & \vdots & \ddots & \vdots \\ a_{n1} & a_{n2} & \cdots & a_{nn} \end{vmatrix} = 0C_{11} + 0C_{12} + \cdots + 0C_{1n} = 0.$$

$\square$

Corollary 3.29 *If A contains two rows which are equal, then $|A| = 0$.*

Proof: To prove this, we will use an inductive argument. If A is a 2×2 matrix with two equal rows, then

$$|A| = \begin{vmatrix} a & b \\ a & b \end{vmatrix} = ab - ab = 0.$$

Now consider a 3×3 matrix with two equal rows. If we expand the determinant by the other row, then each cofactor is a 2×2 determinant with two equal rows, therefore each is zero and so is their sum. For example,

$$|A| = \begin{vmatrix} a & b & c \\ d & e & f \\ a & b & c \end{vmatrix} = -d \begin{vmatrix} b & c \\ b & c \end{vmatrix} + e \begin{vmatrix} a & c \\ a & c \end{vmatrix} - f \begin{vmatrix} a & b \\ a & b \end{vmatrix}$$

$$= 0 + 0 + 0 = 0.$$

Generally, in a similar way, the result for $(n - 1) \times (n - 1)$ matrices implies the result for $n \times n$ matrices. □

Corollary 3.30 *If the cofactors of one row are multiplied by the entries of a different row and added, then the result is 0. That is, if $i \neq j$, then* $a_{j1}C_{i1} + a_{j2}C_{i2} + \cdots + a_{jn}C_{in} = 0.$

Proof: Let

$$A = \begin{pmatrix} a_{11} & a_{12} & \cdots & a_{1n} \\ a_{21} & a_{22} & \cdots & a_{2n} \\ \vdots & \vdots & \ddots & \vdots \\ a_{n1} & a_{n2} & \cdots & a_{nn} \end{pmatrix}.$$

The cofactor expansion of $|A|$ by row i is

$$|A| = a_{i1}C_{i1} + a_{i2}C_{i2} + \cdots + a_{in}C_{in}.$$

Look at the expression

$$a_{j1}C_{i1} + a_{j2}C_{i2} + \cdots + a_{jn}C_{in} \qquad \text{for} \quad i \neq j.$$

This expression is *not* equal to $|A|$, so what is it? It is equal to $|B|$ for some matrix B, but what does the matrix B look like?

In the expression $|B| = a_{j1}C_{i1} + a_{j2}C_{i2} + \cdots + a_{jn}C_{in}$, each cofactor C_{ik}, for $k = 1, \ldots, n$, is made up of entries of the matrix A, omitting the entries from row i. For example, if $i \neq 1$, then C_{i1} is obtained from the matrix resulting in removing row i and column 1 from A, and C_{ik} is obtained by removing row i and column k. So the matrix B will have the same entries as the matrix A except in row i. In the cofactor expansion of a determinant by row i, the entries of row i are the numbers multiplying the cofactors. Therefore, the entries of row i of the matrix B must be $a_{j1}, \ldots, a_{jn}$. Then B has two equal rows, since row i has the same entries as row j. It follows, by Corollary 3.29, that $|B| = 0$, and the result follows. $\square$

Corollary 3.31 *If $A = (a_{ij})$ and if each entry of one of the rows, say row i, can be expressed as a sum of two numbers, $a_{ij} = b_{ij} + c_{ij}$ for $1 \le j \le n$, then $|A| = |B| + |C|$, where B is the matrix A with row i replaced by $b_{i1}, b_{i2}, \cdots b_{in}$ and C is the matrix A with row i replaced by $c_{i1}, c_{i2}, \cdots c_{in}$.*

Proof: First, let's illustrate this with a 3×3 matrix. The corollary states that, for example,

$$|A| = \begin{vmatrix} a & b & c \\ d+p & e+q & f+r \\ g & h & i \end{vmatrix} = \begin{vmatrix} a & b & c \\ d & e & f \\ g & h & i \end{vmatrix} + \begin{vmatrix} a & b & c \\ p & q & r \\ g & h & i \end{vmatrix}$$
$$= |B| + |C|.$$

To show this is true, you just need to use the cofactor expansion by row 2 for each of the determinants. We have

$$\begin{aligned} |A| &= (d+p)C_{21} + (e+q)C_{22} + (f+r)C_{23} \\ &= dC_{21} + eC_{22} + fC_{23} + pC_{21} + qC_{22} + rC_{23} \\ &= |B| + |C|, \end{aligned}$$

where the cofactors C_{21}, C_{22}, C_{23} are exactly the same in each expansion (of the determinants of B and C), since each consists entirely of entries from the matrix A other than those in row i.

The proof for a general matrix A is exactly the same. The cofactor expansion by row i yields

$$\begin{aligned} |A| &= a_{i1}C_{i1} + a_{i2}C_{i2} + \cdots + a_{in}C_{in} \\ &= (b_{i1} + c_{i1})C_{i1} + (b_{i2} + c_{i2})C_{i2} + \cdots + (b_{in} + c_{in})C_{in} \\ &= (b_{i1}C_{i1} + b_{i2}C_{i2} + \cdots + b_{in}C_{in}) \\ &\quad + (c_{i1}C_{i1} + c_{i2}C_{i2} + \cdots + c_{in}C_{in}) \\ &= |B| + |C|. \end{aligned}$$

$\square$

3.3.1 Determinant using row operations

In this section, we take a different approach to evaluating determinants, by making use of row operations.

Definition 3.32 An $n \times n$ matrix A is *upper triangular* if all entries *below* the main diagonal are zero. It is *lower triangular* if all entries *above* the main diagonal are zero.

upper
triangular
matrix
$$\begin{pmatrix} a_{11} & a_{12} & \cdots & a_{1n} \\ 0 & a_{22} & \cdots & a_{2n} \\ \vdots & \vdots & \ddots & \vdots \\ 0 & 0 & \cdots & a_{nn} \end{pmatrix}$$

lower
triangular
matrix
$$\begin{pmatrix} a_{11} & 0 & \cdots & 0 \\ a_{21} & a_{22} & \cdots & 0 \\ \vdots & \vdots & \ddots & \vdots \\ a_{n1} & a_{n2} & \cdots & a_{nn} \end{pmatrix}$$

diagonal
matrix
$$\begin{pmatrix} a_{11} & 0 & \cdots & 0 \\ 0 & a_{22} & \cdots & 0 \\ \vdots & \vdots & \ddots & \vdots \\ 0 & 0 & \cdots & a_{nn} \end{pmatrix}.$$

Suppose we wish to evaluate the determinant of an upper triangular matrix, such as

$$\begin{vmatrix} a_{11} & a_{12} & \cdots & a_{1n} \\ 0 & a_{22} & \cdots & a_{2n} \\ \vdots & \vdots & \ddots & \vdots \\ 0 & 0 & \cdots & a_{nn} \end{vmatrix}.$$

Which row or column should we use for the cofactor expansion? Clearly, the calculations are simplest if we expand by column 1 or row n. Expansion by column 1 gives us

$$|A| = a_{11} \begin{vmatrix} a_{22} & \cdots & a_{2n} \\ \vdots & \ddots & \vdots \\ 0 & \cdots & a_{nn} \end{vmatrix},$$

where the $(n-1) \times (n-1)$ matrix on the right is again upper triangular. Continuing in this way, we see that $|A|$ is just the product of the diagonal entries. The same argument holds true for a matrix which is diagonal or lower triangular, so we have established one more corollary of Theorem 3.19:

Corollary 3.33 *If A is upper triangular, lower triangular or diagonal, then*

$$|A| = a_{11}a_{22} \cdots a_{nn}.$$

A square matrix in row echelon form is upper triangular. If we know how a determinant is affected by a row operation, then this observation will give us an easier way to calculate large determinants. We can use row operations to put the matrix into row echelon form, keep track of any changes and then easily calculate the determinant of the reduced matrix. So how does each row operation affect the value of the determinant? Let's consider each in turn.

The first row operation is:

RO1   multiply a row by a non-zero constant

Suppose the matrix B is obtained from a matrix A by multiplying row i by a non-zero constant α. For example,

$$|A| = \begin{vmatrix} a_{11} & a_{12} & \cdots & a_{1n} \\ a_{21} & a_{22} & \cdots & a_{2n} \\ \vdots & \vdots & \ddots & \vdots \\ a_{n1} & a_{n2} & \cdots & a_{nn} \end{vmatrix}, \qquad |B| = \begin{vmatrix} a_{11} & a_{12} & \cdots & a_{1n} \\ \alpha a_{21} & \alpha a_{22} & \cdots & \alpha a_{2n} \\ \vdots & \vdots & \ddots & \vdots \\ a_{n1} & a_{n2} & \cdots & a_{nn} \end{vmatrix}.$$

If we evaluate $|B|$ using the cofactor expansion by row i, we obtain

$$\begin{aligned} |B| &= \alpha a_{i1} C_{i1} + \alpha a_{i2} C_{i2} + \cdots + \alpha a_{in} C_{in} \\ &= \alpha(a_{i1} C_{i1} + a_{i2} C_{i2} + \cdots + a_{in} C_{in}) \\ &= \alpha |A|. \end{aligned}$$

So we have:

- The effect of multiplying a row of A by α is to multiply $|A|$ by α, $|B| = \alpha |A|$.

When we actually need this, we will use it to factor out a constant α from the determinant as follows:

$$\begin{vmatrix} a_{11} & a_{12} & \cdots & a_{1n} \\ \alpha a_{21} & \alpha a_{22} & \cdots & \alpha a_{2n} \\ \vdots & \vdots & \ddots & \vdots \\ a_{n1} & a_{n2} & \cdots & a_{nn} \end{vmatrix} = \alpha \begin{vmatrix} a_{11} & a_{12} & \cdots & a_{1n} \\ a_{21} & a_{22} & \cdots & a_{2n} \\ \vdots & \vdots & \ddots & \vdots \\ a_{n1} & a_{n2} & \cdots & a_{nn} \end{vmatrix}.$$

The second type of row operation is:

RO2   interchange two rows

This time we will use an inductive proof involving the cofactor expansion. If A is a 2×2 matrix and B is the matrix obtained from A by interchanging the two rows, then

$$|A| = \begin{vmatrix} a & b \\ c & d \end{vmatrix} = ad - bc, \qquad |B| = \begin{vmatrix} c & d \\ a & b \end{vmatrix} = bc - ad,$$

so $|B| = -|A|$.

Now let A be a 3×3 matrix and let B be a matrix obtained from A by interchanging two rows. Then if we expand $|B|$ using a different row, each cofactor contains the determinant of a 2×2 matrix which is a cofactor of A with two rows interchanged, so each will be multiplied by -1, and $|B| = -|A|$. To visualise this, consider for example

$$|A| = \begin{vmatrix} a & b & c \\ d & e & f \\ g & h & i \end{vmatrix}, \qquad |B| = \begin{vmatrix} g & h & i \\ d & e & f \\ a & b & c \end{vmatrix}.$$

Expanding $|A|$ and $|B|$ by row 2, we have

$$|A| = -d \begin{vmatrix} b & c \\ h & i \end{vmatrix} + e \begin{vmatrix} a & c \\ g & i \end{vmatrix} - f \begin{vmatrix} a & b \\ g & h \end{vmatrix}$$

$$|B| = -d \begin{vmatrix} h & i \\ b & c \end{vmatrix} + e \begin{vmatrix} g & i \\ a & c \end{vmatrix} - f \begin{vmatrix} g & h \\ a & b \end{vmatrix} = -|A|$$

since all the 2×2 determinants change sign. In the same way, if this holds for $(n-1) \times (n-1)$ matrices, then it hold for $n \times n$ matrices.

So we have:

- The effect of interchanging two rows of a matrix is to multiply the determinant by -1: $|B| = -|A|$.

Finally, we have the third type of row operation:

RO3 add a multiple of one row to another.

Suppose the matrix B is obtained from the matrix A by replacing row j of A by row j plus k times row i of A, $j \neq i$. For example, consider the case in which B is obtained from A by adding 4 times row 1 of A to row 2. Then

$$|A| = \begin{vmatrix} a_{11} & a_{12} & \cdots & a_{1n} \\ a_{21} & a_{22} & \cdots & a_{2n} \\ \vdots & \vdots & \ddots & \vdots \\ a_{n1} & a_{n2} & \cdots & a_{nn} \end{vmatrix},$$

$$|B| = \begin{vmatrix} a_{11} & a_{12} & \cdots & a_{1n} \\ a_{21} + 4a_{11} & a_{22} + 4a_{12} & \cdots & a_{2n} + 4a_{1n} \\ \vdots & \vdots & \ddots & \vdots \\ a_{n1} & a_{n2} & \cdots & a_{nn} \end{vmatrix}.$$

In general, in a situation like this, we can expand $|B|$ by row j:

$$\begin{aligned}
|B| &= (a_{j1} + ka_{i1})C_{j1} + (a_{j2} + ka_{i2})C_{j2} + \cdots + (a_{jn} + ka_{in})C_{jn} \\
&= a_{j1}C_{j1} + a_{j2}C_{j2} + \cdots + a_{jn}C_{jn} \\
&\quad + k(a_{i1}C_{j1} + a_{i2}C_{j2} + \cdots + a_{in}C_{jn}) \\
&= |A| + 0.
\end{aligned}$$

The last expression in brackets is 0 because it consists of the cofactors of one row multiplied by the entries of another row (see Corollary 3.30). So this row operation does not change the value of $|A|$.

So we see that:

- There is *no change* in the value of the determinant if a multiple of one row is added to another.

We collect these results in the following theorem.

Theorem 3.34 (Effect of a row (column) operation on $|A|$) *All state-ments are true if* row *is replaced by* column.

(RO1)   *If a row is multiplied by a constant α, then $|A|$ changes to $\alpha|A|$.*

(RO2)   *If two rows are interchanged, then $|A|$ changes to $-|A|$.*

(RO3)   *If a multiple of one row is added to another, then there is no change in $|A|$.*

Example 3.35 We can now use row operations to evaluate

$$|A| = \begin{vmatrix} 1 & 2 & -1 & 4 \\ -1 & 3 & 0 & 2 \\ 2 & 1 & 1 & 2 \\ 1 & 4 & 1 & 3 \end{vmatrix}$$

by reducing A to an upper triangular matrix. First, we obtain zeros below the leading one by adding multiples of row 1 to the rows below. The new matrix will have the same determinant as A. So

$$|A| = \begin{vmatrix} 1 & 2 & -1 & 4 \\ 0 & 5 & -1 & 6 \\ 0 & -3 & 3 & -6 \\ 0 & 2 & 2 & -1 \end{vmatrix}.$$

Next, we observe that

$$\begin{vmatrix} 1 & 2 & -1 & 4 \\ 0 & 5 & -1 & 6 \\ 0 & -3 & 3 & -6 \\ 0 & 2 & 2 & -1 \end{vmatrix} = -3 \begin{vmatrix} 1 & 2 & -1 & 4 \\ 0 & 5 & -1 & 6 \\ 0 & 1 & -1 & 2 \\ 0 & 2 & 2 & -1 \end{vmatrix},$$

where we factored -3 from the third row. (We would need to multiply the resulting determinant on the right by -3 in order to put the -3 back into the third row, and to get back a matrix with the same determinant as A.) Next we switch row 2 and row 3, with the effect of changing the sign of the determinant.

$$|A| = 3 \begin{vmatrix} 1 & 2 & -1 & 4 \\ 0 & 1 & -1 & 2 \\ 0 & 5 & -1 & 6 \\ 0 & 2 & 2 & -1 \end{vmatrix}.$$

Next, we use RO3 operations to achieve upper triangular form. These operations result in no change in the value of the determinant. So we have

$$|A| = 3 \begin{vmatrix} 1 & 2 & -1 & 4 \\ 0 & 1 & -1 & 2 \\ 0 & 0 & 4 & -4 \\ 0 & 0 & 4 & -5 \end{vmatrix} = 3 \begin{vmatrix} 1 & 2 & -1 & 4 \\ 0 & 1 & -1 & 2 \\ 0 & 0 & 4 & -4 \\ 0 & 0 & 0 & -1 \end{vmatrix}.$$

Finally, we evaluate the determinant of the upper triangular matrix, obtaining

$$|A| = 3 \begin{vmatrix} 1 & 2 & -1 & 4 \\ 0 & 1 & -1 & 2 \\ 0 & 0 & 4 & -4 \\ 0 & 0 & 0 & -1 \end{vmatrix} = 3(1 \times 1 \times 4 \times (-1)) = -12.$$

A word of caution with row operations on a determinant! What is the change in the value of $|A|$ in the following circumstances:

(1) if R_2 is replaced by $R_2 - 3R_1$?
(2) if R_2 is replaced by $3R_1 - R_2$?

For (1), there is no change, but for (2), the determinant will change sign. Why? Well, $3R_1 - R_2$ is actually two elementary row operations: first, we multiply row 2 by -1 and then we add three times row 1 to it. When performing row operation RO3, to leave the determinant unchanged, you should always add a multiple of another row to the row you are replacing.

Activity 3.36 You can shorten the writing in the above example by expanding the 4×4 determinant using the first column as soon as you have obtained the determinant with zeros under the leading one. You will then be left with a 3×3 determinant to evaluate. Do this. Without looking at the example above, work through the calculations in this way to evaluate

$$|A| = \begin{vmatrix} 1 & 2 & -1 & 4 \\ -1 & 3 & 0 & 2 \\ 2 & 1 & 1 & 2 \\ 1 & 4 & 1 & 3 \end{vmatrix}.$$

3.3.2   The determinant of a product

One very important result concerning determinants can be stated as: 'the determinant of the product of two square matrices is the product of their determinants'. This is the content of the following theorem.

Theorem 3.37 *If A and B are $n \times n$ matrices, then*

$$|AB| = |A||B|.$$

Proof: We will outline the proof of this theorem without filling in all the details. We first prove the theorem in the case when the matrix A is an elementary matrix. We use again the fact established in Section 3.1.1 (page 92) that the matrix obtained by a row operation on the matrix B is equal to the product of the elementary matrix of that row operation times the matrix B.

Let E_1 be an elementary matrix that multiplies a row by a non-zero constant k. Then $E_1 B$ is the matrix B obtained by performing that row operation on B, and by Theorem 3.34, $|E_1 B| = k|B|$. For the same reason, $|E_1| = |E_1 I| = k|I| = k$. Therefore,

$$|E_1 B| = k|B| = |E_1||B|.$$

The argument for the other two types of elementary matrices follows the same steps.

Activity 3.38 Try these. Show that if E_2 is an elementary matrix that switches two rows, then $|E_2 B| = |E_2||B|$, and do the same for an elementary matrix E_3 that adds a multiple of one row to another.

So we assume that the theorem is true when A is any elementary matrix.

Now recall that every matrix is row equivalent to a matrix in reduced row echelon form, so if R denotes the reduced row echelon form of the matrix A, then we can write

$$A = E_r E_{r-1} \ldots E_1 R,$$

where the E_i are elementary matrices. Since A is a square matrix, R is either the identity matrix or a matrix with a row of zeros.

Applying the result for an elementary matrix repeatedly

$$|A| = |E_r E_{r-1} \ldots E_1 R| = |E_r||E_{r-1}| \ldots |E_1||R|,$$

where $|R|$ is either 1 or 0. In fact, since the determinant of an elementary matrix must be non-zero, $|R| = 0$ if and only if $|A| = 0$.

If $R = I$, then by repeated application of the result for elementary matrices, this time with the matrix B,

$$\begin{aligned}
|AB| &= |(E_r E_{r-1} \ldots E_1 I)B| \\
&= |E_r E_{r-1} \ldots E_1 B| \\
&= |E_r||E_{r-1}| \ldots |E_1||B| \\
&= |E_r E_{r-1} \ldots E_1||B| \\
&= |A| \, |B|.
\end{aligned}$$

If $R \neq I$, then

$$|AB| = |E_r E_{r-1} \ldots E_1 R\,B| = |E_r||E_{r-1}| \ldots |E_1||RB|.$$

Since the product matrix RB must also have a row of zeros, $|RB| = 0$. Therefore, $|AB| = 0 = 0|B|$ and the theorem is proved. $\qquad\square$

3.4 Matrix inverse using cofactors

3.4.1 Using determinants to find an inverse

We start with the following characterisation of invertible matrices.

Theorem 3.39 *If A is an $n \times n$ matrix, then A is invertible if and only if $|A| \neq 0$.*

We will give two proofs of Theorem 3.39. The first follows easily from Theorem 3.8. The second is included because it gives us another method to calculate the inverse of a matrix.

First proof of Theorem 3.39
We have already established this theorem indirectly by our arguments in the previous section; we will repeat and collect them here.

By Theorem 3.8, A is invertible if and only if the reduced row echelon form of A is the identity matrix. Let A be any $n \times n$ matrix and let R be the reduced row echelon form of A. Then either R is the identity matrix (in which case A is invertible) and $|R| = 1$, or R is a matrix with a row of zeros (in which case A is not invertible) and $|R| = 0$.

As we have seen, row operations cannot alter the fact that a determinant is zero or non-zero. By performing a row operation, we might be multiplying the determinant by a non-zero constant, or by -1, or not changing the determinant at all. Therefore, we can conclude that $|A| = 0$ if and only if the determinant of its reduced row echelon form, R, is also 0, and $|A| \neq 0$ if and only if $|R| = 1$.

Putting these statements together, $|A| \neq 0$ if and only if the reduced row echelon form of A is the identity; that is (by Theorem 3.8), if and only if A is invertible. □

Second proof of Theorem 3.39

We will now prove Theorem 3.39 directly. Since it is an if and only if statement, we must prove both implications.

First we show that if A is invertible, then $|A| \neq 0$. We assume A^{-1} exists, so that $AA^{-1} = I$. Then taking the determinant of both sides of this equation, $|AA^{-1}| = |I| = 1$. Applying Theorem 3.37 to the product,

$$|AA^{-1}| = |A| \, |A^{-1}| = 1.$$

If the product of two real numbers is non-zero, then neither number can be zero, which proves that $|A| \neq 0$.

As a consequence of this argument, we have the bonus result that

$$|A^{-1}| = \frac{1}{|A|}.$$

We now show the other implication; that is, if $|A| \neq 0$, then A is invertible. To do this, we will construct A^{-1}, and to do this we need some definitions.

Definition 3.40 If A is an $n \times n$ matrix, the *matrix of cofactors of A* is the matrix whose (i, j) entry is C_{ij}, the (i, j) cofactor of A. The *adjoint* (also sometimes called the *adjugate*) of the matrix A is the *transpose* of the matrix of cofactors. That is, the adjoint of A, adj(A), is the matrix

$$\text{adj}(A) = \begin{pmatrix} C_{11} & C_{21} & \cdots & C_{n1} \\ C_{12} & C_{22} & \cdots & C_{n2} \\ \vdots & \vdots & \ddots & \vdots \\ C_{1n} & C_{2n} & \cdots & C_{nn} \end{pmatrix}.$$

Notice that column 1 of this matrix consists of the cofactors of row 1 of A (and row 1 consists of the cofactors of column 1 of A), and similarly for each column and row.

We now multiply the matrix A with its adjoint matrix:

$$A \text{ adj}(A) = \begin{pmatrix} a_{11} & a_{12} & \cdots & a_{1n} \\ a_{21} & a_{22} & \cdots & a_{2n} \\ \vdots & \vdots & \ddots & \vdots \\ a_{n1} & a_{n2} & \cdots & a_{nn} \end{pmatrix} \begin{pmatrix} C_{11} & C_{21} & \cdots & C_{n1} \\ C_{12} & C_{22} & \cdots & C_{n2} \\ \vdots & \vdots & \ddots & \vdots \\ C_{1n} & C_{2n} & \cdots & C_{nn} \end{pmatrix}.$$

Look carefully at what each entry of the product will be.

The $(1, 1)$ entry is $a_{11}C_{11} + a_{12}C_{12} + \cdots + a_{1n}C_{1n}$. This is the cofactor expansion of $|A|$ by row 1.

The $(1,2)$ entry is $a_{11}C_{21} + a_{12}C_{22} + \cdots + a_{1n}C_{2n}$. This consists of the cofactors of row 2 of A multiplied by the entries of row 1, so this is equal to 0 by Corollary 3.30.

Continuing in this way, we see that the entries on the main diagonal of the product are all equal to $|A|$, and all entries off the main diagonal are equal to 0. That is,

$$A \text{ adj}(A) = \begin{pmatrix} |A| & 0 & \cdots & 0 \\ 0 & |A| & \cdots & 0 \\ \vdots & \vdots & \ddots & \vdots \\ 0 & 0 & \cdots & |A| \end{pmatrix} = |A| \, I,$$

since $|A|$ is just a real number, a scalar.

We know $|A| \neq 0$, so we can divide both sides of the equation by $|A|$ to obtain

$$A \left(\frac{1}{|A|} \text{ adj}(A) \right) = I.$$

This implies that A^{-1} exists and is equal to

$$A^{-1} = \frac{1}{|A|} \text{ adj}(A).$$

$\square$

This gives not only a proof of Theorem 3.39, but a useful method to calculate the inverse of a matrix using cofactors.

Example 3.41 Find A^{-1} for the matrix

$$A = \begin{pmatrix} 1 & 2 & 3 \\ -1 & 2 & 1 \\ 4 & 1 & 1 \end{pmatrix}.$$

The first thing to do is to calculate $|A|$ to see if A is invertible. Using the cofactor expansion by row 1,

$$|A| = 1(2-1) - 2(-1-4) + 3(-1-8) = -16 \neq 0.$$

We then calculate the minors: for example

$$M_{11} = \begin{vmatrix} 2 & 1 \\ 1 & 1 \end{vmatrix} = 1,$$

and we can fill in the chart below

$$\begin{array}{lll} M_{11} = 1 & M_{12} = -5 & M_{13} = -9 \\ M_{21} = -1 & M_{22} = -11 & M_{23} = -7 \\ M_{31} = -4 & M_{32} = 4 & M_{33} = 4. \end{array}$$

Next, we change the minors into cofactors, by multiplying by -1 those minors with $i + j$ equal to an odd number. Finally, we transpose the result to form the adjoint matrix, so that

$$A^{-1} = \frac{1}{|A|}\text{adj}(A) = -\frac{1}{16}\begin{pmatrix} 1 & 1 & -4 \\ 5 & -11 & -4 \\ -9 & 7 & 4 \end{pmatrix}.$$

As with all calculations, it is easy to make a mistake. Therefore, having found A^{-1}, the next thing you should do is check your result by showing that $AA^{-1} = I$,

$$-\frac{1}{16}\begin{pmatrix} 1 & 2 & 3 \\ -1 & 2 & 1 \\ 4 & 1 & 1 \end{pmatrix}\begin{pmatrix} 1 & 1 & -4 \\ 5 & -11 & -4 \\ -9 & 7 & 4 \end{pmatrix}$$

$$= -\frac{1}{16}\begin{pmatrix} -16 & 0 & 0 \\ 0 & -16 & 0 \\ 0 & 0 & -16 \end{pmatrix} = I.$$

Activity 3.42 Use this method to find the inverse of the matrix

$$A = \begin{pmatrix} 1 & 2 & 3 \\ 0 & 4 & 0 \\ 5 & 6 & 7 \end{pmatrix}.$$

Check your result.

Remember: The adjoint matrix only contains the *cofactors* of A; the (i, j) entry is the cofactor C_{ji} of A. A common error is to attempt to form the adjoint by multiplying cofactors by entries of the matrix. But the entries of a matrix A only multiply the cofactors when calculating the determinant of A, $|A|$.

3.4.2 Cramer's rule

If A is a square matrix with $|A| \neq 0$, then Cramer's rule gives us an alternative method of solving a system of linear equations $A\mathbf{x} = \mathbf{b}$.

Theorem 3.43 (Cramer's rule) *If A is $n \times n$, $|A| \neq 0$, and $\mathbf{b} \in \mathbb{R}^n$, then the solution $\mathbf{x} = (x_1, x_2, \ldots, x_n)^{\mathrm{T}}$ of the linear system $A\mathbf{x} = \mathbf{b}$ is given by*

$$x_i = \frac{|A_i|}{|A|},$$

where, here, A_i is the matrix obtained from A by replacing the ith column with the vector $\mathbf{b}$.

Before we prove this theorem, let's see how it works.

Example 3.44 Use Cramer's rule to find the solution of the linear system

$$\begin{aligned}
x + 2y + 3z &= 7 \\
-x + 2y + z &= -3 \\
4x + y + z &= 5.
\end{aligned}$$

In matrix form $A\mathbf{x} = \mathbf{b}$, this system is

$$\begin{pmatrix} 1 & 2 & 3 \\ -1 & 2 & 1 \\ 4 & 1 & 1 \end{pmatrix} \begin{pmatrix} x \\ y \\ z \end{pmatrix} = \begin{pmatrix} 7 \\ -3 \\ 5 \end{pmatrix}.$$

We first check that $|A| \neq 0$. This is the same matrix A as in Example 3.41, and we have $|A| = -16$. Then, applying Cramer's rule, we find x by evaluating the determinant of the matrix obtained from A by replacing column 1 with $\mathbf{b}$, and divide this by $|A|$:

$$x = \frac{\begin{vmatrix} 7 & 2 & 3 \\ -3 & 2 & 1 \\ 5 & 1 & 1 \end{vmatrix}}{|A|} = \frac{-16}{-16} = 1.$$

and, in the same way, we obtain y and z:

$$y = \frac{\begin{vmatrix} 1 & 7 & 3 \\ -1 & -3 & 1 \\ 4 & 5 & 1 \end{vmatrix}}{|A|} = \frac{48}{-16} = -3,$$

$$z = \frac{\begin{vmatrix} 1 & 2 & 7 \\ -1 & 2 & -3 \\ 4 & 1 & 5 \end{vmatrix}}{|A|} = \frac{-64}{-16} = 4,$$

which can easily be checked by substitution into the original equations (or by multiplying $A\mathbf{x}$).

We now prove Cramer's rule.

Proof: Since $|A| \neq 0$, A^{-1} exists, and we can solve for $\mathbf{x}$ by multiplying $A\mathbf{x} = \mathbf{b}$ on the left by A^{-1}. Then $\mathbf{x} = A^{-1}\mathbf{b}$:

$$\mathbf{x} = \begin{pmatrix} x_1 \\ x_2 \\ \vdots \\ x_n \end{pmatrix} = \frac{1}{|A|} \begin{pmatrix} C_{11} & C_{21} & \dots & C_{n1} \\ C_{12} & C_{22} & \dots & C_{n2} \\ \vdots & \vdots & \ddots & \vdots \\ C_{1n} & C_{2n} & \dots & C_{nn} \end{pmatrix} \begin{pmatrix} b_1 \\ b_2 \\ \vdots \\ b_n \end{pmatrix}.$$

The entry x_i of the solution is equal to the ith row of this product.

$$x_i = \frac{1}{|A|}(b_1 C_{1i} + b_2 C_{2i} + \dots + b_n C_{ni}).$$

Stare at this expression a moment. The cofactors all come from row i of the adjoint matrix, and they are the cofactors of column i of A, so this looks like a cofactor expansion by column i of a matrix which is identical to A except in column i, where the entries are the components of the vector $\mathbf{b}$. That is, the term in brackets is the cofactor expansion by column i of the matrix A with column i replaced by the vector $\mathbf{b}$; in other words, it is $|A_i|$. $\qquad\square$

To summarise, for a system $A\mathbf{x} = \mathbf{b}$, where A is square and $|A| \neq 0$, to find x_i using Cramer's rule:

(1) replace column i of A by $\mathbf{b}$,
(2) evaluate the determinant of the resulting matrix,
(3) divide by $|A|$.

Cramer's rule is quite an attractive way to solve linear systems of equations, but it should be stressed that it has fairly limited applicability. It only works for square systems, and only for those square systems where the coefficient matrix is invertible. By contrast, the Gaussian elimination method works in all cases: it can be used for non-square systems, square systems in which the matrix is not invertible, systems with infinitely many solutions; and it can also detect when a system is inconsistent.

Activity 3.45 Can you think of another method to obtain the solution to the above example? One way is to use the inverse matrix. Do this. We found A^{-1} in Exercise 3.41. Now use it to find the solution $\mathbf{x}$ of

$$\begin{pmatrix} 1 & 2 & 3 \\ -1 & 2 & 1 \\ 4 & 1 & 1 \end{pmatrix} \begin{pmatrix} x \\ y \\ z \end{pmatrix} = \begin{pmatrix} 7 \\ -3 \\ 5 \end{pmatrix}$$

by calculating $\mathbf{x} = A^{-1}\mathbf{b}$.

3.5 Leontief input–output analysis

In 1973, Wassily Leontief was awarded the Nobel prize in Economics for work he did analysing an economy with many interdependent industries using linear algebra. We present a brief outline of his method here.

Suppose an economy has n interdependent production processes, where the outputs of the n industries are used to run the industries and to satisfy an outside demand. Assume that prices are fixed so that they can be used to measure the output. The problem we wish to solve is to determine the level of output of each industry which will satisfy all demands exactly; that is, both the demands of the other industries and the outside demand. The problem can be described as a system of linear equations, as we shall see by considering the following simple example.

Example 3.46 Suppose there are two industries: water and electricity. Let

$$x_1 = \text{total output of water} \quad (\$ \text{ value})$$
$$x_2 = \text{total output of electricity} \quad (\$ \text{ value}).$$

We can express this as a vector

$$\mathbf{x} = \begin{pmatrix} x_1 \\ x_2 \end{pmatrix}$$

called a *production vector*. Suppose we know that the production of water requires both water and electricity as inputs, and that the production of electricity requires both water and electricity as inputs. Specifically, suppose the following is known:

water production needs $\qquad$ $\left.\begin{array}{c} \$\,0.01 \text{ water} \\ \$\,0.15 \text{ electricity} \end{array}\right\}$
to produce $\$1.00$ water

electricity production needs $\qquad$ $\left.\begin{array}{c} \$\,0.21 \text{ water} \\ \$0.05 \text{ electricity} \end{array}\right\}$
to produce $\$1.00$ electricity.

What is the total water used by the industries? Water is using $\$\,0.01$ for each unit output, so a total of $0.01x_1$; and electricity is using $\$\,0.21$ water for each unit of its output, so a total of $0.21x_2$. The total amount of water used by the industries is therefore $0.01x_1 + 0.21x_2$. In the same way, the total amount of electricity used by the industries is $0.15x_1 + 0.05x_2$. The totals can be expressed as

$$\begin{pmatrix} \text{water} \\ \text{electricity} \end{pmatrix} = \begin{pmatrix} 0.01 & 0.21 \\ 0.15 & 0.05 \end{pmatrix} \begin{pmatrix} x_1 \\ x_2 \end{pmatrix} = C\mathbf{x}.$$

The matrix C is known as a *consumption matrix* or a *technology matrix*.

After the industries have used water and electricity to produce their outputs, how much water and electricity are left to satisfy the outside demand?

Activity 3.47 Think about this before continuing. Write down an expression for the total amount of water which is left after the industries have each used what they need to produce their output. Do the same for electricity.

Let d_1 denote the outside demand for water, and d_2 the demand for electricity. Then in order for the output of these industries to supply the industries and also to satisfy the outside demand exactly, the following equations must be satisfied:

$$\begin{cases} x_1 - 0.01x_1 - 0.21x_2 = d_1 & \text{(water)} \\ x_2 - 0.15x_1 - 0.05x_2 = d_2 & \text{(electricity)}. \end{cases}$$

In matrix notation,

$$\begin{pmatrix} x_1 \\ x_2 \end{pmatrix} - \begin{pmatrix} 0.01 & 0.21 \\ 0.15 & 0.05 \end{pmatrix} \begin{pmatrix} x_1 \\ x_2 \end{pmatrix} = \begin{pmatrix} d_1 \\ d_2 \end{pmatrix},$$

or, $\mathbf{x} - C\mathbf{x} = \mathbf{d}$, where

$$\mathbf{d} = \begin{pmatrix} d_1 \\ d_2 \end{pmatrix}.$$

is the outside *demand vector*. If we use the the fact that $I\mathbf{x} = \mathbf{x}$, where I is the 2×2 identity matrix, then we can rewrite this system in matrix form as

$$I\mathbf{x} - C\mathbf{x} = \mathbf{d}, \qquad \text{or} \qquad (I - C)\mathbf{x} = \mathbf{d}.$$

This is now in the usual matrix form for a system of linear equations. A solution, $\mathbf{x}$, to this system of equations will determine the output levels of each industry required to satisfy all demands exactly.

Now let's look at the general case. Suppose we have an economy with n interdependent industries. If c_{ij} denotes the amount of industry i used by industry j to produce $\$1.00$ of industry j, then the *consumption* or *technology matrix* is $C = (c_{ij})$:

$$C = \begin{pmatrix} c_{11} & c_{12} & \cdots & c_{1n} \\ c_{21} & c_{22} & \cdots & c_{2n} \\ \vdots & \vdots & \ddots & \vdots \\ c_{n1} & c_{n2} & \cdots & c_{nn} \end{pmatrix},$$

where:

- row i lists the amounts of industry i used by each industry
- column j lists the amounts of each industry used by industry j.

If, as before, we denote by $\mathbf{d}$ the $n \times 1$ outside demand vector, then in matrix form the problem we wish to solve is to find the production vector $\mathbf{x}$ such that

$$(I - C)\mathbf{x} = \mathbf{d},$$

a system of n linear equations in n unknowns.

Activity 3.48 Return to the example given above and assume that the public demand for water is $\$627$ and for electricity is $\$4,955$. Find the levels of output which satisfy all demands exactly.

3.6   Learning outcomes

You should now be able to:

- say what is meant by an elementary matrix, and understand how they are used for row operations
- find the inverse of a matrix using row operations
- find the determinant of a square matrix and use it to determine if a matrix is invertible

- find the inverse of a matrix using cofactors
- solve a system of linear equations using Cramer's rule
- say what is meant by the Leontief input–output model and solve input–output analysis problems.

In addition, you should know that:

- There are three methods to solve $A\mathbf{x} = \mathbf{b}$ if A is $n \times n$ and $|A| \neq 0$:
  (1) Gaussian elimination
  (2) find A^{-1}, then calculate $\mathbf{x} = A^{-1}\mathbf{b}$
  (3) Cramer's rule.
- There is one method to solve $A\mathbf{x} = \mathbf{b}$ if A is $m \times n$ and $m \neq n$, or if $|A| = 0$:
  (1) Gaussian elimination.
- There are two methods to find A^{-1}:
  (1) using cofactors for the adjoint matrix
  (2) by row reduction of $(A \mid I)$ to $(I \mid A^{-1})$.
- If A is an $n \times n$ matrix, then the following statements are equivalent (Theorems 3.8 and 3.39):
  (1) A is invertible.
  (2) $A\mathbf{x} = \mathbf{b}$ has a unique solution for any $\mathbf{b} \in \mathbb{R}^n$.
  (3) $A\mathbf{x} = \mathbf{0}$ has only the trivial solution, $\mathbf{x} = \mathbf{0}$.
  (4) the reduced row echelon form of A is I.
  (5) $|A| \neq 0$.

3.7   Comments on activities

Activity 3.2 Only the last matrix is an elementary matrix, representing the operation $R_3 - R_1$ on I. The others each represent two row operations. For example,

$$\begin{pmatrix} 2 & 1 & 0 \\ 0 & 1 & 0 \\ 0 & 0 & 1 \end{pmatrix} = \begin{pmatrix} 1 & 1 & 0 \\ 0 & 1 & 0 \\ 0 & 0 & 1 \end{pmatrix} \begin{pmatrix} 2 & 0 & 0 \\ 0 & 1 & 0 \\ 0 & 0 & 1 \end{pmatrix} = E_2 E_1,$$

where E_1 represents $2R_1$ and E_2 represents $R_1 + R_2$. You should multiply the matrices in the opposite order, $E_1 E_2$, and notice the effect, thinking about the row operations on I.

Activity 3.4 The matrix E is the identity matrix after the row operation $R_2 - 4R_1$ has been performed on it, so the inverse matrix is the identity matrix after $R_2 + 4R_1$,

$$E^{-1} = \begin{pmatrix} 1 & 0 & 0 \\ 4 & 1 & 0 \\ 0 & 0 & 1 \end{pmatrix}.$$

Multiply out EE^{-1} and $E^{-1}E$ as instructed.

Activity 3.11 For the matrix A,

$$(A|I) = \begin{pmatrix} -2 & 1 & 3 & | & 1 & 0 & 0 \\ 0 & -1 & 1 & | & 0 & 1 & 0 \\ 1 & 2 & 0 & | & 0 & 0 & 1 \end{pmatrix} \xrightarrow{R_1 \leftrightarrow R_3} \begin{pmatrix} 1 & 2 & 0 & | & 0 & 0 & 1 \\ 0 & -1 & 1 & | & 0 & 1 & 0 \\ -2 & 1 & 3 & | & 1 & 0 & 0 \end{pmatrix}$$

$$\xrightarrow{R_3 + 2R_1} \begin{pmatrix} 1 & 2 & 0 & | & 0 & 0 & 1 \\ 0 & -1 & 1 & | & 0 & 1 & 0 \\ 0 & 5 & 3 & | & 1 & 0 & 2 \end{pmatrix} \xrightarrow{(-1)R_2} \begin{pmatrix} 1 & 2 & 0 & | & 0 & 0 & 1 \\ 0 & 1 & -1 & | & 0 & -1 & 0 \\ 0 & 5 & 3 & | & 1 & 0 & 2 \end{pmatrix}$$

$$\xrightarrow{R_3 - 5R_2} \begin{pmatrix} 1 & 2 & 0 & | & 0 & 0 & 1 \\ 0 & 1 & -1 & | & 0 & -1 & 0 \\ 0 & 0 & 8 & | & 1 & 5 & 2 \end{pmatrix} \xrightarrow{\frac{1}{8}R_3} \begin{pmatrix} 1 & 2 & 0 & | & 0 & 0 & 1 \\ 0 & 1 & -1 & | & 0 & -1 & 0 \\ 0 & 0 & 1 & | & \frac{1}{8} & \frac{5}{8} & \frac{1}{4} \end{pmatrix}$$

$$\xrightarrow{R_2 + R_3} \begin{pmatrix} 1 & 2 & 0 & | & 0 & 0 & 1 \\ 0 & 1 & 0 & | & \frac{1}{8} & -\frac{3}{8} & \frac{1}{4} \\ 0 & 0 & 1 & | & \frac{1}{8} & \frac{5}{8} & \frac{1}{4} \end{pmatrix} \xrightarrow{R_1 - 2R_2} \begin{pmatrix} 1 & 0 & 0 & | & -\frac{2}{8} & \frac{6}{8} & \frac{1}{2} \\ 0 & 1 & 0 & | & \frac{1}{8} & -\frac{3}{8} & \frac{1}{4} \\ 0 & 0 & 1 & | & \frac{1}{8} & \frac{5}{8} & \frac{1}{4} \end{pmatrix}.$$

So,

$$A^{-1} = \frac{1}{8} \begin{pmatrix} -2 & 6 & 4 \\ 1 & -3 & 2 \\ 1 & 5 & 2 \end{pmatrix}.$$

Now *check* that $AA^{-1} = I$.

When you carry out the row reduction, it is not necessary to always indicate the separation of the two matrices by a line as we have done so far. You just need to keep track of what you are doing.

In the calculation for the inverse of B, we have omitted the line but added a bit of space to make it easier for you to read.

$$(B|I) = \begin{pmatrix} 2 & 1 & 3 & 1 & 0 & 0 \\ 0 & -1 & 1 & 0 & 1 & 0 \\ 1 & 2 & 0 & 0 & 0 & 1 \end{pmatrix} \xrightarrow{R_1 \leftrightarrow R_3} \begin{pmatrix} 1 & 2 & 0 & 0 & 0 & 1 \\ 0 & -1 & 1 & 0 & 1 & 0 \\ 2 & 1 & 3 & 1 & 0 & 0 \end{pmatrix}$$

$$\xrightarrow{R_3 - 2R_1} \begin{pmatrix} 1 & 2 & 0 & 0 & 0 & 1 \\ 0 & -1 & 1 & 0 & 1 & 0 \\ 0 & -3 & 3 & 1 & 0 & -2 \end{pmatrix} \xrightarrow{(-1)R_2} \begin{pmatrix} 1 & 2 & 0 & 0 & 0 & 1 \\ 0 & 1 & -1 & 0 & -1 & 0 \\ 0 & -3 & 3 & 1 & 0 & -2 \end{pmatrix}$$

$$\xrightarrow{R_3 + 3R_2} \begin{pmatrix} 1 & 2 & 0 & 0 & 0 & 1 \\ 0 & 1 & -1 & 0 & -1 & 0 \\ 0 & 0 & 0 & 1 & -3 & -2 \end{pmatrix},$$

which indicates that the matrix B is not invertible; it is not row equivalent to the identity matrix.

Activity 3.16 $C_{13} = 13$.

Activity 3.18 $|M| = -1(8 - 3) - 2(0 - 3) + 1(0 - 2) = -1$.

Activity 3.21 You should either expand by column 1 or row 2. For example, using column 1: $|M| = -1(8 - 3) + 1(6 - 2) = -1$.

Activity 3.36

$$|A| = \begin{vmatrix} 1 & 2 & -1 & 4 \\ 0 & 5 & -1 & 6 \\ 0 & -3 & 3 & -6 \\ 0 & 2 & 2 & -1 \end{vmatrix} = \begin{vmatrix} 5 & -1 & 6 \\ -3 & 3 & -6 \\ 2 & 2 & -1 \end{vmatrix}.$$

At this stage you can expand the 3×3 matrix using a cofactor expansion, or continue a bit more with row operations:

$$|A| = 3\begin{vmatrix} 1 & -1 & 2 \\ 5 & -1 & 6 \\ 2 & 2 & -1 \end{vmatrix} = 3\begin{vmatrix} 1 & -1 & 2 \\ 0 & 4 & -4 \\ 0 & 4 & -5 \end{vmatrix} = 3\begin{vmatrix} 4 & -4 \\ 4 & -5 \end{vmatrix} = 3(-4) = -12.$$

Activity 3.42

$$|A| = -32 \neq 0$$

$$A^{-1} = \frac{1}{|A|}\text{adj}(A) = -\frac{1}{32}\begin{pmatrix} 28 & 4 & -12 \\ 0 & -8 & 0 \\ -20 & 4 & 4 \end{pmatrix} = \frac{1}{8}\begin{pmatrix} -7 & -1 & 3 \\ 0 & 2 & 0 \\ 5 & -1 & -1 \end{pmatrix}.$$

Activity 3.47 The total water output remaining is $x_1 - 0.01x_1 - 0.21x_2$, and the total electricity output left is $x_2 - 0.15x_1 - 0.05x_2$.

Activity 3.48 Solve $(I - C)\mathbf{x} = \mathbf{d}$ by Gaussian elimination, where

$$C = \begin{pmatrix} 0.01 & 0.21 \\ 0.15 & 0.05 \end{pmatrix}, \quad \mathbf{x} = \begin{pmatrix} x_1 \\ x_2 \end{pmatrix}, \quad \mathbf{d} = \begin{pmatrix} 627 \\ 4955 \end{pmatrix}.$$

Reducing the augmented matrix,

$$((I - C)|\mathbf{d}) = \begin{pmatrix} 0.99 & -0.21 & 627 \\ -0.15 & 0.95 & 4955 \end{pmatrix} \rightarrow \begin{pmatrix} 33 & -7 & 20900 \\ -3 & 19 & 99100 \end{pmatrix}$$

$$\begin{pmatrix} 1 & -7/33 & 1900/3 \\ -3 & 19 & 99100 \end{pmatrix} \rightarrow \begin{pmatrix} 1 & -7/33 & 1900/3 \\ 0 & 202/11 & 101000 \end{pmatrix}$$

$$\begin{pmatrix} 1 & -7/33 & 1900/3 \\ 0 & 1 & 5500 \end{pmatrix} \rightarrow \begin{pmatrix} 1 & 0 & 1800 \\ 0 & 1 & 5500 \end{pmatrix}.$$

So, $\quad \mathbf{x} = \begin{pmatrix} 1800 \\ 5500 \end{pmatrix}.$

3.8 Exercises

Exercise 3.1 Use elementary row operations to find any inverses of the following matrices.

$$A = \begin{pmatrix} 1 & 2 & -1 \\ 0 & 1 & 2 \\ 3 & 8 & 1 \end{pmatrix}, \qquad B = \begin{pmatrix} -1 & 2 & 1 \\ 0 & 1 & 2 \\ 3 & 1 & 4 \end{pmatrix}.$$

Let $\mathbf{b} = \begin{pmatrix} 1 \\ 1 \\ 5 \end{pmatrix}$. Find all solutions to $A\mathbf{x} = \mathbf{b}$. Find all solutions to $B\mathbf{x} = \mathbf{b}$.

Is there a vector $\mathbf{d} \in \mathbb{R}^3$ for which $A\mathbf{x} = \mathbf{d}$ is inconsistent? Is there a vector $\mathbf{d} \in \mathbb{R}^3$ for which $B\mathbf{x} = \mathbf{d}$ is inconsistent? In each case, justify your answer and find such a vector $\mathbf{d}$ if one exists.

Exercise 3.2 Use elementary row operations to reduce the matrix

$$A = \begin{pmatrix} 1 & 0 & 2 \\ 0 & 1 & -1 \\ 1 & 4 & -1 \end{pmatrix}$$

to the identity matrix. Hence, write A as a product of elementary matrices.

Use this to evaluate $|A|$ as a product of matrices, then check your answer by evaluating $|A|$ using a cofactor expansion.

Exercise 3.3 Evaluate each of the following determinants using a cofactor expansion along an appropriate row or column.

$$\text{(a)} \begin{vmatrix} 5 & 2 & -4 \\ -3 & 1 & 1 \\ -1 & 7 & 2 \end{vmatrix} \qquad \text{(b)} \begin{vmatrix} 1 & 23 & 6 & -15 \\ 2 & 5 & 0 & 1 \\ 1 & 4 & 0 & 3 \\ 0 & 1 & 0 & 1 \end{vmatrix}.$$

Exercise 3.4 Suppose $w \in \mathbb{R}$ and B is the matrix

$$B = \begin{pmatrix} 2 & 1 & w \\ 3 & 4 & -1 \\ 1 & -2 & 7 \end{pmatrix}.$$

For what values of w is the determinant of B equal to 0?

Exercise 3.5 Evaluate the following determinant using row operations to simplify the calculation:

$$\begin{vmatrix} 5 & 2 & -4 & -2 \\ -3 & 1 & 5 & 1 \\ -4 & 3 & 1 & 3 \\ 2 & 1 & -1 & 1 \end{vmatrix}.$$

Check your answer by evaluating it a second time using column operations.

Exercise 3.6 For which values of λ is the matrix

$$A = \begin{pmatrix} 7 - \lambda & -15 \\ 2 & -4 - \lambda \end{pmatrix}$$

not invertible?

Exercise 3.7 Suppose A is a 3×3 matrix with $|A| = 7$. Find $|2A|$, $|A^2|$, $|2A^{-1}|$, $|(2A)^{-1}|$.

Exercise 3.8 Use the method of the adjoint matrix to find the inverse of each of the following matrices, if it exists.

$$B = \begin{pmatrix} -1 & 2 & 1 \\ 0 & 1 & 2 \\ 3 & 1 & 4 \end{pmatrix}, \qquad C = \begin{pmatrix} 5 & 2 & -1 \\ 1 & 3 & 4 \\ 6 & 5 & 3 \end{pmatrix}.$$

Exercise 3.9 Write out the system of equations $B\mathbf{x} = \mathbf{b}$, where

$$B = \begin{pmatrix} -1 & 2 & 1 \\ 0 & 1 & 2 \\ 3 & 1 & 4 \end{pmatrix}, \qquad \mathbf{x} = \begin{pmatrix} x \\ y \\ z \end{pmatrix}, \qquad \mathbf{b} = \begin{pmatrix} 1 \\ 1 \\ 5 \end{pmatrix}.$$

Find the solution using Cramer's rule.

Exercise 3.10 Consider an economy with three industries,

$$i_1 : \text{water} \qquad i_2 : \text{electricity} \qquad i_3 : \text{gas}$$

interlinked so that the corresponding consumption matrix is

$$C = \begin{pmatrix} 0.2 & 0.3 & 0.2 \\ 0.4 & 0.1 & 0.2 \\ 0 & 0 & 0.1 \end{pmatrix}.$$

Each week the external demands for water, electricity and gas are, respectively,

$$d_1 = \$40,000 \qquad d_2 = \$100,000. \qquad d_3 = \$72,000.$$

(a) How much water, electricity and gas is needed to produce $1 worth of electricity?

(b) What should be the weekly production of each industry in order to satisfy all demands exactly?

Exercise 3.11 The *vector product* or *cross product* of two vectors is defined in $\mathbb{R}^3$ as follows. If

$$\mathbf{a} = \begin{pmatrix} a_1 \\ a_2 \\ a_3 \end{pmatrix}, \quad \mathbf{b} = \begin{pmatrix} b_1 \\ b_2 \\ b_3 \end{pmatrix}, \quad \mathbf{e}_1 = \begin{pmatrix} 1 \\ 0 \\ 0 \end{pmatrix}, \mathbf{e}_2 = \begin{pmatrix} 0 \\ 1 \\ 0 \end{pmatrix}, \mathbf{e}_3 = \begin{pmatrix} 0 \\ 0 \\ 1 \end{pmatrix},$$

then $\mathbf{a} \times \mathbf{b}$ is the vector given by

$$\mathbf{a} \times \mathbf{b} = \begin{vmatrix} \mathbf{e}_1 & \mathbf{e}_2 & \mathbf{e}_3 \\ a_1 & a_2 & a_3 \\ b_1 & b_2 & b_3 \end{vmatrix}$$

$$= (a_2 b_3 - a_3 b_2)\mathbf{e}_1 - (a_1 b_3 - a_3 b_1)\mathbf{e}_2 + (a_1 b_2 - a_2 b_1)\mathbf{e}_3.$$

(That determinant might look odd to you since it has vectors as some of its entries: but really this is just an extension of the earlier notation, and you can take the second equation as the definition of what it means.) The vector $\mathbf{a} \times \mathbf{b}$ is perpendicular to both $\mathbf{a}$ and $\mathbf{b}$ (see part (b)).

(a) Calculate $\mathbf{w} = \mathbf{u} \times \mathbf{v}$ for the vectors $\mathbf{u} = \begin{pmatrix} 1 \\ 2 \\ 3 \end{pmatrix}$ and $\mathbf{v} = \begin{pmatrix} 2 \\ -5 \\ 4 \end{pmatrix}$.

Check that $\mathbf{w}$ is perpendicular to both $\mathbf{u}$ and $\mathbf{v}$.

(b) Show that for general vectors $\mathbf{a}, \mathbf{b}, \mathbf{c} \in \mathbb{R}^3$ the *scalar triple product*, $\langle \mathbf{a}, \mathbf{b} \times \mathbf{c} \rangle$ is given by the following determinant:

$$\langle \mathbf{a}, \mathbf{b} \times \mathbf{c} \rangle = \begin{vmatrix} a_1 & a_2 & a_3 \\ b_1 & b_2 & b_3 \\ c_1 & c_2 & c_3 \end{vmatrix} \qquad (*)$$

Use this and properties of the determinant to show that the vector $\mathbf{b} \times \mathbf{c}$ is perpendicular to both $\mathbf{b}$ and $\mathbf{c}$.

(c) Show that the vectors $\mathbf{a}, \mathbf{b}, \mathbf{c}$ are coplanar (lie in the same plane) if and only if the determinant $(*)$ is equal to 0.

Find the constant t if the vectors

$$\begin{pmatrix} 3 \\ -1 \\ 2 \end{pmatrix}, \begin{pmatrix} t \\ 5 \\ 1 \end{pmatrix}, \begin{pmatrix} -2 \\ 3 \\ 1 \end{pmatrix}$$

are coplanar.

3.9 Problems

Problem 3.1 Use elementary row operations to find inverses of each of the following matrices when the matrix has an inverse.

$$A = \begin{pmatrix} 1 & 2 & 3 \\ 2 & 3 & 0 \\ 0 & 1 & 2 \end{pmatrix} \quad B = \begin{pmatrix} 1 & 2 & 3 \\ 2 & 3 & 0 \\ 0 & 1 & 6 \end{pmatrix} \quad C = \begin{pmatrix} 1 & 0 & 4 & 0 \\ 0 & 1 & 0 & 0 \\ 0 & 0 & 0 & 1 \\ 0 & 0 & 1 & 0 \end{pmatrix}.$$

Is C an elementary matrix? If the answer is 'yes', what operation does it perform? If the answer is 'no', write it as a product of elementary matrices.

Problem 3.2 Given a system of equations $Ax = b$ for several different values of b, it is often more practical to find A^{-1}, if it exists, and then to find the solutions using $x = A^{-1}b$.

Use this method to solve $Ax = b_r$ for the matrix

$$A = \begin{pmatrix} 1 & 2 & 3 \\ 2 & 3 & 0 \\ 0 & 1 & 2 \end{pmatrix},$$

and for each of the following vectors b_r, $r = 1, 2, 3$:

(a) $\quad b_1 = \begin{pmatrix} 1 \\ 0 \\ 3 \end{pmatrix}$;   (b) $\quad b_2 = \begin{pmatrix} 1 \\ 1 \\ 1 \end{pmatrix}$;   (c) $\quad b_3 = \begin{pmatrix} 0 \\ 1 \\ 0 \end{pmatrix}$.

Be certain your solution for A^{-1} is correct before carrying out this problem by checking that $AA^{-1} = I$.

Problem 3.3 Evaluate the following determinants using the cofactor expansion along an appropriate row or column.

(a) $\quad \begin{vmatrix} 2 & 5 & 1 \\ 1 & 0 & 2 \\ 7 & 1 & 1 \end{vmatrix}.$

(b) $\quad \begin{vmatrix} 7 & 5 & 2 & 3 \\ 2 & 0 & 0 & 0 \\ 11 & 2 & 0 & 0 \\ 23 & 57 & 1 & -1 \end{vmatrix}.$

(c) $\quad \begin{vmatrix} 1 & 2 & 1 & 0 \\ 3 & 2 & 1 & 0 \\ 0 & 1 & 6 & 5 \\ 0 & 1 & 1 & 1 \end{vmatrix}.$

(d) $\begin{vmatrix} 0 & 1 & 0 & 0 \\ 1 & 0 & 0 & 0 \\ 0 & 0 & 1 & 0 \\ 0 & 0 & 0 & 1 \end{vmatrix}.$

(e) $\begin{vmatrix} 0 & 0 & 0 & 0 & 0 & 1 \\ 0 & 0 & 0 & 0 & 3 & 2 \\ 0 & 0 & 0 & 2 & 9 & 3 \\ 0 & 0 & 1 & 0 & 7 & 4 \\ 0 & 6 & 9 & 8 & 7 & 5 \\ 1 & 3 & 4 & 2 & 9 & 6 \end{vmatrix}.$

Problem 3.4 Let

$$B = \begin{pmatrix} 3 & t & -2 \\ -1 & 5 & 3 \\ 2 & 1 & 1 \end{pmatrix}.$$

For what values of t is the determinant of B equal to zero?

Problem 3.5 Evaluate the following determinants (use row operations to simplify the calculation).

(a) $\begin{vmatrix} 1 & -4 & 3 & 2 \\ 2 & -7 & 5 & 1 \\ 1 & 2 & 6 & 0 \\ 2 & -10 & 14 & 4 \end{vmatrix}.$

(b) $\begin{vmatrix} 1 & 4 & -1 & 3 & 0 \\ 1 & 7 & 4 & 3 & 8 \\ 2 & 8 & -2 & 6 & 0 \\ 2 & 0 & 5 & 5 & 7 \\ -1 & 9 & 0 & 9 & 2 \end{vmatrix}.$

(c) $\begin{vmatrix} 3 & 3a & 3a^2 \\ 2 & 2b & 2b^2 \\ 1 & c & c^2 \end{vmatrix}.$

Problem 3.6 Consider the matrix

$$A = \begin{pmatrix} 2 - \lambda & 3 \\ 2 & 1 - \lambda \end{pmatrix}, \quad \lambda \in \mathbb{R}.$$

For which values of λ will the matrix equation $A\mathbf{x} = 0$ have non-trivial solutions?

Problem 3.7 Use the method of the adjoint matrix to find the inverse of each of the following matrices, if it exists.

$$A = \begin{pmatrix} 2 & 0 & -3 \\ 0 & 3 & 1 \\ -1 & 4 & 2 \end{pmatrix} \quad B = \begin{pmatrix} 1 & 0 & 2 \\ 2 & 1 & 3 \\ 0 & -1 & 1 \end{pmatrix} \quad C = \begin{pmatrix} 1 & 2 & 0 \\ 0 & 1 & 1 \\ 2 & 1 & -1 \end{pmatrix}.$$

Problem 3.8 Express the following system of equations in matrix form, as $A\mathbf{x} = \mathbf{b}$.

$$\begin{cases} 2x - y + 5z = 2 \\ x + y - 2z = 1 \\ -3x - 2y + z = -7. \end{cases}$$

Solve the system using each of the three matrix methods. Solve it by Gaussian elimination, by using A^{-1}, and by using Cramer's rule. Express your solution in vector form.

Problem 3.9 Use Cramer's rule to find the value of x, y, z for system (a) and to find the value of z for system (b) where a, b are constants, $a \neq \pm b$, $a \neq 2b$.

(a)
$$\begin{cases} x + y + z = 8 \\ 2x + y - z = 3 \\ -x + 2y + z = 3. \end{cases}$$

(b)
$$\begin{cases} ax - ay + bz = a + b \\ bx - by + az = 0 \\ -ax + 2by + 3z = a - b. \end{cases}$$

Problem 3.10 Prove the following statement using either determinants or Theorem 3.12.

> If A and B are $n \times n$ matrices and $(AB)^{-1}$ exists,
> then A and B are invertible.

4

Rank, range and linear equations

In this short chapter, we aim to extend and consolidate what we have learned so far about systems of equations and matrices, and tie together many of the results of the previous chapters. We will intersperse an overview of the previous two chapters with two new concepts, the rank of a matrix and the range of a matrix.

This chapter will serve as a synthesis of what we have learned so far, in anticipation of a return to these topics later.

4.1 The rank of a matrix

4.1.1 The definition of rank

Any matrix A can be reduced to a matrix in reduced row echelon form by elementary row operations. You just have to follow the algorithm and you will obtain first a row-equivalent matrix which is in row echelon form, and then, continuing with the algorithm, a row-equivalent matrix in reduced row echelon form (see Section 3.1.2). Another way to say this is:

- Any matrix A is row-equivalent to a matrix in reduced row echelon form.

There are several ways of defining the *rank* of a matrix, and we shall meet some other (more sophisticated) ways later. All are equivalent. We begin with the following definition:

Definition 4.1 (Rank of a matrix) The *rank*, rank(A), of a matrix A is the number of non-zero rows in a row echelon matrix obtained from A by elementary row operations.

Notice that the definition only requires that the matrix A be put into row echelon form, because by then the number of non-zero rows is determined. By a non-zero row, we simply mean one that contains entries other than 0. Since every non-zero row of a matrix in row echelon form begins with a leading one, this is equivalent to the following definition.

Definition 4.2 The *rank*, rank(A), of a matrix A is the number of leading ones in a row echelon matrix obtained from A by elementary row operations.

Generally, if A is an $m \times n$ matrix, then the number of non-zero rows (the number of leading ones) in a row echelon form of A can certainly be no more than the total number of rows, m. Furthermore, since the leading ones must be in different columns, the number of leading ones in the echelon form can be no more than the total number, n, of columns. Thus, we have:

- For an $m \times n$ matrix A, rank(A) $\leq \min\{m, n\}$, where $\min\{m, n\}$ denotes the smaller of the two integers m and n.

Example 4.3 Consider the matrix

$$M = \begin{pmatrix} 1 & 2 & 1 & 1 \\ 2 & 3 & 0 & 5 \\ 3 & 5 & 1 & 6 \end{pmatrix}.$$

Reducing this using elementary row operations, we have:

$$\begin{pmatrix} 1 & 2 & 1 & 1 \\ 2 & 3 & 0 & 5 \\ 3 & 5 & 1 & 6 \end{pmatrix} \rightarrow \begin{pmatrix} 1 & 2 & 1 & 1 \\ 0 & -1 & -2 & 3 \\ 0 & -1 & -2 & 3 \end{pmatrix} \rightarrow \begin{pmatrix} 1 & 2 & 1 & 1 \\ 0 & 1 & 2 & -3 \\ 0 & 0 & 0 & 0 \end{pmatrix}.$$

This last matrix is in row echelon form and has two non-zero rows (and two leading ones), so the matrix M has rank 2.

Activity 4.4 Show that the matrix

$$B = \begin{pmatrix} 1 & 2 & 1 & 1 \\ 2 & 3 & 0 & 5 \\ 3 & 5 & 1 & 4 \end{pmatrix}$$

has rank 3.

4.1.2 The main theorem again

If a square matrix A of size $n \times n$ has rank n, then its reduced row echelon form has a leading one in every row and (since the leading ones are in different columns) a leading one in every column. Since every column with a leading one has zeros elsewhere, it follows that the reduced echelon form of A must be I, the $n \times n$ identity matrix. Conversely, if the reduced row echelon form of A is I, then, by the definition of rank, A has rank n. The main theoretical result of Chapter 3 is a characterisation of invertible matrices. We can now add to the main theorem, Theorem 3.8, and to Theorem 3.39, one more equivalent statement characterising invertibility. This leads to the following result:

Theorem 4.5 *If A is an $n \times n$ matrix, then the following statements are equivalent.*

- *A^{-1} exists.*
- *$A\mathbf{x} = \mathbf{b}$ has a unique solution for any $\mathbf{b} \in \mathbb{R}^n$.*
- *$A\mathbf{x} = \mathbf{0}$ has only the trivial solution, $\mathbf{x} = \mathbf{0}$.*
- *The reduced echelon form of A is I.*
- *$|A| \neq 0$.*
- *The rank of A is n.*

4.2 Rank and systems of linear equations

4.2.1 General solution and rank

Recall that to solve a system of linear equations using Guassian elimination, we form the augmented matrix and reduce it to echelon form by using elementary row operations. We will look at some examples to review what we learned in Chapter 2, and link this to the concept of the rank of a matrix.

Example 4.6 Consider the system of equations

$$x_1 + 2x_2 + x_3 = 1$$
$$2x_1 + 3x_2 \quad\ = 5$$
$$3x_1 + 5x_2 + x_3 = 4.$$

The augmented matrix is the same as the matrix B in the previous activity. When you reduced B to find the rank, after two steps you found

$$\begin{pmatrix} 1 & 2 & 1 & 1 \\ 2 & 3 & 0 & 5 \\ 3 & 5 & 1 & 4 \end{pmatrix} \rightarrow \begin{pmatrix} 1 & 2 & 1 & 1 \\ 0 & -1 & -2 & 3 \\ 0 & -1 & -2 & 1 \end{pmatrix} \rightarrow \begin{pmatrix} 1 & 2 & 1 & 1 \\ 0 & 1 & 2 & -3 \\ 0 & 0 & 0 & -2 \end{pmatrix}.$$

Thus, the original system of equations is equivalent to the system

$$x_1 + 2x_2 + x_3 = 1$$
$$x_2 + 2x_3 = -3$$
$$0x_1 + 0x_2 + 0x_3 = -2.$$

This system has no solutions, since there are no values of x_1, x_2, x_3 that satisfy the last equation, which reduces to the false statement '$0 = -2$' whatever values we give the unknowns. We deduce, therefore, that the original system has no solutions and we say that it is inconsistent. In this case, there is no reason to reduce the matrix further.

Continuing with our example, the coefficient matrix, A, consists of the first three columns of the augmented matrix, and the row echelon form of A consists of the first three columns of the row echelon form of the augmented matrix:

$$A = \begin{pmatrix} 1 & 2 & 1 \\ 2 & 3 & 0 \\ 3 & 5 & 1 \end{pmatrix} \rightarrow \cdots \rightarrow \begin{pmatrix} 1 & 2 & 1 \\ 0 & 1 & 2 \\ 0 & 0 & 0 \end{pmatrix}.$$

$$(A|\mathbf{b}) = \begin{pmatrix} 1 & 2 & 1 & 1 \\ 2 & 3 & 0 & 5 \\ 3 & 5 & 1 & 4 \end{pmatrix} \rightarrow \cdots \rightarrow \begin{pmatrix} 1 & 2 & 1 & 1 \\ 0 & 1 & 2 & -3 \\ 0 & 0 & 0 & 1 \end{pmatrix}.$$

The rank of the coefficient matrix A is 2, but the rank of the augmented matrix $(A|\mathbf{b})$ is 3.

If, as in Example 4.6, the row reduction of an augmented matrix has a row of the kind $(0\ 0\ \ldots\ 0\ a)$, with $a \neq 0$, then the original system is equivalent to one in which there is an equation

$$0x_1 + 0x_2 + \cdots + 0x_n = a \quad (a \neq 0),$$

which clearly cannot be satisfied by any values of the x_is, and the system is inconsistent. Then the row echelon form of the augmented matrix will have a row of the form $(0\ 0\ \ldots\ 0\ 1)$, and there will be one more leading one than in the row echelon form of the coefficient matrix. Therefore, the rank of the augmented matrix will be greater than the rank of the coefficient matrix. If the system is consistent, there will be no *leading one* in the last column of the augmented matrix and the ranks will be the same. In other words, we have the following result:

- A system $A\mathbf{x} = \mathbf{b}$ is consistent if and only if the rank of the augmented matrix is precisely the same as the rank of the matrix A.

Example 4.7 In contrast, consider the system of equations

$$\begin{aligned}
x_1 + 2x_2 + x_3 &= 1 \\
2x_1 + 3x_2 \phantom{{}+x_3} &= 5 \\
3x_1 + 5x_2 + x_3 &= 6.
\end{aligned}$$

This system has the same coefficient matrix A as in Example 4.6, and the rank of A is 2. The augmented matrix for the system is the matrix M in Example 4.3, which also has rank 2, so this system is consistent.

Activity 4.8 Write down a general solution for this system. Note that since the rank is 2 and there are three columns in A, there is a free variable and therefore there are infinitely many solutions.

Now suppose we have an $m \times n$ matrix A which has rank m. Then there will be a leading one in every row of an echelon form of A, and in this case a system of equations $A\mathbf{x} = \mathbf{b}$ will never be inconsistent. Why? There are two ways to see this. In the first place, if there is a leading one in every row of A, the augmented matrix $(A|\mathbf{b})$ can never have a row of the form $(0\ 0\ \ldots\ 0\ 1)$. Second, the augmented matrix also has m rows (since its size is $m \times (n+1)$), so the rank of $(A|\mathbf{b})$ can never be more than m. So we have the following observation:

- If an $m \times n$ matrix A has rank m the system of linear equations, $A\mathbf{x} = \mathbf{b}$ will be consistent for all $\mathbf{b} \in \mathbb{R}^m$.

Example 4.9 Suppose that

$$B = \begin{pmatrix} 1 & 2 & 1 & 1 \\ 2 & 3 & 0 & 5 \\ 3 & 5 & 1 & 4 \end{pmatrix}$$

is the coefficient matrix of a system of three equations in four unknowns, $B\mathbf{x} = \mathbf{d}$, with $\mathbf{d} \in \mathbb{R}^3$. This matrix B is the same as that of Activity 4.4 and we determined its row echelon form in Example 4.6 (where it was, differently, interpreted as an augmented matrix of a system of three equations in three unknowns):

$$B = \begin{pmatrix} 1 & 2 & 1 & 1 \\ 2 & 3 & 0 & 5 \\ 3 & 5 & 1 & 4 \end{pmatrix} \to \cdots \to \begin{pmatrix} 1 & 2 & 1 & 1 \\ 0 & 1 & 2 & -3 \\ 0 & 0 & 0 & 1 \end{pmatrix}.$$

The matrix B is 3×4 and has rank 3, so as we argued above, the system of equations $B\mathbf{x} = \mathbf{d}$ is always consistent.

Now let's look at the solutions. Any augmented matrix $(B|\mathbf{d})$ will be row equivalent to a matrix in echelon form for which the first four columns are the same as the echelon form of B; that is,

$$(B|\mathbf{d}) \rightarrow \cdots \rightarrow \begin{pmatrix} 1 & 2 & 1 & 1 & p_1 \\ 0 & 1 & 2 & -3 & p_2 \\ 0 & 0 & 0 & 1 & p_3 \end{pmatrix}$$

for some constants p_i, which could be zero. This system will have infinitely many solutions for any $\mathbf{d} \in \mathbb{R}^3$, because the number of columns is greater than the rank of B. There is one column without a leading one, so there is one non-leading variable.

Activity 4.10 If $p_1 = 1$, $p_2 = -2$ and $p_3 = 0$, and

$$\mathbf{x} = (x_1, x_2, x_3, x_4)^T,$$

write down the solution to the given system $B\mathbf{x} = \mathbf{d}$ in vector form, and use it to determine the original vector $\mathbf{d}$.

If we have a consistent system such that the rank r is strictly less than n, the number of unknowns, then as illustrated in Example 4.9, the system in reduced row echelon form (and hence the original one) does not provide enough information to specify the values of $x_1, x_2, \ldots, x_n$ uniquely and we will have infinitely many solutions. Let's consider this in more detail.

Example 4.11 Suppose we are given a system for which the augmented matrix reduces to the row echelon form

$$\begin{pmatrix} 1 & 3 & -2 & 0 & 0 & 0 \\ 0 & 0 & 1 & 2 & 3 & 1 \\ 0 & 0 & 0 & 0 & 1 & 5 \\ 0 & 0 & 0 & 0 & 0 & 0 \end{pmatrix}.$$

Here the rank (number of non-zero rows) is $r = 3$, which is strictly less than the number of unknowns, $n = 5$.

Continuing to reduced row echelon form, we obtain the matrix

$$\begin{pmatrix} 1 & 3 & 0 & 4 & 0 & -28 \\ 0 & 0 & 1 & 2 & 0 & -14 \\ 0 & 0 & 0 & 0 & 1 & 5 \\ 0 & 0 & 0 & 0 & 0 & 0 \end{pmatrix}.$$

Activity 4.12 Verify this. What are the additional two row operations which need to be carried out?

The corresponding system is

$$x_1 + 3x_2 + 4x_4 = -28$$
$$x_3 + 2x_4 = -14$$
$$x_5 = 5.$$

The variables x_1, x_3 and x_5 correspond to the columns with the leading ones and are the *leading variables*. The other variables are the *non-leading variables*.

The form of these equations tells us that we can assign any values to x_2 and x_4, and then the leading variables will be determined. Explicitly, if we give x_2 and x_4 the arbitrary values s and t, where s, t represent any real numbers, the solution is given by

$$x_1 = -28 - 3s - 4t, \quad x_2 = s, \quad x_3 = -14 - 2t, \quad x_4 = t, \quad x_5 = 5.$$

There are infinitely many solutions because the so-called 'free variables' x_2, x_4 can take any values $s, t \in \mathbb{R}$.

Generally, we can describe what happens when the row echelon form has $r < n$ non-zero rows $(0\ 0\ \ldots\ 0\ 1 * * \ldots *)$. If the leading one is in the kth column, it is the coefficient of the variable x_k. So if the rank is r and the leading ones occur in columns $c_1, c_2, \ldots, c_r$, then the general solution to the system can be expressed in a form where the unknowns $x_{c_1}, x_{c_2}, \ldots, x_{c_r}$ (the leading variables) are given in terms of the other $n - r$ unknowns (the non-leading variables), and those $n - r$ unknowns are free to take any values. In Example 4.11, we have $n = 5$ and $r = 3$, and the three variables x_1, x_3, x_5 can be expressed in terms of the $5 - 3 = 2$ free variables x_2, x_4.

If $r = n$, where the number of leading ones r in the echelon form is equal to the number of unknowns n, there is a leading one in every column since the leading ones move to the right as we go down the rows. In this case, a *unique solution* is obtained from the reduced echelon form. In fact, this can be thought of as a special case of the more general one discussed above: since $r = n$, there are $n - r = 0$ free variables, and the solution is therefore unique.

We can now summarise our conclusions thus far concerning a general linear system of m equations in n variables, written as $A\mathbf{x} = \mathbf{b}$, where the coefficient matrix A is an $m \times n$ matrix of rank r:

- If the echelon form of the augmented matrix has a row $(0\ 0\ \ldots\ 0\ 1)$, the original system is inconsistent; it has *no solutions*. In this case, $\text{rank}(A) = r < m$ and $\text{rank}(A|\mathbf{b}) = r + 1$.

- If the echelon form of the augmented matrix has no rows of the above type, the system is consistent, and the general solution involves $n - r$ free variables, where r is the rank of the coefficient matrix. When $r < n$, there are *infinitely many solutions*, but when $r = n$ there are no free variables and so there is a *unique solution*.

A homogeneous system of m equations in n unknowns is always consistent. In this case, the last statement still applies.

- The general solution of a homogeneous system involves $n - r$ free variables, where r is the rank of the coefficient matrix. When $r < n$ there are *infinitely many solutions*, but when $r = n$ there are no free variables and so there is a *unique solution*, namely the *trivial solution*, $\mathbf{x} = \mathbf{0}$.

4.2.2   General solution in vector notation

Continuing with Example 4.11, we found the general solution of the linear system in terms of the two free variables, or *parameters*, s and t. Expressing the solution, $\mathbf{x}$, as a column vector, we have

$$\mathbf{x} = \begin{pmatrix} x_1 \\ x_2 \\ x_3 \\ x_4 \\ x_5 \end{pmatrix} = \begin{pmatrix} -28 - 3s - 4t \\ s \\ -14 - 2t \\ t \\ 5 \end{pmatrix} = \begin{pmatrix} -28 \\ 0 \\ -14 \\ 0 \\ 5 \end{pmatrix} + \begin{pmatrix} -3s \\ s \\ 0 \\ 0 \\ 0 \end{pmatrix} + \begin{pmatrix} -4t \\ 0 \\ -2t \\ t \\ 0 \end{pmatrix}.$$

That is, the general solution is

$$\mathbf{x} = \mathbf{p} + s\mathbf{v}_1 + t\mathbf{v}_2 \quad s, t \in \mathbb{R},$$

where

$$\mathbf{p} = \begin{pmatrix} -28 \\ 0 \\ -14 \\ 0 \\ 5 \end{pmatrix}, \quad \mathbf{v}_1 = \begin{pmatrix} -3 \\ 1 \\ 0 \\ 0 \\ 0 \end{pmatrix}, \quad \mathbf{v}_2 = \begin{pmatrix} -4 \\ 0 \\ -2 \\ 1 \\ 0 \end{pmatrix}.$$

Applying the same method, more generally, to a consistent system of rank r with n unknowns, we can express the general solution of a consistent system $A\mathbf{x} = \mathbf{b}$ in the form

$$\mathbf{x} = \mathbf{p} + a_1\mathbf{v}_1 + a_2\mathbf{v}_2 + \cdots + a_{n-r}\mathbf{v}_{n-r}.$$

Note that if we put all the a_is equal to 0, we get a solution $\mathbf{x} = \mathbf{p}$, which means that $A\mathbf{p} = \mathbf{b}$, so $\mathbf{p}$ is a *particular solution* of the system. Putting $a_1 = 1$ and the remaining a_is equal to 0, we get a solution $\mathbf{x} = \mathbf{p} + \mathbf{v}_1$, which means that $A(\mathbf{p} + \mathbf{v}_1) = \mathbf{b}$. Thus,

$$\mathbf{b} = A(\mathbf{p} + \mathbf{v}_1) = A\mathbf{p} + A\mathbf{v}_1 = \mathbf{b} + A\mathbf{v}_1.$$

Comparing the first and last expressions, we see that $A\mathbf{v}_1 = \mathbf{0}$. Clearly, the same equation holds for $\mathbf{v}_2, \ldots, \mathbf{v}_{n-r}$. So we have proved the following:

- If A is an $m \times n$ matrix of rank r, the general solution of $A\mathbf{x} = \mathbf{b}$ is the sum of:
  - a particular solution $\mathbf{p}$ of the system $A\mathbf{x} = \mathbf{b}$ and
  - a linear combination $a_1\mathbf{v}_1 + a_2\mathbf{v}_2 + \cdots + a_{n-r}\mathbf{v}_{n-r}$ of solutions $\mathbf{v}_1, \mathbf{v}_2, \ldots, \mathbf{v}_{n-r}$ of the homogeneous system $A\mathbf{x} = \mathbf{0}$.
- If A has rank n, then $A\mathbf{x} = \mathbf{0}$ only has the solution $\mathbf{x} = \mathbf{0}$, and so $A\mathbf{x} = \mathbf{b}$ has a unique solution: $\mathbf{p} + \mathbf{0} = \mathbf{p}$.

This is a more precise form of the result of Theorem 2.29, which states that all solutions of a consistent system $A\mathbf{x} = \mathbf{b}$ are of the form $\mathbf{x} = \mathbf{p} + \mathbf{z}$ where $\mathbf{p}$ is any solution of $A\mathbf{x} = \mathbf{b}$ and $\mathbf{z} \in N(A)$, the null space of A (the set of all solutions of $A\mathbf{x} = \mathbf{0}$).

4.3 Range

The *range* of a matrix A is defined as follows:

Definition 4.13 (Range of a matrix) Suppose that A is an $m \times n$ matrix. Then the *range* of A, denoted by $R(A)$, is the subset of $\mathbb{R}^m$ given by

$$R(A) = \{A\mathbf{x} \mid \mathbf{x} \in \mathbb{R}^n\}.$$

That is, the range is the set of all vectors $\mathbf{y} \in \mathbb{R}^m$ of the form $\mathbf{y} = A\mathbf{x}$ for some $\mathbf{x} \in \mathbb{R}^n$.

What is the connection between the range of a matrix A and a system of linear equations $A\mathbf{x} = \mathbf{b}$? If A is $m \times n$, then $\mathbf{x} \in \mathbb{R}^n$ and $\mathbf{b} \in \mathbb{R}^m$. If the system $A\mathbf{x} = \mathbf{b}$ is consistent, then this means that there is a vector $\mathbf{x} \in \mathbb{R}^n$ such that $A\mathbf{x} = \mathbf{b}$, so $\mathbf{b}$ is in the range of A. Conversely, if $\mathbf{b}$ is in the range of A, then the system $A\mathbf{x} = \mathbf{b}$ must have a solution. Therefore, for an $m \times n$ matrix A:

- the range of A, $R(A)$, consists of all vectors $\mathbf{b} \in \mathbb{R}^m$ for which the system of equations $A\mathbf{x} = \mathbf{b}$ is consistent.

Let's look at $R(A)$ from a different point of view. Suppose that the columns of A are $\mathbf{c}_1, \mathbf{c}_2, \ldots, \mathbf{c}_n$, which we can indicate by writing $A = (\mathbf{c}_1\, \mathbf{c}_2\, \ldots\, \mathbf{c}_n)$. If $\mathbf{x} = (\alpha_1, \alpha_2, \ldots, \alpha_n)^T \in \mathbb{R}^n$, then we saw in Chapter 1 (Theorem 1.38) that the product $A\mathbf{x}$ can be expressed as a linear combination of the columns of A, namely

$$A\mathbf{x} = \alpha_1 \mathbf{c}_1 + \alpha_2 \mathbf{c}_2 + \cdots + \alpha_n \mathbf{c}_n.$$

Activity 4.14 This is a good time to convince yourself (again) of this statement. Write out each side using $\mathbf{c}_i = (c_{1i}, c_{2i}, \ldots, c_{mi})^T$ to show that

$$A\mathbf{x} = \alpha_1 \mathbf{c}_1 + \alpha_2 \mathbf{c}_2 + \cdots + \alpha_n \mathbf{c}_n.$$

Try to do this yourself before looking at the solution to this activity. This is a very important result which will be used many times in this text, so make sure you understand how it works.

So, $R(A)$, the set of all matrix products $A\mathbf{x}$, is also the set of all *linear combinations* of the columns of A. For this reason, $R(A)$ is also called the *column space* of A. (We'll discuss this more in the next chapter.)

If $A = (\mathbf{c}_1\, \mathbf{c}_2\, \ldots\, \mathbf{c}_n)$, where $\mathbf{c}_i$ denotes column i of A, then we can write

$$R(A) = \{a_1 \mathbf{c}_1 + a_2 \mathbf{c}_2 + \ldots + a_n \mathbf{c}_n \mid a_1, a_2, \ldots, a_n \in \mathbb{R}\}.$$

Example 4.15 Suppose that

$$A = \begin{pmatrix} 1 & 2 \\ -1 & 3 \\ 2 & 1 \end{pmatrix}.$$

Then for $\mathbf{x} = (\alpha_1, \alpha_2)^T$,

$$A\mathbf{x} = \begin{pmatrix} 1 & 2 \\ -1 & 3 \\ 2 & 1 \end{pmatrix} \begin{pmatrix} \alpha_1 \\ \alpha_2 \end{pmatrix} = \begin{pmatrix} \alpha_1 + 2\alpha_2 \\ -\alpha_1 + 3\alpha_2 \\ 2\alpha_1 + \alpha_2 \end{pmatrix} = \alpha_1 \begin{pmatrix} 1 \\ -1 \\ 2 \end{pmatrix} + \alpha_2 \begin{pmatrix} 2 \\ 3 \\ 1 \end{pmatrix},$$

so

$$R(A) = \left\{ \begin{pmatrix} \alpha_1 + 2\alpha_2 \\ -\alpha_1 + 3\alpha_2 \\ 2\alpha_1 + \alpha_2 \end{pmatrix} \;\middle|\; \alpha_1, \alpha_2 \in \mathbb{R} \right\},$$

or

$$R(A) = \{\alpha_1 \mathbf{c}_1 + \alpha_2 \mathbf{c}_2 \mid \alpha_1, \alpha_2 \in \mathbb{R}\},$$

where $\mathbf{c}_1 = \begin{pmatrix} 1 \\ -1 \\ 2 \end{pmatrix}$ and $\mathbf{c}_2 = \begin{pmatrix} 2 \\ 3 \\ 1 \end{pmatrix}$ are the columns of A.

Again, thinking of the connection with the system of equations $A\mathbf{x} = \mathbf{b}$, we have already shown that $A\mathbf{x} = \mathbf{b}$ is consistent if and only if $\mathbf{b}$ is in the range of A, and we have now shown that $R(A)$ is equal to the set of all linear combinations of the columns of A. Therefore, we can now assert that:

- The system of equations $A\mathbf{x} = \mathbf{b}$ is consistent if and only if $\mathbf{b}$ is a linear combination of the columns of A.

Example 4.16 Consider the following systems of three equations in two unknowns.

$$\begin{cases} x + 2y = 0 \\ -x + 3y = -5 \\ 2x + y = 3 \end{cases} \qquad \begin{cases} x + 2y = 1 \\ -x + 3y = 5 \\ 2x + y = 2. \end{cases}$$

Solving these by Gaussian elimination (or any other method), you will find that the first system is consistent and the second system has no solution. The first system has the unique solution $(x, y)^{\mathrm{T}} = (2, -1)^{\mathrm{T}}$.

Activity 4.17 Do this. Solve each of the above systems.

The coefficient matrix of each of the systems is the same, and is equal to the matrix A in Example 4.15. For the first system,

$$A = \begin{pmatrix} 1 & 2 \\ -1 & 3 \\ 2 & 1 \end{pmatrix}, \quad \mathbf{x} = \begin{pmatrix} x \\ y \end{pmatrix}, \quad \mathbf{b} = \begin{pmatrix} 0 \\ -5 \\ 3 \end{pmatrix}.$$

Checking the solution, you will find that

$$A\mathbf{x} = \begin{pmatrix} 1 & 2 \\ -1 & 3 \\ 2 & 1 \end{pmatrix} \begin{pmatrix} 2 \\ -1 \end{pmatrix} = \begin{pmatrix} 0 \\ -5 \\ 3 \end{pmatrix}$$

or

$$\begin{pmatrix} 0 \\ -5 \\ 3 \end{pmatrix} = 2 \begin{pmatrix} 1 \\ -1 \\ 2 \end{pmatrix} - \begin{pmatrix} 2 \\ 3 \\ 1 \end{pmatrix} = 2\mathbf{c}_1 - \mathbf{c}_2.$$

On the other hand, it is not possible to express the vector $(1, 5, 2)^T$ as a linear combination of the column vectors of A. Trying to do so would lead to precisely the same set of inconsistent equations.

Notice, also, that the homogeneous system $A\mathbf{x} = \mathbf{0}$ has only the trivial solution, and that the only way to express $\mathbf{0}$ as a linear combination of the columns of A is by $0\mathbf{c}_1 + 0\mathbf{c}_2 = \mathbf{0}$.

Activity 4.18 Verify all of the above statements.

4.4   Learning outcomes

You should now be able to:

- explain what is meant by the rank of a matrix
- find the rank of an $m \times n$ matrix A
- explain why a system of linear equations, $A\mathbf{x} = \mathbf{b}$, where A is an $m \times n$ matrix, is consistent if and only if the rank$(A) =$ rank$((A|\mathbf{b}))$; and why if rank$(A) =$ m, then $A\mathbf{x} = \mathbf{b}$ is consistent for all $\mathbf{b} \in \mathbb{R}^m$
- explain why a general solution $\mathbf{x}$ to $A\mathbf{x} = \mathbf{b}$, where A is an $m \times n$ matrix of rank r, is of the form

$$\mathbf{x} = \mathbf{p} + a_1\mathbf{v}_1 + a_2\mathbf{v}_2 + \cdots + a_{n-r}\mathbf{v}_{n-r}, a_i \in \mathbb{R};$$

  specifically why there are $n - r$ arbitrary constants
- explain what is meant by the range of a matrix
- show that if $A = (\mathbf{c}_1 \ \mathbf{c}_2 \ \dots \ \mathbf{c}_n)$, and if $\mathbf{x} = (\alpha_1, \alpha_2, \dots, \alpha_n)^T \in \mathbb{R}^n$, then $A\mathbf{x} = \alpha_1\mathbf{c}_1 + \alpha_2\mathbf{c}_2 + \cdots + \alpha_n\mathbf{c}_n$
- write $\mathbf{b}$ as a linear combination of the columns of A if $A\mathbf{x} = \mathbf{b}$ is consistent
- write $\mathbf{0}$ as a linear combination of the columns of A, and explain when it is possible to do this in some way other than using the trivial solution, $\mathbf{x} = \mathbf{0}$, with all the coefficients in the linear combination equal to 0.

4.5   Comments on activities

Activity 4.8 One more row operation on the row echelon form will obtain a matrix in reduced row echelon form which is row equivalent to the matrix M, from which the solution is found to be

$$\mathbf{x} = \begin{pmatrix} 7 \\ -3 \\ 0 \end{pmatrix} + t \begin{pmatrix} 3 \\ -2 \\ 1 \end{pmatrix}, \quad t \in \mathbb{R}.$$

Activity 4.10 Substitute for p_1, p_2, p_3 in the row echelon form of the augmented matrix and then continue to reduce it to reduced row echelon form. The non-leading variable is x_3. Letting $x_3 = t$, the general solution is

$$\mathbf{x} = \begin{pmatrix} x_1 \\ x_2 \\ x_3 \\ x_4 \end{pmatrix} = \begin{pmatrix} 5 \\ -2 \\ 0 \\ 0 \end{pmatrix} + t \begin{pmatrix} 3 \\ -2 \\ 1 \\ 0 \end{pmatrix} = \mathbf{p} + t\mathbf{v}, \quad t \in \mathbb{R}.$$

Since $B\mathbf{p} = \mathbf{d}$, multiplying $B\mathbf{p}$, you will find that $\mathbf{d} = (1, 4, 5)^{\mathrm{T}}$. (You can check all this by row reducing $(B|\mathbf{d})$.)

Activity 4.14 First write out the matrix product of $A = (c_{ij})$ and $\mathbf{x}$.

$$A\mathbf{x} = \begin{pmatrix} c_{11} & c_{12} & \cdots & c_{1n} \\ c_{21} & c_{22} & \cdots & c_{2n} \\ \vdots & \vdots & \ddots & \vdots \\ c_{m1} & c_{m2} & \cdots & c_{mn} \end{pmatrix} \begin{pmatrix} \alpha_1 \\ \alpha_2 \\ \vdots \\ \alpha_n \end{pmatrix}.$$

The product is $m \times 1$; that is,

$$A\mathbf{x} = \begin{pmatrix} c_{11}\alpha_1 + c_{12}\alpha_2 + \cdots + c_{1n}\alpha_n \\ c_{21}\alpha_1 + c_{22}\alpha_2 + \cdots + c_{2n}\alpha_n \\ \vdots \\ c_{m1}\alpha_1 + c_{m2}\alpha_2 + \cdots + c_{mn}\alpha_n \end{pmatrix}$$

and can be written as a sum of n, $m \times 1$ vectors:

$$A\mathbf{x} = \begin{pmatrix} c_{11}\alpha_1 \\ c_{21}\alpha_1 \\ \vdots \\ c_{m1}\alpha_1 \end{pmatrix} + \begin{pmatrix} c_{12}\alpha_2 \\ c_{22}\alpha_2 \\ \vdots \\ c_{m2}\alpha_2 \end{pmatrix} + \cdots + \begin{pmatrix} c_{1n}\alpha_n \\ c_{2n}\alpha_n \\ \vdots \\ c_{mn}\alpha_n \end{pmatrix}.$$

So,

$$A\mathbf{x} = \alpha_1 \begin{pmatrix} c_{11} \\ c_{21} \\ \vdots \\ c_{m1} \end{pmatrix} + \alpha_2 \begin{pmatrix} c_{12} \\ c_{22} \\ \vdots \\ c_{m2} \end{pmatrix} + \cdots + \alpha_n \begin{pmatrix} c_{1n} \\ c_{2n} \\ \vdots \\ c_{mn} \end{pmatrix}.$$

That is,

$$A\mathbf{x} = \alpha_1 \mathbf{c}_1 + \alpha_2 \mathbf{c}_2 + \cdots + \alpha_n \mathbf{c}_n.$$

All these steps are reversible, so any expression

$$\alpha_1 \mathbf{c}_1 + \alpha_2 \mathbf{c}_2 + \cdots + \alpha_n \mathbf{c}_n$$

can be written in matrix form as $A\mathbf{x}$, where $A = (c_{ij})$ and $\mathbf{x} = (\alpha_1, \alpha_2, \ldots, \alpha_n)^{\mathsf{T}}$.

4.6   Exercises

Exercise 4.1 Solve the following system of equations $A\mathbf{x} = \mathbf{b}$ by reducing the augmented matrix to reduced row echelon form:

$$x_1 + 5x_2 + 3x_3 + 7x_4 + x_5 = 2$$
$$2x_1 + 10x_2 + 3x_3 + 8x_4 + 5x_5 = -5$$
$$x_1 + 5x_2 + x_3 + 3x_4 + 3x_5 = -4.$$

If $r = \text{rank}(A)$ and n is the number of columns of A, show that your solution can be written in the form $\mathbf{x} = \mathbf{p} + a_1\mathbf{v}_1 + \ldots + a_{n-r}\mathbf{v}_{n-r}$ where $a_i \in \mathbb{R}$.

Show also that $A\mathbf{p} = \mathbf{b}$ and that $A\mathbf{v}_i = 0$ for $i = 1, \ldots, n - r$.

Express the vector $\mathbf{b}$ as a linear combination of the columns of the coefficient matrix A. Do the same for the vector $\mathbf{0}$.

Exercise 4.2 Find the rank of the matrix

$$A = \begin{pmatrix} 1 & 0 & 1 & 0 & 2 \\ 2 & 1 & 1 & 1 & 3 \\ 1 & 3 & -1 & 2 & 2 \\ 0 & 3 & -2 & 2 & 0 \end{pmatrix}.$$

Determine $N(A)$, the null space of A, and $R(A)$, the range of A.

Exercise 4.3 Consider the system of linear equations $A\mathbf{x} = \mathbf{b}$ given below, where λ and μ are constants, and

$$A = \begin{pmatrix} 1 & 2 & 0 \\ 5 & 1 & \lambda \\ 1 & -1 & 1 \end{pmatrix}, \quad \mathbf{x} = \begin{pmatrix} x \\ y \\ z \end{pmatrix}, \quad \mathbf{b} = \begin{pmatrix} 2 \\ 7 \\ \mu \end{pmatrix}.$$

Compute the determinant of A, $|A|$.

Determine for which values of λ and μ this system has:

(a)   a unique solution,
(b)   no solutions,
(c)   infinitely many solutions.

In case (a), use Cramer's rule to find the value of z in terms of λ and μ. In case (c), solve the system using row operations and express the solution in vector form, $\mathbf{x} = \mathbf{p} + t\mathbf{v}$.

Exercise 4.4 A system of linear equations $B\mathbf{x} = \mathbf{d}$ is known to have the following general solution:

$$\mathbf{x} = \begin{pmatrix} 1 \\ 0 \\ 2 \\ 0 \end{pmatrix} + s \begin{pmatrix} -3 \\ 1 \\ 0 \\ 0 \end{pmatrix} + t \begin{pmatrix} 1 \\ 0 \\ -1 \\ 1 \end{pmatrix} \quad s, t \in \mathbb{R}.$$

Let $\mathbf{c}_1 = \begin{pmatrix} 1 \\ 1 \\ 2 \end{pmatrix}$ be the first column of B. If $\mathbf{d} = \begin{pmatrix} 3 \\ 5 \\ -2 \end{pmatrix}$, find the matrix B.

Exercise 4.5 Consider the matrix

$$A = \begin{pmatrix} 1 & 2 & 1 \\ 2 & 3 & 0 \\ 3 & 5 & 1 \end{pmatrix}.$$

Find a condition that the components of the vector $\mathbf{b} = (a, b, c)^{\mathrm{T}}$ must satisfy in order for $A\mathbf{x} = \mathbf{b}$ to be consistent. Hence, or otherwise, show that $R(A)$ is a plane in $\mathbb{R}^3$, and write down a Cartesian equation of this plane.

Show that $\mathbf{d} = (1, 5, 6)^{\mathrm{T}}$ is in $R(A)$. Express $\mathbf{d}$ as a linear combination of the columns of A. Is it possible to do this in two different ways? If the answer is yes, then do so; otherwise, justify why this is not possible.

Exercise 4.6 Consider the matrices

$$A = \begin{pmatrix} 1 & 1 & 1 \\ 0 & 1 & -2 \\ 2 & -1 & 8 \\ 3 & 1 & 7 \end{pmatrix} \quad B = \begin{pmatrix} -2 & 3 & -2 & 5 \\ 3 & -6 & 9 & -6 \\ -2 & 9 & -1 & 9 \\ 5 & -6 & 9 & -4 \end{pmatrix} \quad \mathbf{b} = \begin{pmatrix} 4 \\ 1 \\ a \\ b \end{pmatrix}.$$

(a) Find the rank of the matrix A. Find a general solution of $A\mathbf{x} = \mathbf{0}$. Either write down a non-trivial linear combination of the column vectors of A which is equal to the zero vector, $\mathbf{0}$, or justify why this is not possible.

Find all real numbers a and b such that $\mathbf{b} \in R(A)$, where $\mathbf{b}$ is the vector given above. Write down a general solution of $A\mathbf{x} = \mathbf{b}$.

(b) Using row operations, or otherwise, find $|B|$, where B is the matrix given above. What is the rank of B?

Either write down a non-trivial linear combination of the column vectors of B which is equal to the zero vector, $\mathbf{0}$, or justify why this is not possible.

Find all real numbers a and b such that $\mathbf{b} \in R(B)$, the range of B, where $\mathbf{b}$ is the vector given above.

4.7 Problems

Problem 4.1 Solve the following system of equations $A\mathbf{x} = \mathbf{b}$ by reducing the augmented matrix to reduced row echelon form:

$$x_1 - x_2 + x_3 + x_4 + 2x_5 = 4$$
$$-x_1 + x_2 + x_4 - x_5 = -3$$
$$x_1 - x_2 + 2x_3 + 3x_4 + 4x_5 = 7.$$

Show that your solution can be written in the form $\mathbf{x} = \mathbf{p} + s\mathbf{v}_1 + t\mathbf{v}_2$ where $A\mathbf{p} = \mathbf{b}$, $A\mathbf{v}_1 = \mathbf{0}$ and $A\mathbf{v}_2 = \mathbf{0}$.

Problem 4.2 Express the following system of linear equations in matrix form, as $A\mathbf{x} = \mathbf{b}$:

$$x + y + z + w = 3$$
$$y - 2z + 2w = 1$$
$$x + 3z - w = 2.$$

Find the general solution.

(a) Determine $N(A)$, the null space of A.
(b) If $\mathbf{a} = (a, b, c)^\mathsf{T}$, find an equation which a, b, c must satisfy so that $\mathbf{a} \in R(A)$, the range of A.
(c) If $\mathbf{d} = (1, 5, 3)$, determine if the system of equations $A\mathbf{x} = \mathbf{d}$ is consistent, and write down the general solution if it is.

Problem 4.3 Show that the following system of equations is consistent for any $c \in \mathbb{R}$:

$$\begin{cases} x + y - 2z = 1 \\ 2x - y + 2z = 1 \\ cx + z = 0. \end{cases}$$

Solve the system using any matrix method (Cramer's rule, inverse matrix, or Gaussian elimination) and hence write down expressions for x, y, z in terms of c.

Problem 4.4 Consider the following system of equations, where λ is a constant:

$$\begin{cases} 2x + y + z = 3 \\ x - y + 2z = 3 \\ x - 2y + \lambda z = 4. \end{cases}$$

Determine all values of λ, if any, such that this system has:

(1) no solutions;
(2) exactly one solution;
(3) infinitely many solutions.

In case (2) find the solution using either Cramer's rule or an inverse matrix. In case (3) solve the system using Gaussian elimination. Express the solution in vector form.

Problem 4.5 Solve the system of equations $B\mathbf{x} = \mathbf{b}$ using Gaussian elimination, where

$$B = \begin{pmatrix} 1 & 1 & 2 & 1 \\ 2 & 3 & -1 & 1 \\ 1 & 0 & 7 & 2 \end{pmatrix}, \quad \mathbf{b} = \begin{pmatrix} 3 \\ 2 \\ -2 \end{pmatrix}.$$

Show that the vector $\mathbf{b}$ cannot be expressed as a linear combination of the columns of B.

Problem 4.6 A system of linear equations $A\mathbf{x} = \mathbf{d}$ is known to have the following solution:

$$\mathbf{x} = \begin{pmatrix} 1 \\ 2 \\ 0 \\ -1 \\ 0 \end{pmatrix} + s \begin{pmatrix} 2 \\ 1 \\ 1 \\ 0 \\ 0 \end{pmatrix} + t \begin{pmatrix} 1 \\ 1 \\ 0 \\ -1 \\ 1 \end{pmatrix}, \quad s, t \in \mathbb{R}.$$

Assume that A is an $m \times n$ matrix. Let $\mathbf{c}_1, \mathbf{c}_2, \ldots, \mathbf{c}_n$ denote the columns of A.

Answer each of the following questions *or*, if there is insufficient information to answer a question, say so.

(1) What number is n?
(2) What number is m?
(3) What (number) is the rank of A?
(4) Describe the null space $N(A)$.
(5) Write down an expression for $\mathbf{d}$ as a linear combination of the columns of A.

(6)   Write down a non-trivial linear combination of the columns c_i which is equal to $\mathbf{0}$.

Problem 4.7 Let

$$A = \begin{pmatrix} 3 & 1 & 5 & 9 & -1 \\ 1 & 0 & 1 & 2 & -1 \\ -2 & 1 & 0 & -1 & 2 \\ 1 & 1 & 3 & 5 & 0 \end{pmatrix}, \quad \mathbf{b} = \begin{pmatrix} 11 \\ 5 \\ -8 \\ 4 \end{pmatrix}.$$

Solve the system of equations, $A\mathbf{x} = \mathbf{b}$, using Gaussian elimination. Express your solution in vector form, as $\mathbf{x} = \mathbf{p} + a_1\mathbf{v}_1 + \cdots + a_n\mathbf{v}_n$, and verify that $k = n - r$ where r is the rank of A. What is n?

If possible, express $\mathbf{b}$ as a linear combination of the column vectors of A in two different ways.

5

Vector spaces

In this chapter, we study the important theoretical concept of a vector space. This, and the related concepts to be explored in the subsequent chapters, will enable us to extend and to understand more deeply what we've already learned about matrices and linear equations, and lead us to new and important ways to apply linear algebra. There is, necessarily, a bit of a step upwards in the level of 'abstraction', but it is worth the effort in order to help our fundamental understanding.

5.1   Vector spaces

5.1.1   Definition of a vector space

We know that vectors of $\mathbb{R}^n$ can be added together and that they can be 'scaled' by real numbers. That is, for every $\mathbf{x}, \mathbf{y} \in \mathbb{R}^n$ and every $\alpha \in \mathbb{R}$, it makes sense to talk about $\mathbf{x} + \mathbf{y}$ and $\alpha \mathbf{x}$. Furthermore, these operations of addition and multiplication by a *scalar* (that is, multiplication by a real number) behave and interact 'sensibly' in that, for example,

$$\alpha(\mathbf{x} + \mathbf{y}) = \alpha \mathbf{x} + \alpha \mathbf{y},$$
$$\alpha(\beta \mathbf{x}) = (\alpha \beta)\mathbf{x},$$
$$\mathbf{x} + \mathbf{y} = \mathbf{y} + \mathbf{x},$$

and so on.

But it is not only vectors in $\mathbb{R}^n$ that can be added and multiplied by scalars. There are other sets of objects for which this is possible. Consider the set F of all functions from $\mathbb{R}$ to $\mathbb{R}$. Then any two of these functions can be added; given $f, g \in F$, we simply define the function

$f + g$ by

$$(f + g)(x) = f(x) + g(x).$$

Also, for any $\alpha \in \mathbb{R}$, the function αf is given by

$$(\alpha f)(x) = \alpha f(x).$$

These operations of addition and scalar multiplication are sometimes known as *pointwise addition* and *pointwise scalar multiplication*. This might seem a bit abstract, but think about what the functions $x + x^2$ and $2x$ represent: the former is the function x plus the function x^2, and the latter is the function x multiplied by the scalar 2. So this is just a different way of looking at something with which you are already familiar. It turns out that F and its rules for addition and multiplication by a scalar satisfy the same key properties as the set of vectors in $\mathbb{R}^n$ with its addition and scalar multiplication. We refer to a set with an addition and scalar multiplication which behave appropriately as a *vector space*. We now give the formal definition of a vector space.

Definition 5.1 (Vector space) A (real) *vector space* V is a non-empty set equipped with an addition operation and a scalar multiplication operation such that for all $\alpha, \beta \in \mathbb{R}$ and all $\mathbf{u}, \mathbf{v}, \mathbf{w} \in V$:

1. $\mathbf{u} + \mathbf{v} \in V$ (*closure* under addition).
2. $\mathbf{u} + \mathbf{v} = \mathbf{v} + \mathbf{u}$ (the *commutative* law for addition).
3. $\mathbf{u} + (\mathbf{v} + \mathbf{w}) = (\mathbf{u} + \mathbf{v}) + \mathbf{w}$ (the *associative* law for addition).
4. there is a single member $\mathbf{0}$ of V, called the *zero vector*, such that for all $\mathbf{v} \in V, \mathbf{v} + \mathbf{0} = \mathbf{v}$.
5. for every $\mathbf{v} \in V$ there is an element $\mathbf{w} \in V$ (usually written as $-\mathbf{v}$), called the *negative* of $\mathbf{v}$, such that $\mathbf{v} + \mathbf{w} = \mathbf{0}$.
6. $\alpha \mathbf{v} \in V$ (*closure* under scalar multiplication).
7. $\alpha(\mathbf{u} + \mathbf{v}) = \alpha\mathbf{u} + \alpha\mathbf{v}$ (*distributive* law).
8. $(\alpha + \beta)\mathbf{v} = \alpha\mathbf{v} + \beta\mathbf{v}$ (*distributive* law).
9. $\alpha(\beta\mathbf{v}) = (\alpha\beta)\mathbf{v}$ (*associative* law for scalar multiplication).
10. $1\mathbf{v} = \mathbf{v}$.

This list of properties, called axioms, in the definition is the shortest possible number which will enable any vector space V to behave the way we would like it to behave with respect to addition and scalar multiplication; that is, like $\mathbb{R}^n$. Other properties which we would expect to be true follow from those listed in the definition. For instance, we can see that $0\mathbf{x} = \mathbf{0}$ for all $\mathbf{x}$, as follows:

$$0\mathbf{x} = (0 + 0)\mathbf{x} = 0\mathbf{x} + 0\mathbf{x}$$

by axiom 8; so, adding the negative $-0\mathbf{x}$ of $0\mathbf{x}$ to each side,

$$\mathbf{0} = 0\mathbf{x} + (-0\mathbf{x}) = (0\mathbf{x} + 0\mathbf{x}) + (-0\mathbf{x}) = 0\mathbf{x} + (0\mathbf{x} + (-0\mathbf{x}))$$
$$= 0\mathbf{x} + \mathbf{0} = 0\mathbf{x}$$

by axioms 5, 3, 5 again, and 4. The proof may seem a bit contrived, but just remember the result:

$$0\mathbf{x} = \mathbf{0}.$$

This would be easy to show in $\mathbb{R}^n$ because we know what the vector $\mathbf{0}$ looks like, namely $\mathbf{0} = (0, 0, \ldots, 0)^{\mathrm{T}}$. But because we want to show it is true in any vector space, V, we have to derive this property directly from the definition. Once we've established a result, we can use it to prove other properties which hold in a vector space V.

Activity 5.2 Prove that for any vector $\mathbf{x}$ in a vector space V,

$$(-1)\mathbf{x} = -\mathbf{x},$$

the negative of the vector $\mathbf{x}$, using a similar argument with $0 = 1 + (-1)$.
If you're feeling confident, show that $\alpha\mathbf{0} = \mathbf{0}$ for any $\alpha \in \mathbb{R}$.

Note that the definition of a vector space says nothing at all about multiplying together two vectors, or an inner product. The only operations with which the definition is concerned are addition and scalar multiplication.

A vector space as we have defined it is called a *real vector space*, to emphasise that the 'scalars' α, β and so on are real numbers rather than (say) complex numbers. There is a notion of *complex vector space*, where the scalars are complex numbers, which we shall cover in Chapter 13.

In the discussions that follow, be aware of whether we are talking about $\mathbb{R}^n$ or about an abstract vector space V.

5.1.2 Examples

Example 5.3 The set $\mathbb{R}^n$ is a vector space with the usual way of adding and scalar multiplying vectors.

Example 5.4 The set $V = \{\mathbf{0}\}$ consisting only of the zero vector is a vector space, with addition defined by $\mathbf{0} + \mathbf{0} = \mathbf{0}$, and scalar multiplication defined by $\alpha\mathbf{0} = \mathbf{0}$ for all $\alpha \in \mathbb{R}$.

Example 5.5 The set F of functions from $\mathbb{R}$ to $\mathbb{R}$ with pointwise addition and scalar multiplication (described earlier in this section) is a vector space. Note that the zero vector in this space is the function that maps every real number to 0 – that is, the *identically zero* function.

Activity 5.6 Show that all 10 axioms of a vector space are satisfied. In particular, if the function f is a vector in this space, what is the vector $-f$?

Example 5.7 The set of $m \times n$ matrices with real entries is a vector space, with the usual addition and scalar multiplication of matrices. The 'zero vector' in this vector space is the zero $m \times n$ matrix, which has all entries equal to 0.

Example 5.8 The set S of all infinite sequences of real numbers, $\mathbf{y} = \{y_1, y_2, \ldots, y_n, \ldots\}$, $y_i \in \mathbb{R}$, is a vector space. We can also use the notation $\mathbf{y} = \{y_n\}$, $n \geq 1$ for a sequence. For example, the sequence $\mathbf{y} = \{1, 2, 4, 8, 16, 32, \ldots\}$ can also be represented as $\{y_n\}$ with $y_t = 2^t$, $t = 0, 1, 2, 3, \ldots$.

The operation of addition of sequences is by adding components. If $\mathbf{y}, \mathbf{z} \in S$,

$$\mathbf{y} = \{y_1, y_2, \ldots, y_n, \ldots\}, \mathbf{z} = \{z_1, z_2, \ldots, z_n, \ldots\},$$

then

$$\mathbf{y} + \mathbf{z} = \{y_1 + z_1, y_2 + z_2, \ldots, y_n + z_n, \ldots\}.$$

Multiplication by a scalar $\alpha \in \mathbb{R}$ is defined in a similar way, by

$$\alpha \mathbf{y} = \{\alpha y_1, \alpha y_2, \ldots, \alpha y_n, \ldots\}.$$

These operations satisfy all the requirements for S to be a vector space. The sum and scalar multiple of an infinite sequence as defined above is again an infinite sequence. The zero vector is the sequence consisting entirely of zeros, and the negative of $\mathbf{y} = \{y_n\}$ is $-\mathbf{y} = \{-y_n\}$. The remaining axioms are satisfied because the components of a sequence are real numbers. For example, using the notation $\mathbf{y} = \{y_n\}$, $\mathbf{z} = \{z_n\}$, $n \geq 1$,

$$\mathbf{y} + \mathbf{z} = \{y_n + z_n\} = \{z_n + y_n\} = \mathbf{z} + \mathbf{y}.$$

Activity 5.9 If it is not immediately clear to you that all ten axioms are satisfied, then try to write down proofs for some of them.

The following example concerns a subset of $\mathbb{R}^3$.

Example 5.10 Let W be the set of all vectors in $\mathbb{R}^3$ with the third entry equal to 0; that is,

$$W = \left\{ \begin{pmatrix} x \\ y \\ 0 \end{pmatrix} \;\middle|\; x, y \in \mathbb{R} \right\}.$$

Then W is a vector space with the usual addition and scalar multiplication. To verify this, we need only check that W is non-empty and closed under addition and scalar multiplication. Why is this so? The axioms 2, 3, 7, 8, 9, 10 will hold for vectors in W because they hold for all vectors in $\mathbb{R}^3$, and if W is closed under addition and scalar multiplication, then all linear combinations: of vectors in W are still in W. Furthermore, if we can show that W is closed under scalar multiplication, then for any particular $\mathbf{v} \in W$, $0\mathbf{v} = \mathbf{0} \in W$ and $(-1)\mathbf{v} = -\mathbf{v} \in W$. So we simply need to check that $W \neq \emptyset$ (W is non-empty), that if $\mathbf{u}, \mathbf{v} \in W$, then $\mathbf{u} + \mathbf{v} \in W$, and if $\alpha \in \mathbb{R}$ and $\mathbf{v} \in W$, then $\alpha\mathbf{v} \in W$. Each of these is easy to check.

Activity 5.11 Verify that $W \neq \emptyset$, and that for $\mathbf{u}, \mathbf{v} \in W$ and $\alpha \in \mathbb{R}$, $\mathbf{u} + \mathbf{v} \in W$ and $\alpha\mathbf{v} \in W$.

5.1.3 Linear combinations

For vectors $\mathbf{v}_1, \mathbf{v}_2, \ldots, \mathbf{v}_k$ in a vector space V, the vector

$$\mathbf{v} = a_1\mathbf{v}_1 + a_2\mathbf{v}_2 + \cdots + a_k\mathbf{v}_k$$

is known as a *linear combination* of the vectors $\mathbf{v}_1, \mathbf{v}_2, \ldots, \mathbf{v}_k$. The scalars a_i are called *coefficients*. The structure of a vector space is designed for us to work with linear combinations of vectors.

Example 5.12 Suppose we want to express the vector $\mathbf{w} = (2, -5)^\mathsf{T}$ in $\mathbb{R}^2$ as a linear combination of the vectors $\mathbf{v}_1 = (1, 2)^\mathsf{T}$ and $\mathbf{v}_2 = (1, -1)^\mathsf{T}$. Then we solve the system of linear equations given by the components of the vector equation

$$\begin{pmatrix} 2 \\ -5 \end{pmatrix} = \alpha \begin{pmatrix} 1 \\ 2 \end{pmatrix} + \beta \begin{pmatrix} 1 \\ -1 \end{pmatrix}$$

to obtain $\alpha = -1$ and $\beta = 3$. Then $\mathbf{w} = -\mathbf{v}_1 + 3\mathbf{v}_2$, which is easily checked by performing the scalar multiplication and addition:

$$\begin{pmatrix} 2 \\ -5 \end{pmatrix} = - \begin{pmatrix} 1 \\ 2 \end{pmatrix} + 3 \begin{pmatrix} 1 \\ -1 \end{pmatrix}.$$

Activity 5.13 On a graph, sketch the vectors $\mathbf{v}_1$ and $\mathbf{v}_2$ and then sketch the vector $\mathbf{w}$ as a linear combination of these. Sketch also the vector $\mathbf{x} = \frac{1}{2}\mathbf{v}_1 + \mathbf{v}_2$. Do you think you can reach any point on your piece of paper as a linear combination of $\mathbf{v}_1$ and $\mathbf{v}_2$?

Example 5.14 If F is the vector space of functions from $\mathbb{R}$ to $\mathbb{R}$, then the function $f : x \mapsto 2x^2 + 3x + 4$ can be expressed as a linear combination of three simpler functions, $f = 2g + 3h + 4k$, where $g : x \mapsto x^2, h : x \mapsto x$ and $k : x \mapsto 1$.

5.2   Subspaces

5.2.1   Definition of a subspace

Example 5.10 is informative. Arguing as we did there, if V is a vector space and $W \subseteq V$ is non-empty and closed under scalar multiplication and addition, then W too is a vector space (and we do not need to verify that all the other axioms hold). The formal definition of a *subspace* is as follows:

Definition 5.15 (Subspace) A *subspace* W of a vector space V is a non-empty subset of V that is itself a vector space under the same operations of addition and scalar multiplication as V.

The discussion given in Example 5.10 justifies the following important result:

Theorem 5.16 *Suppose V is a vector space. Then a non-empty subset W of V is a subspace if and only if both the following hold:*

- *for all $\mathbf{u}, \mathbf{v} \in W$, $\mathbf{u} + \mathbf{v} \in W$*
  (*that is, W is closed under addition*),
- *for all $\mathbf{v} \in W$ and $\alpha \in \mathbb{R}$, $\alpha \mathbf{v} \in W$*
  (*that is, W is closed under scalar multiplication*).

Activity 5.17 Write out a proof of this theorem, following the discussion in example 5.10.

5.2.2 Examples

Example 5.18 In $\mathbb{R}^2$, the lines $y = 2x$ and $y = 2x + 1$ can be defined as the sets of vectors,

$$S = \left\{ \begin{pmatrix} x \\ y \end{pmatrix} \,\middle|\, y = 2x, x \in \mathbb{R} \right\},$$

$$U = \left\{ \begin{pmatrix} x \\ y \end{pmatrix} \,\middle|\, y = 2x + 1, x \in \mathbb{R} \right\}.$$

Each vector in one of the sets is the position vector of a point on that line. We will show that the set S is a subspace of $\mathbb{R}^2$, and that the set U is not a subspace of $\mathbb{R}^2$.

If $\mathbf{v} = \begin{pmatrix} 1 \\ 2 \end{pmatrix}$ and $\mathbf{p} = \begin{pmatrix} 0 \\ 1 \end{pmatrix}$, these sets can equally well be expressed as

$$S = \{\mathbf{x} \mid \mathbf{x} = t\mathbf{v}, \ t \in \mathbb{R}\}, \qquad U = \{\mathbf{x} \mid \mathbf{x} = \mathbf{p} + t\mathbf{v}, \ t \in \mathbb{R}\}.$$

Activity 5.19 Show that the two descriptions of S describe the same set of vectors.

To show S is a subspace, we need to show that it is non-empty, and we need to show that it is closed under addition and closed under scalar multiplication using *any* vectors in S and *any* scalar in $\mathbb{R}$. We'll use the second set of definitions, so our line is the set of vectors

$$S = \{\mathbf{x} \mid \mathbf{x} = t\mathbf{v}, \ t \in \mathbb{R}\}, \qquad \mathbf{v} = \begin{pmatrix} 1 \\ 2 \end{pmatrix}.$$

The set S is non-empty, since $\mathbf{0} = 0\mathbf{v} \in S$.

Let $\mathbf{u}, \mathbf{w}$ be any vectors in S and let $\alpha \in \mathbb{R}$. Then

$$\mathbf{u} = s \begin{pmatrix} 1 \\ 2 \end{pmatrix} \qquad \mathbf{w} = t \begin{pmatrix} 1 \\ 2 \end{pmatrix} \qquad \text{for some } s, t \in \mathbb{R}.$$

- closure under addition:

$$\mathbf{u} + \mathbf{w} = s \begin{pmatrix} 1 \\ 2 \end{pmatrix} + t \begin{pmatrix} 1 \\ 2 \end{pmatrix} = (s + t) \begin{pmatrix} 1 \\ 2 \end{pmatrix} \in S \ (\text{since } s + t \in \mathbb{R}).$$

- closure under scalar multiplication:

$$\alpha \mathbf{u} = \alpha \left(s \begin{pmatrix} 1 \\ 2 \end{pmatrix} \right) = (\alpha s) \begin{pmatrix} 1 \\ 2 \end{pmatrix} \in S \ (\text{since } \alpha s \in \mathbb{R}).$$

This shows that S is a subspace of $\mathbb{R}^2$. $\qquad\qquad\square$

To show U is not a subspace, any *one* of the three following statements (counterexamples) will suffice:

1. $\mathbf{0} \notin U$.

2. U is not closed under addition:

$$\begin{pmatrix} 0 \\ 1 \end{pmatrix} \in U, \quad \begin{pmatrix} 1 \\ 3 \end{pmatrix} \in U, \quad \text{but} \quad \begin{pmatrix} 0 \\ 1 \end{pmatrix} + \begin{pmatrix} 1 \\ 3 \end{pmatrix} = \begin{pmatrix} 1 \\ 4 \end{pmatrix} \notin U,$$

    since $4 \neq 2(1) + 1$.

3. U is not closed under scalar multiplication:

$$\begin{pmatrix} 0 \\ 1 \end{pmatrix} \in U, \quad 2 \in \mathbb{R}, \quad \text{but} \quad 2 \begin{pmatrix} 0 \\ 1 \end{pmatrix} = \begin{pmatrix} 0 \\ 2 \end{pmatrix} \notin U.$$

□

Activity 5.20 Show that $\mathbf{0} \notin U$. Why does this suffice to show that U is not a subspace?

The line $y = 2x + 1$ is an example of an *affine subset*, a 'translation' of a subspace.

It is useful to visualise what is happening here by looking at the graphs of the lines $y = 2x$ and $y = 2x + 1$. Sketch $y = 2x$ and sketch the position vector of any point on the line. You will find that the vector lies along the line, so any scalar multiple of that position vector will also lie along the line, as will the sum of any two such position vectors. These position vectors are all still in the set S. Now sketch the line $y = 2x + 1$. First, notice that it does not contain the origin. Now sketch the position vector of any point on the line. You will find that the position vector does *not* lie along the line, but goes from the origin up to the point on the line. If you scalar multiply this vector by any constant $\alpha \neq 1$, it will be the position vector of a point which is not on the line, so the resulting vector will not be in U. The same is true if you add together the position vectors of two points on the line. So U is not a subspace.

Activity 5.21 Do these two sketches as described above.

If V is any vector space and $\mathbf{v} \in V$, then the set

$$S = \{\alpha \mathbf{v} \mid \alpha \in \mathbb{R}\}$$

is a subspace of V. If $\mathbf{v} \neq \mathbf{0}$, then the set S defines a *line* through the origin in V.

Activity 5.22 Show this. Let $\mathbf{v}$ be any non-zero vector in a vector space V and show that the set

$$S = \{\alpha \mathbf{v} \mid \alpha \in \mathbb{R}\}$$

is a subspace.

Example 5.23 If V is a vector space, then V is a subspace of V.

Example 5.24 If V is a vector space, then the set $\{0\}$ is a subspace of V. The set $\{0\}$ is *not empty*, it contains one vector, namely the zero vector. It is a subspace because $0 + 0 = 0$ and $\alpha 0 = 0$ for any $\alpha \in \mathbb{R}$.

5.2.3 Deciding if a subset is a subspace

Given any subset S of a vector space V, how do you decide if it is a subspace? First, look carefully at the definition of S: what is the requirement for a vector in V to be in the subset S? Check that $0 \in S$. If $0 \notin S$, then you know immediately that S is not a subspace.

If $0 \in S$, then using some vectors in the subset, see if adding them and scalar multiplying them will give you another vector in S.

To *prove* that S *is a subspace*, you will need to verify that it is closed under addition and closed under scalar multiplication for *any* vectors in S. (To represent a general vector in $\mathbb{R}^n$, you will need to use letters to represent the vector and possibly its components.) You will need to show that the sum of two general vectors and the multiple of a general vector by any scalar, say $\alpha \in \mathbb{R}$, also satisfy the definition of S.

To *prove* a set S *is not a subspace*, you only need to find *one counterexample*: either two particular vectors for which the sum does not satisfy the definition of S, or a vector for which some scalar multiple does not satisfy the definition of S. (For a particular vector in $\mathbb{R}^n$, use numbers.)

Activity 5.25 Write down a general vector (using letters) and a particular vector (using numbers) for each of the following subsets. Show that one of the sets is a subspace of $\mathbb{R}^3$ and the other is not:

$$S_1 = \left\{ \begin{pmatrix} x \\ x^2 \\ 0 \end{pmatrix} \,\middle|\, x \in \mathbb{R} \right\}, \qquad S_2 = \left\{ \begin{pmatrix} x \\ 2x \\ 0 \end{pmatrix} \,\middle|\, x \in \mathbb{R} \right\}.$$

There is an alternative characterisation of a subspace. We have seen that a subspace is a non-empty subset W of a vector space that is closed under addition and scalar multiplication, meaning that if $\mathbf{u}, \mathbf{v} \in W$ and $\alpha \in \mathbb{R}$, then both $\mathbf{u} + \mathbf{v}$ and $\alpha \mathbf{v}$ are in W. Now, it is fairly easy to see that the following equivalent property characterises when W will be a subspace:

Theorem 5.26 *A non-empty subset W of a vector space is a subspace if and only if for all $\mathbf{u}, \mathbf{v} \in W$ and all $\alpha, \beta \in \mathbb{R}$, we have $\alpha \mathbf{u} + \beta \mathbf{v} \in W$.*

That is, W is a subspace if it is non-empty and *closed under linear combination*.

Activity 5.27 If it is not already obvious to you, show that the property given above is equivalent to closure under addition and scalar multiplication.

To summarise, here is how you would prove that a subset W of a vector space is a subspace:

- Prove that W is non-empty. Usually the easiest way is to show that $\mathbf{0} \in W$.
- Prove that W is closed under addition: if $\mathbf{u}, \mathbf{v}$ are *any* vectors in W, then $\mathbf{u} + \mathbf{v} \in W$.
- Prove that W is closed under scalar multiplication: if $\mathbf{v}$ is *any* vector in W and α is *any* real number, then $\alpha \mathbf{v} \in W$.

Alternatively, by Theorem 5.26, you could do the following:

- Prove that W is non-empty. Usually the easiest way is to show that $\mathbf{0} \in W$.
- Prove that W is closed under linear combinations: if $\mathbf{u}, \mathbf{v}$ are *any* vectors in W, and α, β are *any* real numbers, then $\alpha \mathbf{u} + \beta \mathbf{v} \in W$.

In doing either of these, your arguments have to be *general*: you need $\mathbf{u}, \mathbf{v}, \alpha$ (and β, in the second approach) to be *arbitrary*. Simply showing these statements for some particular vectors or numbers is not enough. On the other hand, if you want to show that a set is *not* a subspace, then as we've noted above, it suffices to show how some of these properties fail for *particular* choices of vectors or scalars.

5.2.4 Null space and range of a matrix

Suppose that A is an $m \times n$ matrix. Then the null space $N(A)$, the set of solutions to the homogeneous linear system $A\mathbf{x} = \mathbf{0}$, is a subspace of $\mathbb{R}^n$.

Theorem 5.28 *For any $m \times n$ matrix A, $N(A)$ is a subspace of $\mathbb{R}^n$.*

Proof: Since A is $m \times n$, the set $N(A)$ is a subset of $\mathbb{R}^n$. To prove it is a subspace, we have to verify that $N(A) \neq \emptyset$, and that if $\mathbf{u}, \mathbf{v} \in N(A)$ and $\alpha \in \mathbb{R}$, then $\mathbf{u} + \mathbf{v} \in N(A)$ and $\alpha \mathbf{u} \in N(A)$. Since $A\mathbf{0} = \mathbf{0}$, $\mathbf{0} \in N(A)$ and hence $N(A) \neq \emptyset$. Suppose $\mathbf{u}, \mathbf{v} \in N(A)$. Then, to show $\mathbf{u} + \mathbf{v} \in N(A)$ and $\alpha \mathbf{u} \in N(A)$, we must show that $\mathbf{u} + \mathbf{v}$ and $\alpha \mathbf{u}$ are solutions of $A\mathbf{x} = \mathbf{0}$. We have

$$A(\mathbf{u} + \mathbf{v}) = A\mathbf{u} + A\mathbf{v} = \mathbf{0} + \mathbf{0} = \mathbf{0}$$

and

$$A(\alpha\mathbf{u}) = \alpha(A\mathbf{u}) = \alpha\mathbf{0} = \mathbf{0},$$

so we have shown what was needed. $\square$

The null space is the set of solutions to the *homogeneous* linear system. If we instead consider the set of solutions S to a general system $A\mathbf{x} = \mathbf{b}$, S is *not* a subspace of $\mathbb{R}^n$ if $\mathbf{b} \neq \mathbf{0}$ (that is, if the system is not homogeneous). This is because $\mathbf{0}$ does not belong to S. However, as we saw in Chapter 2 (Theorem 2.29), there is a relationship between S and $N(A)$: if $\mathbf{x}_0$ is any solution of $A\mathbf{x} = \mathbf{b}$, then

$$S = \{\mathbf{x}_0 + \mathbf{z} \mid \mathbf{z} \in N(A)\},$$

which we may write as $\mathbf{x}_0 + N(A)$. S is an affine subset, a translation of the subspace $N(A)$.

Definition 5.29 (Affine subset) If W is a subspace of a vector space V and $\mathbf{x} \in V$, then the set $\mathbf{x} + W$ defined by

$$\mathbf{x} + W = \{\mathbf{x} + \mathbf{w} \mid \mathbf{w} \in W\}$$

is said to be an *affine subset* of V.

In general, an affine subset is not a subspace, although every subspace is an affine subset, as we can see by taking $\mathbf{x} = \mathbf{0}$.

Recall that the *range* of an $m \times n$ matrix is

$$R(A) = \{A\mathbf{x} \mid \mathbf{x} \in \mathbb{R}^n\}.$$

Theorem 5.30 *For any $m \times n$ matrix A, $R(A)$ is a subspace of $\mathbb{R}^m$.*

Proof: Since A is $m \times n$, the set $R(A)$ consists of $m \times 1$ vectors, so it is a subset of $\mathbb{R}^m$. It is non-empty since $A\mathbf{0} = \mathbf{0} \in R(A)$. We need to show that if $\mathbf{u}, \mathbf{v} \in R(A)$, then $\mathbf{u} + \mathbf{v} \in R(A)$, and for any $\alpha \in \mathbb{R}$, $\alpha\mathbf{v} \in R(A)$. So suppose $\mathbf{u}, \mathbf{v} \in R(A)$. Then for some $\mathbf{y}_1, \mathbf{y}_2 \in \mathbb{R}^n$, $\mathbf{u} = A\mathbf{y}_1$, $\mathbf{v} = A\mathbf{y}_2$. We need to show that $\mathbf{u} + \mathbf{v} = A\mathbf{y}$ for some $\mathbf{y}$. Well,

$$\mathbf{u} + \mathbf{v} = A\mathbf{y}_1 + A\mathbf{y}_2 = A(\mathbf{y}_1 + \mathbf{y}_2),$$

so we may take $\mathbf{y} = \mathbf{y}_1 + \mathbf{y}_2$ to see that, indeed, $\mathbf{u} + \mathbf{v} \in R(A)$. Next,

$$\alpha\mathbf{v} = \alpha(A\mathbf{y}_1) = A(\alpha\mathbf{y}_1),$$

so $\alpha\mathbf{v} = A\mathbf{y}$ for some $\mathbf{y}$ (namely, $\mathbf{y} = \alpha\mathbf{y}_1$) and hence $\alpha\mathbf{v} \in R(A)$. Therefore, $R(A)$ is a subspace of $\mathbb{R}^m$. $\square$

5.3 Linear span

Recall that by a *linear combination* of vectors $v_1, v_2, \ldots, v_k$ we mean a vector of the form

$$v = \alpha_1 v_1 + \alpha_2 v_2 + \cdots + \alpha_k v_k,$$

for some constants $\alpha_i \in \mathbb{R}$. If we add together two vectors of this form, we get another linear combination of the vectors $v_1, v_2, \ldots, v_k$. The same is true of any scalar multiple of v.

Activity 5.31 Show this; show that if $v = \alpha_1 v_1 + \alpha_2 v_2 + \cdots + \alpha_k v_k$ and $w = \beta_1 v_1 + \beta_2 v_2 + \cdots + \beta_k v_k$, then $v + w$ and $sv, s \in \mathbb{R}$, are also linear combinations of the vectors $v_1, v_2, \ldots, v_k$.

The set of all linear combinations of a given set of vectors of a vector space V forms a subspace, and we give it a special name.

Definition 5.32 (Linear span) Suppose that V is a vector space and that $v_1, v_2, \ldots, v_k \in V$. The *linear span* of $X = \{v_1, \ldots, v_k\}$ is the set of all linear combinations of the vectors $v_1, \ldots, v_k$, denoted by $\mathrm{Lin}(X)$ or $\mathrm{Lin}\{v_1, v_2, \ldots, v_k\}$. That is,

$$\mathrm{Lin}\{v_1, v_2, \ldots, v_k\} = \{\alpha_1 v_1 + \cdots + \alpha_k v_k \mid \alpha_1, \alpha_2, \ldots, \alpha_k \in \mathbb{R}\}.$$

Theorem 5.33 *If $X = \{v_1, \ldots, v_k\}$ is a set of vectors of a vector space V, then $\mathrm{Lin}(X)$ is a subspace of V. It is the smallest subspace containing the vectors $v_1, v_2, \ldots, v_k$.*

Proof: The set $\mathrm{Lin}(X)$ is non-empty, since

$$0 = 0v_1 + \cdots + 0v_k \in \mathrm{Lin}(X).$$

If you have carefully carried out Activity 5.31 above, then you have shown that $\mathrm{Lin}(X)$ is closed under addition and scalar multiplication. Therefore, it is a subspace of V. Furthermore, any vector space which contains the vectors $v_1, v_2, \ldots, v_k$ must also contain all linear combinations of these vectors, so it must contain $\mathrm{Lin}(X)$. That is, $\mathrm{Lin}(X)$ is the smallest subspace of V containing $v_1, v_2, \ldots, v_k$. $\square$

The subspace $\mathrm{Lin}(X)$ is also known as the *subspace spanned by* the set $X = \{v_1, \ldots, v_k\}$, or, simply, as the *span* of $\{v_1, v_2, \ldots, v_k\}$. If $V = \mathrm{Lin}(X)$, then we say that the set $\{v_1, v_2, \ldots, v_k\}$ *spans* the vector space V.

If we know that a set of vectors $\{v_1, v_2, \ldots, v_k\}$ spans a vector space V, then we know that any vector $w \in V$ can be expressed in some way as a linear combination of the vectors $v_1, v_2, \ldots, v_k$. This gives us a lot of information about the vector space V.

5.3.1 Row space and column space of a matrix

If A is an $m \times n$ matrix, then the columns of A are vectors in $\mathbb{R}^m$ and the rows of A are row vectors, $1 \times n$ matrices. When written as a column – that is, when transposed – a row gives an $n \times 1$ matrix; that is, a vector in $\mathbb{R}^n$. (Recall, from Chapter 1, that, by a vector, we mean a column vector.) We define the row space of A to be the linear span of its rows, when written as vectors, and the column space to be the linear span of its columns.

Definition 5.34 (Column space) If A is an $m \times n$ matrix, and if $\mathbf{c}_1, \mathbf{c}_2, \ldots, \mathbf{c}_n$ denote the columns of A, then the column space of A, $CS(A)$, is

$$CS(A) = \text{Lin}\{\mathbf{c}_1, \mathbf{c}_2, \ldots, \mathbf{c}_n\}.$$

The column space of an $m \times n$ matrix A is a subspace of $\mathbb{R}^m$.

Definition 5.35 (Row space) If A is an $m \times n$ matrix, and if $\mathbf{r}_1, \mathbf{r}_2, \ldots, \mathbf{r}_n$ denote the rows of A written as vectors, then the row space of A, $RS(A)$, is

$$RS(A) = \text{Lin}\{\mathbf{r}_1, \mathbf{r}_2, \ldots, \mathbf{r}_m\}.$$

The row space of an $m \times n$ matrix A is a subspace of $\mathbb{R}^n$.

We should just add a note of clarification. Our approach to the definition of row space is slightly different from that found in some other texts. It is perfectly valid to think of the set of row vectors, by which we mean $1 \times n$ matrices, as a vector space in an entirely analogous way to $\mathbb{R}^n$, with the corresponding addition and scalar multiplication. This is simply a different 'version' of $\mathbb{R}^n$, populated by row vectors rather than column vectors. Then the row space could be defined as the linear span of the rows of the matrix, and is a subspace of this vector space. However, for our purposes, we prefer not to have to work with two versions of $\mathbb{R}^n$, and nor do we want (as some are content to do) to make no distinction between rows and columns. It is because we want the row space to be a subspace of Euclidean space as we understand it (which entails working with column vectors) that we have defined row space in the way we have.

In the previous chapter, we observed that the range, $R(A)$, of an $m \times n$ matrix A is equal to the set of all linear combinations of its columns. (See Section 4.3.) That is, $R(A)$ is equal to the linear span of the columns of A, so

$$R(A) = CS(A).$$

Therefore, the range of A and the column space of A are precisely the same subspace of $\mathbb{R}^m$, although their original definitions are different.

On the other hand, the row space of A is a subspace of $\mathbb{R}^n$. Although the notations are similar, it is important not to confuse the row space of a matrix, $RS(A)$, with the range of a matrix $R(A) = CS(A)$.

We have seen one more vector space associated with a matrix, namely the null space, $N(A)$, which is also a subspace of $\mathbb{R}^n$.

Recall that two vectors in $\mathbb{R}^n$ are orthogonal if and only if their inner product is equal to 0. The null space of a matrix A and the row space of A are orthogonal subspaces of $\mathbb{R}^n$, meaning that every vector in the null space is orthogonal to every vector in the row space. Why is this true? A vector $\mathbf{x}$ is in $N(A)$ if and only if $A\mathbf{x} = \mathbf{0}$. Look at the ith component of the product $A\mathbf{x}$. This is just the inner product of $\mathbf{r}_i$ with $\mathbf{x}$, where $\mathbf{r}_i$ is the ith row of A written as a vector. But, since $A\mathbf{x} = \mathbf{0}$, it must be true that $\langle \mathbf{r}_i, \mathbf{x} \rangle = 0$ for each i. Since any $\mathbf{r} \in RS(A)$ is some linear combination of the spanning vectors $\mathbf{r}_1, \ldots, \mathbf{r}_m$, the inner product $\langle \mathbf{r}, \mathbf{x} \rangle$ equals zero for any $\mathbf{r} \in RS(A)$ and any $\mathbf{x} \in N(A)$. We restate this important fact:

- If A is an $m \times n$ matrix, then for any $\mathbf{r} \in RS(A)$ and any $\mathbf{x} \in N(A)$, $\langle \mathbf{r}, \mathbf{x} \rangle = 0$; that is, $\mathbf{r}$ and $\mathbf{x}$ are orthogonal.

Activity 5.36 Show that if $\{\mathbf{r}_1, \mathbf{r}_2, \ldots, \mathbf{r}_m\}$ is any set of vectors in $\mathbb{R}^n$, and $\mathbf{x} \in \mathbb{R}^n$ is such that $\langle \mathbf{r}_i, \mathbf{x} \rangle = 0$ for $i = 1, \ldots, m$, then $\langle \mathbf{r}, \mathbf{x} \rangle = 0$ for any linear combination $\mathbf{r} = a_1\mathbf{r}_1 + a_2\mathbf{r}_2 + \cdots + a_m\mathbf{r}_m$.

5.3.2   Lines and planes in $\mathbb{R}^3$

What is the set $\mathrm{Lin}\{\mathbf{v}\}$, the linear span of a single non-zero vector $\mathbf{v} \in \mathbb{R}^n$? Since the set is defined by

$$\mathrm{Lin}\{\mathbf{v}\} = \{\alpha\mathbf{v} \mid \alpha \in \mathbb{R}\},$$

we have already seen that $\mathrm{Lin}\{\mathbf{v}\}$ defines a line through the origin in $\mathbb{R}^n$. In fact, in Activity 5.22 you proved directly that this is a subspace for any vector space, V.

In Chapter 1 (Section 1.11), we saw that a plane in $\mathbb{R}^3$ can be defined either as the set of all vectors $\mathbf{x} = (x, y, z)^\mathrm{T}$ whose components satisfy a single Cartesian equation, $ax + by + cz = d$, or as the set of all vectors $\mathbf{x}$ which satisfy a vector equation with two parameters, $\mathbf{x} = \mathbf{p} + s\mathbf{v} + t\mathbf{w}, s, t \in \mathbb{R}$, where $\mathbf{v}$ and $\mathbf{w}$ are non-parallel vectors and $\mathbf{p}$ is the position vector of a point on the plane. These definitions are equivalent, as it is possible to go from one representation of a given plane to the other.

If $d = 0$, the plane contains the origin; so, taking $\mathbf{p} = \mathbf{0}$, a plane through the origin is the set of vectors

$$\{\mathbf{x} \mid \mathbf{x} = s\mathbf{v} + t\mathbf{w}, \ s, t \in \mathbb{R}\}.$$

Since this is the linear span, $\text{Lin}\{\mathbf{v}, \mathbf{w}\}$, of two vectors in $\mathbb{R}^3$, a plane through the origin is a subspace of $\mathbb{R}^3$.

Let's look at a specific example.

Example 5.37 Let S be the set given by

$$S = \left\{ \begin{pmatrix} x \\ y \\ z \end{pmatrix} \ \middle| \ 3x - 2y + z = 0 \right\}.$$

Then for $\mathbf{x} \in S$,

$$\mathbf{x} = \begin{pmatrix} x \\ y \\ z \end{pmatrix} = \begin{pmatrix} x \\ y \\ 2y - 3x \end{pmatrix} = \begin{pmatrix} x \\ 0 \\ -3x \end{pmatrix} + \begin{pmatrix} 0 \\ y \\ 2y \end{pmatrix}$$

$$= x \begin{pmatrix} 1 \\ 0 \\ -3 \end{pmatrix} + y \begin{pmatrix} 0 \\ 1 \\ 2 \end{pmatrix}.$$

That is, S can be expressed as the set

$$S = \{\mathbf{x} \mid \mathbf{x} = s\mathbf{v}_1 + t\mathbf{v}_2, \ s, t \in \mathbb{R}\},$$

where $\mathbf{v}_1, \mathbf{v}_2$ are the vectors $\mathbf{v}_1 = (1, 0, -3)^\text{T}$, $\mathbf{v}_2 = (0, 1, 2)^\text{T}$. Since S is the linear span of two vectors, it is a subspace of $\mathbb{R}^3$. Of course, you can show directly that S is a subspace by showing it is non-empty, and closed under addition and scalar multiplication.

If $d \neq 0$, then the plane is not a subspace. It is an affine subset, a translation of a subspace.

Activity 5.38 Show this in general, as follows. If a, b, c are real numbers, not all zero, show that the set

$$S = \left\{ \begin{pmatrix} x \\ y \\ z \end{pmatrix} \ \middle| \ ax + by + cz = d \right\}$$

is a subspace if $d = 0$ by showing that S is non-empty and that it is closed under addition and scalar multiplication. Show, however, that if $d \neq 0$, the set S is *not* a subspace.

In the same way as for planes in $\mathbb{R}^3$, any hyperplane in $\mathbb{R}^n$ which contains the origin is a subspace of $\mathbb{R}^n$. You can show this directly, exactly as in the activity above, or you can show it is the linear span of $n - 1$ vectors in $\mathbb{R}^n$.

5.4   Learning outcomes

You should now be able to:

- explain what is meant by a vector space and a subspace
- prove that a given set is a vector space
- decide whether or not a subset of a vector space is a subspace
- prove that a subset is a subspace or show by a counterexample that it is not a subspace
- explain what is meant by the linear span of a set of vectors
- explain what is meant by the column space and the row space of a matrix
- explain why the range of a matrix is equal to the column space of the matrix
- explain why the row space of a matrix is orthogonal to the null space of the matrix.

5.5   Comments on activities

Activity 5.2 For any $\mathbf{x}$,

$$\mathbf{0} = 0\mathbf{x} = (1 + (-1))\mathbf{x} = 1\mathbf{x} + (-1)\mathbf{x} = \mathbf{x} + (-1)\mathbf{x},$$

so adding the negative $-\mathbf{x}$ of $\mathbf{x}$ to each side, and using axioms 3 and 4 of the definition of a vector space,

$$-\mathbf{x} = -\mathbf{x} + \mathbf{0} = -\mathbf{x} + \mathbf{x} + (-1)\mathbf{x} = (-1)\mathbf{x},$$

which proves that $-\mathbf{x} = (-1)\mathbf{x}$.

To show that $\alpha\mathbf{0} = \mathbf{0}$ for any $\alpha \in \mathbb{R}$, let $\mathbf{u}$ denote any vector in V. Then

$$\alpha\mathbf{0} + \alpha\mathbf{u} = \alpha(\mathbf{0} + \mathbf{u}) = \alpha\mathbf{u}.$$

Why? This follows from axioms 7 and 4 of the definition of a vector space. Now add $-\alpha\mathbf{u}$ to both sides of the equation, to obtain the result that $\alpha\mathbf{0} = \mathbf{0}$. Which axioms are you using to deduce this?

Activity 5.6 The axioms are not hard to check. For example, to check axiom 2, let $f, g \in F$. Then the function $f + g$ is given by

$$(f + g)(x) = f(x) + g(x) = g(x) + f(x) = (g + f)(x),$$

since real numbers commute. But this means $f + g = g + f$. We omit the details of the other axioms; they are all straightforward and follow from the properties of real numbers. The negative of a function f is the function $-f$ given by $(-f)(x) = -(f(x))$ for all x.

Activity 5.9 Just write out each sequence in the shorter form, $\mathbf{y} = \{y_n\}$ and use the properties of real numbers.

Activity 5.11 Clearly, $W \neq \emptyset$ since $\mathbf{0} \in W$. Suppose

$$\mathbf{u} = \begin{pmatrix} x \\ y \\ 0 \end{pmatrix}, \quad \mathbf{v} = \begin{pmatrix} x' \\ y' \\ 0 \end{pmatrix} \in W,$$

and that $\alpha \in \mathbb{R}$. Then

$$\mathbf{u} + \mathbf{v} = \begin{pmatrix} x + x' \\ y + y' \\ 0 \end{pmatrix} \in W \quad \text{and} \quad \alpha \mathbf{v} = \begin{pmatrix} \alpha x \\ \alpha y \\ 0 \end{pmatrix} \in W,$$

as required.

Activity 5.13 Do the sketches as instructed. Yes, you can reach any point in $\mathbb{R}^2$ as a linear combination of these vectors. Why? Because you can always solve the system of linear equations resulting from the vector equation $\mathbf{x} = \alpha \mathbf{v}_1 + \beta \mathbf{v}_2$ for α and β (since the determinant of the coefficient matrix is non-zero).

Activity 5.17 Since this is an 'if and only if' statement, you must prove it both ways.

If W is a subspace, then certainly it is closed under addition and scalar multiplication.

Now suppose that W is a non-empty subset of a vector space V, which is closed under the addition and scalar multiplication defined on V, so that axioms 1 and 6 are satisfied for W under these operations. W is non-empty, so there is a vector $\mathbf{v} \in W$. Since W is closed under scalar multiplication, then also $\mathbf{0} = 0\mathbf{v} \in W$ and $(-1)\mathbf{v} = \mathbf{v} \in W$ for any $\mathbf{v} \in W$. The remainder of the axioms are satisfied in W since they are true in V, and any vector in W is also in V (and any linear combination of vectors in W remains in W).

Activity 5.19 This follows from

$$\mathbf{x} = \begin{pmatrix} x \\ y \end{pmatrix} = t \begin{pmatrix} 1 \\ 2 \end{pmatrix} = \begin{pmatrix} t \\ 2t \end{pmatrix} \iff \left. \begin{array}{l} x = t \\ y = 2t \end{array} \right\} \iff y = 2x.$$

Activity 5.20 The vector $\mathbf{0}$ is not in the set U as

$$\mathbf{0} = \begin{pmatrix} 0 \\ 0 \end{pmatrix} \neq \begin{pmatrix} 0 \\ 1 \end{pmatrix} + t \begin{pmatrix} 1 \\ 2 \end{pmatrix} \qquad \text{for any } t \in \mathbb{R},$$

so axiom 4 of the definition of a vector space is not satisfied.

Activity 5.22 Note first that S is non-empty because $\mathbf{0} \in S$. Suppose that $\mathbf{x}, \mathbf{y} \in S$. (Why are we carefully not using the usual symbols $\mathbf{u}$ and $\mathbf{v}$? It is because $\mathbf{v}$ is representing a particular vector and is used in the definition of the set S.) Suppose also that $\beta \in \mathbb{R}$. Now, because $\mathbf{x}$ and $\mathbf{y}$ belong to S, there are $\alpha, \alpha' \in \mathbb{R}$ such that $\mathbf{x} = \alpha\mathbf{v}$ and $\mathbf{y} = \alpha'\mathbf{v}$. Then,

$$\mathbf{x} + \mathbf{y} = \alpha\mathbf{v} + \alpha'\mathbf{v} = (\alpha + \alpha')\mathbf{v},$$

which is in S since it is a scalar multiple of $\mathbf{v}$. Also,

$$\beta\mathbf{x} = \beta(\alpha\mathbf{v}) = (\beta\alpha)\mathbf{v} \in S$$

and it follows that S is a subspace.

Activity 5.25 A general vector in S_1 is of the form $\begin{pmatrix} a \\ a^2 \\ 0 \end{pmatrix}$, $a \in \mathbb{R}$, and one particular vector, taking $x = 1$, is $\begin{pmatrix} 1 \\ 1 \\ 0 \end{pmatrix}$. A general vector in S_2 is of the form $\begin{pmatrix} a \\ 2a \\ 0 \end{pmatrix}$, $a \in \mathbb{R}$, and one particular vector, taking $x = 1$, is $\begin{pmatrix} 1 \\ 2 \\ 0 \end{pmatrix}$.

Each of these subsets contains the zero vector, $\mathbf{0}$.

The set S_1 is not a subspace. To show this, you need to find *one counterexample*, one or two particular vectors in S_1 which do not satisfy the closure properties. For example,

$$\begin{pmatrix} 1 \\ 1 \\ 0 \end{pmatrix} \in S_1 \quad \text{but} \quad \begin{pmatrix} 1 \\ 1 \\ 0 \end{pmatrix} + \begin{pmatrix} 1 \\ 1 \\ 0 \end{pmatrix} = \begin{pmatrix} 2 \\ 2 \\ 0 \end{pmatrix} \notin S_1.$$

The set S_2 is a subspace. You need to show it is closed under addition and scalar multiplication using general vectors. Let $\mathbf{u}, \mathbf{v} \in S_2$, $\alpha \in \mathbb{R}$.

Then

$$\mathbf{u} = \begin{pmatrix} a \\ 2a \\ 0 \end{pmatrix} \quad \text{and} \quad \mathbf{v} = \begin{pmatrix} b \\ 2b \\ 0 \end{pmatrix}, \quad \text{for some } a, b \in \mathbb{R}.$$

Taking the sum and scalar multiple,

$$\mathbf{u} + \mathbf{v} = \begin{pmatrix} a \\ 2a \\ 0 \end{pmatrix} + \begin{pmatrix} b \\ 2b \\ 0 \end{pmatrix} = \begin{pmatrix} a+b \\ 2(a+b) \\ 0 \end{pmatrix} \in S_2 \quad \text{and}$$

$$\alpha\mathbf{u} = \begin{pmatrix} \alpha a \\ 2(\alpha a) \\ 0 \end{pmatrix} \in S_2,$$

which proves that S_2 is a subspace.

Activity 5.27 Let $\mathbf{u}, \mathbf{v} \in W$. If $\alpha\mathbf{u} + \beta\mathbf{v} \in W$ for all $\alpha, \beta \in \mathbb{R}$, then taking, first, $\alpha = \beta = 1$ and, second, $\beta = 0$, we have $\mathbf{u} + \mathbf{v} \in W$ and $\alpha\mathbf{u} \in W$ for all $\alpha \in \mathbb{R}$. Conversely, if W is closed under addition and scalar multiplication, then (by closure under scalar multiplication) $\alpha\mathbf{u} \in W$ and $\beta\mathbf{v} \in W$ for all $\alpha, \beta \in \mathbb{R}$, and it follows, by closure under addition, that $\alpha\mathbf{u} + \beta\mathbf{v} \in W$.

Activity 5.31 Any two such vectors will be of the form

$$\mathbf{v} = \alpha_1\mathbf{v}_1 + \alpha_2\mathbf{v}_2 + \cdots + \alpha_k\mathbf{v}_k$$

and

$$\mathbf{v}' = \alpha_1'\mathbf{v}_1 + \alpha_2'\mathbf{v}_2 + \cdots + \alpha_k'\mathbf{v}_k$$

and we will have

$$\mathbf{v} + \mathbf{v}' = (\alpha_1 + \alpha_1')\mathbf{v}_1 + (\alpha_2 + \alpha_2')\mathbf{v}_2 + \cdots + (\alpha_k + \alpha_k')\mathbf{v}_k,$$

which is a linear combination of the vectors $\mathbf{v}_1, \mathbf{v}_2, \ldots, \mathbf{v}_k$. Also,

$$\begin{aligned} \alpha\mathbf{v} &= \alpha(\alpha_1\mathbf{v}_1 + \alpha_2\mathbf{v}_2 + \cdots + \alpha_k\mathbf{v}_k) \\ &= (\alpha\alpha_1)\mathbf{v}_1 + (\alpha\alpha_2)\mathbf{v}_2 + \cdots + (\alpha\alpha_k)\mathbf{v}_k \end{aligned}$$

is a linear combination of the vectors $\mathbf{v}_1, \mathbf{v}_2, \ldots, \mathbf{v}_k$.

Activity 5.36 Using properties of the inner product of two vectors in $\mathbb{R}^n$,

$$\begin{aligned} \langle \mathbf{r}, \mathbf{x} \rangle &= \langle a_1\mathbf{r}_1 + a_2\mathbf{r}_2 + \cdots + a_m\mathbf{r}_m, \mathbf{x} \rangle \\ &= a_1\langle \mathbf{r}_1, \mathbf{x} \rangle + a_2\langle \mathbf{r}_2, \mathbf{x} \rangle + \cdots + a_m\langle \mathbf{r}_m, \mathbf{x} \rangle. \end{aligned}$$

Since $\langle \mathbf{r}_i, \mathbf{x} \rangle = 0$ for each vector $\mathbf{r}_i$, we can conclude that also $\langle \mathbf{r}, \mathbf{x} \rangle = 0$.

Activity 5.38 It is easy to see that $S \neq \emptyset$ in either case: just list one vector in each set. For example, $(0, 0, 0)^{\mathsf{T}} \in S$ if $d = 0$, and, assuming $a \neq 0$, $(d/a, 0, 0) \in S$ if $d \neq 0$ (or even if $d = 0$).

Suppose $d = 0$. Let $\mathbf{u}, \mathbf{v} \in S$ and $\alpha \in \mathbb{R}$. Then

$$\mathbf{u} = \begin{pmatrix} x \\ y \\ z \end{pmatrix}, \quad \mathbf{v} = \begin{pmatrix} x' \\ y' \\ z' \end{pmatrix},$$

where $ax + by + cz = 0$ and $ax' + by' + cz' = 0$. Consider $\mathbf{u} + \mathbf{v}$. This equals

$$\begin{pmatrix} x + x' \\ y + y' \\ z + z' \end{pmatrix} = \begin{pmatrix} X \\ Y \\ Z \end{pmatrix}$$

and we want to show this belongs to S. Now, this is the case, because

$$\begin{aligned} aX + bY + cZ &= a(x + x') + b(y + y') + c(z + z') \\ &= (ax + by + cz) + (ax' + by' + cz') \\ &= 0 + 0 \\ &= 0, \end{aligned}$$

and, similarly, it can be shown that for any $\alpha \in \mathbb{R}$, $\alpha \mathbf{v} \in S$. So in this case S is closed under addition and scalar multiplication and is therefore a subspace.

If $d \neq 0$, the simple statement that $\mathbf{0}$ does not satisfy the equation means that in this case S is not a subspace.

(However, you can see why closure fails when d is not 0; for then, choosing any two particular vectors for $\mathbf{u}$ and $\mathbf{v}$, if $\mathbf{u} + \mathbf{v} = (X, Y, Z)^{\mathsf{T}}$, then $aX + bY + cZ$ will equal $2d$, which will not be the same as d. So we will not have $\mathbf{u} + \mathbf{v} \in S$. Similarly, we can see that $\alpha \mathbf{v}$ will not be in S if $\alpha \neq 1$.)

5.6   Exercises

Exercise 5.1 Which of the following sets are subspaces of $\mathbb{R}^3$?

$$S_1 = \left\{ \begin{pmatrix} x \\ y \\ z \end{pmatrix} \,\middle|\, z = y = 3x \right\}, \quad S_2 = \left\{ \begin{pmatrix} x \\ y \\ z \end{pmatrix} \,\middle|\, z + y = 3x \right\},$$

$$S_3 = \left\{ \begin{pmatrix} x \\ y \\ z \end{pmatrix} \,\middle|\, zy = 3x \right\}, \quad S_4 = \left\{ \begin{pmatrix} x \\ y \\ z \end{pmatrix} \,\middle|\, xyz = 0 \right\}.$$

Provide proofs or counterexamples to justify your answers.

Exercise 5.2 Suppose A is an $n \times n$ matrix and $\lambda \in \mathbb{R}$ is a fixed constant. Show that the set

$$S = \{\mathbf{x} \mid A\mathbf{x} = \lambda\mathbf{x}\}$$

is a subspace of $\mathbb{R}^n$.

Exercise 5.3 Consider the following vectors

$$\mathbf{v}_1 = \begin{pmatrix} -1 \\ 0 \\ 1 \end{pmatrix}, \quad \mathbf{v}_2 = \begin{pmatrix} 1 \\ 2 \\ 3 \end{pmatrix}, \quad \mathbf{u} = \begin{pmatrix} -1 \\ 2 \\ 5 \end{pmatrix}, \quad \mathbf{w} = \begin{pmatrix} 1 \\ 2 \\ 5 \end{pmatrix}.$$

(a) Show that $\mathbf{u}$ can be expressed as a linear combination of $\mathbf{v}_1$ and $\mathbf{v}_2$, and write down the linear combination; but that $\mathbf{w}$ cannot be expressed as a linear combination of $\mathbf{v}_1$ and $\mathbf{v}_2$.

(b) What subspace of $\mathbb{R}^3$ is given by $\mathrm{Lin}\{\mathbf{v}_1, \mathbf{v}_2, \mathbf{u}\}$? What subspace of $\mathbb{R}^3$ is given by $\mathrm{Lin}\{\mathbf{v}_1, \mathbf{v}_2, \mathbf{w}\}$?

(c) Show that the set $\{\mathbf{v}_1, \mathbf{v}_2, \mathbf{u}, \mathbf{w}\}$ spans $\mathbb{R}^3$. Show also that any vector $\mathbf{b} \in \mathbb{R}^3$ can be expressed as a linear combination of $\mathbf{v}_1, \mathbf{v}_2, \mathbf{u}, \mathbf{w}$ in infinitely many ways.

Exercise 5.4 If $\mathbf{v}, \mathbf{w} \in \mathbb{R}^n$, explain the difference between the sets

$$A = \{\mathbf{v}, \mathbf{w}\} \quad \text{and} \quad B = \mathrm{Lin}\{\mathbf{v}, \mathbf{w}\}.$$

Exercise 5.5 Let F be the vector space of all functions from $\mathbb{R} \to \mathbb{R}$ with pointwise addition and scalar multiplication. Let n be a fixed positive integer and let $\mathcal{P}_n$ be the set of all real polynomial functions of degree at most n; that is, $\mathcal{P}_n$ consists of all functions of the form

$$f(x) = a_0 + a_1 x + a_2 x^2 + \cdots + a_n x^n, \quad \text{where } a_0, a_1, \ldots, a_n \in \mathbb{R}.$$

Prove that $\mathcal{P}_n$ is a subspace of F, under the usual pointwise addition and scalar multiplication for real functions. Find a finite set of functions which spans $\mathcal{P}_n$.

Exercise 5.6 Let U and W be subspaces of a vector space V.

(a) Show that $U \cap W$ is a subspace of V.

(b) Show that $U \cup W$ is not a subspace of V unless $U \subseteq W$ or $W \subseteq U$.

Recall the intersection of the two sets U and W is defined as

$$U \cap W = \{\mathbf{x} : \mathbf{x} \in U \text{ and } \mathbf{x} \in W\}$$

and that the union of the two sets U and W is defined as

$$U \cup W = \{\mathbf{x} : \mathbf{x} \in U \text{ or } \mathbf{x} \in W\}.$$

Give a simple example of sets U and W in $\mathbb{R}^3$ illustrating that $U \cap W$ is a subspace, but for which $U \cup W$ is not.

5.7 Problems

Problem 5.1 Which of the following are subspaces of $\mathbb{R}^3$?

$$S_1 = \left\{ \begin{pmatrix} x \\ y \\ z \end{pmatrix} \,\middle|\, x + y + z = 0 \right\}, \quad S_2 = \left\{ \begin{pmatrix} x \\ y \\ z \end{pmatrix} \,\middle|\, x^2 + y^2 + z^2 = 1 \right\},$$

$$S_3 = \left\{ \begin{pmatrix} x \\ y \\ z \end{pmatrix} \,\middle|\, x = 0 \right\}, \quad S_4 = \left\{ \begin{pmatrix} x \\ y \\ z \end{pmatrix} \,\middle|\, xy = 0 \right\},$$

$$S_5 = \left\{ \begin{pmatrix} x \\ y \\ z \end{pmatrix} \,\middle|\, x = 0 \text{ and } y = 0 \right\} = \left\{ \begin{pmatrix} 0 \\ 0 \\ z \end{pmatrix} \,\middle|\, z \in \mathbb{R} \right\}.$$

Provide proofs or counterexamples to justify your answers. Describe the sets geometrically.

Problem 5.2 Suppose

$$\mathbf{u} = \begin{pmatrix} 2 \\ -1 \\ 1 \end{pmatrix} \quad \text{and} \quad \mathbf{v} = \begin{pmatrix} -1 \\ 1 \\ 3 \end{pmatrix}.$$

Determine which of the vectors below are in $\text{Lin}\{\mathbf{u}, \mathbf{v}\}$, and for each such vector, express it as a linear combination of $\mathbf{u}$ and $\mathbf{v}$:

$$\mathbf{a} = \begin{pmatrix} 3 \\ -2 \\ 4 \end{pmatrix}, \quad \mathbf{b} = \begin{pmatrix} 0 \\ 0 \\ 0 \end{pmatrix}, \quad \mathbf{c} = \begin{pmatrix} 7 \\ -5 \\ -7 \end{pmatrix}.$$

Problem 5.3 Let

$$S_1 = \left\{ \begin{pmatrix} 1 \\ 2 \\ 3 \end{pmatrix}, \begin{pmatrix} 1 \\ 0 \\ -1 \end{pmatrix}, \begin{pmatrix} 0 \\ 1 \\ 1 \end{pmatrix}, \begin{pmatrix} 1 \\ 1 \\ 0 \end{pmatrix} \right\},$$

$$S_2 = \left\{ \begin{pmatrix} 1 \\ 0 \\ -1 \end{pmatrix}, \begin{pmatrix} 2 \\ 1 \\ 3 \end{pmatrix}, \begin{pmatrix} 1 \\ 2 \\ 9 \end{pmatrix} \right\}, \quad S_3 = \left\{ \begin{pmatrix} 1 \\ 1 \\ 1 \end{pmatrix}, \begin{pmatrix} 2 \\ 0 \\ 1 \end{pmatrix} \right\}.$$

Show that the set S_1 spans $\mathbb{R}^3$, but any vector $\mathbf{v} \in \mathbb{R}^3$ can be written as a linear combination of the vectors in S_1 in infinitely many ways. Show that S_2 and S_3 do not span $\mathbb{R}^3$.

Problem 5.4

(a) Solve the following equation to find the coefficients α and β by finding A^{-1}:

$$\begin{pmatrix} 2 \\ -5 \end{pmatrix} = \alpha \begin{pmatrix} 1 \\ 2 \end{pmatrix} + \beta \begin{pmatrix} 1 \\ -1 \end{pmatrix} = \begin{pmatrix} 1 & 1 \\ 2 & -1 \end{pmatrix} \begin{pmatrix} \alpha \\ \beta \end{pmatrix} = A\mathbf{x}.$$

(b) Show that $\mathrm{Lin}\{\mathbf{w}_1, \mathbf{w}_2\} = \mathrm{Lin}\left\{ \begin{pmatrix} 1 \\ 2 \end{pmatrix}, \begin{pmatrix} 1 \\ -1 \end{pmatrix} \right\} = \mathbb{R}^2$. That is, show *any* vector $\mathbf{b} \in \mathbb{R}^2$ can be expressed as a linear combination of $\mathbf{w}_1$ and $\mathbf{w}_2$ by solving $\mathbf{b} = A\mathbf{x}$ for $\mathbf{x}$:

$$\begin{pmatrix} b_1 \\ b_2 \end{pmatrix} = \alpha \begin{pmatrix} 1 \\ 2 \end{pmatrix} + \beta \begin{pmatrix} 1 \\ -1 \end{pmatrix} = \begin{pmatrix} 1 & 1 \\ 2 & -1 \end{pmatrix} \begin{pmatrix} \alpha \\ \beta \end{pmatrix} = A\mathbf{x}.$$

(c) Show, in general, that if $\mathbf{v}$ and $\mathbf{w}$ are non-zero vectors in $\mathbb{R}^2$, with $\mathbf{v} = (a, c)^{\mathrm{T}}$ and $\mathbf{w} = (b, d)^{\mathrm{T}}$, then

$$\mathrm{Lin}\{\mathbf{v}, \mathbf{w}\} = \mathbb{R}^2 \iff \mathbf{v} \neq t\mathbf{w} \text{ for any } t \in \mathbb{R} \iff \begin{vmatrix} a & b \\ c & d \end{vmatrix} \neq 0.$$

Problem 5.5 Let F be the vector space of all functions from $\mathbb{R} \to \mathbb{R}$ with pointwise addition and scalar multiplication. (See Example 5.5.)

(a) Which of the following sets are subspaces of F?

$$S_1 = \{f \in F \mid f(0) = 1\}, \qquad S_2 = \{f \in F \mid f(1) = 0\}.$$

(b) (For readers who have studied calculus) Show that the set

$$S_3 = \{f \in F \mid f \text{ is differentiable and } f' - f = 0\}$$

is a subspace of F.

Problem 5.6 Let $M_2(\mathbb{R})$ denote the set of all 2×2 matrices with real entries. Show that $M_2(\mathbb{R})$ is a vector space under the usual matrix addition and scalar multiplication.

Which of the following subsets are subspaces of $M_2(\mathbb{R})$?

$$W_1 = \left\{ \begin{pmatrix} a & 0 \\ 0 & b \end{pmatrix} \mid a, b \in \mathbb{R} \right\}, \qquad W_2 = \left\{ \begin{pmatrix} a & 1 \\ 1 & b \end{pmatrix} \mid a, b \in \mathbb{R} \right\},$$

$$W_3 = \left\{ \begin{pmatrix} a^2 & 0 \\ 0 & b^2 \end{pmatrix} \mid a, b \in \mathbb{R} \right\}.$$

Justify your answers.

Problem 5.7 If V is a vector space, show that for all $\alpha \in \mathbb{R}$ and $\mathbf{u} \in V$,

$$\alpha \mathbf{u} = \mathbf{0} \text{ if and only if } \alpha = 0 \text{ or } \mathbf{u} = \mathbf{0}.$$

6

Linear independence, bases and dimension

In this chapter, we look into the structure of vector spaces, developing the concept of a basis. This will enable us to understand more about a given vector space, and know precisely what we mean by its dimension. In particular, you should then have a clear understanding of the statement that $\mathbb{R}^n$ is an n-dimensional vector space.

6.1 Linear independence

Linear independence is a central idea in the theory of vector spaces. If $\{v_1, v_2, \ldots, v_k\}$ is a set of vectors in a vector space V, then the vector equation

$$\alpha_1 v_1 + \alpha_2 v_2 + \cdots + \alpha_r v_k = 0$$

always has the trivial solution, $\alpha_1 = \alpha_2 = \cdots = \alpha_k = 0$.

If this is the only possible solution of the vector equation, then we say that the vectors $v_1, v_2, \ldots, v_k$ are *linearly independent*. If there are numbers $\alpha_1, \alpha_2, \ldots, \alpha_k$, *not all zero*, such that

$$\alpha_1 v_1 + \alpha_2 v_2 + \cdots + \alpha_k v_k = 0,$$

then the vectors are not linearly independent; we say they are *linearly dependent*. In this case, the left-hand side is termed a *non-trivial linear combination* of the vectors $\{v_1, v_2, \ldots, v_k\}$.

So the vectors $\{v_1, v_2, \ldots, v_k\}$ are linearly independent if no non-trivial linear combination of them is equal to the zero vector, or, equivalently, if whenever

$$\alpha_1 x_1 + \alpha_2 x_2 + \cdots + \alpha_k x_k = 0,$$

then, necessarily, $\alpha_1 = \alpha_2 = \cdots = \alpha_k = 0$.

We state the formal definitions now.

Definition 6.1 (Linear independence) Let V be a vector space and $\mathbf{v}_1, \ldots, \mathbf{v}_k \in V$. Then $\mathbf{v}_1, \mathbf{v}_2, \ldots, \mathbf{v}_k$ are *linearly independent* (or *form a linearly independent set*) if and only if the vector equation

$$\alpha_1 \mathbf{v}_1 + \alpha_2 \mathbf{v}_2 + \cdots + \alpha_k \mathbf{v}_k = \mathbf{0}$$

has the unique solution

$$\alpha_1 = \alpha_2 = \cdots = \alpha_k = 0;$$

that is, if and only if no non-trivial linear combination of the vectors equals the zero vector.

Definition 6.2 (Linear dependence) Let V be a vector space and $\mathbf{v}_1, \mathbf{v}_2, \ldots, \mathbf{v}_k \in V$. Then $\mathbf{v}_1, \mathbf{v}_2, \ldots, \mathbf{v}_k$ are *linearly dependent* (or *form a linearly dependent set*) if and only if there are real numbers $\alpha_1, \alpha_2, \ldots, \alpha_k$, *not all zero*, such that

$$\alpha_1 \mathbf{v}_1 + \alpha_2 \mathbf{v}_2 + \cdots + \alpha_k \mathbf{v}_k = \mathbf{0};$$

that is, if and only if some non-trivial linear combination of the vectors is equal to the zero vector.

Example 6.3 In $\mathbb{R}^2$, the vectors

$$\mathbf{v} = \begin{pmatrix} 1 \\ 2 \end{pmatrix} \quad \text{and} \quad \mathbf{w} = \begin{pmatrix} 1 \\ -1 \end{pmatrix}$$

are linearly independent. Why? Well, suppose we have a linear combination of these vectors which is equal to the zero vector:

$$\alpha \begin{pmatrix} 1 \\ 2 \end{pmatrix} + \beta \begin{pmatrix} 1 \\ -1 \end{pmatrix} = \begin{pmatrix} 0 \\ 0 \end{pmatrix}.$$

Then this vector equation holds if and only if

$$\begin{cases} \alpha + \beta = 0 \\ 2\alpha - \beta = 0. \end{cases}$$

This homogeneous linear system has only the trivial solution, $\alpha = 0$, $\beta = 0$, so the vectors are linearly independent.

Activity 6.4 Show that the vectors

$$\mathbf{p} = \begin{pmatrix} 1 \\ -2 \end{pmatrix} \quad \text{and} \quad \mathbf{q} = \begin{pmatrix} -2 \\ 4 \end{pmatrix}$$

are linearly dependent by writing down a non-trivial linear combination which is equal to the zero vector.

Example 6.5 In $\mathbb{R}^3$, the following vectors are linearly dependent:

$$\mathbf{v}_1 = \begin{pmatrix} 1 \\ 2 \\ 3 \end{pmatrix}, \quad \mathbf{v}_2 = \begin{pmatrix} 2 \\ 1 \\ 5 \end{pmatrix}, \quad \mathbf{v}_3 = \begin{pmatrix} 4 \\ 5 \\ 11 \end{pmatrix}.$$

This is because

$$2\mathbf{v}_1 + \mathbf{v}_2 - \mathbf{v}_3 = \mathbf{0}.$$

(Check this!).

Note that this can also be written as $\mathbf{v}_3 = 2\mathbf{v}_1 + \mathbf{v}_2$.

This example illustrates the following general result. Try to prove it yourself before looking at the proof.

Theorem 6.6 *The set $\{\mathbf{v}_1, \mathbf{v}_2, \ldots, \mathbf{v}_k\} \subseteq V$ is linearly dependent if and only if at least one vector $\mathbf{v}_i$ is a linear combination of the other vectors.*

Proof: Since this is an 'if and only if' statement, we must prove it both ways. If $\{\mathbf{v}_1, \mathbf{v}_2, \ldots, \mathbf{v}_k\}$ is linearly dependent, the equation

$$\alpha_1 \mathbf{v}_1 + \alpha_2 \mathbf{v}_2 + \cdots + \alpha_r \mathbf{v}_k = \mathbf{0}$$

has a solution with some $\alpha_i \neq 0$. Then we can solve for the vector $\mathbf{v}_i$:

$$\mathbf{v}_i = -\frac{\alpha_1}{\alpha_i}\mathbf{v}_1 - \frac{\alpha_2}{\alpha_i}\mathbf{v}_2 - \cdots - \frac{\alpha_{i-1}}{\alpha_i}\mathbf{v}_{i-1} - \frac{\alpha_{i+1}}{\alpha_i}\mathbf{v}_{i+1} - \cdots - \frac{\alpha_k}{\alpha_i}\mathbf{v}_k,$$

which expresses $\mathbf{v}_i$ as a linear combination of the other vectors in the set.

If $\mathbf{v}_i$ is a linear combination of the other vectors, say

$$\mathbf{v}_i = \beta_1 \mathbf{v}_1 + \cdots + \beta_{i-1}\mathbf{v}_{i-1} + \beta_{i+1}\mathbf{v}_{i+1} + \cdots + \beta_k \mathbf{v}_k,$$

then

$$\beta_1 \mathbf{v}_1 + \cdots + \beta_{i-1}\mathbf{v}_{i-1} - \mathbf{v}_i + \beta_{i+1}\mathbf{v}_{i+1} + \cdots + \beta_k \mathbf{v}_k = \mathbf{0}$$

is a non-trivial linear combination of the vectors that is equal to the zero vector, since the coefficient of $\mathbf{v}_i$ is $-1 \neq 0$. Therefore, the vectors are linearly dependent. $\qquad\qquad\square$

Theorem 6.6 has the following consequence.

Corollary 6.7 *Two vectors are linearly dependent if and only if at least one vector is a scalar multiple of the other.*

Example 6.8 The vectors

$$\mathbf{v}_1 = \begin{pmatrix} 1 \\ 2 \\ 3 \end{pmatrix}, \quad \mathbf{v}_2 = \begin{pmatrix} 2 \\ 1 \\ 5 \end{pmatrix},$$

in the example above are linearly independent, since neither is a scalar multiple of the other.

If V is any vector space, and $\{v_1, v_2, \ldots, v_k\} \subset V$, then the set $\{v_1, v_2, \ldots, v_k, 0\}$ is linearly dependent. Why? Well, we can write

$$0v_1 + 0v_2 + \cdots + 0v_k + a0 = 0,$$

where $a \neq 0$ is any real number (for example, let $a = 1$). This is a non-trivial linear combination equal to the zero vector. Therefore, we have shown the following.

Theorem 6.9 *In a vector space V, a non-empty set of vectors which contains the zero vector is linearly dependent.*

6.1.1 Uniqueness of linear combinations

There is an important property of linearly independent sets of vectors which holds for any vector space V.

Theorem 6.10 *If $v_1, v_2, \ldots, v_m$ are linearly independent vectors in V and if*

$$a_1v_1 + a_2v_2 + \cdots + a_mv_m = b_1v_1 + b_2v_2 + \cdots + b_mv_m,$$

then

$$a_1 = b_1, \quad a_2 = b_2, \quad \ldots, \quad a_m = b_m.$$

Activity 6.11 Prove this. Use the fact that

$$a_1v_1 + a_2v_2 + \cdots + a_mv_m = b_1v_1 + b_2v_2 + \cdots + b_mv_m$$

if and only if

$$(a_1 - b_1)v_1 + (a_2 - b_2)v_2 + \cdots + (a_m - b_m)v_m = 0.$$

What does this theorem say about $x = c_1v_1 + c_2v_2 + \cdots + c_mv_m$? (Pause for a moment and think about this before you continue reading.)

It says that *if* a vector x can be expressed as a linear combination of linearly independent vectors, then this can be done in only one way: the linear combination is *unique*.

6.1.2 Testing for linear independence in $\mathbb{R}^n$

Given k vectors $v_1, \ldots, v_k \in \mathbb{R}^n$, the vector expression

$$\alpha_1v_1 + \alpha_2v_2 + \cdots + \alpha_kv_k$$

equals $A\mathbf{x}$, where A is the $n \times k$ matrix whose columns are the vectors $\mathbf{v}_1, \mathbf{v}_2, \ldots, \mathbf{v}_k$ and $\mathbf{x}$ is the vector, $\mathbf{x} = (\alpha_1, \alpha_2, \ldots, \alpha_k)^{\mathrm{T}}$. (This is by Theorem 1.38.) So the equation

$$\alpha_1 \mathbf{v}_1 + \alpha_2 \mathbf{v}_2 + \cdots + \alpha_k \mathbf{v}_k = \mathbf{0}$$

is equivalent to the matrix equation $A\mathbf{x} = \mathbf{0}$, which is a homogeneous system of n linear equations in k unknowns. Then, the question of whether or not a set of vectors in $\mathbb{R}^n$ is linearly independent can be answered by looking at the solutions of the homogeneous system $A\mathbf{x} = \mathbf{0}$. We state this practical relationship as the following theorem:

Theorem 6.12 *The vectors* $\mathbf{v}_1, \mathbf{v}_2, \ldots, \mathbf{v}_k$ *in* $\mathbb{R}^n$ *are linearly dependent if and only if the linear system* $A\mathbf{x} = \mathbf{0}$ *has a solution other than* $\mathbf{x} = \mathbf{0}$, *where* A *is the matrix* $A = (\mathbf{v}_1 \; \mathbf{v}_2 \; \cdots \; \mathbf{v}_k)$. *Equivalently, the vectors are linearly independent precisely when the only solution to the system is* $\mathbf{x} = \mathbf{0}$.

If the vectors are linearly dependent, then any solution $\mathbf{x} \neq \mathbf{0}$, $\mathbf{x} = (\alpha_1, \alpha_2, \ldots, \alpha_k)^{\mathrm{T}}$ of the system $A\mathbf{x} = \mathbf{0}$ will directly give a non-trivial linear combination of the vectors that equals the zero vector, using the relationship that $A\mathbf{x} = \alpha_1 \mathbf{v}_1 + \alpha \mathbf{v}_2 + \cdots + \alpha_k \mathbf{v}_k$.

Example 6.13 The vectors

$$\mathbf{v}_1 = \begin{pmatrix} 1 \\ 2 \end{pmatrix}, \quad \mathbf{v}_2 = \begin{pmatrix} 1 \\ -1 \end{pmatrix}, \quad \mathbf{v}_3 = \begin{pmatrix} 2 \\ -5 \end{pmatrix}$$

are linearly dependent. To show this, and to find a linear dependence relationship, we solve $A\mathbf{x} = \mathbf{0}$ by reducing the coefficient matrix A to reduced row echelon form:

$$A = \begin{pmatrix} 1 & 1 & 2 \\ 2 & -1 & -5 \end{pmatrix} \to \cdots \to \begin{pmatrix} 1 & 0 & -1 \\ 0 & 1 & 3 \end{pmatrix}.$$

There is one non-leading variable, so the general solution is

$$\mathbf{x} = \begin{pmatrix} t \\ -3t \\ t \end{pmatrix} \quad t \in \mathbb{R}.$$

In particular, taking $t = 1$, and using the relationship

$$A\mathbf{x} = t\mathbf{v}_1 - 3t\mathbf{v}_2 + t\mathbf{v}_3 = \mathbf{0},$$

we have that

$$\begin{pmatrix} 1 \\ 2 \end{pmatrix} - 3 \begin{pmatrix} 1 \\ -1 \end{pmatrix} + \begin{pmatrix} 2 \\ -5 \end{pmatrix} = \begin{pmatrix} 0 \\ 0 \end{pmatrix},$$

which is a non-trivial linear combination equal to the zero vector.

Activity 6.14 This is the method used to find the linear combination given in Example 6.5 for the vectors $\mathbf{v}_1, \mathbf{v}_2, \mathbf{v}_3$,

$$\mathbf{v}_1 = \begin{pmatrix} 1 \\ 2 \\ 3 \end{pmatrix}, \quad \mathbf{v}_2 = \begin{pmatrix} 2 \\ 1 \\ 5 \end{pmatrix}, \quad \mathbf{v}_3 = \begin{pmatrix} 4 \\ 5 \\ 11 \end{pmatrix}.$$

Find the solution of $a_1\mathbf{v}_1 + a_2\mathbf{v}_2 + +a_3\mathbf{v}_3 = \mathbf{0}$ to obtain a linear dependence relation.

Continuing with this line of thought, we know from our experience of solving linear systems with row operations that the system $A\mathbf{x} = \mathbf{0}$ will have precisely the one solution $\mathbf{x} = \mathbf{0}$ if and only if we obtain from the $n \times k$ matrix A an echelon matrix in which there are k leading ones. That is, if and only if $\text{rank}(A) = k$. (Make sure you recall why this is true.) Thus, we have the following result:

Theorem 6.15 *Suppose* $\mathbf{v}_1, \ldots, \mathbf{v}_k \in \mathbb{R}^n$. *The set* $\{\mathbf{v}_1, \ldots, \mathbf{v}_k\}$ *is linearly independent if and only if the* $n \times k$ *matrix* $A = (\mathbf{v}_1 \ \mathbf{v}_2 \ \cdots \ \mathbf{v}_k)$ *has rank* k.

But the rank is always at most the number of rows, so we certainly need to have $k \leq n$. Also, there is a set of n linearly independent vectors in $\mathbb{R}^n$. In fact, there are infinitely many such sets, but an obvious one is

$$\{\mathbf{e}_1, \mathbf{e}_2, \ldots, \mathbf{e}_n\},$$

where $\mathbf{e}_i$ is the vector with every entry equal to 0 except for the ith entry, which is 1. That is,

$$\mathbf{e}_1 = \begin{pmatrix} 1 \\ 0 \\ \vdots \\ 0 \end{pmatrix}, \quad \mathbf{e}_2 = \begin{pmatrix} 0 \\ 1 \\ \vdots \\ 0 \end{pmatrix}, \quad \ldots, \quad \mathbf{e}_n = \begin{pmatrix} 0 \\ 0 \\ \vdots \\ 1 \end{pmatrix}.$$

This set of vectors is known as the *standard basis* of $\mathbb{R}^n$.

Activity 6.16 Show that the set of vectors

$$\{\mathbf{e}_1, \mathbf{e}_2, \ldots, \mathbf{e}_n\},$$

in $\mathbb{R}^n$ is linearly independent.

Therefore, we have established the following result:

Theorem 6.17 *The maximum size of a linearly independent set of vectors in* $\mathbb{R}^n$ *is* n.

So any set of more than n vectors in $\mathbb{R}^n$ is linearly *dependent*. On the other hand, it should not be imagined that any set of n or fewer is linearly independent; that is not true.

Example 6.18 In $\mathbb{R}^4$, which of the following sets of vectors are linearly independent?

$$L_1 = \left\{ \begin{pmatrix} 1 \\ 0 \\ -1 \\ 0 \end{pmatrix}, \begin{pmatrix} 1 \\ 2 \\ 9 \\ 2 \end{pmatrix}, \begin{pmatrix} 2 \\ 1 \\ 3 \\ 1 \end{pmatrix}, \begin{pmatrix} 0 \\ 0 \\ 1 \\ 0 \end{pmatrix}, \begin{pmatrix} 2 \\ 5 \\ 9 \\ 1 \end{pmatrix} \right\},$$

$$L_2 = \left\{ \begin{pmatrix} 1 \\ 0 \\ -1 \\ 0 \end{pmatrix}, \begin{pmatrix} 1 \\ 2 \\ 9 \\ 2 \end{pmatrix} \right\}, \qquad L_3 = \left\{ \begin{pmatrix} 1 \\ 0 \\ -1 \\ 0 \end{pmatrix}, \begin{pmatrix} 1 \\ 2 \\ 9 \\ 2 \end{pmatrix}, \begin{pmatrix} 2 \\ 1 \\ 3 \\ 1 \end{pmatrix} \right\},$$

$$L_4 = \left\{ \begin{pmatrix} 1 \\ 0 \\ -1 \\ 0 \end{pmatrix}, \begin{pmatrix} 1 \\ 2 \\ 9 \\ 2 \end{pmatrix}, \begin{pmatrix} 2 \\ 1 \\ 3 \\ 1 \end{pmatrix}, \begin{pmatrix} 0 \\ 0 \\ 1 \\ 0 \end{pmatrix} \right\}.$$

Try this yourself before reading the answers.

The set L_1 is linearly dependent because it consists of five vectors in $\mathbb{R}^4$. The set L_2 is linearly independent because neither vector is a scalar multiple of the other. To see that the set L_3 is linearly dependent, write the vectors as the columns of a matrix A and reduce A to echelon form to find that the rank of A is 2. This means that there is a non-trivial linear combination of the vectors which is equal to $\mathbf{0}$, or, equivalently, that one of the vectors is a linear combination of the other two. The last set, L_4, contains the set L_3 and is therefore also linearly dependent, since it is still true that one of the vectors is a linear combination of the others.

Activity 6.19 For the set L_3 above, find the solution of the corresponding homogeneous system $A\mathbf{x} = \mathbf{0}$, where A is the matrix whose columns are the vectors of L_3. Use the solution to write down a non-trivial linear combination of the vectors that is equal to the zero vector. Express one of the vectors as a linear combination of the other two.

6.1.3 Linear independence and span

As we have just seen in Example 6.18, if we have a linearly dependent set of vectors, and if we add to the set another vector, then the set is still linearly dependent, because it is still true that at least one of the vectors is a linear combination of the others. This is true whether or not the vector we add is a linear combination of the vectors already in the set.

On the other hand, if we have a linearly independent set of vectors, and if we add another vector to the set, then the new set may or may not be linearly independent, depending on the vector we add to the set. The following is a very useful result which tells us that if we have a linearly independent set of vectors and add to the set a vector which is not a linear combination of those vectors, then the new set is still linearly independent. (Clearly, if we were to add to the set a vector which is a linear combination of the vectors in the set, then the new set would be linearly dependent by Theorem 6.6.)

Theorem 6.20 *If* $S = \{v_1, v_2, \ldots, v_k\}$ *is a linearly independent set of vectors in a vector space* V *and if* $w \in V$ *is not in the linear span of* S, $w \notin \mathrm{Lin}(S)$, *then the set of vectors* $\{v_1, v_2, \ldots, v_k, w\}$ *is linearly independent.*

Proof: To show that the set $\{v_1, v_2, \ldots, v_k, w\}$ is linearly independent, we need to show that the vector equation

$$a_1 v_1 + a_2 v_2 + \cdots + a_k v_k + b w = 0$$

has only the trivial solution. If $b \neq 0$, then we can solve the vector equation for w and hence express w as a linear combination of the vectors in S, which would contradict the assumption that $w \notin \mathrm{Lin}(S)$. Therefore, we must have $b = 0$. But that leaves the expression

$$a_1 v_1 + a_2 v_2 + \cdots + a_k v_k = 0,$$

and since S is linearly independent, all the coefficients a_i must be 0. Hence the set $\{v_1, v_2, \ldots, v_k, w\}$ is linearly independent. $\square$

Now suppose we have a set of vectors $S = \{v_1, v_2, \ldots, v_k\}$ which spans a vector space V, so $V = \mathrm{Lin}(S)$. If the set of vectors is linearly independent, and if we remove a vector, say v_i, from the set, then the smaller set of $k-1$ vectors cannot span V, because v_i (which belongs to V) is not a linear combination of the remaining vectors. On the other hand, if the set is linearly dependent, then some vector, say v_i, is a linear combination of the others; we can safely remove it and the set of $k-1$ vectors will still span V.

6.1.4   Linear independence and span in $\mathbb{R}^n$

If $S = \{\mathbf{v}_1, \mathbf{v}_2, \ldots, \mathbf{v}_k\}$ is a set of vectors in $\mathbb{R}^n$, then we have seen that the questions of whether or not the set S spans $\mathbb{R}^n$ or is linearly independent can be answered by looking at the matrix A whose columns are the vectors $\mathbf{v}_1, \mathbf{v}_2, \ldots, \mathbf{v}_k$.

The set S spans $\mathbb{R}^n$ if we can show that the system of linear equations $A\mathbf{x} = \mathbf{v}$ has a solution for all $\mathbf{v} \in \mathbb{R}^n$; that is, if the system of equations $A\mathbf{x} = \mathbf{v}$ is consistent for all $\mathbf{v} \in \mathbb{R}^n$. We looked at this in Section 4.2 (page 135): if the $n \times k$ matrix A has rank n, then S will span $\mathbb{R}^n$. So we must have $k \geq n$.

In Section 6.1.2 (Theorem 6.12), we saw that the set $S = \{\mathbf{v}_1, \mathbf{v}_2, \ldots, \mathbf{v}_k\}$ is linearly independent if and only if the system of equations $A\mathbf{x} = \mathbf{0}$ has a unique solution, namely the trivial solution, so if and only if the matrix A has rank k. Therefore, we must have $k \leq n$.

So to do both – to span $\mathbb{R}^n$ *and* to be linearly independent – the set S must have precisely n vectors. If we have a set of n vectors $\{\mathbf{v}_1, \mathbf{v}_2, \ldots, \mathbf{v}_n\}$ in $\mathbb{R}^n$, then the matrix A whose columns are the vectors $\mathbf{v}_1, \mathbf{v}_2, \ldots, \mathbf{v}_n$ is a square $n \times n$ matrix. Therefore, to decide if they span $\mathbb{R}^n$ or if they are linearly independent, we only need to evaluate $|A|$.

Example 6.21   The set of vectors $\{\mathbf{v}_1, \mathbf{v}_2, \mathbf{w}\}$, where

$$\mathbf{v}_1 = \begin{pmatrix} 1 \\ 2 \\ 3 \end{pmatrix}, \quad \mathbf{v}_2 = \begin{pmatrix} 2 \\ 1 \\ 5 \end{pmatrix}, \quad \mathbf{w} = \begin{pmatrix} 4 \\ 5 \\ 1 \end{pmatrix}$$

is linearly independent. The vector equation $a_1\mathbf{v}_1 + a_2\mathbf{v}_2 + a_3\mathbf{w} = \mathbf{0}$ has only the trivial solution, since

$$|A| = \begin{vmatrix} 1 & 2 & 4 \\ 2 & 1 & 5 \\ 3 & 5 & 1 \end{vmatrix} = 30 \neq 0.$$

To emphasise how the properties of linear independence and span work together in $\mathbb{R}^n$, we will prove the following result, which shows explicitly that a linearly independent set of n vectors in $\mathbb{R}^n$ also *spans* $\mathbb{R}^n$.

Theorem 6.22   *If $\mathbf{v}_1, \mathbf{v}_2, \ldots, \mathbf{v}_n$ are linearly independent vectors in $\mathbb{R}^n$, then for any $\mathbf{x}$ in $\mathbb{R}^n$, $\mathbf{x}$ can be written as a unique linear combination of $\mathbf{v}_1, \ldots, \mathbf{v}_n$.*

Proof: Because $\mathbf{v}_1, \ldots, \mathbf{v}_n$ are linearly independent, the $n \times n$ matrix

$$A = (\mathbf{v}_1 \ \mathbf{v}_2 \ \ldots \ \mathbf{v}_n)$$

has rank$(A) = n$. (In other words, A reduces to the $n \times n$ identity matrix.) By Theorem 4.5, the system $A\mathbf{z} = \mathbf{x}$ has a unique solution for any $\mathbf{x} \in \mathbb{R}^n$. But let's spell it out. Since there is a leading one in every row of the reduced echelon form of A, we can find a solution to $A\mathbf{z} = \mathbf{x}$, so *any* vector $\mathbf{x}$ can be expressed in the form

$$\mathbf{x} = A\mathbf{z} = (\mathbf{v}_1 \ \mathbf{v}_2 \ \cdots \ \mathbf{v}_n) \begin{pmatrix} \alpha_1 \\ \alpha_2 \\ \vdots \\ \alpha_n \end{pmatrix},$$

where we have written $\mathbf{z}$ as $(\alpha_1, \alpha_2, \ldots, \alpha_n)^{\mathrm{T}}$. Expanding this matrix product, we have that any $\mathbf{x} \in \mathbb{R}^n$ can be expressed as a linear combination

$$\mathbf{x} = \alpha_1 \mathbf{v}_1 + \alpha_2 \mathbf{v}_2 + \cdots + \alpha_n \mathbf{v}_n,$$

as required. This linear combination is unique since the vectors are linearly independent (or, because there is a leading one in every column of the echelon matrix, so there are no free variables). $\square$

It follows from this theorem that if we have a set of n linearly independent vectors in $\mathbb{R}^n$, then the set of vectors also spans $\mathbb{R}^n$. So any vector in $\mathbb{R}^n$ can be expressed *in exactly one way* as a linear combination of the n vectors. We say that the n vectors form a *basis* of $\mathbb{R}^n$. This is the subject of the next section.

6.2 Bases

Consider a set of vectors $\{\mathbf{v}_1, \mathbf{v}_2, \ldots, \mathbf{v}_k\}$ in a vector space V. We have seen two concepts associated with this set. If the set $\{\mathbf{v}_1, \mathbf{v}_2, \ldots, \mathbf{v}_k\}$ spans V, then any vector $\mathbf{x} \in V$ can be expressed as a linear combination of the vectors $\mathbf{v}_1, \mathbf{v}_2, \ldots, \mathbf{v}_k$. If the set $\{\mathbf{v}_1, \mathbf{v}_2, \ldots, \mathbf{v}_k\}$ is linearly independent and if a vector $\mathbf{x} \in V$ *can* be expressed as a linear combination of the vectors $\mathbf{v}_1, \mathbf{v}_2, \ldots, \mathbf{v}_k$, then this expression is unique.

If a set of vectors $\{\mathbf{v}_1, \mathbf{v}_2, \ldots, \mathbf{v}_k\}$ has both properties – if it spans V and it is linearly independent – then *every* vector $\mathbf{v} \in V$ can be expressed as a *unique* linear combination of $\mathbf{v}_1, \mathbf{v}_2, \ldots, \mathbf{v}_k$. This gives us the important concept of a *basis*.

Definition 6.23 (Basis) Let V be a vector space. Then the subset $B = \{\mathbf{v}_1, \mathbf{v}_2, \ldots, \mathbf{v}_n\}$ of V is said to be a *basis* for (or of) V if:

(1) B is a linearly independent set of vectors, *and*
(2) B spans V; that is, $V = \mathrm{Lin}(B)$.

An alternative characterisation of a basis can be given. The set $B = \{\mathbf{v}_1, \mathbf{v}_2, \ldots, \mathbf{v}_n\}$ is a basis of V if every vector in V can be expressed *in exactly one way* as a linear combination of the vectors in B. The set B spans V if and only if each vector in V is a linear combination of the vectors in B; and B is linearly independent if and only if any linear combination of vectors in B is unique. We have therefore shown:

Theorem 6.24 $B = \{\mathbf{v}_1, \mathbf{v}_2, \ldots, \mathbf{v}_n\}$ *is a basis of V if and only if any* $\mathbf{v} \in V$ *is a* unique *linear combination of* $\mathbf{v}_1, \mathbf{v}_2, \ldots, \mathbf{v}_n$.

Example 6.25 The vector space $\mathbb{R}^n$ has the basis $\{\mathbf{e}_1, \mathbf{e}_2, \ldots, \mathbf{e}_n\}$ where $\mathbf{e}_i$ is (as earlier) the vector with every entry equal to 0 except for the ith entry, which is 1. It is clear that the vectors are linearly independent (as you showed in Activity 6.16 on page 177), and it is easy to see that they span the whole of $\mathbb{R}^n$, since for any $\mathbf{x} = (x_1, x_2, \ldots, x_n)^{\mathrm{T}} \in \mathbb{R}^n$, $\mathbf{x} = x_1\mathbf{e}_1 + x_2\mathbf{e}_2 + \cdots + x_n\mathbf{e}_n$. That is,

$$\mathbf{x} = \begin{pmatrix} x_1 \\ x_2 \\ \vdots \\ x_n \end{pmatrix} = x_1 \begin{pmatrix} 1 \\ 0 \\ \vdots \\ 0 \end{pmatrix} + x_2 \begin{pmatrix} 0 \\ 1 \\ \vdots \\ 0 \end{pmatrix} + \cdots + x_n \begin{pmatrix} 0 \\ 0 \\ \vdots \\ 1 \end{pmatrix}.$$

The basis $\{\mathbf{e}_1, \mathbf{e}_2, \ldots, \mathbf{e}_n\}$ is the *standard basis* of $\mathbb{R}^n$.

Example 6.26 We will find a basis of the subspace of $\mathbb{R}^3$ given by

$$W = \left\{ \begin{pmatrix} x \\ y \\ z \end{pmatrix} \;\middle|\; x + y - 3z = 0 \right\}.$$

If $\mathbf{x} = (x, y, z)^{\mathrm{T}}$ is any vector in W, then its components must satisfy $y = -x + 3z$, and we can express $\mathbf{x}$ as

$$\mathbf{x} = \begin{pmatrix} x \\ y \\ z \end{pmatrix} = \begin{pmatrix} x \\ -x + 3z \\ z \end{pmatrix} = x \begin{pmatrix} 1 \\ -1 \\ 0 \end{pmatrix} + z \begin{pmatrix} 0 \\ 3 \\ 1 \end{pmatrix}$$

$$= x\mathbf{v} + z\mathbf{w} \quad (x, z \in \mathbb{R}).$$

This shows that the set $\{\mathbf{v}, \mathbf{w}\}$ spans W. The set is also linearly independent. Why? Because of the positions of the zeros and ones, if $\alpha\mathbf{v} + \beta\mathbf{w} = \mathbf{0}$, then, necessarily, $\alpha = 0$ and $\beta = 0$.

Example 6.27 The set

$$S = \left\{ \begin{pmatrix} 1 \\ 2 \end{pmatrix}, \begin{pmatrix} 1 \\ -1 \end{pmatrix} \right\}$$

is a basis of $\mathbb{R}^2$. We can show this either using Theorem 6.22, or by showing that it spans $\mathbb{R}^2$ and is linearly independent, or, equivalently, that any vector $\mathbf{b} \in \mathbb{R}^2$ is a *unique* linear combination of these two vectors. We will do the latter. Writing the vectors as the columns of a matrix A, we find that $|A| \neq 0$, so this is true by Theorem 4.5.

As in the above example, we can show that n vectors in $\mathbb{R}^n$ are a basis of $\mathbb{R}^n$ by writing them as the columns of a matrix A and invoking Theorem 4.5. Turning this around, we can see that if $A = (\mathbf{v}_1 \; \mathbf{v}_2 \; \ldots \; \mathbf{v}_n)$ is an $n \times n$ matrix with $\text{rank}(A) = n$, then the columns of A are a basis of $\mathbb{R}^n$. Indeed, by Theorem 4.5, the system $A\mathbf{z} = \mathbf{x}$ will have a unique solution for any $\mathbf{x} \in \mathbb{R}^n$, so any vector $\mathbf{x} \in \mathbb{R}^n$ can be written as a unique linear combination of the column vectors. We therefore have two more equivalent statements to add to Theorem 4.5, resulting in the following extended version of that result:

Theorem 6.28 *If A is an $n \times n$ matrix, then the following statements are equivalent:*

- A^{-1} *exists.*
- $A\mathbf{x} = \mathbf{b}$ *has a unique solution for any $\mathbf{b} \in \mathbb{R}^n$.*
- $A\mathbf{x} = \mathbf{0}$ *has only the trivial solution, $\mathbf{x} = \mathbf{0}$.*
- *The reduced echelon form of A is I.*
- $|A| \neq 0$.
- *The rank of A is n.*
- *The column vectors of A are a basis of $\mathbb{R}^n$.*
- *The rows of A (written as vectors) are a basis of $\mathbb{R}^n$.*

The last statement can be seen from the facts that $|A^{\mathrm{T}}| = |A|$ and the rows of A are the columns of A^{T}. This theorem provides an easy way to determine if a set of n vectors is a basis of $\mathbb{R}^n$. We simply write the n vectors as the columns of a matrix and evaluate its determinant.

Example 6.29 The vector space $\text{Lin}\{\mathbf{v}_1, \mathbf{v}_2, \mathbf{w}\}$, where

$$\mathbf{v}_1 = \begin{pmatrix} 1 \\ 2 \\ 3 \end{pmatrix}, \quad \mathbf{v}_2 = \begin{pmatrix} 2 \\ 1 \\ 5 \end{pmatrix}, \quad \mathbf{w} = \begin{pmatrix} 4 \\ 5 \\ 1 \end{pmatrix}$$

is $\mathbb{R}^3$, and the set of vectors $\{\mathbf{v}_1, \mathbf{v}_2, \mathbf{w}\}$ is a basis. Why? In Example 6.21, we showed that the set $\{\mathbf{v}_1, \mathbf{v}_2, \mathbf{w}\}$ is linearly independent, and since it contains three vectors in $\mathbb{R}^3$, it is a basis of $\mathbb{R}^3$. (In fact, we showed that $|A| \neq 0$, where A is the matrix whose column vectors are $\mathbf{v}_1, \mathbf{v}_2, \mathbf{w}$.)

What about the vector space $U = \text{Lin}\{v_1, v_2, v_3\}$, where v_1, v_2, v_3 are the following vectors from Example 6.5?

$$v_1 = \begin{pmatrix} 1 \\ 2 \\ 3 \end{pmatrix}, \quad v_2 = \begin{pmatrix} 2 \\ 1 \\ 5 \end{pmatrix}, \quad v_3 = \begin{pmatrix} 4 \\ 5 \\ 11 \end{pmatrix}.$$

This set of vectors is linearly dependent since $v_3 = 2v_1 + v_2$, so we know that $v_3 \in \text{Lin}\{v_1, v_2\}$. Therefore, $\text{Lin}\{v_1, v_2\} = \text{Lin}\{v_1, v_2, v_3\}$. Furthermore, $\{v_1, v_2\}$ is linearly independent since neither vector is a scalar multiple of the other, so this space is the linear span of two linearly independent vectors in $\mathbb{R}^3$ and is therefore a plane. The set $\{v_1, v_2\}$ is a basis of U. A parametric equation of this plane is given by

$$x = \begin{pmatrix} x \\ y \\ z \end{pmatrix} = sv_1 + tv_2 = s\begin{pmatrix} 1 \\ 2 \\ 3 \end{pmatrix} + t\begin{pmatrix} 2 \\ 1 \\ 5 \end{pmatrix}, \qquad s, t \in \mathbb{R},$$

and we could find a Cartesian equation by eliminating the variables s and t from the component equations. But there is a much simpler way. The vector x belongs to U if and only if x can be expressed as a linear combination of v_1 and v_2, as in the equation above; that is, if and only if x, v_1, v_2 are linearly dependent. This will be the case if and only if we have

$$|A| = \begin{vmatrix} 1 & 2 & x \\ 2 & 1 & y \\ 3 & 5 & z \end{vmatrix} = 0.$$

Expanding this determinant by column 3, we obtain

$$|A| = 7x + y - 3z = 0.$$

This is the equation of the plane.

Activity 6.30　Carry out the calculation of the determinant. Then verify that $7x + y - 3z = 0$ is the equation of the plane by checking that the vectors v_1, v_2, v_3 each satisfy this equation.

Another way to look at a basis is as a smallest spanning set of vectors.

Theorem 6.31　*If V is a vector space, then a smallest spanning set is a basis of V.*

Proof: Suppose we have a set of vectors S and we know that S is a smallest spanning set for V, so $V = \text{Lin}(S)$. If S is linearly independent, then it is a basis. So can S be linearly dependent? If S is linearly dependent, then there is a vector $v \in S$ which is a linear combination of the other vectors in S. But this means that we can remove $\mathbf{v}$ from S, and the remaining smaller set of vectors will still span V since any linear combination of the vectors in S will also be a linear combination of this smaller set. But we assumed thât S was a smallest spanning set, so this is not possible. S must be linearly independent and therefore S is a basis of V. $\qquad\qquad\square$

6.3 Coordinates

What is the importance of a basis? If $S = \{\mathbf{v}_1, \mathbf{v}_2, \ldots, \mathbf{v}_n\}$ is a basis of a vector space V, then any vector $\mathbf{v} \in V$ can be expressed *uniquely* as $\mathbf{v} = \alpha_1 \mathbf{v}_1 + \alpha_2 \mathbf{v}_2 + \cdots + \alpha_n \mathbf{v}_n$. The real numbers $\alpha_1, \alpha_2, \cdots, \alpha_n$ are the *coordinates* of $\mathbf{v}$ with respect to the basis, S.

Definition 6.32 (Coordinates) If $S = \{\mathbf{v}_1, \mathbf{v}_2, \ldots, \mathbf{v}_n\}$ is a basis of a vector space V and $\mathbf{v} = \alpha_1 \mathbf{v}_1 + \alpha_2 \mathbf{v}_2 + \cdots + \alpha_n \mathbf{v}_n$, then the real numbers $\alpha_1, \alpha_2, \cdots, \alpha_n$ are the *coordinates* of $\mathbf{v}$ with respect to the basis, S. We use the notation

$$[\mathbf{v}]_S = \begin{bmatrix} \alpha_1 \\ \alpha_2 \\ \vdots \\ \alpha_n \end{bmatrix}_S$$

to denote the *coordinate vector* of $\mathbf{v}$ in the basis S.

Example 6.33 The sets $B = \{\mathbf{e}_1, \mathbf{e}_2\}$ and $S = \{\mathbf{v}_1, \mathbf{v}_2\}$, where

$$B = \left\{ \begin{pmatrix} 1 \\ 0 \end{pmatrix}, \begin{pmatrix} 0 \\ 1 \end{pmatrix} \right\} \quad \text{and} \quad S = \left\{ \begin{pmatrix} 1 \\ 2 \end{pmatrix}, \begin{pmatrix} 1 \\ -1 \end{pmatrix} \right\}$$

are each a basis of $\mathbb{R}^2$. The coordinates of the vector $\mathbf{v} = (2, -5)^{\mathrm{T}}$ in each basis are given by the coordinate vectors,

$$[\mathbf{v}]_B = \begin{bmatrix} 2 \\ -5 \end{bmatrix}_B \quad \text{and} \quad [\mathbf{v}]_S = \begin{bmatrix} -1 \\ 3 \end{bmatrix}_S .$$

In the standard basis, the coordinates of $\mathbf{v}$ are precisely the components of the vector $\mathbf{v}$, so we just write the standard coordinates as

$$\mathbf{v} = \begin{pmatrix} 2 \\ -5 \end{pmatrix} .$$

In the basis S, the components of $\mathbf{v}$ arise from the observation that

$$\mathbf{v} = -1 \begin{pmatrix} 1 \\ 2 \end{pmatrix} + 3 \begin{pmatrix} 1 \\ -1 \end{pmatrix} = \begin{pmatrix} 2 \\ -5 \end{pmatrix}, \quad \text{so} \quad [\mathbf{v}]_S = \begin{bmatrix} -1 \\ 3 \end{bmatrix}_S .$$

Activity 6.34 For the example above, sketch the vector $\mathbf{v}$ on graph paper and show it as the sum of the vectors given by each of the linear combinations: $\mathbf{v} = 2\mathbf{e}_1 - 5\mathbf{e}_2$ and $\mathbf{v} = -1\mathbf{v}_1 + 3\mathbf{v}_2$.

Activity 6.35 In Example 6.26, we showed that a basis of the plane W,

$$W = \left\{ \begin{pmatrix} x \\ y \\ z \end{pmatrix} \middle| \; x + y - 3z = 0 \right\},$$

is given by the set of vectors $B = \{\mathbf{v}, \mathbf{w}\}$, where $\mathbf{v} = (1, -1, 0)^{\mathrm{T}}$ and $\mathbf{w} = (0, 3, 1)^{\mathrm{T}}$. Show that the vector $\mathbf{y} = (5, 1, 2)^{\mathrm{T}}$ belongs to W and find its coordinates in the basis B.

6.4   Dimension

6.4.1   Definition of dimension

A fundamental fact concerning vector spaces is that if a vector space V has a finite basis, meaning a basis consisting of a finite number of vectors, then all bases of V contain precisely the same number of vectors.

In order to prove this, we first need to establish the following result.

Theorem 6.36 *Let V be a vector space with a basis*

$$B = \{\mathbf{v}_1, \mathbf{v}_2, \ldots, \mathbf{v}_n\}$$

of n vectors. Then any set of $n + 1$ vectors is linearly dependent.

This fact is easily established for $\mathbb{R}^n$, since it is a direct consequence of Theorem 6.17 on page 177. But we will show directly that any set of $n + 1$ vectors in $\mathbb{R}^n$ is linearly dependent, because the proof will indicate to us how to prove the theorem for any vector space V.

If $S = \{\mathbf{w}_1, \mathbf{w}_2, \ldots, \mathbf{w}_{n+1}\} \subset \mathbb{R}^n$, and if A is the $n \times (n + 1)$ matrix whose columns are the vectors $\mathbf{w}_1, \mathbf{w}_2, \ldots, \mathbf{w}_{n+1}$, then the homogeneous system of equations $A\mathbf{x} = \mathbf{0}$ will have infinitely many solutions. Indeed, since the reduced row echelon form of A can have at most n leading ones, there will always be a free variable, and hence infinitely many solutions. Therefore, the set S of $n + 1$ vectors is linearly dependent. Using these ideas, we can now prove the theorem in general for any vector space V with a basis of n vectors.

Proof: Let $S = \{\mathbf{w}_1, \mathbf{w}_2, \ldots, \mathbf{w}_{n+1}\}$ be any set of $n + 1$ vectors in V. Then each of the vectors $\mathbf{w}_i$ can be expressed as a unique linear combination of the vectors in the basis B. Let

$$\mathbf{w}_i = a_{1,i}\mathbf{v}_1 + a_{2,i}\mathbf{v}_2 + \cdots + a_{n,i}\mathbf{v}_n.$$

Now consider any linear combination of the vectors $\mathbf{w}_1, \mathbf{w}_2, \ldots, \mathbf{w}_{n+1}$ such that

$$b_1\mathbf{w}_1 + b_2\mathbf{w}_2 + \cdots + b_{n+1}\mathbf{w}_{n+1} = \mathbf{0}.$$

Substituting for the vectors $\mathbf{w}_i$ in the linear combination, we obtain

$$b_1(a_{1,1}\mathbf{v}_1 + a_{2,1}\mathbf{v}_2 + \cdots + a_{n,1}\mathbf{v}_n) + \cdots$$
$$+ b_{n+1}(a_{1,n+1}\mathbf{v}_1 + a_{2,n+1}\mathbf{v}_2 + \cdots + a_{n,n+1}\mathbf{v}_n) = \mathbf{0}$$

Now comes the tricky bit. We rewrite the linear combination by collecting all the terms which multiply each of the vectors $\mathbf{v}_i$. We have

$$(b_1 a_{1,1} + b_2 a_{1,2} + \cdots + b_{n+1} a_{1,n+1})\mathbf{v}_1 + \cdots$$
$$+ (b_1 a_{n,1} + b_2 a_{n,2} + \cdots + b_{n+1} a_{n,n+1})\mathbf{v}_n = \mathbf{0}.$$

But since the set $B = \{\mathbf{v}_1, \mathbf{v}_2, \ldots, \mathbf{v}_n\}$ is a basis, all the coefficients must be equal to 0. This gives us a homogeneous system of n linear equations in the $n + 1$ unknowns $b_1, b_2, \ldots, b_{n+1}$,

$$\begin{cases} b_1 a_{1,1} + b_2 a_{1,2} + \cdots + b_{n+1} a_{1,n+1} = 0 \\ b_1 a_{2,1} + b_2 a_{2,2} + \cdots + b_{n+1} a_{2,n+1} = 0 \\ \quad\quad\quad\quad\quad\quad\quad\quad\vdots \\ b_1 a_{n,1} + b_2 a_{n,2} + \cdots + b_{n+1} a_{n,n+1} = 0, \end{cases}$$

which must therefore have a non-trivial solution. So there are constants $b_1, b_2, \ldots, b_{n+1}$, not all zero, such that

$$b_1\mathbf{w}_1 + b_2\mathbf{w}_2 + \cdots + b_{n+1}\mathbf{w}_{n+1} = \mathbf{0}.$$

This proves that the set of vectors $S = \{\mathbf{w}_1, \mathbf{w}_2, \ldots, \mathbf{w}_{n+1}\}$ is linearly dependent. □

Using this result, it is now a simple matter to prove the following theorem, which states that all bases of a vector space with a finite basis are the same size; that is, they have the same number of vectors.

Theorem 6.37 *Suppose that a vector space V has a finite basis consisting of r vectors. Then any basis of V consists of exactly r vectors.*

Proof: Suppose V has a basis $B = \{\mathbf{v}_1, \mathbf{v}_2, \ldots, \mathbf{v}_r\}$ consisting of r vectors and a basis $S = \{\mathbf{w}_1, \mathbf{w}_2, \ldots, \mathbf{w}_s\}$ consisting of s vectors. By Theorem 6.36, we must have $s \leq r$ since B is a basis, and so any set of $r + 1$ vectors would be linearly dependent and therefore not a basis. In the same way, since S is a basis, any set of $s + 1$ vectors would be linearly dependent, so $r \leq s$. Therefore, $r = s$.   □

This enables us to define exactly what we mean by the dimension of a vector space V.

Definition 6.38 (Dimension) The number k of vectors in a finite basis of a vector space V is the *dimension* of V, and is denoted $\dim(V)$. The vector space $V = \{\mathbf{0}\}$ is defined to have dimension 0.

A vector space which has a finite basis – that is, a basis consisting of a finite number of vectors – is said to be *finite-dimensional*. Not all vector spaces are finite-dimensional. If a vector space does not have a basis consisting of a finite number of vectors, then it is said to be *infinite-dimensional*.

Example 6.39 We already know $\mathbb{R}^n$ has a basis of size n; for example, the standard basis consists of n vectors. So $\mathbb{R}^n$ has dimension n. (This is reassuring, since it is often referred to as n-dimensional Euclidean space.)

Example 6.40 A plane in $\mathbb{R}^3$ is a two-dimensional subspace. It can be expressed as the linear span of a set of two linearly independent vectors. A line in $\mathbb{R}^n$ is a one-dimensional subspace. A hyperplane in $\mathbb{R}^n$ is an $(n - 1)$-dimensional subspace of $\mathbb{R}^n$.

Example 6.41 The vector space F of real functions with pointwise addition and scalar multiplication (see Example 5.5) has no finite basis. It is an infinite-dimensional vector space. The set S of real-valued sequences (of Example 5.8) is also an infinite-dimensional vector space.

If we know the dimension of a finite-dimensional vector space V, then we know how many vectors we need for a basis. If we have the correct number of vectors for a basis and we know *either* that the vectors span V *or* that they are linearly independent, then we can conclude that both must be true and they form a basis. This is shown in the following theorem. That is, if we know the dimension is k and we have a set of k vectors, then we do not need to show both. We only need to show either that the set is linearly independent or that it spans V.

Theorem 6.42 *Let V be a finite-dimensional vector space of dimension k. Then:*

- *k is the largest size of a linearly independent set of vectors in V. Furthermore, any set of k linearly independent vectors is necessarily a basis of V;*
- *k is the smallest size of a spanning set of vectors for V. Furthermore, any set of k vectors that spans V is necessarily a basis.*

Thus, $k = \dim(V)$ is the largest possible size of a linearly independent set of vectors in V, and the smallest possible size of a spanning set of vectors (a set of vectors whose linear span is V). We have already proved part of this theorem for $\mathbb{R}^n$ as Theorem 6.22.

Proof: If V has dimension k, then every basis of V contains precisely k vectors. Now suppose we have any set $S = \{\mathbf{w}_1, \mathbf{w}_2, \ldots, \mathbf{w}_k\}$ of k linearly independent vectors. If S does not span V, then there must be some vector $\mathbf{v} \in V$ which cannot be expressed as a linear combination of $\mathbf{w}_1, \mathbf{w}_2, \ldots, \mathbf{w}_k$, and if we add this vector $\mathbf{v}$ to the set, then by Theorem 6.20 the set of vectors $\{\mathbf{w}_1, \mathbf{w}_2, \ldots, \mathbf{w}_k, \mathbf{v}\}$ would still be linearly independent. But we have already shown in Theorem 6.36 that k is the maximum size of a linearly independent set of vectors; any set of $k + 1$ vectors is linearly dependent. Therefore, such a vector $\mathbf{v}$ cannot exist. The set S must span V, and so S is a basis of V.

To prove the next part of this theorem, suppose we have any set $S = \{\mathbf{w}_1, \mathbf{w}_2, \ldots, \mathbf{w}_k\}$ of k vectors which spans V. If the set is linearly dependent, then one of the vectors can be expressed as a linear combination of the others. In this case, we could remove it from the set and we would have a set of $k - 1$ vectors which still spans V. This would imply the existence of a basis of V with at most $k - 1$ vectors, since either the new set of $k - 1$ vectors is linearly independent, or we could repeat the process until we arrive at some subset which both spans and is linearly independent. But every basis of V has precisely k vectors, by Theorem 6.37. Therefore, the set S must be linearly independent, and S is a basis of V. This argument also shows that it is not possible for fewer than k vectors to span S. $\qquad\square$

Example 6.43 We know (from Example 6.26) that the plane W in $\mathbb{R}^3$,

$$W = \left\{ \begin{pmatrix} x \\ y \\ z \end{pmatrix} \;\middle|\; x + y - 3z = 0 \right\},$$

has dimension 2, because we found a basis for it consisting of two vectors. If we choose *any* set of two linearly independent vectors

in W, then that set will be a basis of W. For example, the vectors
$\mathbf{v}_1 = (1, 2, 1)^T$ and $\mathbf{v}_2 = (3, 0, 1)^T$ are linearly independent (why?),
so by the Theorem 6.42, $S = \{\mathbf{v}_1, \mathbf{v}_2\}$ is a basis of W.

6.4.2   Dimension and bases of subspaces

Suppose that W is a subspace of the finite-dimensional vector space
V. Any set of linearly independent vectors in W is also a linearly
independent set in V.

Activity 6.44 Prove this last statement.

Now, the dimension of W is the largest size of a linearly independent
set of vectors in W, so there is a set of $\dim(W)$ linearly independent
vectors in V. But then this means that $\dim(W) \le \dim(V)$, since the
largest possible size of a linearly independent set in V is $\dim(V)$. There
is another important relationship between bases of W and V: this is that
any basis of W can be extended to one of V. The following result states
this precisely:

Theorem 6.45 *Suppose that V is a finite-dimensional vector space and
that W is a subspace of V. Then $\dim(W) \le \dim(V)$. Furthermore, if
$\{\mathbf{w}_1, \mathbf{w}_2, \dots, \mathbf{w}_r\}$ is a basis of W, then there are $s = \dim(V) - \dim(W)$
vectors $\mathbf{v}_1, \mathbf{v}_2, \dots, \mathbf{v}_s \in V$ such that $\{\mathbf{w}_1, \mathbf{w}_2, \dots, \mathbf{w}_r, \mathbf{v}_1, \mathbf{v}_2, \dots, \mathbf{v}_s\}$ is
a basis of V. (In the case $W = V$, the basis of W is already a basis
of V.) That is, we can obtain a basis of the whole space V by adding
certain vectors of V to any basis of W.*

Proof: If $\{\mathbf{w}_1, \mathbf{w}_2, \dots, \mathbf{w}_r\}$ is a basis of W, then the set of vectors is a
linearly independent set of vectors in V. If the set spans V, then it is a
basis of V, and $W = V$. If not, there is a vector $\mathbf{v}_1 \in V$, which cannot
be expressed as a linear combination of the vectors $\mathbf{w}_1, \mathbf{w}_2, \dots, \mathbf{w}_r$.
Then the set of vectors $\{\mathbf{w}_1, \mathbf{w}_2, \dots, \mathbf{w}_r, \mathbf{v}_1\}$ is a linearly independent
set of vectors in V by Theorem 6.20. Continuing in this way, we can
find vectors $\mathbf{v}_2, \dots, \mathbf{v}_s \in V$ until the linearly independent set of vectors
$\{\mathbf{w}_1, \mathbf{w}_2, \dots, \mathbf{w}_r, \mathbf{v}_1, \mathbf{v}_2, \dots, \mathbf{v}_s\}$ spans V and is therefore a basis. This
must occur when $r + s = \dim(V)$, so $\dim(W) \le \dim(V)$. $\square$

Example 6.46 The plane W in $\mathbb{R}^3$,

$$W = \{\mathbf{x} \mid x + y - 3z = 0\},$$

has a basis consisting of the vectors $\mathbf{v}_1 = (1, 2, 1)^T$ and $\mathbf{v}_2 = (3, 0, 1)^T$.

Let $\mathbf{v}_3$ be any vector which is *not* in this plane. For example, the vector $\mathbf{v}_3 = (1, 0, 0)^{\mathrm{T}}$ is not in W, since its components do not satisfy the equation. Then the set $S = \{\mathbf{v}_1, \mathbf{v}_2, \mathbf{v}_3\}$ is a basis of $\mathbb{R}^3$. Why?

Activity 6.47 Answer this question. Why can you conclude that S is a basis of $\mathbb{R}^3$?

6.5 Basis and dimension in $\mathbb{R}^n$

6.5.1 Row space, column space and null space

We have met three important subspaces associated with an $m \times n$ matrix A:

- the row space is the linear span of the rows of the matrix (when they are written as vectors) and is a subspace of $\mathbb{R}^n$,
- the null space is the set of all solutions of $A\mathbf{x} = \mathbf{0}$ and is also a subspace of $\mathbb{R}^n$,
- the column space, or range of the matrix, is the linear span of the column vectors of A and is a subspace of $\mathbb{R}^m$.

In Chapter 12, we will meet a fourth subspace associated with A, namely $N(A^{\mathrm{T}})$, but these are, for now, the three main ones.

In order to find a basis for each of these three spaces, we put the matrix A into reduced row echelon form.

To understand how and why this works, we will first work carefully through an example.

Example 6.48 Let A be the 4×5 matrix

$$A = \begin{pmatrix} 1 & 2 & 1 & 1 & 2 \\ 0 & 1 & 2 & 1 & 4 \\ -1 & 3 & 9 & 1 & 9 \\ 0 & 1 & 2 & 0 & 1 \end{pmatrix}.$$

Then the row space of A, $RS(A)$ is the linear span of the transposed rows:

$$RS(A) = \mathrm{Lin}\left\{ \begin{pmatrix} 1 \\ 2 \\ 1 \\ 1 \\ 2 \end{pmatrix}, \begin{pmatrix} 0 \\ 1 \\ 2 \\ 1 \\ 4 \end{pmatrix}, \begin{pmatrix} -1 \\ 3 \\ 9 \\ 1 \\ 9 \end{pmatrix}, \begin{pmatrix} 0 \\ 1 \\ 2 \\ 0 \\ 1 \end{pmatrix} \right\}.$$

The null space, $N(A)$, is the set of all solutions of $A\mathbf{x} = \mathbf{0}$.

Whereas the row space and the null space of A are subspaces of $\mathbb{R}^5$, the column space is a subspace of $\mathbb{R}^4$. Recall that the column space of a matrix, $CS(A)$, is the same as the range of the matrix, $R(A)$. Although the original definitions of the range and the column space are different, we saw in Section 5.3.1 that they are precisely the same set of vectors, namely the linear span of the columns of A:

$$R(A) = CS(A) = \mathrm{Lin}\left\{\begin{pmatrix}1\\0\\-1\\0\end{pmatrix}, \begin{pmatrix}2\\1\\3\\1\end{pmatrix}, \begin{pmatrix}1\\2\\9\\2\end{pmatrix}, \begin{pmatrix}1\\1\\1\\0\end{pmatrix}, \begin{pmatrix}2\\4\\9\\1\end{pmatrix}\right\}.$$

We put the matrix A into reduced row echelon form using elementary row operations. Each one of these row operations involves replacing one row of the matrix with a linear combination of that row and another row: for example, our first step will be to replace row 3 with 'row 3 + row 1'. We will let R denote the matrix which is the reduced row echelon form of A.

$$A \longrightarrow \cdots \longrightarrow R = \begin{pmatrix} 1 & 0 & -3 & 0 & -3 \\ 0 & 1 & 2 & 0 & 1 \\ 0 & 0 & 0 & 1 & 3 \\ 0 & 0 & 0 & 0 & 0 \end{pmatrix}.$$

Activity 6.49 Carry out the row reduction of A to obtain the matrix R.

The row space of the matrix R is the linear span of the rows of R (written as vectors) and it is clear that a basis for this is given by the non-zero rows. Why is this? Each of these rows begins with a leading one, and since the rows below have zeros beneath the leading ones of the rows above, the set

$$\left\{\begin{pmatrix}1\\0\\-3\\0\\-3\end{pmatrix}, \begin{pmatrix}0\\1\\2\\0\\1\end{pmatrix}, \begin{pmatrix}0\\0\\0\\1\\3\end{pmatrix}\right\}$$

is linearly independent and is therefore a basis of $RS(R)$.

Activity 6.50 Validate this argument: use the definition of linear independence to show that this set of vectors is linearly independent.

But $RS(R) = RS(A)$. Why? Each of the rows of R is a linear combination of the rows of A, obtained by performing the row operations. Therefore, $RS(R) \subseteq RS(A)$. But each of these row operations is reversible, so each of the rows of A can be obtained as a linear combination of the rows of R, so $RS(A) \subseteq RS(R)$. Therefore, $RS(A) = RS(R)$ and we have found a basis for the row space of A. It is the set of non-zero rows (written as vectors) in the reduced row echelon form of A. These are the rows with a leading one.

In this example, the row space of A is a three-dimensional subspace of $\mathbb{R}^5$, where $\text{rank}(A) = 3$.

To find a basis of the null space of A, we write down the general solution of $A\mathbf{x} = \mathbf{0}$, with $\mathbf{x} = (x_1, x_2, x_3, x_4, x_5)^{\mathrm{T}}$. Looking at R, we see that the non-leading variables, corresponding to the columns without a leading one, are x_3 and x_5. If we set $x_3 = s$ and $x_5 = t$, where $s, t \in \mathbb{R}$ represent any real numbers, then the solution is

$$\begin{pmatrix} x_1 \\ x_2 \\ x_3 \\ x_4 \\ x_5 \end{pmatrix} = \begin{pmatrix} 3s + 3t \\ -2s - t \\ s \\ -3t \\ t \end{pmatrix} = s \begin{pmatrix} 3 \\ -2 \\ 1 \\ 0 \\ 0 \end{pmatrix} + t \begin{pmatrix} 3 \\ -1 \\ 0 \\ -3 \\ 1 \end{pmatrix}$$

$$= s\mathbf{v}_1 + t\mathbf{v}_2, \qquad s, t \in \mathbb{R}.$$

Activity 6.51 Check this solution. Check the steps indicated and check that $A\mathbf{v}_1 = \mathbf{0}$. $A\mathbf{v}_2 = \mathbf{0}$.

The set of vectors $\{\mathbf{v}_1, \mathbf{v}_2\}$ is a basis of the null space, $N(A)$. Why? They span the null space since every solution of $A\mathbf{x} = \mathbf{0}$ can be expressed as a linear combination of $\mathbf{v}_1, \mathbf{v}_2$, and they are linearly independent because of the positions of the zeros and ones as the third and fifth components; the only linear combination $s\mathbf{v}_1 + t\mathbf{v}_2$ which is equal to the zero vector is given by $s = t = 0$. This is not an accident. The assignment of the arbitrary parameters, s to x_3 and t to x_5, ensures that the vector $\mathbf{v}_1$ will have a '1' as its third component and a '0' as its fifth component, and vice versa for $\mathbf{v}_2$.

In this example, the null space of A is a two-dimensional subspace of $\mathbb{R}^5$. Here $2 = n - r$, where $n = 5$ is the number of columns of A and $r = 3$ is the rank of the matrix. Since A has r leading ones, there are $n - r$ non-leading variables, and each of these determines one of the basis vectors for $N(A)$.

Finding a basis of the column space of A from the reduced row echelon form is not as obvious. The columns of R are very different from the columns of A. In particular, any column with a leading one has zeros elsewhere, so the columns with leading ones in the reduced row echelon form of any matrix are vectors of the standard basis of $\mathbb{R}^m$. In our example, these are $\mathbf{e}_1, \mathbf{e}_2, \mathbf{e}_3 \in \mathbb{R}^4$, which are in columns 1, 2 and 4 of R. But we can use this information to deduce which of the column vectors of A are linearly independent, and by finding a linearly independent subset of the spanning set, we obtain a basis.

Let c_1, c_2, c_3, c_4, c_5 denote the columns of A. Then, if we choose the columns corresponding to the leading ones in the reduced row echelon form of A, namely c_1, c_2, c_4, these three vectors are linearly independent. Why? Because if we row reduce the matrix consisting of these three columns,

$$\begin{pmatrix} 1 & 2 & 1 \\ 0 & 1 & 1 \\ -1 & 3 & 1 \\ 0 & 1 & 0 \end{pmatrix} \rightarrow \cdots \rightarrow \begin{pmatrix} 1 & 0 & 0 \\ 0 & 1 & 0 \\ 0 & 0 & 1 \\ 0 & 0 & 0 \end{pmatrix},$$

the reduced row echelon form will have a leading one in every column. On the other hand, if we include either of the vectors c_3 or c_5, then this will no longer be true, and the set will be linearly dependent.

Activity 6.52 Make sure you understand why this is true.

Since $\{c_1, c_2, c_4\}$ is linearly independent, but $\{c_1, c_2, c_4, c_3\}$ is linearly dependent, we know that c_3 can be expressed as a linear combination of c_1, c_2, c_4. The same applies to c_5, so

$$CS(A) = \text{Lin}\{c_1, c_2, c_3, c_4, c_5\} = \text{Lin}\{c_1, c_2, c_4\}$$

and we have found a basis, namely $\{c_1, c_2, c_4\}$.

So in our example, the range, or column space of A is a three-dimensional subspace of $\mathbb{R}^4$, where $\text{rank}(A) = 3$. Our basis consists of those columns of A which correspond to the columns of the reduced row echelon form with the leading ones.

The same considerations can be applied to any $m \times n$ matrix A, as we shall see in the next section.

If we are given k vectors $v_1, v_2, \ldots, v_k$ in $\mathbb{R}^n$ and we want to find a basis for the linear span $V = \text{Lin}\{v_1, v_2, \ldots, v_k\}$, then we have a choice of how to do this using matrices. The point is that the k vectors might not form a linearly independent set (and hence they are not a basis). One method to obtain a basis for V is to write the spanning set of vectors as the rows of a $k \times n$ matrix and find a basis of the row space. In this case, we will obtain a simplified set of vectors for the basis (in the sense that there will be leading ones and zeros in the vectors), which are linear combinations of the original vectors. Alternatively, we can write the spanning set of vectors as the columns of an $n \times k$ matrix and find a basis of the column space. This will consist of a subset of the original spanning set, namely those vectors in the spanning set which correspond to the columns with the leading ones in the reduced echelon form.

6.5.2 The rank–nullity theorem

We have seen that the range and null space of an $m \times n$ matrix are subspaces of $\mathbb{R}^m$ and $\mathbb{R}^n$, respectively (Section 5.2.4). Their dimensions are so important that they are given special names.

Definition 6.53 (Rank and nullity) The *rank* of a matrix A is

$$\text{rank}(A) = \dim(R(A))$$

and the *nullity* is

$$\text{nullity}(A) = \dim(N(A)).$$

We have, of course, already used the word 'rank', so it had better be the case that the usage just given coincides with the earlier one. Fortunately, it does. If you look again at how we obtained a basis of the column space, or range, in Example 6.48 in the previous section, you will see the correspondence between the basis vectors and the leading ones in the reduced row echelon form. This connection is the content of the following theorem.

Theorem 6.54 *Suppose that A is an $m \times n$ matrix with columns $c_1, c_2, \ldots, c_n$, and that the reduced row echelon form obtained from A has leading ones in columns $i_1, i_2, \ldots, i_r$. Then a basis for $R(A)$ is*

$$B = \{c_{i_1}, c_{i_2}, \ldots, c_{i_r}\}.$$

Note that the basis is formed from columns of A, *not* columns of the echelon matrix: the basis consists of those columns of A *corresponding to* the leading ones in the reduced row echelon form.

Proof: Any solution $x = (\alpha_1, \alpha_2, \ldots, \alpha_n)$ of $Ax = 0$ gives a linear combination of the columns of A which is equal to the zero vector,

$$0 = \alpha_1 c_1 + \alpha_2 c_2 + \cdots + \alpha_n c_n.$$

If R denotes the reduced echelon form of A, and if $c_1', c_2', \ldots, c_n'$ denote the columns of R, then exactly the same relationship holds:

$$0 = \alpha_1 c_1' + \alpha_2 c_2' + \cdots + \alpha_n c_n'.$$

In fact, we use R to obtain the solution $x = (\alpha_1, \alpha_2, \ldots, \alpha_n)$. So the linear dependence relations are the same for the columns of both matrices. This means that the linearly independent columns of A correspond precisely to the linearly independent columns of R. Which columns of R are linearly independent? The columns which contain the leading ones. Why? (Think about this before continuing to read.)

The form of the reduced row echelon matrix R is such that the columns with the leading ones are $\mathbf{c}'_{i_1} = \mathbf{e}_1$, $\mathbf{c}'_{i_2} = \mathbf{e}_2$, $\ldots$, $\mathbf{c}'_{i_r} = \mathbf{e}_r$, where $\mathbf{e}_1, \mathbf{e}_2, \ldots, \mathbf{e}_r$ are the first r vectors of the standard basis of $\mathbb{R}^m$. These vectors are linearly independent. Furthermore, since R has r leading ones, the matrix has precisely r non-zero rows, so any other column vector of R (corresponding to a column without a leading one) is of the form

$$\mathbf{c}'_j = \begin{pmatrix} \alpha_1 \\ \vdots \\ \alpha_r \\ 0 \\ \vdots \\ 0 \end{pmatrix} = \alpha_1 \mathbf{e}_1 + \cdots + \alpha_r \mathbf{e}_r.$$

This gives a linear dependence relationship:

$$\alpha_1 \mathbf{e}_1 + \cdots + \alpha_r \mathbf{e}_r - \mathbf{c}'_j = \mathbf{0}.$$

The same linear dependence relationship holds for the columns of A, so that

$$\alpha_1 \mathbf{c}_{i_1} + \cdots + \alpha_r \mathbf{c}_{i_r} - \mathbf{c}_j = \mathbf{0}.$$

This implies that the set B spans $R(A)$. Since the only linear combination of the vectors $\mathbf{e}_1, \mathbf{e}_2, \ldots, \mathbf{e}_r$ which is equal to the zero vector is the one with all coefficients equal to zero, the same is true for the vectors $\mathbf{c}_{i_1}, \mathbf{c}_{i_2}, \ldots, \mathbf{c}_{i_r}$. Therefore these vectors are linearly independent, and the set B is a basis of $R(A)$.   □

Although the row space and the column space of an $m \times n$ matrix may be subspaces of different Euclidean spaces, $RS(A) \subseteq \mathbb{R}^n$ and $CS(A) \subseteq \mathbb{R}^m$, it turns out that these spaces have the same dimension. Try to see why this might be true by looking again at the example in the previous section, before reading the proof of the following theorem.

Theorem 6.55 *If A is an $m \times n$ matrix, then*

$$\dim(RS(A)) = \dim(CS(A)) = \text{rank}(A).$$

Proof: By Theorem 6.54, the dimension of the column space, or range, is equal to the number of leading ones in the reduced row echelon form of A.

If R denotes the reduced row echelon form of A, then $RS(A) = RS(R)$ and a basis of this space is given by the non-zero rows of R; that is, the rows with the leading ones. The reason this works is that: (i) row operations are such that, at any stage in the procedure, the row space of the reduced matrix is equal to the row space of the original matrix (since the rows of the reduced matrix are linear combinations of the original rows), and (ii) the non-zero rows of an echelon matrix are linearly independent (since each has a one in a position where the rows below it all have a zero).

Therefore, the dimension of the row space is also equal to the number of leading ones in R; that is,

$$\dim(RS(A)) = \dim(CS(A)) = \dim(R(A)) = \text{rank}(A).$$

$\square$

Example 6.56 Let B be the matrix

$$B = \begin{pmatrix} 1 & 1 & 2 & 1 \\ 2 & 0 & 1 & 1 \\ 9 & -1 & 3 & 4 \end{pmatrix}.$$

The reduced row echelon form of the matrix is (verify this!)

$$E = \begin{pmatrix} 1 & 0 & \frac{1}{2} & \frac{1}{2} \\ 0 & 1 & \frac{3}{2} & \frac{1}{2} \\ 0 & 0 & 0 & 0 \end{pmatrix}.$$

The leading ones in this echelon matrix are in the first and second columns, so a basis for $R(B)$ can be obtained by taking the first and second columns of B. (Note: 'columns of B', *not* of the echelon matrix!) Therefore, a basis for $R(B)$ is

$$\left\{ \begin{pmatrix} 1 \\ 2 \\ 9 \end{pmatrix}, \begin{pmatrix} 1 \\ 0 \\ -1 \end{pmatrix} \right\}.$$

A basis of the row space of B consists of the two non-zero rows of the reduced matrix or, alternatively, the first two rows of the original matrix (written as vectors):

$$\left\{ \begin{pmatrix} 1 \\ 0 \\ \frac{1}{2} \\ \frac{1}{2} \end{pmatrix}, \begin{pmatrix} 0 \\ 1 \\ \frac{3}{2} \\ \frac{1}{2} \end{pmatrix} \right\} \quad \text{or} \quad \left\{ \begin{pmatrix} 1 \\ 1 \\ 2 \\ 1 \end{pmatrix}, \begin{pmatrix} 2 \\ 0 \\ 1 \\ 1 \end{pmatrix} \right\}.$$

Note that the column space is a two-dimensional subspace of $\mathbb{R}^3$ (a plane) and the row space is a two-dimensional subspace of $\mathbb{R}^4$. The columns of B and E satisfy the same linear dependence relations, which can be easily read from the reduced echelon form of the matrix,

$$c_3 = \frac{1}{2}c_1 + \frac{3}{2}c_2, \qquad c_4 = \frac{1}{2}c_1 + \frac{1}{2}c_2.$$

Activity 6.57 Check that the columns of B satisfy these same linear dependence relations.

There is a very important relationship between the rank and nullity of a matrix, known as the *rank–nullity theorem* or *dimension theorem* for matrices. This theorem states that

$$\dim(R(A)) + \dim(N(A)) = n,$$

where n is the number of columns of the matrix A.

We have already seen some indications of this result in our considerations of linear systems (Section 4.2). Recall that if an $m \times n$ matrix A has rank r, then the general solution to the system $A\mathbf{x} = \mathbf{0}$ involves $n - r$ 'free parameters'. Specifically, the general solution takes the form

$$\mathbf{x} = s_1\mathbf{v}_1 + s_2\mathbf{v}_2 + \cdots + s_{n-r}\mathbf{v}_{n-r}, \qquad s_i \in \mathbb{R},$$

where $\mathbf{v}_1, \mathbf{v}_2, \ldots, \mathbf{v}_{n-r}$ are themselves solutions of the system $A\mathbf{x} = \mathbf{0}$. But the set of solutions of $A\mathbf{x} = \mathbf{0}$ is precisely the null space $N(A)$. Thus, the null space is spanned by the $n - r$ vectors $\mathbf{v}_1, \mathbf{v}_2, \ldots, \mathbf{v}_{n-r}$, and so its dimension is at most $n - r$. In fact, it turns out that its dimension is precisely $n - r$. That is,

$$\text{nullity}(A) = n - \text{rank}(A).$$

To see this, we need to show that the vectors $\mathbf{v}_1, \mathbf{v}_2, \ldots, \mathbf{v}_{n-r}$ are linearly independent. Because of the way in which these vectors arise (look at the examples we worked through), it will be the case that for each of them there is some position where that vector will have an entry equal to 1 and the entry in that same position for all the other vectors will be 0. From this we can see that no non-trivial linear combination of them can be the zero vector, so they are linearly independent. We have therefore proved the following central result:

Theorem 6.58 (Rank–nullity theorem) *For an $m \times n$ matrix A,*

$$\text{rank}(A) + \text{nullity}(A) = n.$$

Activity 6.59 Find a basis of the null space of the matrix B in Example 6.56 and verify the rank–nullity theorem:

$$B = \begin{pmatrix} 1 & 1 & 2 & 1 \\ 2 & 0 & 1 & 1 \\ 9 & -1 & 3 & 4 \end{pmatrix}.$$

Use the null space basis vectors to obtain the linear dependence relations between the columns of B (which we found earlier using the columns of R).

This is a good time to recall that the row space, $RS(A)$, and the null space, $N(A)$, of an $m \times n$ matrix A are orthogonal subspaces of $\mathbb{R}^n$. As we saw in Section 5.3.1, any vector in one of the spaces is orthogonal to any vector in the other. Therefore, the only vector in both spaces is the zero vector; that is, the intersection $RS(A) \cap N(A) = \{\mathbf{0}\}$.

Activity 6.60 Prove this last statement.

6.6   Learning outcomes

You should now be able to:

- explain what is meant by linear independence and linear dependence
- determine whether a given set of vectors is linearly independent or linearly dependent, and in the latter case, find a non-trivial linear combination of the vectors which equals the zero vector
- explain what is meant by a basis
- find a basis for a linear span
- find a basis for the null space, range and row space of a matrix from its reduced row echelon form
- explain what it means for a vector space to be finite-dimensional and what is meant by the dimension of a finite-dimensional vector space
- explain how rank and nullity are defined, and the relationship between them (the rank–nullity theorem).

6.7   Comments on activities

Activity 6.4 Since $2\mathbf{p} - \mathbf{q} = \mathbf{0}$, the vectors are linearly dependent.

Activity 6.11 As noted,

$$a_1\mathbf{v}_1 + a_2\mathbf{v}_2 + \cdots + a_m\mathbf{v}_m = b_1\mathbf{v}_1 + b_2\mathbf{v}_2 + \cdots + b_m\mathbf{v}_m$$

if and only if

$$(a_1 - b_1)\mathbf{v}_1 + (a_2 - b_2)\mathbf{v}_2 + \cdots + (a_m - b_m)\mathbf{v}_m = \mathbf{0}.$$

But since the vectors are linearly independent, this can be true only if $a_1 - b_1 = 0$, $a_2 - b_2 = 0$, and so on. That is, for each i, we must have $a_i = b_i$.

Activity 6.14 We need to find constants a_i such that

$$a_1 \begin{pmatrix} 1 \\ 2 \\ 3 \end{pmatrix} + a_2 \begin{pmatrix} 2 \\ 1 \\ 5 \end{pmatrix} + a_3 \begin{pmatrix} 4 \\ 5 \\ 11 \end{pmatrix} = \begin{pmatrix} 0 \\ 0 \\ 0 \end{pmatrix}.$$

This is equivalent to the matrix equation $A\mathbf{x} = \mathbf{0}$, where A is the matrix whose columns are these vectors. Row reducing A, we find

$$A = \begin{pmatrix} 1 & 2 & 4 \\ 2 & 1 & 5 \\ 3 & 5 & 11 \end{pmatrix} \rightarrow \begin{pmatrix} 1 & 0 & 2 \\ 0 & 1 & 1 \\ 0 & 0 & 0 \end{pmatrix}.$$

Setting the non-leading variable equal to t, the general solution of $A\mathbf{x} = \mathbf{0}$ is

$$\mathbf{x} = t \begin{pmatrix} -2 \\ -1 \\ 1 \end{pmatrix}, \qquad t \in \mathbb{R}.$$

Taking $t = 1$, then $A\mathbf{x} = -2\mathbf{v}_1 - \mathbf{v}_2 + \mathbf{v}_3 = \mathbf{0}$, which is a linear dependence relation.

Activity 6.16 Looking at the components of the vector equation

$$a_1 \begin{pmatrix} 1 \\ 0 \\ \vdots \\ 0 \end{pmatrix} + a_2 \begin{pmatrix} 0 \\ 1 \\ \vdots \\ 0 \end{pmatrix} + \cdots + a_n \begin{pmatrix} 0 \\ 0 \\ \vdots \\ 1 \end{pmatrix} = \begin{pmatrix} 0 \\ 0 \\ \vdots \\ 0 \end{pmatrix},$$

you can see that the positions of the ones and zeros in the vectors lead to the equations $a_1 = 0$ from the first component, $a_2 = 0$ from the second component and so on, so that $a_i = 0$ $(1 \le i \le n)$ is the only possible solution and the vectors are linearly independent. (Alternatively, the matrix $A = (\mathbf{e}_1, \mathbf{e}_2, \ldots, \mathbf{e}_n)$ is the $n \times n$ identity matrix, so the only solution to $A\mathbf{z} = \mathbf{0}$ is the trivial solution, proving that the vectors are linearly independent.)

Activity 6.19 The general solution to the system is

$$\mathbf{x} = \begin{pmatrix} x \\ y \\ z \end{pmatrix} = t \begin{pmatrix} -3/2 \\ -1/2 \\ 1 \end{pmatrix}, \qquad t \in \mathbb{R}.$$

Taking $t = -1$, for instance, and multiplying out the equation $A\mathbf{x} = \mathbf{0}$, we see that

$$\frac{3}{2} \begin{pmatrix} 1 \\ 0 \\ -1 \\ 0 \end{pmatrix} + \frac{1}{2} \begin{pmatrix} 1 \\ 2 \\ 9 \\ 2 \end{pmatrix} - \begin{pmatrix} 2 \\ 1 \\ 3 \\ 1 \end{pmatrix} = \mathbf{0},$$

and hence

$$
\begin{pmatrix} 2 \\ 1 \\ 3 \\ 1 \end{pmatrix} = \frac{3}{2} \begin{pmatrix} 1 \\ 0 \\ -1 \\ 0 \end{pmatrix} + \frac{1}{2} \begin{pmatrix} 1 \\ 2 \\ 9 \\ 2 \end{pmatrix}.
$$

Activity 6.35 The vector **y** belongs to W since it satisfies the equation of the plane, $(5) + (1) - 3(2) = 0$. For the coordinates in the basis B, you need to solve the vector equation

$$
\begin{pmatrix} 5 \\ 1 \\ 2 \end{pmatrix} = \alpha \begin{pmatrix} 1 \\ -1 \\ 0 \end{pmatrix} + \beta \begin{pmatrix} 0 \\ 3 \\ 1 \end{pmatrix}
$$

for constants α and β. Because of the positions of the zeros and ones in this basis, this can be done by inspection. From row 1 (equating the first components), we must have $\alpha = 5$, and from row 3, we must have $\beta = 2$. Checking the middle row, $1 = 5(-1) + 2(3)$. Therefore,

$$
[\mathbf{y}]_B = \begin{bmatrix} 5 \\ 2 \end{bmatrix}_B.
$$

Activity 6.44 If $S = \{\mathbf{w}_1, \mathbf{w}_2, \ldots, \mathbf{w}_r\}$ is a linearly independent set of vectors in W, then we can state that the only linear combination

$$
a_1 \mathbf{w}_1 + a_2 \mathbf{w}_2 + \cdots + a_r \mathbf{w}_r = \mathbf{0}
$$

is the trivial one, with all $a_i = 0$. But all the vectors in W are also in V, and this statement still holds true, so S is a linearly independent set of vectors in V.

Activity 6.47 The set S is linearly independent since $\mathbf{v}_3 \notin \mathrm{Lin}\{\mathbf{v}_1, \mathbf{v}_2\}$, and it contains precisely $3 = \dim(\mathbb{R}^3)$ vectors.

Activity 6.50 Consider any linear combination of these vectors which is equal to the zero vector,

$$
a_1 \begin{pmatrix} 1 \\ 0 \\ -3 \\ 0 \\ -3 \end{pmatrix} + a_2 \begin{pmatrix} 0 \\ 1 \\ 2 \\ 0 \\ 1 \end{pmatrix} + a_3 \begin{pmatrix} 0 \\ 0 \\ 0 \\ 1 \\ 3 \end{pmatrix} = \begin{pmatrix} 0 \\ 0 \\ 0 \\ 0 \\ 0 \end{pmatrix}.
$$

Clearly, the only solution is the trivial one, $a_1 = a_2 = a_3 = 0$.

This will happen for any set of non-zero rows from the reduced row echelon form of a matrix. Since the vectors arise as the non-zero rows of a matrix in reduced row echelon form, each vector contains a leading one as it first non-zero component, and all the other vectors have zeros in those positions. So in this example the equations relating the first, second and fourth components (the positions of the leading ones) tell us that all coefficients are zero and the vectors are linearly independent.

Activity 6.59 A general solution of the system of equations $Bx = 0$ is

$$\mathbf{x} = s_1 \begin{pmatrix} -\frac{1}{2} \\ -\frac{3}{2} \\ 1 \\ 0 \end{pmatrix} + s_2 \begin{pmatrix} -\frac{1}{2} \\ -\frac{1}{2} \\ 0 \\ 1 \end{pmatrix} = s_1\mathbf{u}_1 + s_2\mathbf{u}_2.$$

The set $\{\mathbf{u}_1, \mathbf{u}_2\}$ is a basis of the null space of B, so $\dim(N(B)) = 2$. From the example, $\text{rank}(B) = 2$. The matrix B has $n = 4$ columns:

$$\text{rank}(B) + \text{nullity}(B) = 2 + 2 = 4 = n.$$

The basis vectors of the null space give the same linear dependence relations between the column vectors as those given in the example. Since $A\mathbf{u}_1 = 0$ and $A\mathbf{u}_2 = 0$,

$$A\mathbf{u}_1 = -\frac{1}{2}\mathbf{c}_1 - \frac{3}{2}\mathbf{c}_2 + \mathbf{c}_3 = 0 \quad \text{and} \quad A\mathbf{u}_2 = -\frac{1}{2}\mathbf{c}_1 - \frac{1}{2}\mathbf{c}_2 + \mathbf{c}_4 = 0.$$

Activity 6.60 Let A be an $m \times n$ matrix. If $\mathbf{x} \in RS(A)$, then $\langle \mathbf{x}, \mathbf{v} \rangle = 0$ for all $\mathbf{v} \in N(A)$, and if $\mathbf{x} \in N(A)$, then $\langle \mathbf{w}, \mathbf{x} \rangle = 0$ for all $\mathbf{w} \in RS(A)$. Therefore, if $\mathbf{x}$ is in both, we must have $\langle \mathbf{x}, \mathbf{x} \rangle = \|\mathbf{x}\|^2 = 0$. But this implies that $\mathbf{x}$ is the zero vector; that is, $RS(A) \cap N(A) = \{\mathbf{0}\}$.

6.8 Exercises

Exercise 6.1 Show that the vectors $\mathbf{x}_1, \mathbf{x}_2, \mathbf{x}_3$ given below are linearly independent:

$$\mathbf{x}_1 = \begin{pmatrix} 2 \\ 1 \\ -1 \end{pmatrix}, \quad \mathbf{x}_2 = \begin{pmatrix} 3 \\ 4 \\ 6 \end{pmatrix}, \quad \mathbf{x}_3 = \begin{pmatrix} -2 \\ 3 \\ 2 \end{pmatrix}.$$

Express the vector

$$\mathbf{v} = \begin{pmatrix} -5 \\ 7 \\ -2 \end{pmatrix}$$

as a linear combination of them.

Exercise 6.2 Let

$$\mathbf{x}_1 = \begin{pmatrix} 2 \\ 3 \\ 5 \end{pmatrix}, \quad \mathbf{x}_2 = \begin{pmatrix} 1 \\ 1 \\ 2 \end{pmatrix}, \quad \mathbf{v} = \begin{pmatrix} a \\ b \\ c \end{pmatrix}.$$

Find a vector $\mathbf{x}_3$ such that $\{\mathbf{x}_1, \mathbf{x}_2, \mathbf{x}_3\}$ is a linearly independent set of vectors.

Find a condition that a, b, c must satisfy for the set of vectors $\{\mathbf{x}_1, \mathbf{x}_2, \mathbf{v}\}$ to be linearly dependent.

Exercise 6.3 Using the definition of linear independence, show that any non-empty subset of a linearly independent set of vectors is linearly independent.

Exercise 6.4 Let $S = \{\mathbf{v}_1, \mathbf{v}_2, \ldots, \mathbf{v}_n\}$ be a set of n vectors in $\mathbb{R}^n$ and let A be the matrix whose columns are the vectors $\mathbf{v}_1, \mathbf{v}_2, \ldots, \mathbf{v}_n$. Explain why the set S is linearly independent if and only if $|A| \neq 0$.

Exercise 6.5 Show that the following vectors are linearly dependent by finding a non-trivial linear combination of the vectors that equals the zero vector.

$$\begin{pmatrix} 1 \\ 2 \\ 1 \\ 2 \end{pmatrix}, \begin{pmatrix} 0 \\ -1 \\ 3 \\ 4 \end{pmatrix}, \begin{pmatrix} 4 \\ -11 \\ 5 \\ -1 \end{pmatrix}, \begin{pmatrix} 9 \\ 2 \\ 1 \\ -3 \end{pmatrix}.$$

Exercise 6.6 Prove that if $n > m$, then any set of n vectors in $\mathbb{R}^m$ is linearly dependent.

Exercise 6.7 Let A be any matrix. Let $\mathbf{v}_1$ and $\mathbf{v}_2$ be two non-zero vectors and suppose that $A\mathbf{v}_1 = 2\mathbf{v}_1$ and $A\mathbf{v}_2 = 5\mathbf{v}_2$. Prove that $\{\mathbf{v}_1, \mathbf{v}_2\}$ is linearly independent. (Hint: Assume $\alpha_1\mathbf{v}_1 + \alpha_2\mathbf{v}_2 = \mathbf{0}$. Multiply this equation through by A to get a second equation for $\mathbf{v}_1$ and $\mathbf{v}_2$. Then solve the two equations simultaneously.)

Can you generalise this result?

Exercise 6.8 Consider the sets

$$U = \left\{ \begin{pmatrix} -1 \\ 0 \\ 1 \end{pmatrix}, \begin{pmatrix} 1 \\ 2 \\ 3 \end{pmatrix}, \begin{pmatrix} -1 \\ 2 \\ 5 \end{pmatrix} \right\}, \quad W = \left\{ \begin{pmatrix} -1 \\ 0 \\ 1 \end{pmatrix}, \begin{pmatrix} 1 \\ 2 \\ 3 \end{pmatrix}, \begin{pmatrix} 1 \\ 2 \\ 5 \end{pmatrix} \right\}.$$

What subspace of $\mathbb{R}^3$ is $\mathrm{Lin}(U)$? $\mathrm{Lin}(W)$? Find a basis for each subspace and show that one of them is a plane in $\mathbb{R}^3$. Find a Cartesian equation for the plane.

Exercise 6.9 Write down a basis for the xz-plane in $\mathbb{R}^3$.

Exercise 6.10 Let B be the set of vectors $B = \{v_1, v_2, v_3\}$, where $v_1 = (1, 1, 0)^T$, $v_2 = (-4, 0, 3)^T$, $v_3 = (3, 5, 1)^T$. Show that B is a basis of $\mathbb{R}^3$.

Let $w = (-1, 7, 5)^T$ and $e_1 = (1, 0, 0)^T$. Find the coordinates of w and e_1 with respect to the basis B.

Exercise 6.11 Let V be a vector space with a basis $B = \{v_1, v_2, \ldots, v_n\}$. Show that for any $u, w \in V$,

$$[\alpha u + \beta w]_B = \alpha [u]_B + \beta [w]_B.$$

Exercise 6.12 Consider the matrix

$$A = \begin{pmatrix} 1 & 2 & -1 & 3 \\ 2 & 3 & 0 & 1 \\ -4 & -5 & -2 & 3 \end{pmatrix}.$$

Find a basis of the row space of A, $RS(A)$, and the column space, $CS(A)$. State why $CS(A)$ is a plane in $\mathbb{R}^3$, and find a Cartesian equation of this plane.

State the rank–nullity theorem (the dimension theorem for matrices), ensuring that you define each term. Use it to determine the dimension of the null space, $N(A)$.

For what real values of a is the vector

$$b(a) = \begin{pmatrix} -1 \\ a \\ a^2 \end{pmatrix}$$

in the range of A? Write down any vectors in $R(A)$ of this form.

Exercise 6.13 A matrix A is said to have *full column rank* if and only if the columns of A are linearly independent. If A is an $m \times k$ matrix with full column rank, show that:

(1) $A^T A$ is a symmetric $k \times k$ matrix,
(2) $A^T A$ is invertible.

Then verify the above results for the matrix $M = \begin{pmatrix} 1 & -2 \\ 3 & 0 \\ 1 & 1 \end{pmatrix}$.

Exercise 6.14 Let B be an $m \times k$ matrix whose row space, $RS(B)$, is a plane in $\mathbb{R}^3$ with Cartesian equation $4x - 5y + 3z = 0$.

From the given information, can you determine either m or k for the matrix B? If it is possible, do so.

Can you determine the null space of B? If so, write down a general solution of $Bx = 0$.

Exercise 6.15 Let S be the vector space of all infinite sequences of real numbers. Let W be the subset which consists of sequences for which all entries beyond the third are zero. Show that W is a subspace of S of dimension 3.

6.9 Problems

Problem 6.1 Determine which of the following sets of vectors are linearly independent.

$$L_1 = \left\{ \begin{pmatrix} 1 \\ 2 \end{pmatrix}, \begin{pmatrix} 1 \\ 3 \end{pmatrix} \right\}, \qquad L_2 = \left\{ \begin{pmatrix} 1 \\ 2 \end{pmatrix}, \begin{pmatrix} 1 \\ 3 \end{pmatrix}, \begin{pmatrix} 4 \\ 5 \end{pmatrix} \right\},$$

$$L_3 = \left\{ \begin{pmatrix} 0 \\ 0 \end{pmatrix}, \begin{pmatrix} 1 \\ 2 \end{pmatrix} \right\}, \qquad L_4 = \left\{ \begin{pmatrix} 1 \\ 2 \\ 0 \end{pmatrix}, \begin{pmatrix} 2 \\ 7 \\ 0 \end{pmatrix}, \begin{pmatrix} 3 \\ 5 \\ 0 \end{pmatrix} \right\},$$

$$L_5 = \left\{ \begin{pmatrix} 1 \\ 2 \\ 0 \end{pmatrix}, \begin{pmatrix} 3 \\ 0 \\ -1 \end{pmatrix}, \begin{pmatrix} 4 \\ 1 \\ 2 \end{pmatrix} \right\}.$$

Problem 6.2 Which of the following sets of vectors in $\mathbb{R}^4$ are linearly independent?

$$S_1 = \left\{ \begin{pmatrix} 1 \\ 2 \\ 1 \\ 3 \end{pmatrix}, \begin{pmatrix} 2 \\ 0 \\ -1 \\ 2 \end{pmatrix} \right\}, \qquad S_2 = \left\{ \begin{pmatrix} 1 \\ 2 \\ 1 \\ 3 \end{pmatrix}, \begin{pmatrix} 2 \\ 0 \\ -1 \\ 2 \end{pmatrix}, \begin{pmatrix} 1 \\ 1 \\ 1 \\ 2 \end{pmatrix} \right\},$$

$$S_3 = \left\{ \begin{pmatrix} 1 \\ 2 \\ 1 \\ 3 \end{pmatrix}, \begin{pmatrix} 2 \\ 0 \\ -1 \\ 2 \end{pmatrix}, \begin{pmatrix} 4 \\ 4 \\ 1 \\ 8 \end{pmatrix} \right\},$$

$$S_4 = \left\{ \begin{pmatrix} 1 \\ 2 \\ 1 \\ 3 \end{pmatrix}, \begin{pmatrix} 2 \\ 0 \\ -1 \\ 2 \end{pmatrix}, \begin{pmatrix} 4 \\ 4 \\ 1 \\ 8 \end{pmatrix}, \begin{pmatrix} 1 \\ 1 \\ 1 \\ 2 \end{pmatrix} \right\}.$$

Problem 6.3 Show that the following set of vectors is linearly dependent

$$S = \{ \mathbf{v}_1, \mathbf{v}_2, \mathbf{v}_3, \mathbf{v}_4, \mathbf{v}_5 \} = \left\{ \begin{pmatrix} 1 \\ 2 \\ 1 \end{pmatrix}, \begin{pmatrix} 2 \\ 0 \\ -1 \end{pmatrix}, \begin{pmatrix} 4 \\ 4 \\ 1 \end{pmatrix}, \begin{pmatrix} 1 \\ 1 \\ 1 \end{pmatrix}, \begin{pmatrix} 3 \\ 5 \\ 0 \end{pmatrix} \right\}.$$

Write down a largest subset W of S which is a linearly independent set of vectors. Express any of the vectors which are in S but not in W as a linear combination of the vectors in W.

Problem 6.4 What does it mean to say that a set of vectors $\{v_1, v_2, \ldots, v_n\}$ is *linearly dependent?*

Given the following vectors

$$v_1 = \begin{pmatrix} 1 \\ 2 \\ 0 \\ -1 \end{pmatrix}, \quad v_2 = \begin{pmatrix} 1 \\ 1 \\ 1 \\ 1 \end{pmatrix}, \quad v_3 = \begin{pmatrix} 4 \\ 5 \\ 3 \\ 2 \end{pmatrix}, \quad v_4 = \begin{pmatrix} 5 \\ 5 \\ 2 \\ 2 \end{pmatrix},$$

show that $\{v_1, v_2, v_3, v_4\}$ is linearly dependent.

Is it possible to express v_4 as a linear combination of the other vectors? If so, do this. If not, explain why not. What about the vector v_3?

Problem 6.5 For each of the sets S_i of vectors given below, find a basis of the vector space $\text{Lin}(S_i)$ and state its dimension. Describe geometrically any sets $\text{Lin}(S_i)$ which are proper subspaces of a Euclidean space $\mathbb{R}^n$, giving Cartesian equations for any lines and planes.

$$S_1 = \left\{ \begin{pmatrix} 1 \\ 2 \end{pmatrix}, \begin{pmatrix} 2 \\ 3 \end{pmatrix} \right\}, \quad S_2 = \left\{ \begin{pmatrix} 1 \\ -1 \end{pmatrix}, \begin{pmatrix} 0 \\ 0 \end{pmatrix}, \begin{pmatrix} 2 \\ -2 \end{pmatrix}, \begin{pmatrix} -3 \\ 3 \end{pmatrix} \right\},$$

$$S_3 = \left\{ \begin{pmatrix} 1 \\ 0 \\ -1 \end{pmatrix}, \begin{pmatrix} 2 \\ 1 \\ 3 \end{pmatrix}, \begin{pmatrix} 1 \\ 2 \\ 9 \end{pmatrix} \right\},$$

$$S_4 = \left\{ \begin{pmatrix} 1 \\ 2 \\ 1 \\ 3 \end{pmatrix}, \begin{pmatrix} 2 \\ 0 \\ -1 \\ 2 \end{pmatrix}, \begin{pmatrix} 4 \\ 4 \\ 1 \\ 8 \end{pmatrix}, \begin{pmatrix} 1 \\ 1 \\ 1 \\ 2 \end{pmatrix} \right\}.$$

Problem 6.6 Which of the following sets are a basis for $\mathbb{R}^3$? (State reasons for your answers.)

$$S_1 = \left\{ \begin{pmatrix} 1 \\ 2 \\ 3 \end{pmatrix}, \begin{pmatrix} 2 \\ 1 \\ 0 \end{pmatrix}, \begin{pmatrix} 4 \\ 1 \\ 0 \end{pmatrix}, \begin{pmatrix} 7 \\ 2 \\ 1 \end{pmatrix} \right\}, \quad S_2 = \left\{ \begin{pmatrix} 1 \\ 0 \\ 1 \end{pmatrix}, \begin{pmatrix} 1 \\ -1 \\ 1 \end{pmatrix} \right\},$$

$$S_3 = \left\{ \begin{pmatrix} 2 \\ 1 \\ 1 \end{pmatrix}, \begin{pmatrix} 1 \\ 2 \\ -1 \end{pmatrix}, \begin{pmatrix} 3 \\ 3 \\ 0 \end{pmatrix} \right\}, \quad S_4 = \left\{ \begin{pmatrix} 2 \\ 1 \\ 1 \end{pmatrix}, \begin{pmatrix} 1 \\ 2 \\ -1 \end{pmatrix}, \begin{pmatrix} 3 \\ 3 \\ 1 \end{pmatrix} \right\}.$$

Problem 6.7 Find the coordinates of the vector $(1, 2, 1)^T$ with respect to each of the following bases for $\mathbb{R}^3$:

$$B_1 = \left\{ \begin{pmatrix} 1 \\ 0 \\ 0 \end{pmatrix}, \begin{pmatrix} 0 \\ 1 \\ 0 \end{pmatrix}, \begin{pmatrix} 0 \\ 0 \\ 1 \end{pmatrix} \right\}, \quad B_2 = \left\{ \begin{pmatrix} 1 \\ 1 \\ 1 \end{pmatrix}, \begin{pmatrix} 1 \\ -1 \\ 0 \end{pmatrix}, \begin{pmatrix} 2 \\ -3 \\ -3 \end{pmatrix} \right\}.$$

Problem 6.8 Find a basis for each of the following subspaces of $\mathbb{R}^3$.

(a) The plane $x - 2y + z = 0$;
(b) The yz-plane.

Problem 6.9 Prove that the set

$$H = \left\{ \begin{pmatrix} 2t \\ t \\ 3t \end{pmatrix} : t \in \mathbb{R} \right\}$$

is a subspace of $\mathbb{R}^3$.

Show that every vector $\mathbf{w} \in H$ is a unique linear combination of the vectors

$$\mathbf{v}_1 = \begin{pmatrix} 1 \\ 0 \\ -1 \end{pmatrix} \quad \text{and} \quad \mathbf{v}_2 = \begin{pmatrix} 0 \\ 1 \\ 5 \end{pmatrix}.$$

Is $\{\mathbf{v}_1, \mathbf{v}_2\}$ a basis of the subspace H? If yes, state why. If no, write down a basis of H. State the dimension of H.

Let G be the subspace $G = \text{Lin}\{\mathbf{v}_1, \mathbf{v}_2\}$. Is $\{\mathbf{v}_1, \mathbf{v}_2\}$ a basis of G? Why or why not?

State a geometric description of each of the subspaces H and G. What is the relationship between them?

Problem 6.10 Find the general solution of each of the following systems of equations in the form $\mathbf{x} = \mathbf{p} + \alpha_1 \mathbf{s}_1 + \cdots + \alpha_{n-r} \mathbf{s}_{n-r}$ where $\mathbf{p}$ is a particular solution of the system and $\{\mathbf{s}_1, \ldots, \mathbf{s}_{n-r}\}$ is a basis for the null space of the coefficient matrix.

$$A\mathbf{x} = \mathbf{b}_1: \qquad\qquad\qquad\qquad B\mathbf{x} = \mathbf{d}_1:$$

$$\begin{cases} x_1 + x_2 + x_3 + x_4 = 4 \\ 2x_1 + x_3 - x_4 = 2 \\ 2x_2 + x_3 + 3x_4 = 6 \end{cases} \qquad \begin{cases} x_1 + 2x_2 - x_3 - x_4 = 3 \\ x_1 - x_2 - 2x_3 - x_4 = 1 \\ 2x_1 + x_2 - x_3 = 3. \end{cases}$$

Find the set of all $\mathbf{b} \in \mathbb{R}^3$ such that $A\mathbf{x} = \mathbf{b}$ is consistent.

Find the set of all $\mathbf{d} \in \mathbb{R}^3$ such that $B\mathbf{x} = \mathbf{d}$ is consistent.

Problem 6.11 Let

$$A = \begin{pmatrix} 1 & -2 & 1 & 1 & 2 \\ -1 & 3 & 0 & 2 & -2 \\ 0 & 1 & 1 & 3 & 4 \\ 1 & 2 & 5 & 13 & 5 \end{pmatrix}.$$

Find a basis for the range of A. Find a basis of the row space and the null space of A. Verify the rank–nullity theorem for the matrix A.

Problem 6.12 Find a basis of the row space, a basis of the range, and a basis of the null space of the matrix

$$B = \begin{pmatrix} 1 & 2 & 1 & 3 & 0 \\ 0 & 1 & 1 & 1 & -1 \\ 1 & 3 & 2 & 0 & 1 \end{pmatrix}.$$

Find the rank of B and verify the rank–nullity theorem.

Let $\mathbf{b} = \mathbf{c}_1 + \mathbf{c}_5$, the sum of the first and last column of the matrix B. Without solving the system, use the information you have obtained to write down a general solution of the system of equations $B\mathbf{x} = \mathbf{b}$.

Problem 6.13 Find the rank of the matrix

$$A = \begin{pmatrix} 1 & 0 & 1 \\ 1 & 1 & 2 \\ 0 & -1 & -1 \end{pmatrix}.$$

Find a basis of the row space and a basis of the column space of A. Show that $RS(A)$ and $CS(A)$ are each planes in $\mathbb{R}^3$. Find Cartesian equations for these planes and hence show that they are two different subspaces.

Find the null space of A and verify the rank–nullity theorem. Show that the basis vectors of the null space are orthogonal to the basis vectors of the row space of A.

Without solving the equations, determine if the systems of equations $A\mathbf{x} = \mathbf{b}_1$ and $A\mathbf{x} = \mathbf{b}_2$ are consistent, where

$$\mathbf{b}_1 = \begin{pmatrix} 1 \\ -1 \\ 2 \end{pmatrix} \quad \text{and} \quad \mathbf{b}_2 = \begin{pmatrix} 2 \\ 1 \\ 3 \end{pmatrix}.$$

If the system is consistent, then find the general solution. If possible, express each of $\mathbf{b}_1$ and $\mathbf{b}_2$ as a linear combination of the columns of A.

Problem 6.14 A portion of the matrix A and the reduced row echelon form of A are shown below:

$$A = \begin{pmatrix} 1 & 4 & * & * \\ 2 & -1 & * & * \\ 3 & 2 & * & * \end{pmatrix} \rightarrow \cdots \rightarrow \begin{pmatrix} 1 & 0 & -1 & 5 \\ 0 & 1 & 3 & 2 \\ 0 & 0 & 0 & 0 \end{pmatrix}.$$

Find a basis of the row space of A, $RS(A)$, a basis of the range of A, $R(A)$, and a basis of the null space of A, $N(A)$.

Let $\mathbf{b} = (9, 0, a)^{\mathrm{T}}$ for some $a \in \mathbb{R}$. The matrix equation $A\mathbf{x} = \mathbf{b}$ represents how many equations in how many unknowns? Find the value of a for which the system of equations $A\mathbf{x} = \mathbf{b}$ is consistent.

Find, if possible, the missing columns of A.

Problem 6.15 Let B be a 3×4 matrix whose null space is

$$N(B) = \left\{ \mathbf{x} = t \begin{pmatrix} 1 \\ 2 \\ 3 \\ 4 \end{pmatrix} \middle| t \in \mathbb{R} \right\}.$$

Determine the rank of B. Find the range of B, $R(B)$.

Consider the row space of B, $RS(B)$. Show that the vector $\mathbf{v}_1 = (4, 0, 0, -1)^{\mathrm{T}}$ is in $RS(B)$. Extend $\{\mathbf{v}_1\}$ to a basis of $RS(B)$, and justify that your set of vectors is a basis.

Problem 6.16 If A is an $m \times n$ matrix and B is an $n \times p$ matrix, show that

$$R(AB) \subseteq R(A) \quad \text{and} \quad N(B) \subseteq N(AB).$$

Deduce that

$$\text{rank}(AB) \leq \min\{\text{rank}(A), \text{rank}(B)\}.$$

7

Linear transformations and change of basis

We now turn our attention to special types of functions between vector spaces known as linear transformations. We will look at the matrix representations of linear transformations between Euclidean vector spaces, and discuss the concept of similarity of matrices. These ideas will then be employed to investigate change of basis and change of coordinates. This material provides the fundamental theoretical underpinning for the technique of diagonalisation, which has many applications, as we shall see later.

7.1   Linear transformations

A function from one vector space V to a vector space W is a rule which assigns to every vector $\mathbf{v} \in V$ a unique vector $\mathbf{w} \in W$. If this function between vector spaces is *linear*, then it is known as a *linear transformation*, (or *linear mapping* or *linear function*).

Definition 7.1 (Linear transformation) Suppose that V and W are vector spaces. A function $T : V \to W$ is *linear* if for all $\mathbf{u}, \mathbf{v} \in V$ and all $\alpha \in \mathbb{R}$:

1. $T(\mathbf{u} + \mathbf{v}) = T(\mathbf{u}) + T(\mathbf{v})$, and
2. $T(\alpha \mathbf{u}) = \alpha T(\mathbf{u})$.

A *linear transformation* is a linear function between vector spaces.

A linear transformation of a vector space V to itself, $T : V \to V$ is often known as a *linear operator*.

If T is linear, then for all $\mathbf{u}, \mathbf{v} \in V$ and $\alpha, \beta \in \mathbb{R}$,

$$T(\alpha\mathbf{u} + \beta\mathbf{v}) = \alpha T(\mathbf{u}) + \beta T(\mathbf{v}).$$

This single condition implies the two in the definition, and is implied by them.

Activity 7.2 Prove that this single condition is equivalent to the two of the definition.

So a linear transformation maps linear combinations of vectors to the same linear combinations of the image vectors. In this sense, it preserves the 'linearity' of a vector space.

In particular, if $T : V \to W$, then for $\mathbf{0} \in V$, $T(\mathbf{0}) = \mathbf{0} \in W$. That is, a linear transformation from V to W maps the zero vector in V to the zero vector in W. This can be seen in a number of ways. For instance, take any $\mathbf{x} \in V$. Then $T(\mathbf{0}) = T(0\mathbf{x}) = 0T(\mathbf{x}) = \mathbf{0}$.

7.1.1 Examples

Example 7.3 To get an idea of what a linear mapping might look like, let us look first at $\mathbb{R}$. What mappings $F : \mathbb{R} \to \mathbb{R}$ are linear?

The function $F_1(x) = px$ for any $p \in \mathbb{R}$ is a linear transformation, since for any $x, y \in \mathbb{R}, \alpha, \beta \in \mathbb{R}$, we have

$$F_1(\alpha x + \beta y) = p(\alpha x + \beta y) = \alpha(px) + \beta(py) = \alpha F_1(x) + \beta F_1(y).$$

But neither of the functions $F_2(x) = px + q$, (for $p, q \in \mathbb{R}, q \neq 0$) or $F_3(x) = x^2$ is linear.

Activity 7.4 Show this. Use a specific example of real numbers to show that neither of these functions satisfies the property

$$T(x + y) = T(x) + T(y) \text{ for all } x, y \in \mathbb{R}.$$

Example 7.5 Suppose that A is an $m \times n$ matrix. Let T be the function given by $T(\mathbf{x}) = A\mathbf{x}$ for $\mathbf{x} \in \mathbb{R}^n$. That is, T is simply multiplication by A. Then T is a linear transformation, $T : \mathbb{R}^n \to \mathbb{R}^m$. This is easily checked, as follows: first,

$$T(\mathbf{u} + \mathbf{v}) = A(\mathbf{u} + \mathbf{v}) = A\mathbf{u} + A\mathbf{v} = T(\mathbf{u}) + T(\mathbf{v}),$$

and, second,

$$T(\alpha\mathbf{u}) = A(\alpha\mathbf{u}) = \alpha A\mathbf{u} = \alpha T(\mathbf{u}).$$

So the two 'linearity' conditions are satisfied. We call T the linear transformation *corresponding to* A, and sometimes denote it by T_A to identify it as such.

Example 7.6 (More complicated) Let us take $V = \mathbb{R}^n$ and take W to be the vector space of all functions $f : \mathbb{R} \to \mathbb{R}$ (with pointwise addition and scalar multiplication). Define a function $T : \mathbb{R}^n \to W$ as follows:

$$T(\mathbf{u}) = T \begin{pmatrix} u_1 \\ u_2 \\ \vdots \\ u_n \end{pmatrix} = p_{u_1, u_2, \ldots, u_n} = p_{\mathbf{u}},$$

where $p_{\mathbf{u}} = p_{u_1, u_2, \ldots, u_n}$ is the polynomial function given by

$$p_{u_1, u_2, \ldots, u_n}(x) = u_1 x + u_2 x^2 + u_3 x^3 + \cdots + u_n x^n.$$

Then T is a linear transformation. To check this, we need to verify that

$$T(\mathbf{u} + \mathbf{v}) = T(\mathbf{u}) + T(\mathbf{v}) \quad \text{and} \quad T(\alpha \mathbf{u}) = \alpha T(\mathbf{u}).$$

Now, $T(\mathbf{u} + \mathbf{v}) = p_{\mathbf{u}+\mathbf{v}}$, $T(\mathbf{u}) = p_{\mathbf{u}}$, and $T(\mathbf{v}) = p_{\mathbf{v}}$, so we need to check that $p_{\mathbf{u}+\mathbf{v}} = p_{\mathbf{u}} + p_{\mathbf{v}}$. This is in fact true, since, for all x,

$$\begin{aligned}
p_{\mathbf{u}+\mathbf{v}}(x) &= p_{u_1+v_1, \ldots, u_n+v_n} \\
&= (u_1 + v_1)x + \cdots + (u_n + v_n)x^n \\
&= (u_1 x + \cdots + u_n x^n) + (v_1 x + \cdots + v_n x^n) \\
&= p_{\mathbf{u}}(x) + p_{\mathbf{v}}(x) \\
&= (p_{\mathbf{u}} + p_{\mathbf{v}})(x).
\end{aligned}$$

The fact that for *all* x, $p_{\mathbf{u}+\mathbf{v}}(x) = (p_{\mathbf{u}} + p_{\mathbf{v}})(x)$ means that the functions $p_{\mathbf{u}+\mathbf{v}}$ and $p_{\mathbf{u}} + p_{\mathbf{v}}$ are identical. The proof that $T(\alpha \mathbf{u}) = \alpha T(\mathbf{u})$ is similar.

Activity 7.7 Do this. Prove that $T(\alpha \mathbf{u}) = \alpha T(\mathbf{u})$.

7.1.2   Linear transformations and matrices

In this section, we consider only linear transformations from $\mathbb{R}^n$ to $\mathbb{R}^m$ for some m and n. But much of what we say can be extended to linear transformations mapping from any finite-dimensional vector space to any other finite-dimensional vector space.

We have seen that any $m \times n$ matrix A defines a linear transformation $T : \mathbb{R}^n \to \mathbb{R}^m$ given by $T(\mathbf{v}) = A\mathbf{v}$.

There is a reverse connection: for every linear transformation $T : \mathbb{R}^n \to \mathbb{R}^m$, there is a matrix A such that $T(\mathbf{v}) = A\mathbf{v}$. In this context, we sometimes denote the matrix by A_T in order to identify it as the matrix corresponding to T. (Note that in the expression A_T, T refers to a linear transformation. This should not be confused with A^T, the transpose of the matrix A.)

Theorem 7.8 *Suppose that $T : \mathbb{R}^n \to \mathbb{R}^m$ is a linear transformation. Let $\{\mathbf{e}_1, \mathbf{e}_2, \ldots, \mathbf{e}_n\}$ denote the standard basis of $\mathbb{R}^n$ and let A be the matrix whose columns are the vectors $T(\mathbf{e}_1), T(\mathbf{e}_2), \ldots, T(\mathbf{e}_n)$: that is,*

$$A = (T(\mathbf{e}_1) \ T(\mathbf{e}_2) \ \ldots \ T(\mathbf{e}_n)).$$

Then, for every $\mathbf{x} \in \mathbb{R}^n$, $T(\mathbf{x}) = A\mathbf{x}$.

Proof: Let $\mathbf{x} = (x_1, x_2, \ldots, x_n)^T$ be any vector in $\mathbb{R}^n$. Then

$$\begin{pmatrix} x_1 \\ x_2 \\ \vdots \\ x_n \end{pmatrix} = x_1 \begin{pmatrix} 1 \\ 0 \\ \vdots \\ 0 \end{pmatrix} + x_2 \begin{pmatrix} 0 \\ 1 \\ \vdots \\ 0 \end{pmatrix} + \cdots + x_n \begin{pmatrix} 0 \\ 0 \\ \vdots \\ 1 \end{pmatrix}$$

$$= x_1 \mathbf{e}_1 + x_2 \mathbf{e}_2 + \cdots + x_n \mathbf{e}_n.$$

Then by the linearity properties of T we have

$$\begin{aligned} T(\mathbf{x}) &= T(x_1 \mathbf{e}_1 + x_2 \mathbf{e}_2 + \cdots + x_n \mathbf{e}_n) \\ &= T(x_1 \mathbf{e}_1) + T(x_2 \mathbf{e}_2) + \cdots + T(x_n \mathbf{e}_n) \\ &= x_1 T(\mathbf{e}_1) + x_2 T(\mathbf{e}_2) + \cdots + x_n T(\mathbf{e}_n). \end{aligned}$$

But this is just a linear combination of the columns of A, so we have (by Theorem 1.38),

$$T(\mathbf{x}) = (T(\mathbf{e}_1) \ T(\mathbf{e}_2) \ \ldots \ T(\mathbf{e}_n))\mathbf{x} = A\mathbf{x},$$

exactly as we wanted.　　　　　　　　　　　　　　　　　　　□

Thus, to each matrix A there corresponds a linear transformation T_A, and to each linear transformation T there corresponds a matrix A_T. Note that the matrix A we found was determined by using the standard basis in both vector spaces; later in this chapter we will generalise this to use other bases.

Example 7.9 Let $T : \mathbb{R}^3 \to \mathbb{R}^3$ be the linear transformation given by

$$T \begin{pmatrix} x \\ y \\ z \end{pmatrix} = \begin{pmatrix} x + y + z \\ x - y \\ x + 2y - 3z \end{pmatrix}.$$

We can find the image of a vector, say $\mathbf{u} = (1, 2, 3)^{\mathrm{T}}$, by substituting its components into the definition, so that, for example, $T(\mathbf{u}) = (6, -1, -4)^{\mathrm{T}}$.

To find the matrix of this linear transformation, we need the images of the standard basis vectors. We have

$$T(\mathbf{e}_1) = \begin{pmatrix} 1 \\ 1 \\ 1 \end{pmatrix}, \qquad T(\mathbf{e}_2) = \begin{pmatrix} 1 \\ -1 \\ 2 \end{pmatrix}, \qquad T(\mathbf{e}_3) = \begin{pmatrix} 1 \\ 0 \\ -3 \end{pmatrix}.$$

The matrix representing T is $A = (T(\mathbf{e}_1) \; T(\mathbf{e}_2) \; T(\mathbf{e}_3))$, which is

$$A = \begin{pmatrix} 1 & 1 & 1 \\ 1 & -1 & 0 \\ 1 & 2 & -3 \end{pmatrix}.$$

Notice that the entries of the matrix A are just the coefficients of x, y, z in the definition of T.

Activity 7.10 Calculate the matrix product $A\mathbf{u}$ for the vector $\mathbf{u} = (1, 2, 3)^{\mathrm{T}}$ and the matrix A above, and observe that this has exactly the same effect as substituting the components.

7.1.3   Linear transformations on $\mathbb{R}^2$

Linear transformations from $\mathbb{R}^2$ to $\mathbb{R}^2$ have the advantage that we can 'observe' them as mappings from one copy of the Cartesian plane to another. For example, we can visualise a reflection in the x axis, which is given by

$$T : \begin{pmatrix} x \\ y \end{pmatrix} \mapsto \begin{pmatrix} x \\ -y \end{pmatrix},$$

with matrix

$$A = \begin{pmatrix} 1 & 0 \\ 0 & -1 \end{pmatrix}.$$

We know that linear transformations preserve linear combinations, and we can interpret this geometrically by saying that lines are mapped to lines and parallelograms are mapped to parallelograms. Because

Figure 7.1 A
rotation

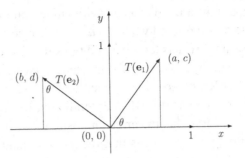

we know that any linear transformation $T : \mathbb{R}^2 \to \mathbb{R}^2$ corresponds to
multiplication by a matrix A, we can describe the effects of these on the
plane. As another example, consider the linear transformation

$$T(\mathbf{x}) = \begin{pmatrix} 2 & 0 \\ 0 & 3 \end{pmatrix} \begin{pmatrix} x \\ y \end{pmatrix}.$$

This has the effect of stretching the plane away from the origin by a
factor of 2 in the x direction and by a factor of 3 in the y direction. If
we look at the effect of this linear transformation on the parallelogram
whose sides are the vectors $\mathbf{e}_1$ and $\mathbf{e}_2$ (a unit square), we find that the
image is a parallelogram (a rectangle) whose corresponding sides are
$2\mathbf{e}_1$ and $3\mathbf{e}_2$. (In this sense, the linear transformation can be described
as an 'enlargement'.)

What about a rotation? If we 'rotate' the plane about the origin
anticlockwise by an angle θ, the unit square with sides $\mathbf{e}_1$ and $\mathbf{e}_2$ will be
rotated. To find the matrix A which represents this linear transformation,
we need to find the images of the standard basis vectors $\mathbf{e}_1$ and $\mathbf{e}_2$. Let

$$T(\mathbf{e}_1) = \begin{pmatrix} a \\ c \end{pmatrix}, \qquad T(\mathbf{e}_2) = \begin{pmatrix} b \\ d \end{pmatrix},$$

so that

$$A = \begin{pmatrix} a & b \\ c & d \end{pmatrix}.$$

We want to determine the coordinates a, c and b, d. It is helpful to draw
a diagram of $\mathbb{R}^2$ such as Figure 7.1, with the images $T(\mathbf{e}_1)$ and $T(\mathbf{e}_2)$
after rotation anticlockwise by an angle θ, $0 < \theta < \frac{\pi}{2}$.

The vectors

$$T(\mathbf{e}_1) = \begin{pmatrix} a \\ c \end{pmatrix} \quad \text{and} \quad T(\mathbf{e}_2) = \begin{pmatrix} b \\ d \end{pmatrix}$$

are orthogonal and each has length 1 since they are the rotated standard
basis vectors. We drop a perpendicular from the point (a, c) to the

x axis, forming a right-angled triangle with angle θ at the origin. Since the x coordinate of the rotated vector is a and the y coordinate is c, the side opposite the angle θ has length c and the side adjacent to the angle θ has length a. The hypotenuse of this triangle (which is the rotated unit vector $\mathbf{e}_1$) has length equal to 1. We therefore have $a = \cos \theta$ and $c = \sin \theta$. Similarly, we can drop the perpendicular from the point (b, d) to the x axis and observe that the angle opposite the x axis is equal to θ. Again, basic trigonometry tells us that the x coordinate is $b = -\sin \theta$ (it has length $\sin \theta$ and is in the negative x direction), and the height is $d = \cos \theta$. Therefore,

$$A = \begin{pmatrix} a & b \\ c & d \end{pmatrix} = \begin{pmatrix} \cos \theta & -\sin \theta \\ \sin \theta & \cos \theta \end{pmatrix}$$

is the matrix of rotation anticlockwise by an angle θ. Although we have shown this using an angle $0 < \theta < \frac{\pi}{2}$, the argument can be extended to any angle θ.

Example 7.11 If $\theta = \frac{\pi}{4}$, then rotation anticlockwise by $\frac{\pi}{4}$ radians is given by the matrix

$$B = \begin{pmatrix} \cos \frac{\pi}{4} & -\sin \frac{\pi}{4} \\ \sin \frac{\pi}{4} & \cos \frac{\pi}{4} \end{pmatrix} = \begin{pmatrix} \frac{1}{\sqrt{2}} & -\frac{1}{\sqrt{2}} \\ \frac{1}{\sqrt{2}} & \frac{1}{\sqrt{2}} \end{pmatrix}.$$

Activity 7.12 Confirm this by sketching the vectors $\mathbf{e}_1$ and $\mathbf{e}_2$ and the image vectors

$$T(\mathbf{e}_1) = \begin{pmatrix} \frac{1}{\sqrt{2}} \\ \frac{1}{\sqrt{2}} \end{pmatrix} \quad \text{and} \quad T(\mathbf{e}_2) = \begin{pmatrix} -\frac{1}{\sqrt{2}} \\ \frac{1}{\sqrt{2}} \end{pmatrix}.$$

What is the matrix of the linear transformation which is a rotation anticlockwise by π radians? What is the matrix of the linear transformation which is a reflection in the line $y = x$? Think about what each of these two transformations does to the standard basis vectors $\mathbf{e}_1$ and $\mathbf{e}_2$ and find these matrices.

7.1.4   Identity and zero linear transformations

If V is a vector space, we can define a linear transformation $T : V \to V$ by $T(\mathbf{v}) = \mathbf{v}$, called the *identity* linear transformation.

If $V = \mathbb{R}^n$, the matrix of this linear transformation is I, the $n \times n$ identity matrix.

There is also a linear transformation $T : V \to W$ defined by $T(\mathbf{v}) = \mathbf{0}$, mapping every vector in V to the zero vector in W.

If $V = \mathbb{R}^n$ and $W = \mathbb{R}^m$, the matrix of this linear transformation is an $m \times n$ matrix consisting entirely of zeros.

7.1.5 Composition and combinations of linear transformations

The *composition* of linear transformations is again a linear transformation. If $T : V \to W$ and $S : W \to U$, then ST is the linear transformation given by

$$ST(\mathbf{v}) = S(T(\mathbf{v})) = S(\mathbf{w}) = \mathbf{u},$$

where $\mathbf{w} = T(\mathbf{v})$. Note that ST means do T and then do S; that is, $V \xrightarrow{T} W \xrightarrow{S} U$. (For ST, work from the inside, out.)

If $T : \mathbb{R}^n \to \mathbb{R}^m$ and $S : \mathbb{R}^m \to \mathbb{R}^p$, then in terms of matrices,

$$ST(\mathbf{v}) = S(T(\mathbf{v})) = S(A_T\mathbf{v}) = A_S A_T\mathbf{v}.$$

That is, $A_{ST} = A_S A_T$; the matrix of the composition is obtained by matrix multiplication of the matrices of the linear transformations. The order is important. Composition of linear transformations, like multiplication of matrices, is not commutative.

Activity 7.13 What are the sizes of the matrices A_S and A_T? Show that the sizes of these matrices indicate in what order they should be multiplied (and therefore in what order the composition of the linear transformations is written).

A *linear combination* of linear transformations is again a linear transformation. If $S, T : V \longrightarrow W$ are linear transformations between the same vector spaces, then $S + T$ and αS, $\alpha \in \mathbb{R}$, are linear transformations, and therefore so is $\alpha S + \beta T$ for any $\alpha, \beta \in \mathbb{R}$.

Activity 7.14 If you have any doubts about why any of the linear transformations mentioned in this section are linear transformations, try to prove that they are by showing the linearity conditions.

For example, the composition ST is a linear transformation because

$$\begin{aligned}
ST(\alpha\mathbf{x} + \beta\mathbf{y}) &= S(T(\alpha\mathbf{x} + \beta\mathbf{y})) \\
&= S(\alpha T(\mathbf{x}) + \beta T(\mathbf{y})) \\
&= \alpha S(T(\mathbf{x})) + \beta S(T(\mathbf{y})) \\
&= \alpha ST(\mathbf{x}) + \beta ST(\mathbf{y}).
\end{aligned}$$

7.1.6   Inverse linear transformations

If V and W are finite-dimensional vector spaces of the same dimension, then the *inverse* of a linear transformation $T : V \rightarrow W$ is the linear transformation $T^{-1} : W \rightarrow V$ such that

$$T^{-1}(T(\mathbf{v})) = \mathbf{v}.$$

In $\mathbb{R}^n$, if T^{-1} exists, then its matrix satisfies

$$T^{-1}(T(\mathbf{v})) = A_{T^{-1}} A_T \mathbf{v} = I\mathbf{v}.$$

That is, T^{-1} exists if and only if $(A_T)^{-1}$ exists, and $(A_T)^{-1} = A_{T^{-1}}$.

Activity 7.15 What result about inverse matrices is being used here in order to make this conclusion?

Example 7.16 In $\mathbb{R}^2$, the inverse of rotation anticlockwise by an angle θ is rotation clockwise by the same angle. Thinking of clockwise rotation by θ as anticlockwise rotation by an angle $-\theta$, the matrix of rotation clockwise by θ is given by,

$$A_{T^{-1}} = \begin{pmatrix} \cos(-\theta) & -\sin(-\theta) \\ \sin(-\theta) & \cos(-\theta) \end{pmatrix} = \begin{pmatrix} \cos\theta & \sin\theta \\ -\sin\theta & \cos\theta \end{pmatrix}.$$

This is easily checked:

$$A_{T^{-1}} A_T = \begin{pmatrix} \cos\theta & \sin\theta \\ -\sin\theta & \cos\theta \end{pmatrix} \begin{pmatrix} \cos\theta & -\sin\theta \\ \sin\theta & \cos\theta \end{pmatrix} = \begin{pmatrix} 1 & 0 \\ 0 & 1 \end{pmatrix}.$$

Activity 7.17 Check this by multiplying the matrices.

Example 7.18 Is there an inverse to the linear transformation in Example 7.9,

$$T \begin{pmatrix} x \\ y \\ z \end{pmatrix} = \begin{pmatrix} x + y + z \\ x - y \\ x + 2y - 3z \end{pmatrix} ?$$

We found

$$A = \begin{pmatrix} 1 & 1 & 1 \\ 1 & -1 & 0 \\ 1 & 2 & -3 \end{pmatrix}.$$

Since $|A| = 9$, the matrix is invertible, and T^{-1} is given by the matrix

$$A^{-1} = \frac{1}{9} \begin{pmatrix} 3 & 5 & 1 \\ 3 & -4 & 1 \\ 3 & -1 & -2 \end{pmatrix}.$$

That is,

$$T^{-1} \begin{pmatrix} u \\ v \\ w \end{pmatrix} = \begin{pmatrix} \frac{1}{3}u + \frac{5}{9}v + \frac{1}{9}w \\ \frac{1}{3}u - \frac{4}{9}v + \frac{1}{9}w \\ \frac{1}{3}u - \frac{1}{9}v + -\frac{2}{9}w \end{pmatrix}.$$

Activity 7.19 Check that $T^{-1}(\mathbf{w})$ for $\mathbf{w} = T(\mathbf{u}) = (6, -1, -4)^{\mathsf{T}}$ is $\mathbf{u} = (1, 2, 3)^{\mathsf{T}}$ (see Example 7.9). Also check that $T^{-1}T = I$.

7.1.7 Linear transformations from V to W

Theorem 7.20 *Let V be a finite-dimensional vector space and let T be a linear transformation from V to a vector space W. Then T is completely determined by what it does to a basis of V.*

Proof: Let $\dim(V) = n$, and let $B = \{\mathbf{v}_1, \mathbf{v}_2, \ldots, \mathbf{v}_n\}$ be a basis of V. Then any $\mathbf{v} \in V$ can be uniquely expressed as a linear combination of these basis vectors: $\mathbf{v} = a_1\mathbf{v}_1 + a_2\mathbf{v}_2 + \cdots + a_n\mathbf{v}_n$.
  Then

$$T(\mathbf{v}) = T(a_1\mathbf{v}_1 + a_2\mathbf{v}_2 + \cdots + a_n\mathbf{v}_n)$$
$$= a_1 T(\mathbf{v}_1) + a_2 T(\mathbf{v}_2) + \cdots + a_n T(\mathbf{v}_n).$$

That is, if $\mathbf{v} \in V$ is expressed as a linear combination of the basis vectors, then the image $T(\mathbf{v})$ is the same linear combination of the images of the basis vectors. Therefore, if we know T on the basis vectors, we know it for all $\mathbf{v} \in V$. □

If both V and W are finite-dimensional vector spaces, then this result allows us to find a matrix which corresponds to the linear transformation T. The matrix will depend on the basis of V and the basis of W. Suppose V has $\dim(V) = n$ and basis $B = \{\mathbf{v}_1, \mathbf{v}_2, \ldots, \mathbf{v}_n\}$, and that W has $\dim(W) = m$ and basis $S = \{\mathbf{w}_1, \mathbf{w}_2, \ldots, \mathbf{w}_m\}$. Then the coordinate vector of a vector $\mathbf{v} \in V$ is denoted by $[\mathbf{v}]_B$, and the coordinate vector of a vector $T(\mathbf{v}) \in W$ is denoted $[T(\mathbf{v})]_S$. By working with the coordinates of these vectors (rather than the vectors themselves), we can find a matrix such that $[T(\mathbf{v})]_S = A[\mathbf{v}]_B$.
  Using the result above we can write,

$$[T(\mathbf{v})]_S = a_1[T(\mathbf{v}_1)]_S + a_2[T(\mathbf{v}_2)]_S + \cdots + a_n[T(\mathbf{v}_n)]_S$$
$$= ([T(\mathbf{v}_1)]_S \quad [T(\mathbf{v}_2)]_S \quad \cdots \quad [T(\mathbf{v}_n)]_S)[\mathbf{v}]_B,$$

where $[\mathbf{v}]_B = (a_1, a_2, \cdots, a_n)^{\mathsf{T}}$ is the coordinate matrix of the vector $\mathbf{v}$ in the basis B, and $[T(\mathbf{v}_i)]_S$ are the coordinates of the image vectors in the basis S (see Exercise 6.11). That is, A is the matrix whose columns are the images $T(\mathbf{v}_i)$ expressed in the basis of W. Then we have $[T(\mathbf{v})]_S = A[\mathbf{v}]_B$.

7.2   Range and null space

7.2.1   Definitions of range and null space

Just as we have the range and null space of a matrix, so we have the range and null space of a linear transformation, defined as follows:

Definition 7.21 (Range and null space) Suppose that T is a linear transformation from a vector space V to a vector space W. Then the *range*, $R(T)$, of T is

$$R(T) = \{T(\mathbf{v}) \mid \mathbf{v} \in V\},$$

and the *null space*, $N(T)$, of T is

$$N(T) = \{\mathbf{v} \in V \mid T(\mathbf{v}) = \mathbf{0}\},$$

where $\mathbf{0}$ denotes the zero vector of W.

The null space is also called the *kernel*, and may be denoted $\ker(T)$ in some texts.

The range and null space of a linear transformation $T : V \to W$ are subspaces of W and V, respectively.

Activity 7.22 Prove this. Try this yourself before looking at the answer to this activity.

Of course, for any $m \times n$ matrix, A, if T is the linear transformation $T(\mathbf{x}) = A\mathbf{x}$, then $R(T) = R(A)$ and $N(T) = N(A)$. The definitions of the subspaces are the same since $T : \mathbb{R}^n \to \mathbb{R}^m$ and for all $\mathbf{x} \in \mathbb{R}^n$, if $T(\mathbf{x}) = A\mathbf{x}$, we have:

$$R(T) = \{T(\mathbf{x}) \mid \mathbf{x} \in \mathbb{R}^n\} = \{A\mathbf{x} \mid \mathbf{x} \in \mathbb{R}^n\} = R(A) \subseteq \mathbb{R}^m,$$

and

$$N(T) = \{\mathbf{x} \in \mathbb{R}^n \mid T(\mathbf{x}) = \mathbf{0}\} = \{\mathbf{x} \in \mathbb{R}^n \mid A\mathbf{x} = \mathbf{0}\} = N(A) \subseteq \mathbb{R}^n.$$

Example 7.23 We find the null space and range of the linear transformation $S : \mathbb{R}^2 \to \mathbb{R}^4$ given by

$$S\begin{pmatrix} x \\ y \end{pmatrix} = \begin{pmatrix} x+y \\ x \\ x-y \\ y \end{pmatrix}.$$

The matrix of the linear transformation is

$$A_S = \begin{pmatrix} 1 & 1 \\ 1 & 0 \\ 1 & -1 \\ 0 & 1 \end{pmatrix}.$$

Observe that this matrix has rank 2 (by having two linearly independent columns, or you could alternatively see this by putting it into row echelon form), so that $N(S) = \{0\}$, the subspace of $\mathbb{R}^2$ consisting of only the zero vector. This can also be seen directly from the fact that

$$\begin{pmatrix} x+y \\ x \\ x-y \\ y \end{pmatrix} = \begin{pmatrix} 0 \\ 0 \\ 0 \\ 0 \end{pmatrix} \iff x = 0, \ y = 0 \iff \begin{pmatrix} x \\ y \end{pmatrix} = \begin{pmatrix} 0 \\ 0 \end{pmatrix}.$$

The range, $R(S)$, is the two-dimensional subspace of $\mathbb{R}^4$ with basis given by the column vectors of A_S.

7.2.2 Rank–nullity theorem for linear transformations

If V and W are both finite-dimensional, then so are $R(T)$ and $N(T)$, and we refer to their dimensions as the *rank* and *nullity* of the linear transformation, respectively.

Definition 7.24 (Rank and nullity of a linear transformation) The *rank* of a linear transformation T is

$$\text{rank}(T) = \dim(R(T))$$

and the *nullity* is

$$\text{nullity}(T) = \dim(N(T)).$$

As for matrices, there is a strong link between these two dimensions known as the *Rank–nullity theorem* or the *Dimension theorem* for linear transformations. Here we are concerned with subspaces of any vector spaces V and W (not just Euclidean spaces).

Theorem 7.25 (Rank–nullity theorem linear transformations)

Suppose that T is a linear transformation from the finite-dimensional vector space V to the vector space W. Then

$$\text{rank}(T) + \text{nullity}(T) = \dim(V).$$

(Note that this result holds even if W is not finite-dimensional.)

Proof: Assume that $\dim(V) = n$ and $\text{nullity}(T) = k$. We need to show that $\text{rank}(T) = n - k$. Let $\{v_1, v_2, \ldots, v_k\}$ be a basis of the null space, $N(T)$. As $N(T)$ is a subspace of V, we can extend this basis to a basis of V, $\{v_1, v_2, \ldots, v_k, v_{k+1}, \ldots, v_n\}$ (by Theorem 6.45). For any $v \in V$, we have $v = a_1 v_1 + a_2 v_2 + \cdots + a_n v_n$. Then,

$$
\begin{aligned}
T(v) &= a_1 T(v_1) + \cdots + a_k T(v_k) + a_{k+1} T(v_{k+1}) + \cdots + a_n T(v_n) \\
&= a_1 0 + \cdots + a_k 0 + a_{k+1} T(v_{k+1}) + \cdots + a_n T(v_n) \\
&= a_{k+1} T(v_{k+1}) + \cdots + a_n T(v_n),
\end{aligned}
$$

since $T(v_i) = 0$ for $i = 1, \ldots, k$ (because $v_i \in N(A)$). Hence the vectors $\{T(v_{k+1}), \cdots, T(v_n)\}$ span the range, $R(T)$. If they are a basis of $R(T)$, then $\text{rank}(T) = n - k$. So it only remains to show that they are linearly independent.

If there is a linear combination of the vectors equal to the zero vector,

$$b_{k+1} T(v_{k+1}) + \cdots + b_n T(v_n) = T(b_{k+1} v_{k+1} + \cdots + b_n v_n) = 0,$$

then the vector $b_{k+1} v_{k+1} + \cdots + b_n v_n$ is in the null space of T, and can be written as a linear combination of the basis vectors of $N(T)$,

$$b_{k+1} v_{k+1} + \cdots + b_n v_n = b_1 v_1 + \cdots + b_k v_k.$$

Rearranging, we have

$$b_1 v_1 + \cdots + b_k v_k - b_{k+1} v_{k+1} - \cdots - b_n v_n = 0.$$

But $\{v_1, v_2, \ldots, v_k, v_{k+1}, \ldots, v_n\}$ is a basis of V, and hence all coefficients b_i must be 0. This shows that $\{T(v_{k+1}), \cdots, T(v_n)\}$ are linearly independent and the theorem is proved. □

For an $m \times n$ matrix A, if $T(x) = Ax$, then T is a linear transformation from $V = \mathbb{R}^n$ to $W = \mathbb{R}^m$, and we have $\text{rank}(T) = \text{rank}(A)$ and $\text{nullity}(T) = \text{nullity}(A)$. So this theorem is the same as the earlier result that

$$\text{rank}(A) + \text{nullity}(A) = n.$$

Here n is the dimension of $\mathbb{R}^n = V$ (which, of course, is the same as the number of columns of A).

Example 7.26 Is it possible to construct a linear transformation $T : \mathbb{R}^3 \to \mathbb{R}^3$ with

$$N(T) = \left\{ t \begin{pmatrix} 1 \\ 2 \\ 3 \end{pmatrix} : t \in \mathbb{R} \right\}, \qquad R(T) = xy\text{-plane}?$$

A linear transformation $T : \mathbb{R}^3 \to \mathbb{R}^3$ must satisfy the rank–nullity theorem with $n = 3$:

$$\text{nullity}(T) + \text{rank}(T) = 3.$$

Since the dimension of the null space of T is 1 and the dimension of $R(T)$ is 2, the rank–nullity theorem is satisfied, so at this stage, we certainly can't rule out the possibility that such a linear transformation exists. (Of course, if it was not satisfied, we would know straight away that we couldn't have a linear transformation of the type suggested.)

To find a linear transformation T with $N(T)$ and $R(T)$ as above, we construct a matrix A_T, which must be 3×3 since $T : \mathbb{R}^3 \to \mathbb{R}^3$. Note that if $R(A_T) = R(T)$ is the xy-plane, then the column vectors of A_T must be linearly dependent and include a basis for this plane. You can take any two linearly independent vectors in the xy-plane to be the first two columns of the matrix, and the third column must be a linear combination of the first two. The linear dependency condition they must satisfy is revealed by the basis of the null space.

For example, we may take the first two column vectors to be the standard basis vectors, $\mathbf{c}_1 = \mathbf{e}_1$, and $\mathbf{c}_2 = \mathbf{e}_2$. Then if $\mathbf{v}$ is the null space basis vector, we must have $A_T \mathbf{v} = \mathbf{0}$. This means

$$A_T \mathbf{v} = (\, \mathbf{c}_1 \quad \mathbf{c}_2 \quad \mathbf{c}_3 \,) \begin{pmatrix} 1 \\ 2 \\ 3 \end{pmatrix} = 1\,\mathbf{c}_1 + 2\,\mathbf{c}_2 + 3\,\mathbf{c}_3 = \mathbf{0}.$$

Therefore, we must have $\mathbf{c}_3 = -\frac{1}{3}\mathbf{c}_1 - \frac{2}{3}\mathbf{c}_2$. So one possible linear transformation satisfying these conditions is given by the matrix

$$A_T = \begin{pmatrix} 1 & 0 & -\frac{1}{3} \\ 0 & 1 & -\frac{2}{3} \\ 0 & 0 & 0 \end{pmatrix}.$$

7.3 Coordinate change

In this section, we shall limit our discussion to $\mathbb{R}^n$ for some n, but much of what we say can be extended to any finite-dimensional vector space V.

Suppose that the vectors $\mathbf{v}_1, \mathbf{v}_2, \ldots, \mathbf{v}_n$ form a basis B for $\mathbb{R}^n$. Then, as we have seen, any $\mathbf{x} \in \mathbb{R}^n$ can be written in exactly one way as a linear combination,

$$\mathbf{x} = \alpha_1 \mathbf{v}_1 + \alpha_2 \mathbf{v}_2 + \cdots + \alpha_n \mathbf{v}_n,$$

of the vectors in the basis, and the vector

$$[\mathbf{x}]_B = \begin{bmatrix} \alpha_1 \\ \alpha_2 \\ \vdots \\ \alpha_n \end{bmatrix}_B$$

is called the *coordinate vector* of $\mathbf{x}$ with respect to the basis $B = \{\mathbf{v}_1, \mathbf{v}_2, \ldots, \mathbf{v}_n\}$.

One very straightforward observation is that the coordinate vector of any $\mathbf{x} \in \mathbb{R}^n$ with respect to the standard basis is just $\mathbf{x}$ itself. This is because if $\mathbf{x} = (x_1, x_2, \ldots, x_n)^T$, then

$$\mathbf{x} = x_1 \mathbf{e}_1 + x_2 \mathbf{e}_2 + \cdots + x_n \mathbf{e}_n.$$

What is less immediately obvious is how to find the coordinates of a vector $\mathbf{x}$ with respect to a basis other than the standard one.

7.3.1   Change of coordinates from standard to basis B

To find the coordinates of a vector with respect to a basis $B = \{\mathbf{v}_1, \mathbf{v}_2, \ldots, \mathbf{v}_n\}$, we need to solve the system of linear equations

$$a_1 \mathbf{v}_1 + a_2 \mathbf{v}_2 + \cdots + a_n \mathbf{v}_n = \mathbf{x},$$

which, in matrix form, is

$$\mathbf{x} = (\mathbf{v}_1 \ \mathbf{v}_2 \ \ldots \ \mathbf{v}_n)\mathbf{a}$$

with $\mathbf{a} = (a_1, a_2, \ldots, a_n)^T = [\mathbf{x}]_B$. In other words, if we let P be the matrix whose columns are the basis vectors (in order),

$$P = (\mathbf{v}_1 \ \mathbf{v}_2 \ \ldots \ \mathbf{v}_n),$$

then for any $\mathbf{x} \in \mathbb{R}^n$,

$$\mathbf{x} = P[\mathbf{x}]_B.$$

The matrix P is invertible because its columns form a basis. So we can also write

$$[\mathbf{x}]_B = P^{-1}\mathbf{x}.$$

Definition 7.27 (Transition matrix) If $B = \{\mathbf{v}_1, \mathbf{v}_2, \ldots, \mathbf{v}_n\}$ is a basis of $\mathbb{R}^n$, the matrix

$$P = (\mathbf{v}_1 \ \mathbf{v}_2 \ \ldots \ \mathbf{v}_n),$$

whose columns are the basis vectors in B, is called the *transition matrix from B coordinates to standard coordinates*. Then, the matrix P^{-1} is the transition matrix from standard coordinates to coordinates in the basis B.

In order to emphasise the connection of a transition matrix P with the corresponding basis B, we will sometimes denote the matrix by P_B.

Example 7.28 Let B be the following set of vectors of $\mathbb{R}^3$:

$$B = \left\{ \begin{pmatrix} 1 \\ 2 \\ -1 \end{pmatrix}, \begin{pmatrix} 2 \\ -1 \\ 4 \end{pmatrix}, \begin{pmatrix} 3 \\ 2 \\ 1 \end{pmatrix} \right\}.$$

To show that B is a basis, we can write the vectors as the columns of a matrix P,

$$P = \begin{pmatrix} 1 & 2 & 3 \\ 2 & -1 & 2 \\ -1 & 4 & 1 \end{pmatrix},$$

then evaluate the determinant. We have $|P| = 4 \neq 0$ so B is a basis of $\mathbb{R}^3$.

If we are given the B coordinates of a vector $\mathbf{v}$, say

$$[\mathbf{v}]_B = \begin{bmatrix} 4 \\ 1 \\ -5 \end{bmatrix}_B,$$

then we can find its standard coordinates either directly as a linear combination of the basis vectors,

$$\mathbf{v} = 4 \begin{pmatrix} 1 \\ 2 \\ -1 \end{pmatrix} + \begin{pmatrix} 2 \\ -1 \\ 4 \end{pmatrix} - 5 \begin{pmatrix} 3 \\ 2 \\ 1 \end{pmatrix} = \begin{pmatrix} -9 \\ -3 \\ -5 \end{pmatrix},$$

or by using the matrix P,

$$\mathbf{v} = P[\mathbf{v}]_B = \begin{pmatrix} 1 & 2 & 3 \\ 2 & -1 & 2 \\ -1 & 4 & 1 \end{pmatrix} \begin{bmatrix} 4 \\ 1 \\ -5 \end{bmatrix}_B = \begin{pmatrix} -9 \\ -3 \\ -5 \end{pmatrix},$$

which, of course, amounts to the same thing.

To find the B coordinates of a vector $\mathbf{x}$, say $\mathbf{x} = (5, 7, -3)^T$, we need to find constants a_1, a_2, a_3 such that

$$\begin{pmatrix} 5 \\ 7 \\ -3 \end{pmatrix} = a_1 \begin{pmatrix} 1 \\ 2 \\ -1 \end{pmatrix} + a_2 \begin{pmatrix} 2 \\ -1 \\ 4 \end{pmatrix} + a_3 \begin{pmatrix} 3 \\ 2 \\ 1 \end{pmatrix}.$$

We can do this either using Gaussian elimination to solve the system $P\mathbf{a} = \mathbf{x}$ for $\mathbf{a} = (a_1, a_2, a_3)^T$ or by using the inverse matrix, P^{-1}, to find

$$[\mathbf{x}]_B = P^{-1}\mathbf{x} = \begin{bmatrix} 1 \\ -1 \\ 2 \end{bmatrix}_B.$$

We can check the result as follows:

$$\mathbf{x} = 1 \begin{pmatrix} 1 \\ 2 \\ -1 \end{pmatrix} + (-1) \begin{pmatrix} 2 \\ -1 \\ 4 \end{pmatrix} + 2 \begin{pmatrix} 3 \\ 2 \\ 1 \end{pmatrix} = \begin{pmatrix} 5 \\ 7 \\ -3 \end{pmatrix}.$$

Activity 7.29 Check all the calculations in this example. Find P^{-1} and use it to find $[\mathbf{x}]_B$.

Activity 7.30 Continuing with this example, what are the B coordinates of the basis vectors

$$\mathbf{v}_1 = \begin{pmatrix} 1 \\ 2 \\ -1 \end{pmatrix}, \qquad \mathbf{v}_2 = \begin{pmatrix} 2 \\ -1 \\ 4 \end{pmatrix}, \qquad \mathbf{v}_3 = \begin{pmatrix} 3 \\ 2 \\ 1 \end{pmatrix}?$$

7.3.2 Change of basis as a linear transformation

If P is the transition matrix from coordinates in a basis B of $\mathbb{R}^n$ to standard coordinates, then considered as the matrix of a linear transformation, $T(\mathbf{x}) = P\mathbf{x}$, the linear transformation actually maps the standard basis vectors, $\mathbf{e}_i$, to the new basis vectors, $\mathbf{v}_i$. That is, $T(\mathbf{e}_i) = \mathbf{v}_i$.

Example 7.31 Suppose we wish to change basis in $\mathbb{R}^2$ by a rotation of the axes $\frac{\pi}{4}$ radians anticlockwise. What are the coordinates of a vector with respect to this new basis, $B = \{\mathbf{v}_1, \mathbf{v}_2\}$?

The matrix of the linear transformation which performs this rotation is given by

$$\begin{pmatrix} \cos \frac{\pi}{4} & -\sin \frac{\pi}{4} \\ \sin \frac{\pi}{4} & \cos \frac{\pi}{4} \end{pmatrix} = \begin{pmatrix} \frac{1}{\sqrt{2}} & -\frac{1}{\sqrt{2}} \\ \frac{1}{\sqrt{2}} & \frac{1}{\sqrt{2}} \end{pmatrix},$$

and the column vectors of the matrix are the new basis vectors, $\mathbf{v}_1 = T(\mathbf{e}_1)$ and $\mathbf{v}_2 = T(\mathbf{e}_2)$, since these are the images of the standard basis vectors. So the matrix is also the transition matrix from B coordinates to standard coordinates:

$$P = \begin{pmatrix} \frac{1}{\sqrt{2}} & -\frac{1}{\sqrt{2}} \\ \frac{1}{\sqrt{2}} & \frac{1}{\sqrt{2}} \end{pmatrix},$$

and we have $\mathbf{v} = P[\mathbf{v}]_B$. Then the coordinates of a vector with respect to the new basis are given by $[\mathbf{v}]_B = P^{-1}\mathbf{v}$. The inverse of rotation anticlockwise is rotation clockwise, so we have,

$$P^{-1} = \begin{pmatrix} \cos(-\frac{\pi}{4}) & -\sin(-\frac{\pi}{4}) \\ \sin(-\frac{\pi}{4}) & \cos(-\frac{\pi}{4}) \end{pmatrix} = \begin{pmatrix} \cos \frac{\pi}{4} & \sin \frac{\pi}{4} \\ -\sin \frac{\pi}{4} & \cos \frac{\pi}{4} \end{pmatrix}$$

$$= \begin{pmatrix} \frac{1}{\sqrt{2}} & \frac{1}{\sqrt{2}} \\ -\frac{1}{\sqrt{2}} & \frac{1}{\sqrt{2}} \end{pmatrix}.$$

Suppose we want the new coordinates of a vector, say $\mathbf{x} = (1, 1)^T$. Then we have

$$[\mathbf{x}]_B = P^{-1}\mathbf{x} = \begin{pmatrix} \frac{1}{\sqrt{2}} & \frac{1}{\sqrt{2}} \\ -\frac{1}{\sqrt{2}} & \frac{1}{\sqrt{2}} \end{pmatrix} \begin{pmatrix} 1 \\ 1 \end{pmatrix} = \begin{bmatrix} \sqrt{2} \\ 0 \end{bmatrix}_B.$$

From a different viewpoint, we could have noticed that

$$\mathbf{x} = \begin{pmatrix} 1 \\ 1 \end{pmatrix} = \sqrt{2} \begin{pmatrix} \frac{1}{\sqrt{2}} \\ \frac{1}{\sqrt{2}} \end{pmatrix} = \sqrt{2}\,\mathbf{v}_1,$$

so that

$$[\mathbf{x}]_B = \sqrt{2} \begin{bmatrix} 1 \\ 0 \end{bmatrix}_B = \begin{bmatrix} \sqrt{2} \\ 0 \end{bmatrix}_B.$$

7.3.3 Change of coordinates from basis B to basis B'

Given a basis B of $\mathbb{R}^n$ with transition matrix P_B, and another basis B' with transition matrix $P_{B'}$, how do we change from coordinates in the basis B to coordinates in the basis B'?

The answer is quite simple. First, we change from B coordinates to standard coordinates using $\mathbf{v} = P_B[\mathbf{v}]_B$ and then change from standard coordinates to B' coordinates using $[\mathbf{v}]_{B'} = P_{B'}^{-1}\mathbf{v}$. That is,

$$[\mathbf{v}]_{B'} = P_{B'}^{-1}P_B[\mathbf{v}]_B.$$

The matrix $M = P_{B'}^{-1}P_B$ is the transition matrix from B coordinates to B' coordinates.

In practice, the easiest way to obtain the matrix M is as the product of the two transition matrices, $M = P_{B'}^{-1}P_B$. But let's look more closely at the matrix M. If the basis B is the set of vectors $B = \{\mathbf{v}_1, \mathbf{v}_2, \dots, \mathbf{v}_n\}$, then these are the columns of the transition matrix, $P_B = (\mathbf{v}_1 \ \mathbf{v}_2 \ \dots \ \mathbf{v}_n)$. Looking closely at the columns of the product matrix,

$$M = P_{B'}^{-1}P_B = P_{B'}^{-1}(\mathbf{v}_1 \ \mathbf{v}_2 \ \dots \ \mathbf{v}_n) = (P_{B'}^{-1}\mathbf{v}_1 \ P_{B'}^{-1}\mathbf{v}_2 \ \dots \ P_{B'\mathbf{v}_n}^{-1});$$

that is, each column of the matrix M is obtained by multiplying the matrix $P_{B'}^{-1}$ by the corresponding column of P_B. But $P_{B'}^{-1}\mathbf{v}_i$ is just the B' coordinates of the vector $\mathbf{v}_i$, so the matrix M is given by

$$M = ([\mathbf{v}_1]_{B'} \ [\mathbf{v}_2]_{B'} \ \dots \ [\mathbf{v}_n]_{B'}).$$

We have therefore established the following result:

Theorem 7.32 *If B and B' are two bases of $\mathbb{R}^n$, with*

$$B = \{\mathbf{v}_1, \mathbf{v}_2, \dots, \mathbf{v}_n\},$$

then the transition matrix from B coordinates to B' coordinates is given by

$$M = ([\mathbf{v}_1]_{B'} \ [\mathbf{v}_2]_{B'} \ \dots \ [\mathbf{v}_n]_{B'}).$$

Activity 7.33 The above proof used the following fact about matrix multiplication. If A is an $m \times n$ matrix and B is an $n \times p$ matrix with column vectors $\mathbf{b}_1, \mathbf{b}_2, \dots, \mathbf{b}_p$, then the product AB is the $m \times p$ matrix whose columns are $A\mathbf{b}_1, A\mathbf{b}_2, \dots, A\mathbf{b}_p$; that is,

$$AB = A(\mathbf{b}_1 \ \mathbf{b}_2 \ \dots \ \mathbf{b}_p) = (A\mathbf{b}_1 \ A\mathbf{b}_2 \ \dots \ A\mathbf{b}_p).$$

Why is this correct?

Example 7.34 Each of the sets of vectors

$$B = \left\{\begin{pmatrix} 1 \\ 2 \end{pmatrix}, \begin{pmatrix} -1 \\ 1 \end{pmatrix}\right\}, \qquad S = \left\{\begin{pmatrix} 3 \\ 1 \end{pmatrix}, \begin{pmatrix} 5 \\ 2 \end{pmatrix}\right\}$$

is a basis of $\mathbb{R}^2$, since if

$$P = \begin{pmatrix} 1 & -1 \\ 2 & 1 \end{pmatrix}, \qquad Q = \begin{pmatrix} 3 & 5 \\ 1 & 2 \end{pmatrix},$$

then their determinants, $|P| = 3$ and $|Q| = 1$, are non-zero. P is the transition matrix from B coordinates to standard and Q is the transition matrix from S coordinates to standard.

Suppose you are given a vector $\mathbf{x} \in \mathbb{R}^2$ with

$$[\mathbf{x}]_B = \begin{bmatrix} 4 \\ -1 \end{bmatrix}_B.$$

How do you find the coordinates of $\mathbf{x}$ in the basis S? There are two approaches you can take.

First, you can find the standard coordinates of $\mathbf{x}$,

$$\mathbf{x} = 4 \begin{pmatrix} 1 \\ 2 \end{pmatrix} - \begin{pmatrix} -1 \\ 1 \end{pmatrix} = \begin{pmatrix} 1 & -1 \\ 2 & 1 \end{pmatrix} \begin{pmatrix} 4 \\ -1 \end{pmatrix} = \begin{pmatrix} 5 \\ 7 \end{pmatrix}$$

and then find the S coordinates using Q^{-1}

$$[\mathbf{x}]_S = Q^{-1}\mathbf{x} = \begin{pmatrix} 2 & -5 \\ -1 & 3 \end{pmatrix} \begin{pmatrix} 5 \\ 7 \end{pmatrix} = \begin{bmatrix} -25 \\ 16 \end{bmatrix}_S.$$

Alternatively, you can calculate the transition matrix M from B coordinates to S coordinates. Using $\mathbf{v} = P[\mathbf{v}]_B$ and $\mathbf{v} = Q[\mathbf{v}]_S$, we have $[\mathbf{v}]_S = Q^{-1}P[\mathbf{v}]_B$, so

$$M = Q^{-1}P = \begin{pmatrix} 2 & -5 \\ -1 & 3 \end{pmatrix} \begin{pmatrix} 1 & -1 \\ 2 & 1 \end{pmatrix} = \begin{pmatrix} -8 & -7 \\ 5 & 4 \end{pmatrix}.$$

Then

$$[\mathbf{x}]_S = \begin{pmatrix} -8 & -7 \\ 5 & 4 \end{pmatrix} \begin{pmatrix} 4 \\ -1 \end{pmatrix} = \begin{bmatrix} -25 \\ 16 \end{bmatrix}_S.$$

Note that the columns of M are the S coordinates of the basis B vectors (which would be another way to find M).

Activity 7.35 Check the calculations in this example. In particular, check the S coordinates of the vector $\mathbf{x}$, and check that the columns of M are the basis B vectors in S coordinates.

7.4 Change of basis and similarity

7.4.1 Change of basis and linear transformations

We have already seen that if T is a linear transformation from $\mathbb{R}^n$ to $\mathbb{R}^m$, then there is a corresponding matrix A such that $T(\mathbf{x}) = A\mathbf{x}$ for all $\mathbf{x}$. The matrix A is given by

$$A = (T(\mathbf{e}_1) \; T(\mathbf{e}_2) \; \dots \; T(\mathbf{e}_n)).$$

This matrix is obtained using the standard basis in both $\mathbb{R}^n$ and in $\mathbb{R}^m$. Now suppose that B is a basis of $\mathbb{R}^n$ and B' a basis of $\mathbb{R}^m$, and suppose we want to know the coordinates $[T(\mathbf{x})]_{B'}$ of $T(\mathbf{x})$ with respect to B', given the coordinates $[\mathbf{x}]_B$ of $\mathbf{x}$ with respect to B. Is there a matrix M such that

$$[T(\mathbf{x})]_{B'} = M[\mathbf{x}]_B$$

for all $\mathbf{x}$? Indeed there is, as the following result shows.

Theorem 7.36 *Suppose that $B = \{\mathbf{v}_1, \ldots, \mathbf{v}_n\}$ and $B' = \{\mathbf{v}'_1, \ldots, \mathbf{v}'_m\}$ are bases of $\mathbb{R}^n$ and $\mathbb{R}^m$ and that $T : \mathbb{R}^n \to \mathbb{R}^m$ is a linear transformation. Let $M = A_{[B,B']}$ be the $m \times n$ matrix with the ith column equal to $[T(\mathbf{v}_i)]_{B'}$, the coordinate vector of $T(\mathbf{v}_i)$ with respect to the basis B'. Then for all $\mathbf{x}$, $[T(\mathbf{x})]_{B'} = M[\mathbf{x}]_B$.*

The matrix $M = A_{[B,B']}$ is the matrix which represents T with respect to the bases B and B'.

Proof: In order to prove this theorem, let's look at the stages of transition which occur from changing basis from B to standard, performing the linear transformation in standard coordinates and then changing to the basis B'.

Let A be the matrix representing T in standard coordinates, and let P_B and $P_{B'}$ be, respectively, the transition matrix from B coordinates to standard coordinates in $\mathbb{R}^n$ and the transition matrix from B' coordinates to standard coordinates in $\mathbb{R}^m$. (So P_B is an $n \times n$ matrix having the basis vectors of B as columns, and $P_{B'}$ is an $m \times m$ matrix having the basis vectors of B' as columns.) Then we know that for any $\mathbf{x} \in \mathbb{R}^n$, $\mathbf{x} = P_B[\mathbf{x}]_B$; and, similarly, for any $\mathbf{u} \in \mathbb{R}^m$, $\mathbf{u} = P_{B'}[\mathbf{u}]_{B'}$, so $[\mathbf{u}]_{B'} = P_{B'}^{-1}\mathbf{u}$.

We want to find a matrix M such that $[T(\mathbf{x})]_{B'} = M[\mathbf{x}]_B$. If we start with a vector $\mathbf{x}$ in B coordinates, then $\mathbf{x} = P_B[\mathbf{x}]_B$ will give us the standard coordinates. We can then perform the linear transformation on $\mathbf{x}$ using the matrix A,

$$T(\mathbf{x}) = A\mathbf{x} = AP_B[\mathbf{x}]_B,$$

giving us the image vector $T(\mathbf{x})$ in standard coordinates in $\mathbb{R}^m$. To obtain the B' coordinates of this vector, all we have to do is multiply on the left by the matrix $P_{B'}^{-1}$; that is,

$$[T(\mathbf{x})]_{B'} = P_{B'}^{-1}T(\mathbf{x}).$$

Then substituting what we found for $T(\mathbf{x})$ in standard coordinates,

$$[T(\mathbf{x})]_{B'} = P_{B'}^{-1}AP_B[\mathbf{x}]_B.$$

Since this is true for any $\mathbf{x} \in \mathbb{R}^n$, we conclude that

$$M = P_{B'}^{-1} A P_B$$

is the matrix of the linear transformation in the new bases.

This, in fact, is the easiest way to calculate the matrix M. But let's take a closer look at the columns of $M = P_{B'}^{-1} A P_B$. We have $P_B = (\mathbf{v}_1 \ \mathbf{v}_2 \ \ldots \ \mathbf{v}_n)$, so

$$A P_B = A(\mathbf{v}_1 \ \mathbf{v}_2 \ \ldots \ \mathbf{v}_n) = (A\mathbf{v}_1 \ A\mathbf{v}_2 \ \ldots \ A\mathbf{v}_n).$$

But $A\mathbf{v}_i = T(\mathbf{v}_i)$, so

$$A P_B = (T(\mathbf{v}_1) \ T(\mathbf{v}_2) \ \ldots \ T(\mathbf{v}_n)).$$

Then $M = P_{B'}^{-1} A P_B$, so

$$M = P_{B'}^{-1} (T(\mathbf{v}_1) \ T(\mathbf{v}_2) \ \ldots \ T(\mathbf{v}_n))$$
$$= (P_{B'}^{-1}(T(\mathbf{v}_1)) \ P_{B'}^{-1}(T(\mathbf{v}_2)) \ \ldots \ P_{B'}^{-1}(T(\mathbf{v}_n)));$$

that is, $M = ([T(\mathbf{v}_1)]_{B'} \ [T(\mathbf{v}_2)]_{B'} \ldots [T(\mathbf{v}_n)]_{B'})$ and the theorem is proved. $\qquad\square$

Thus, if we *change the basis* from the standard bases of $\mathbb{R}^n$ and $\mathbb{R}^m$, the matrix representation of the linear transformation T changes.

7.4.2 Similarity

A particular case of this Theorem 7.36 is so important it is worth stating separately. It corresponds to the case in which $m = n$ and $B' = B$.

Theorem 7.37 *Suppose that $T : \mathbb{R}^n \to \mathbb{R}^n$ is a linear transformation and that $B = \{\mathbf{x}_1, \mathbf{x}_2, \ldots, \mathbf{x}_n\}$ is a basis of $\mathbb{R}^n$. Let A be the matrix corresponding to T in standard coordinates, so that $T(\mathbf{x}) = A\mathbf{x}$. Let*

$$P = (\mathbf{x}_1 \ \mathbf{x}_2 \ \ldots \ \mathbf{x}_n)$$

be the matrix whose columns are the vectors of B. Then for all $\mathbf{x} \in \mathbb{R}^n$,

$$[T(\mathbf{x})]_B = P^{-1} A P [\mathbf{x}]_B.$$

In other words, $A_{[B,B]} = P^{-1} A P$ is the matrix representing T in the basis B. The relationship between the matrices $A_{[B,B]}$ and A is a central one in the theory of linear algebra. The matrix $A_{[B,B]}$ performs the *same* linear transformation as the matrix A, only $A_{[B,B]}$ describes it in terms of the basis B rather than in standard coordinates. This likeness of effect inspires the following definition.

Definition 7.38 (Similarity) We say that the square matrix C is *similar* to the matrix A if there is an invertible matrix P such that $C = P^{-1}AP$.

Note that 'similar' has a very precise meaning here: it doesn't mean that the matrices somehow 'look like' each other (as normal use of the word similar would suggest), but that they represent the same linear transformation in different bases.

Similarity defines an equivalence relation on matrices. Recall that an equivalence relation satisfies three properties; it is reflexive, symmetric and transitive (see Section 3.1.2). For similarity, this means that:

- a matrix A is similar to itself (reflexive),
- if C is similar to A, then A is similar to C (symmetric), and
- if D is similar to C, and C to A, then D is similar to A (transitive).

Activity 7.39 Prove these! (Note that we have purposely not used the letter B here to denote a matrix, since we used it in the previous discussion to denote a set of n vectors which form a basis of $\mathbb{R}^n$.)

Because the relationship is symmetric, we usually just say that A and C are *similar matrices*, meaning one is similar to the other, and we can express this either as $C = P^{-1}AP$ or $A = Q^{-1}CQ$ for invertible matrices P and Q (in which case $Q = P^{-1}$).

As we shall see in subsequent chapters, this relationship can be used to great advantage if the new basis B is chosen carefully.

Let's look at some examples to see why we might want to change basis from standard coordinates to another basis of $\mathbb{R}^n$.

Example 7.40 You may have seen graphs of conic sections, and you may know, for example, that the set of points $(x, y) \in \mathbb{R}^2$ such that $x^2 + y^2 = 1$ is a circle of radius one centered at the origin, and that, similarly, the set of points $x^2 + 4y^2 = 4$ is an ellipse.

These equations are said to be in *standard form* and as such they are easy to sketch. For example, to sketch a graph of the ellipse in $\mathbb{R}^2$, all you need to do is note that if $x = 0$, then $y = \pm 1$, and if $y = 0$, then $x = \pm 2$; mark these four points on a set of coordinates axes and connect them in an ellipse (see Figure 7.2).

Suppose, however, we want to graph the set of points $(x, y) \in \mathbb{R}^2$ which satisfy the equation $5x^2 + 5y^2 - 6xy = 2$. It turns out that this, too, is an ellipse, but this is not obvious, and sketching the graph is far from easy. So suppose we are told to do the following: change the

Figure 7.2 The ellipse $x^2 + 4y^2 = 4$

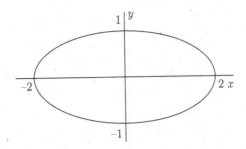

basis in $\mathbb{R}^2$ by rotating the axes by $\frac{\pi}{4}$ radians anticlockwise, express the equation in the new coordinates and then sketch it.

Let's see what happens. The linear transformation which accomplishes this rotation has matrix

$$A_T = \begin{pmatrix} \cos\theta & -\sin\theta \\ \sin\theta & \cos\theta \end{pmatrix} = \begin{pmatrix} \frac{1}{\sqrt{2}} & -\frac{1}{\sqrt{2}} \\ \frac{1}{\sqrt{2}} & \frac{1}{\sqrt{2}} \end{pmatrix} = P,$$

where the columns of P are the new basis vectors. We'll call the new basis B and denote the new coordinates of a vector $\mathbf{v}$ by X and Y; more precisely, $[\mathbf{v}]_B = (X, Y)^T$. Then,

$$\mathbf{v} = P[\mathbf{v}]_B \quad \Longleftrightarrow \quad \begin{pmatrix} x \\ y \end{pmatrix} = \begin{pmatrix} \frac{1}{\sqrt{2}} & -\frac{1}{\sqrt{2}} \\ \frac{1}{\sqrt{2}} & \frac{1}{\sqrt{2}} \end{pmatrix} \begin{pmatrix} X \\ Y \end{pmatrix}.$$

To see the effect of this change of basis, we can substitute

$$x = \frac{1}{\sqrt{2}}X - \frac{1}{\sqrt{2}}Y \quad \text{and} \quad y = \frac{1}{\sqrt{2}}X + \frac{1}{\sqrt{2}}Y$$

into the equation $5x^2 + 5y^2 - 6xy = 2$ and collect terms. The result in the new coordinates is the equation

$$X^2 + 4Y^2 = 1.$$

Activity 7.41 Carry out the substitution for x and y to obtain the new equation.

So how do we sketch this? The new coordinate axes are obtained from the standard ones by rotating the plane by $\frac{\pi}{4}$ radians anticlockwise. So we first sketch these new X, Y axes and then sketch $X^2 + 4Y^2 = 1$ on the new axes as described above (by marking out the points on the X axis where $Y = 0$, and the points on the Y axis where $X = 0$ and connecting them in an ellipse). See Figure 7.3.

We will look at these ideas again later in Chapter 11. Now let's look at a different kind of example.

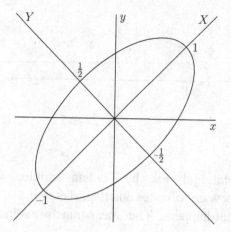

Figure 7.3 The ellipse $5x^2 + 5y^2 - 6xy = 2$

Example 7.42 Suppose we are given the linear transformation $T : \mathbb{R}^2 \to \mathbb{R}^2$,

$$T \begin{pmatrix} x \\ y \end{pmatrix} = \begin{pmatrix} x + 3y \\ -x + 5y \end{pmatrix}$$

and we are asked to describe the effect of this linear transformation on the xy-plane. At this point there isn't much we can say (other than perhaps sketch a unit square and see what happens to it). So suppose we are told to change the basis in $\mathbb{R}^2$ to a new basis

$$B = \{\mathbf{v}_1 \mathbf{v}_2\} = \left\{ \begin{pmatrix} 1 \\ 1 \end{pmatrix}, \begin{pmatrix} 3 \\ 1 \end{pmatrix} \right\}$$

and then to find the matrix of the linear transformation in this basis. Call this matrix C. We have just seen that $C = P^{-1}AP$, where A is the matrix of the linear transformation in standard coordinates and P is the transition matrix from B coordinates to standard:

$$A = \begin{pmatrix} 1 & 3 \\ -1 & 5 \end{pmatrix}, \quad P = \begin{pmatrix} 1 & 3 \\ 1 & 1 \end{pmatrix}.$$

Then

$$C = P^{-1}AP = \frac{1}{2} \begin{pmatrix} -1 & 3 \\ 1 & -1 \end{pmatrix} \begin{pmatrix} 1 & 3 \\ -1 & 5 \end{pmatrix} \begin{pmatrix} 1 & 3 \\ 1 & 1 \end{pmatrix} = \begin{pmatrix} 4 & 0 \\ 0 & 2 \end{pmatrix}.$$

Activity 7.43 Check this calculation by multiplying the matrices.

So what does this tell us? The B coordinates of the B basis vectors are

$$[\mathbf{v}_1]_B = \begin{bmatrix} 1 \\ 0 \end{bmatrix}_B \quad \text{and} \quad [\mathbf{v}_2]_B = \begin{bmatrix} 0 \\ 1 \end{bmatrix}_B,$$

so in B coordinates the linear transformation can be described as a stretch in the direction of $\mathbf{v}_1$ by a factor of 4 and a stretch in the direction of $\mathbf{v}_2$ by a factor of 2:

$$[T(\mathbf{v}_1)]_B = \begin{pmatrix} 4 & 0 \\ 0 & 2 \end{pmatrix} \begin{bmatrix} 1 \\ 0 \end{bmatrix}_B = \begin{bmatrix} 4 \\ 0 \end{bmatrix}_B = 4[\mathbf{v}_1]_B,$$

and, similarly, $[T(\mathbf{v}_2)]_B = 2[\mathbf{v}_2]_B$. But the effect of T is the same no matter what basis is being used to describe it; it is only the matrices which change. So this statement must be true even in standard coordinates; that is, we must have

$$A\mathbf{v}_1 = 4\mathbf{v}_1 \quad \text{and} \quad A\mathbf{v}_2 = 2\mathbf{v}_2.$$

Activity 7.44 Check this. Show that $A\mathbf{v}_1 = 4\mathbf{v}_1$ and $A\mathbf{v}_2 = 2\mathbf{v}_2$.

In each of these examples, we were told the basis to use in $\mathbb{R}^2$ in order to solve the questions we posed. So you should now be asking a question: 'How did we know which basis would work for each of these examples?' We shall begin to discover the answer in the next chapter.

7.5 Learning outcomes

You should now be able to:

- explain what is meant by a linear transformation and be able to prove a given mapping is linear
- explain what is meant by the range and null space, and rank and nullity of a linear transformation
- explain the rank–nullity theorem (the dimension theorem) for linear transformations and be able to apply it
- explain the two-way relationship between matrices and linear transformations
- find the matrix representation of a transformation with respect to two given bases
- change between different bases of a vector space
- explain what it means to say that two square matrices are similar.

7.6 Comments on activities

Activity 7.2 To show the condition is equivalent to the other two, we need to prove two things: first, that the two conditions imply this one and, second, that this single condition implies the other two. So suppose the two conditions of the definition hold and suppose that $\mathbf{u}, \mathbf{v} \in V$

and $\alpha, \beta \in \mathbb{R}$. Then we have $T(\alpha\mathbf{u}) = \alpha T(\mathbf{u})$ and $T(\beta\mathbf{v}) = \beta T(\mathbf{v})$, by property 2, and, by property 1, we then have

$$T(\alpha\mathbf{u} + \beta\mathbf{v}) = T(\alpha\mathbf{u}) + T(\beta\mathbf{v}) = \alpha T(\mathbf{u}) + \beta T(\mathbf{v}),$$

as required. On the other hand, suppose that for all $\mathbf{u}, \mathbf{v} \in V$ and $\alpha, \beta \in \mathbb{R}$, we have $T(\alpha\mathbf{u} + \beta\mathbf{v}) = \alpha T(\mathbf{u}) + \beta T(\mathbf{v})$. Then property 1 follows on taking $\alpha = \beta = 1$ and property 2 follows on taking $\beta = 0$.

Activity 7.4 You just need to use one specific example to show this. For example, let $x = 3$ and $y = 4$. Then

$$F_2(3 + 4) = 7p + q \neq F_2(3) + F_2(4) = (3p + q) + (4p + q)$$
$$= 7p + 2q$$

since $q \neq 0$. Similarly, $F_3(3 + 4) = 49$ but $F_3(3) + F_3(4) = 25$, so F_3 is not linear. (Of course, you can conclude F_2 is not linear since $F_2(0) = q \neq 0$.)

Activity 7.7 $T(\alpha\mathbf{u}) = p_{\alpha\mathbf{u}}$, and $T(\mathbf{u}) = p_\mathbf{u}$, so we need to check that $p_{\alpha\mathbf{u}} = \alpha p_\mathbf{u}$. Now, for all x,

$$p_{\alpha\mathbf{u}}(x) = p_{\alpha u_1, \alpha u_2, \dots, \alpha u_n}(x)$$
$$= (\alpha u_1)x + (\alpha u_2)x^2 + \cdots + (\alpha u_n)x^n$$
$$= \alpha(u_1 x + u_2 x^2 + \cdots + u_n x^n)$$
$$= \alpha p_\mathbf{u}(x),$$

as required.

Activity 7.12 Rotation by π radians is given by the matrix A below, whereas reflection in the line $y = x$ is given by the matrix C:

$$A = \begin{pmatrix} -1 & 0 \\ 0 & -1 \end{pmatrix}, \qquad C = \begin{pmatrix} 0 & 1 \\ 1 & 0 \end{pmatrix}.$$

Activity 7.13 Since A_T is $m \times n$ and A_S is $p \times m$, the matrices can be multiplied in the order $A_S A_T$ and not necessarily in the other order, unless $p = n$. Therefore, the composite linear transformation ST is defined. But, in any case, you should still remember that this means: first do T and then do S.

Activity 7.15 The result that if $AB = I$, then $A = B^{-1}$ and $B = A^{-1}$ (Theorem 3.12).

Activity 7.19 You can check that $AA^{-1} = I$, or you can substitute $u = x + y + z$, $v = x - y$, and $w = x + 2y - 3z$ into the formula for T^{-1} to see that you get back the vector $(x, y, z)^{\mathsf{T}}$.

Activity 7.22 This is very similar to the proofs in Chapter 5 that, for a matrix A, $R(A)$ and $N(A)$ are subspaces.

First, we show $R(T)$ is a subspace of W. Note that it is non-empty since $T(\mathbf{0}) = \mathbf{0}$ and hence it contains $\mathbf{0}$. (The fact that $T(\mathbf{0}) = \mathbf{0}$ can be seen in a number of ways. For instance, take any $\mathbf{x} \in V$. Then $T(\mathbf{0}) = T(0\mathbf{x}) = 0T(\mathbf{x}) = \mathbf{0}$.) We need to show that if $\mathbf{u}, \mathbf{v} \in R(T)$, then $\mathbf{u} + \mathbf{v} \in R(T)$, and, for any $\alpha \in \mathbb{R}$, $\alpha\mathbf{v} \in R(T)$. Suppose $\mathbf{u}, \mathbf{v} \in R(T)$. Then for some $\mathbf{y}_1, \mathbf{y}_2 \in V$, $\mathbf{u} = T(\mathbf{y}_1)$, $\mathbf{v} = T(\mathbf{y}_2)$. Now,

$$\mathbf{u} + \mathbf{v} = T(\mathbf{y}_1) + T(\mathbf{y}_2) = T(\mathbf{y}_1 + \mathbf{y}_2),$$

and so $\mathbf{u} + \mathbf{v} \in R(T)$. Next,

$$\alpha\mathbf{v} = \alpha(T(\mathbf{y}_1)) = T(\alpha\mathbf{y}_1),$$

so $\alpha\mathbf{v} \in R(A)$.

Now consider $N(T)$. It is non-empty because the fact that $T(\mathbf{0}) = \mathbf{0}$ shows $\mathbf{0} \in N(T)$. Suppose $\mathbf{u}, \mathbf{v} \in N(A)$ and $\alpha \in \mathbb{R}$. Then to show $\mathbf{u} + \mathbf{v} \in N(T)$ and $\alpha\mathbf{u} \in N(T)$, we must show that $T(\mathbf{u} + \mathbf{v}) = \mathbf{0}$ and $T(\alpha\mathbf{u}) = \mathbf{0}$. We have

$$T(\mathbf{u} + \mathbf{v}) = T(\mathbf{u}) + T(\mathbf{v}) = \mathbf{0} + \mathbf{0} = \mathbf{0}$$

and

$$T(\alpha\mathbf{u}) = \alpha(T(\mathbf{u})) = \alpha\mathbf{0} = \mathbf{0},$$

so we have shown what we needed.

Activity 7.29 Once you have found P^{-1}, check that it is correct; check that $PP^{-1} = I$. We have

$$[\mathbf{x}]_B = P^{-1}\mathbf{x} = \frac{1}{4}\begin{pmatrix} -9 & 10 & 7 \\ -4 & 4 & 4 \\ 7 & -6 & -5 \end{pmatrix}\begin{pmatrix} 5 \\ 7 \\ -3 \end{pmatrix}.$$

Activity 7.30 The B coordinates are

$$[\mathbf{v}_1]_B = \begin{bmatrix} 1 \\ 0 \\ 0 \end{bmatrix}_B, \quad [\mathbf{v}_2]_B = \begin{bmatrix} 0 \\ 1 \\ 0 \end{bmatrix}_B, \quad [\mathbf{v}_3]_B = \begin{bmatrix} 0 \\ 0 \\ 1 \end{bmatrix}_B$$

since, for example, $\mathbf{v}_1 = 1\mathbf{v}_1 + 0\mathbf{v}_2 + 0\mathbf{v}_3$. This emphasises the fact that the order of the vectors in $B = \{\mathbf{v}_1, \mathbf{v}_2, \mathbf{v}_3\}$ is important, and that it must be the same order for the columns of the matrix P.

Activity 7.33 This is just how matrix multiplication works. For example, let's look closely at the first column, $\mathbf{c}_1$, of the product matrix.

$$
\begin{pmatrix} a_{11} & \cdots & a_{1n} \\ a_{21} & \cdots & a_{2n} \\ \vdots & \ddots & \vdots \\ a_{m1} & \cdots & a_{mn} \end{pmatrix}
\begin{pmatrix} b_{11} & \cdots & b_{1p} \\ b_{21} & \cdots & b_{2p} \\ \vdots & \ddots & \vdots \\ b_{np} & \cdots & b_{nn} \end{pmatrix}
=
\begin{pmatrix} c_{11} & \cdots & \cdots \\ c_{21} & \cdots & \cdots \\ \vdots & \ddots & \ddots \\ c_{m1} & \cdots & \cdots \end{pmatrix}.
$$

$$\quad\quad\quad\quad\quad\quad\quad \underset{\mathbf{b}_1}{\uparrow} \quad\quad\quad\quad\quad\quad\quad \underset{\mathbf{c}_1}{\uparrow}$$

Each entry c_{i1} of column 1 of the product is obtained by taking the inner product of the ith row of A (regarded as a vector) with column 1 of B, so c_{i1} is the same as the ith entry of $A\mathbf{b}_1$. This is true for any of the columns of AB, so that $AB = (A\mathbf{b}_1, A\mathbf{b}_2, \ldots, A\mathbf{b}_p)$.

Activity 7.35 We have, for $\mathbf{x}$,

$$-25 \begin{pmatrix} 3 \\ 1 \end{pmatrix} + 16 \begin{pmatrix} 5 \\ 2 \end{pmatrix} = \begin{pmatrix} 5 \\ 7 \end{pmatrix} \quad \text{so} \quad [\mathbf{x}]_S = \begin{bmatrix} -25 \\ 16 \end{bmatrix}_S,$$

and for the basis B vectors,

$$-8 \begin{pmatrix} 3 \\ 1 \end{pmatrix} + 5 \begin{pmatrix} 5 \\ 2 \end{pmatrix} = \begin{pmatrix} 1 \\ 2 \end{pmatrix} \quad \text{and} \quad -7 \begin{pmatrix} 3 \\ 1 \end{pmatrix} + 4 \begin{pmatrix} 5 \\ 2 \end{pmatrix} = \begin{pmatrix} -1 \\ 1 \end{pmatrix},$$

so these are the correct coordinates in the basis S. You could also check all of these by using $[\mathbf{v}]_S = Q^{-1}\mathbf{v}$.

Activity 7.39 Let I denote the $n \times n$ identity matrix. Then $A = I^{-1}AI$, which shows that A is similar to itself.

If C is similar to A, then there is an invertible matrix P such that $C = P^{-1}AP$. But then $A = PCP^{-1} = (P^{-1})CP^{-1}$, so A is similar to C.

If D is similar to C, then there is an invertible matrix Q such that $D = Q^{-1}CQ$. If C is similar to A, then there is an invertible matrix P such that $C = P^{-1}AP$. Then

$$D = Q^{-1}CQ = Q^{-1}P^{-1}APQ = (PQ)^{-1}A(PQ),$$

so D is similar to A.

Activity 7.44 We have

$$A\mathbf{v}_1 = \begin{pmatrix} 1 & 3 \\ -1 & 5 \end{pmatrix} \begin{pmatrix} 1 \\ 1 \end{pmatrix} = \begin{pmatrix} 4 \\ 4 \end{pmatrix} = 4\mathbf{v}_1,$$

$$A\mathbf{v}_2 = \begin{pmatrix} 1 & 3 \\ -1 & 5 \end{pmatrix} \begin{pmatrix} 3 \\ 1 \end{pmatrix} = \begin{pmatrix} 6 \\ 2 \end{pmatrix} = 2\mathbf{v}_2.$$

7.7 Exercises

Exercise 7.1 Find bases for the null space and range of the linear transformation $T : \mathbb{R}^3 \to \mathbb{R}^3$ given by

$$T \begin{pmatrix} x_1 \\ x_2 \\ x_3 \end{pmatrix} = \begin{pmatrix} x_1 + x_2 + 2x_3 \\ x_1 + x_3 \\ 2x_1 + x_2 + 3x_3 \end{pmatrix}.$$

Verify the rank–nullity theorem. Is T invertible?

Exercise 7.2 Let S and T be the linear transformations from $\mathbb{R}^2$ to $\mathbb{R}^2$ given by the matrices

$$A_S = \begin{pmatrix} 0 & 1 \\ 1 & 0 \end{pmatrix}, \qquad A_T = \begin{pmatrix} 0 & 1 \\ -1 & 0 \end{pmatrix}.$$

Sketch the vectors $\mathbf{e}_1$ and $\mathbf{e}_2$ in the xy-plane, and sketch the unit square. Describe the effect of S in words, and illustrate it using the unit square by adding the images $S(\mathbf{e}_1)$ and $S(\mathbf{e}_2)$ to your sketch (and filling in the image of the unit square). Do the same for the linear transformation T.

Now consider the composed linear transformations ST and TS. Illustrate the effect of ST and TS using the unit square (by first performing one linear transformation and then the other). Then calculate their matrices to check that $ST \neq TS$.

You should also check that your matrix for ST matches the images $ST(\mathbf{e}_1)$ and $ST(\mathbf{e}_2)$ in your sketch, and do the same for TS.

Exercise 7.3 Consider the vectors

$$\mathbf{v}_1 = \begin{pmatrix} 1 \\ 0 \\ 1 \end{pmatrix}, \quad \mathbf{v}_2 = \begin{pmatrix} -1 \\ 1 \\ 2 \end{pmatrix}, \quad \mathbf{v}_3 = \begin{pmatrix} 0 \\ 1 \\ 5 \end{pmatrix} \quad \text{and} \quad \mathbf{u} = \begin{pmatrix} 1 \\ 2 \\ 3 \end{pmatrix}.$$

Show that $B = \{\mathbf{v}_1, \mathbf{v}_2, \mathbf{v}_3\}$ is a basis of $\mathbb{R}^3$. Find the B coordinates of $\mathbf{u}$ and hence express $\mathbf{u}$ as a linear combination of $\mathbf{v}_1, \mathbf{v}_2, \mathbf{v}_3$.

A linear transformation $S : \mathbb{R}^3 \to \mathbb{R}^3$ is known to have the following effect

$$S(\mathbf{v}_1) = \mathbf{e}_1 \qquad S(\mathbf{v}_2) = \mathbf{e}_2 \qquad S(\mathbf{v}_3) = \mathbf{e}_3,$$

where e_1, e_2, e_3, are the standard basis vectors in $\mathbb{R}^3$. Using properties of linear transformations, find $S(\mathbf{u})$.

Find, if possible, the null space of S and the range of S.

Write down the corresponding matrix A_S.

Exercise 7.4 Show that the rank–nullity theorem for linear transformations does not rule out the possibility that there exists a linear transformation $T : \mathbb{R}^3 \rightarrow \mathbb{R}^2$, whose null space, $N(T)$, consists of vectors $\mathbf{x} = (x, y, z)^T \in \mathbb{R}^3$ with $x = y = z$ and whose range, $R(T)$, is $\mathbb{R}^2$.

Suppose, further, that we require that T maps $e_1, e_2 \in \mathbb{R}^3$ to the standard basis vectors in $\mathbb{R}^2$. Find a matrix A_T such that the linear transformation $T(\mathbf{x}) = A_T\mathbf{x}$ is as required. Write down an expression for $T(\mathbf{x})$ as a vector in $\mathbb{R}^2$ in terms of x, y, z.

Exercise 7.5 If S and T are the linear transformations given in the previous two exercises, decide which composed linear transformation, ST or TS, is defined, and find its corresponding matrix.

Exercise 7.6 Let $\{e_1, e_2, e_3, e_4\}$ be the standard basis of $\mathbb{R}^4$, and let $\mathbf{v}_1, \mathbf{v}_2, \mathbf{v}_3, \mathbf{x}$ be the following vectors in $\mathbb{R}^3$ (where x, y, z are constants):

$$\mathbf{v}_1 = \begin{pmatrix} 1 \\ 0 \\ -1 \end{pmatrix}, \quad \mathbf{v}_2 = \begin{pmatrix} 2 \\ 1 \\ 2 \end{pmatrix}, \quad \mathbf{v}_3 = \begin{pmatrix} 5 \\ 1 \\ -1 \end{pmatrix}, \quad \mathbf{x} = \begin{pmatrix} x \\ y \\ z \end{pmatrix}.$$

Let T be a linear transformation, $T : \mathbb{R}^4 \rightarrow \mathbb{R}^3$, given by

$$T(e_1) = \mathbf{v}_1, \qquad T(e_2) = \mathbf{v}_2, \qquad T(e_3) = \mathbf{v}_3, \qquad T(e_4) = \mathbf{x}.$$

(i) Suppose the vector $\mathbf{x}$ is such that the linear transformation T has

$$\dim(R(T)) = \dim(N(T)).$$

Write down a condition that the components of $\mathbf{x}$ must satisfy for this to happen. Find a basis of $R(T)$ in this case.

(ii) Suppose the vector $\mathbf{x}$ is such that the linear transformation has

$$\dim(N(T)) = 1.$$

Write down a condition that the components of $\mathbf{x}$ must satisfy for this to happen. Find a basis of $N(T)$ in this case.

Exercise 7.7 Determine for what values of the constant λ, the vectors

$$\mathbf{v}_1 = \begin{pmatrix} 1 \\ 3 \\ -5 \end{pmatrix}, \quad \mathbf{v}_2 = \begin{pmatrix} 1 \\ -1 \\ 1 \end{pmatrix}, \quad \mathbf{v}_3 = \begin{pmatrix} 2 \\ 0 \\ \lambda \end{pmatrix}$$

form a basis of $\mathbb{R}^3$.

Let $\mathbf{b} = (2, 0, 1)^T$ and $\mathbf{s} = (2, 0, 3)^T$. Deduce that each of the sets

$$B = \{\mathbf{v}_1, \mathbf{v}_2, \mathbf{b}\} \quad \text{and} \quad S = \{\mathbf{v}_1, \mathbf{v}_2, \mathbf{s}\}$$

is a basis of $\mathbb{R}^3$. Find the transition matrix P from S coordinates to B coordinates.

If $[\mathbf{w}]_S = \begin{bmatrix} 1 \\ 2 \\ 2 \end{bmatrix}_S$, find $[\mathbf{w}]_B$.

Exercise 7.8 Consider the plane W in $\mathbb{R}^3$,

$$W = \left\{ \begin{pmatrix} x \\ y \\ z \end{pmatrix} \;\middle|\; x - 2y + 3z = 0 \right\}.$$

Show that each of the sets

$$S = \left\{ \begin{pmatrix} 2 \\ 1 \\ 0 \end{pmatrix}, \begin{pmatrix} -3 \\ 0 \\ 1 \end{pmatrix} \right\} \qquad B = \left\{ \begin{pmatrix} -1 \\ 1 \\ 1 \end{pmatrix}, \begin{pmatrix} 1 \\ 2 \\ 1 \end{pmatrix} \right\}$$

is a basis of W.

Show that the vector $\mathbf{v} = (5, 7, 3)^T$ is in W and find its coordinates, $[\mathbf{v}]_S$, in the basis S.

Find a transition matrix M from coordinates in the basis B to coordinates in the basis S; that is,

$$[\mathbf{x}]_S = M[\mathbf{x}]_B.$$

Use this to find $[\mathbf{v}]_B$ for the vector $\mathbf{v} = (5, 7, 3)^T$ and check your answer.

Exercise 7.9 Suppose that $T : \mathbb{R}^2 \to \mathbb{R}^3$ is the linear transformation given by

$$T \begin{pmatrix} x_1 \\ x_2 \end{pmatrix} = \begin{pmatrix} x_2 \\ -5x_1 + 13x_2 \\ -7x_1 + 16x_2 \end{pmatrix}.$$

Find the matrix $A_{[B,B']}$ of T with respect to the bases

$$B = \left\{ \begin{pmatrix} 3 \\ 1 \end{pmatrix}, \begin{pmatrix} 5 \\ 2 \end{pmatrix} \right\} \quad \text{and} \quad B' = \left\{ \begin{pmatrix} 1 \\ 0 \\ -1 \end{pmatrix}, \begin{pmatrix} -1 \\ 2 \\ 2 \end{pmatrix}, \begin{pmatrix} 0 \\ 1 \\ 2 \end{pmatrix} \right\}.$$

Exercise 7.10 (For readers who have studied calculus) Consider the vector space F of functions $f : \mathbb{R} \to \mathbb{R}$ with pointwise addition and scalar multiplication. The symbol $C^\infty(\mathbb{R})$ denotes the set of all functions with continuous derivatives of all orders. Examples of such functions are polynomials, e^x, $\cos x$, $\sin x$. Show that $C^\infty(\mathbb{R})$ is a subspace of

F. Show also that differentiation, $D : f \to f'$, is a linear operator on $C^\infty(\mathbb{R})$.

7.8   Problems

Problem 7.1 Suppose T and S are linear transformations with respective matrices:

$$A_T = \begin{pmatrix} \frac{1}{\sqrt{2}} & -\frac{1}{\sqrt{2}} \\ \frac{1}{\sqrt{2}} & \frac{1}{\sqrt{2}} \end{pmatrix}, \qquad A_S = \begin{pmatrix} -1 & 0 \\ 0 & 1 \end{pmatrix}.$$

(a)   Sketch the effects of T and S on the standard basis, and hence on the unit square with sides e_1, e_2. Describe T and S in words.

(b)   Illustrate ST and TS using the unit square. Then calculate their matrices to check that $ST \neq TS$.

Problem 7.2 Consider the vectors

$$v_1 = \begin{pmatrix} 1 \\ 0 \\ 1 \end{pmatrix}, \ v_2 = \begin{pmatrix} 1 \\ 1 \\ 3 \end{pmatrix}, \ v_3 = \begin{pmatrix} 0 \\ 0 \\ 1 \end{pmatrix},$$

$$w_1 = \begin{pmatrix} 1 \\ 1 \\ 1 \end{pmatrix}, \ w_2 = \begin{pmatrix} 1 \\ -1 \\ 0 \end{pmatrix}, \ w_3 = \begin{pmatrix} 0 \\ 1 \\ -1 \end{pmatrix}.$$

(a)   Show that each of the sets $B = \{v_1, v_2, v_3\}$ and $\widehat{B} = \{w_1, w_2, w_3\}$ is a basis of $\mathbb{R}^3$.

(b)   Write down the matrix A_T of the linear transformation T given by $T(e_1) = v_1$, $T(e_2) = v_2$, $T(e_3) = v_3$, where $\{e_1, e_2, e_3\} \subset \mathbb{R}^3$ is the standard basis.
Express $T(x)$ for $x = (x, y, z)^T$ as a vector in $\mathbb{R}^3$ (in terms of x, y and z).

(c)   Write down the matrix A_S of the linear transformation S given by $S(v_1) = e_1$, $S(v_2) = e_2$, $S(v_3) = e_3$. What is the relationship between S and T?

(d)   Write down the matrix A_R of the linear transformation R given by $R(e_1) = w_1$, $R(e_2) = w_2$, $R(e_3) = w_3$.

(e)   Is RS defined? What does this linear transformation do to v_1, v_2 and v_3? Find the matrix A_{RS} and use it to check your answer.

Problem 7.3 For each of the following linear transformations, find a basis for the null space of T, $N(T)$, and a basis for the range of T, $R(T)$. Verify the rank–nullity theorem in each case. If any of the linear transformations are invertible, find the inverse, T^{-1}.

(a) $T : \mathbb{R}^2 \to \mathbb{R}^3$ given by $T\begin{pmatrix} x \\ y \end{pmatrix} = \begin{pmatrix} x + 2y \\ 0 \\ 0 \end{pmatrix}.$

(b) $T : \mathbb{R}^3 \to \mathbb{R}^3$ given by $T\begin{pmatrix} x \\ y \\ z \end{pmatrix} = \begin{pmatrix} x + y + z \\ y + z \\ z \end{pmatrix}.$

(c) $T : \mathbb{R}^3 \to \mathbb{R}^3$ given by $T\begin{pmatrix} x \\ y \\ z \end{pmatrix} = \begin{pmatrix} 1 & 1 & 0 \\ 0 & 1 & 1 \\ -1 & 0 & 1 \end{pmatrix} \begin{pmatrix} x \\ y \\ z \end{pmatrix}.$

Problem 7.4 Consider the vectors

$$\mathbf{v}_1 = \begin{pmatrix} 1 \\ 2 \\ 0 \\ -1 \end{pmatrix}, \quad \mathbf{v}_2 = \begin{pmatrix} 1 \\ 1 \\ 1 \\ 1 \end{pmatrix}, \quad \mathbf{v}_3 = \begin{pmatrix} 4 \\ 5 \\ 3 \\ 2 \end{pmatrix}, \quad \mathbf{w} = \begin{pmatrix} 5 \\ 5 \\ 2 \\ 2 \end{pmatrix}.$$

Let $\{\mathbf{e}_1, \mathbf{e}_2, \mathbf{e}_3\}$ be the standard basis in $\mathbb{R}^3$ and let $T : \mathbb{R}^3 \to \mathbb{R}^4$ be the linear transformation defined by

$$T(\mathbf{e}_1) = \mathbf{v}_1, \quad T(\mathbf{e}_2) = \mathbf{v}_2, \quad T(\mathbf{e}_3) = \mathbf{v}_3.$$

Write down the matrix A_T such that $T(\mathbf{x}) = A_T\mathbf{x}$.

What is the dimension of $R(T)$, the range of T? Is the vector $\mathbf{w}$ in $R(T)$? Justify your answers.

State the rank–nullity theorem for linear transformations and use it to determine the dimension of the null space of T, $N(T)$.

Find a basis of $N(T)$.

Problem 7.5 If any of the linear transformations T_i given below can be defined, write down a matrix A_{T_i} such that $T_i(\mathbf{x}) = A_{T_i}\mathbf{x}$. Otherwise, state why T_i is not defined.

$T_1 : \mathbb{R}^3 \to \mathbb{R}^2$, where the null space of T_1 is the x axis and the range of T_1 is the line $y = x$.

$T_2 : \mathbb{R}^2 \to \mathbb{R}^3$, such that $N(T_2) = \{0\}$ and

$$R(T_2) = \mathrm{Lin}\left\{ \begin{pmatrix} 1 \\ 0 \\ 1 \end{pmatrix}, \begin{pmatrix} 1 \\ 1 \\ 1 \end{pmatrix}, \begin{pmatrix} 2 \\ 1 \\ 2 \end{pmatrix} \right\}.$$

$T_3 : \mathbb{R}^2 \to \mathbb{R}^3$, where the null space of T_3 is the line $y = 2x$ and the range of T_3 is the line

$$-\frac{x}{4} = \frac{y}{5} = z.$$

Problem 7.6 Let V and W be the subspaces

$$V = \text{Lin} \left\{ \begin{pmatrix} 1 \\ 0 \\ -1 \end{pmatrix}, \begin{pmatrix} 1 \\ -1 \\ 0 \end{pmatrix} \right\} \subset \mathbb{R}^3,$$

$$W = \text{Lin} \left\{ \begin{pmatrix} 1 \\ 3 \\ 1 \\ 0 \end{pmatrix}, \begin{pmatrix} -2 \\ 5 \\ 0 \\ 1 \end{pmatrix} \right\} \subset \mathbb{R}^4.$$

Consider the possibility of a linear transformation T and a linear transformation S such that

$$T : \mathbb{R}^a \to \mathbb{R}^b \quad \text{with} \quad N(T) = V \quad \text{and} \quad R(T) = W;$$

$$S : \mathbb{R}^c \to \mathbb{R}^d \quad \text{with} \quad N(S) = W \quad \text{and} \quad R(S) = V.$$

Show that one of these, S or T, cannot exist. Then find a matrix A_S or A_T representing the other linear transformation (with respect to the standard basis in each Euclidean space).

Check your answer by row reducing the matrix and finding the null space and range of the linear transformation.

Problem 7.7 Consider the linear transformations $T : \mathbb{R}^2 \to \mathbb{R}^3$ and $S : \mathbb{R}^3 \to \mathbb{R}^3$ with matrices

$$A_T = \begin{pmatrix} 1 & -3 \\ -2 & 6 \\ -1 & 3 \end{pmatrix}, \qquad A_S = \begin{pmatrix} 1 & 0 & 1 \\ -1 & 1 & 0 \\ 0 & 2 & 1 \end{pmatrix}.$$

Find the null space of T, $N(T)$, and the range of T, $R(T)$. Describe each subspace or write down a basis. Do the same for the null space of S, $N(S)$, and the range of S, $R(S)$.

Which linear transformation is defined: ST or TS?

Deduce the null space of the composed linear transformation.

Use the rank–nullity theorem to find the dimension of the range of the composed linear transformation.

Problem 7.8 Let

$$A = \begin{pmatrix} 5 & 9 & -1 & 15 \\ -3 & 1 & 7 & -4 \\ 1 & 2 & 0 & 3 \end{pmatrix}, \qquad d = \begin{pmatrix} 4 \\ 4 \\ 1 \end{pmatrix}.$$

(a) Let T denote the linear transformation $T(x) = Ax$. Find a basis of the null space of T, $N(T)$. Find the range of T, $R(T)$. Show that $d \in R(T)$. Find all vectors x such that $T(x) = d$.

(b)   Let S be a linear transformation, $S : \mathbb{R}^3 \to \mathbb{R}^2$, such that the range of S is $\mathbb{R}^2$ and the null space of S is the subspace:

$$N(S) = \{\mathbf{x} \mid \mathbf{x} = t\mathbf{d}, \ t \in \mathbb{R}\},$$

where $\mathbf{d}$ is the vector given above.

Consider the composition ST of linear transformations S and T. Deduce the range of ST from the ranges of T and S. Then use the rank–nullity theorem to determine the dimension of $N(ST)$, the null space of ST.

Find a basis of $N(ST)$.

Problem 7.9 Let $\mathbf{c}_1$, $\mathbf{c}_2$, $\mathbf{c}_3$ denote the columns of the matrix P, where

$$P = \begin{pmatrix} 1 & 0 & 1 \\ 1 & 1 & 2 \\ 3 & -2 & 4 \end{pmatrix}.$$

Find the determinant of P. Why can you deduce that the set $B = \{\mathbf{c}_1, \mathbf{c}_2, \mathbf{c}_3\}$ is a basis of $\mathbb{R}^3$?

If $\mathbf{w} = (1, 1, 0)^{\mathrm{T}}$ in standard coordinates, find the coordinates of $\mathbf{w}$ in the basis B, $[\mathbf{w}]_B$.

Find the vector $\mathbf{v}$ in standard coordinates if $[\mathbf{v}]_B = \begin{bmatrix} 6 \\ -1 \\ -2 \end{bmatrix}_B$.

Problem 7.10 Show that each of the following sets B and $\widehat{B}$ is a basis of $\mathbb{R}^3$:

$$B = \left\{ \begin{pmatrix} 1 \\ 0 \\ 1 \end{pmatrix}, \begin{pmatrix} 1 \\ 1 \\ 3 \end{pmatrix}, \begin{pmatrix} 0 \\ 0 \\ 1 \end{pmatrix} \right\} \quad \text{and} \quad \widehat{B} = \left\{ \begin{pmatrix} 1 \\ 1 \\ 1 \end{pmatrix}, \begin{pmatrix} 1 \\ -1 \\ 0 \end{pmatrix}, \begin{pmatrix} 0 \\ 1 \\ -1 \end{pmatrix} \right\}.$$

Write down the transition matrix P from B coordinates to standard coordinates. Write down the transition matrix Q from $\widehat{B}$ coordinates to standard coordinates.

Find the transition matrix from $\widehat{B}$ coordinates to B coordinates.

If $[\mathbf{x}]_{\widehat{B}} = \begin{bmatrix} 2 \\ 1 \\ 3 \end{bmatrix}_{\widehat{B}}$, find $[\mathbf{x}]_B$.

Problem 7.11

(a)   Change the basis in $\mathbb{R}^2$ by a rotation of the axes through an angle of $\pi/6$ *clockwise*: First write down the matrix of the linear transformation which accomplishes this rotation and then write down the new basis vectors, $\mathbf{v}_1$ and $\mathbf{v}_2$ (which are the images of $\mathbf{e}_1$ and $\mathbf{e}_2$).

Let $B = \{v_1, v_2\}$ be the new basis. Write down the transition matrix P from B coordinates to standard coordinates.

(b)   The curve C is given in standard coordinates, x, y, by the equation $3x^2 + 2\sqrt{3}xy + 5y^2 = 6$. Find the equation of the curve in the new B coordinates, (X, Y).

Use this information to sketch the curve C in the xy-plane.

Problem 7.12 Suppose

$$M = \left\{ \begin{pmatrix} 2 \\ 1 \end{pmatrix}, \begin{pmatrix} -1 \\ 1 \end{pmatrix} \right\}, \quad v = \begin{pmatrix} -1 \\ 2 \end{pmatrix}, \quad T\begin{pmatrix} x \\ y \end{pmatrix} = \begin{pmatrix} 7x - 2y \\ -x + 8y \end{pmatrix}.$$

(a)   Show that M is a basis of $\mathbb{R}^2$. Write down the transition matrix from M coordinates to standard coordinates. Find $[v]_M$, the M coordinates of the vector v.

(b)   Write down the matrix A of the linear transformation

$$T : \mathbb{R}^2 \to \mathbb{R}^2$$

with respect to the standard basis.

Find the matrix of T in M coordinates. Call it D.

Describe geometrically the effect of the transformation T as a map from $\mathbb{R}^2 \to \mathbb{R}^2$.

(c)   Find the image of $[v]_M$ using the matrix D.

Check your answer using standard coordinates.

8

Diagonalisation

One of the most useful techniques in applications of matrices and linear algebra is *diagonalisation*. This relies on the topic of *eigenvalues and eigenvectors*, and is related to change of basis. We will learn how to find eigenvalues and eigenvectors of an $n \times n$ matrix, how to diagonalise a matrix when it is possible to do so and also how to recognise when it is not possible. We shall see in the next chapter how useful a technique diagonalisation is.

All matrices in this chapter are square $n \times n$ matrices with real entries, so all vectors will be in $\mathbb{R}^n$ for some n.

8.1   Eigenvalues and eigenvectors

8.1.1   Definition of eigenvalues and eigenvectors

The first important ideas we need are those of eigenvalues and their corresponding eigenvectors.

Definition 8.1 Suppose that A is a square matrix. The number λ is said to be an *eigenvalue* of A if for some *non-zero* vector $\mathbf{x}$,

$$A\mathbf{x} = \lambda\mathbf{x}.$$

Any *non-zero* vector $\mathbf{x}$ for which this equation holds is called an *eigenvector for eigenvalue* λ or an *eigenvector of A corresponding to eigenvalue* λ.

8.1.2   Finding eigenvalues and eigenvectors

To determine whether λ is an eigenvalue of A, we need to determine whether there are any non-zero solutions $\mathbf{x}$ to the matrix equation

$A\mathbf{x} = \lambda\mathbf{x}$. Note that the matrix equation $A\mathbf{x} = \lambda\mathbf{x}$ is not of the standard form, since the right-hand side is not a fixed vector $\mathbf{b}$, but depends explicitly on $\mathbf{x}$. However, we can rewrite it in standard form. Note that $\lambda\mathbf{x} = \lambda I\mathbf{x}$, where I is, as usual, the identity matrix. So, the equation is equivalent to $A\mathbf{x} = \lambda I\mathbf{x}$, or $A\mathbf{x} - \lambda I\mathbf{x} = \mathbf{0}$, which is equivalent to $(A - \lambda I)\mathbf{x} = \mathbf{0}$.

Now, a square linear system $B\mathbf{x} = \mathbf{0}$ has solutions other than $\mathbf{x} = \mathbf{0}$ precisely when $|B| = 0$. Therefore, taking $B = A - \lambda I$, λ is an eigenvalue if and only if the determinant of the matrix $A - \lambda I$ is zero. This determinant, $p(\lambda) = |A - \lambda I|$ is a polynomial of degree n in the variable λ.

Definition 8.2 (Characteristic polynomial and equation) The polynomial $|A - \lambda I|$ is known as the *characteristic polynomial* of A, and the equation $|A - \lambda I| = 0$ is called the *characteristic equation* of A.

To find the eigenvalues, we solve the characteristic equation $|A - \lambda I| = 0$. Let us illustrate with a 2×2 example.

Example 8.3 Let

$$A = \begin{pmatrix} 7 & -15 \\ 2 & -4 \end{pmatrix}.$$

Then

$$A - \lambda I = \begin{pmatrix} 7 & -15 \\ 2 & -4 \end{pmatrix} - \lambda \begin{pmatrix} 1 & 0 \\ 0 & 1 \end{pmatrix} = \begin{pmatrix} 7 - \lambda & -15 \\ 2 & -4 - \lambda \end{pmatrix}$$

and the characteristic polynomial is

$$\begin{aligned} |A - \lambda I| &= \begin{vmatrix} 7 - \lambda & -15 \\ 2 & -4 - \lambda \end{vmatrix} \\ &= (7 - \lambda)(-4 - \lambda) + 30 \\ &= \lambda^2 - 3\lambda - 28 + 30 \\ &= \lambda^2 - 3\lambda + 2. \end{aligned}$$

So the eigenvalues are the solutions of $\lambda^2 - 3\lambda + 2 = 0$. To solve this for λ, we could use either the formula for the solutions to a quadratic equation, or simply observe that the characteristic polynomial factorises. We have $(\lambda - 1)(\lambda - 2) = 0$ with solutions $\lambda = 1$ and $\lambda = 2$. Hence the eigenvalues of A are 1 and 2, and these are the *only* eigenvalues of A.

To find an eigenvector for each eigenvalue λ, we have to find a nontrivial solution to $(A - \lambda I)\mathbf{x} = \mathbf{0}$, meaning a solution other than the zero vector. (We stress the fact that eigenvectors cannot be the zero vector because this is a mistake many students make.) This is easy, since for a particular value of λ, all we need to do is solve a simple linear system. We illustrate by finding the eigenvectors for the matrix of Example 8.3.

Example 8.4 We find the eigenvectors of

$$A = \begin{pmatrix} 7 & -15 \\ 2 & -4 \end{pmatrix}.$$

We have seen that the eigenvalues are 1 and 2. To find the eigenvectors for eigenvalue 1, we solve the system $(A - I)\mathbf{x} = \mathbf{0}$. We do this by putting the coefficient matrix $A - I$ into reduced echelon form.

$$(A - I) = \begin{pmatrix} 6 & -15 \\ 2 & -5 \end{pmatrix} \longrightarrow \cdots \longrightarrow \begin{pmatrix} 1 & -\frac{5}{2} \\ 0 & 0 \end{pmatrix}.$$

This system has solutions

$$\mathbf{v} = t\begin{pmatrix} 5 \\ 2 \end{pmatrix}, \quad \text{for any} \quad t \in \mathbb{R}.$$

There are infinitely many eigenvectors for 1: for each $t \neq 0$, $\mathbf{v}$ is an eigenvector of A corresponding to $\lambda = 1$. But be careful not to think that you can choose $t = 0$; for then $\mathbf{v}$ becomes the zero vector, and this is never an eigenvector, simply by definition. To find the eigenvectors for 2, we solve $(A - 2I)\mathbf{x} = \mathbf{0}$ by reducing the coefficient matrix,

$$(A - 2I) = \begin{pmatrix} 5 & -15 \\ 2 & -6 \end{pmatrix} \longrightarrow \cdots \longrightarrow \begin{pmatrix} 1 & -3 \\ 0 & 0 \end{pmatrix}.$$

Setting the non-leading variable equal to t, we obtain the solutions

$$\mathbf{v} = t\begin{pmatrix} 3 \\ 1 \end{pmatrix}, \quad t \in \mathbb{R}.$$

Any non-zero scalar multiple of the vector $(3, 1)^{\mathrm{T}}$ is an eigenvector of A for eigenvalue 2.

Note that, in this example, each system of equations is simple enough to be solved directly. For example, if $\mathbf{x} = (x_1, x_2)^{\mathrm{T}}$, the system $(A - 2I)\mathbf{x} = \mathbf{0}$ consists of the equations

$$5x_1 - 15x_2 = 0, \qquad 2x_1 - 6x_2 = 0.$$

Clearly, both equations are equivalent to $x_1 = 3x_2$. If we set $x_2 = t$ for any real number t, then we obtain the eigenvectors for $\lambda = 2$ as before. However, we prefer to use row operations. There are two reasons for this. The first reason is that the system of equations may not be as simple as the one just given, particularly for an $n \times n$ matrix where $n > 2$. The second reason is that putting the matrix $A - \lambda I$ into echelon form provides a useful check on the eigenvalue. If $|A - \lambda I| = 0$, the echelon form of $A - \lambda I$ must have a row of zeros, so the system $(A - \lambda I)\mathbf{x} = \mathbf{0}$ has a non-trivial solution. If we have reduced the matrix $(A - \lambda_0 I)$ for some supposed eigenvalue λ_0 and do not obtain a zero row, we know immediately that there is an error, either in the row reduction or in the choice of λ_0, and we can go back and correct it.

We now give two examples with 3×3 matrices.

Example 8.5 Suppose that

$$A = \begin{pmatrix} 4 & 0 & 4 \\ 0 & 4 & 4 \\ 4 & 4 & 8 \end{pmatrix}.$$

Let's find the eigenvalues of A and corresponding eigenvectors for each eigenvalue.

To find the eigenvalues, we solve $|A - \lambda I| = 0$. Now,

$$
\begin{aligned}
|A - \lambda I| &= \begin{vmatrix} 4 - \lambda & 0 & 4 \\ 0 & 4 - \lambda & 4 \\ 4 & 4 & 8 - \lambda \end{vmatrix} \\
&= (4 - \lambda) \begin{vmatrix} 4 - \lambda & 4 \\ 4 & 8 - \lambda \end{vmatrix} + 4 \begin{vmatrix} 0 & 4 - \lambda \\ 4 & 4 \end{vmatrix} \\
&= (4 - \lambda)((4 - \lambda)(8 - \lambda) - 16) + 4(-4(4 - \lambda)) \\
&= (4 - \lambda)((4 - \lambda)(8 - \lambda) - 16) - 16(4 - \lambda).
\end{aligned}
$$

We notice that each of the two terms in this expression has $4 - \lambda$ as a factor, so instead of expanding everything, we take $4 - \lambda$ out as a common factor, obtaining

$$
\begin{aligned}
|A - \lambda I| &= (4 - \lambda)((4 - \lambda)(8 - \lambda) - 16 - 16) \\
&= (4 - \lambda)(32 - 12\lambda + \lambda^2 - 32) \\
&= (4 - \lambda)(\lambda^2 - 12\lambda) \\
&= (4 - \lambda)\lambda(\lambda - 12).
\end{aligned}
$$

It follows that the eigenvalues are 4, 0, 12. (The characteristic polynomial will not always factorise so easily. Here it was simple because of the common factor $4 - \lambda$. The next example is more difficult.)

To find an eigenvector for 4, we have to solve the equation $(A - 4I)\mathbf{x} = \mathbf{0}$ for $\mathbf{x} = (x_1, x_2, x_3)^{\mathrm{T}}$. Using row operations, we have

$$\begin{pmatrix} 0 & 0 & 4 \\ 0 & 0 & 4 \\ 4 & 4 & 4 \end{pmatrix} \longrightarrow \cdots \longrightarrow \begin{pmatrix} 1 & 1 & 0 \\ 0 & 0 & 1 \\ 0 & 0 & 0 \end{pmatrix}.$$

Thus, $x_3 = 0$ and setting the free variable $x_2 = t$, the solutions are

$$\mathbf{x} = t \begin{pmatrix} -1 \\ 1 \\ 0 \end{pmatrix}, \qquad t \in \mathbb{R}.$$

So the eigenvectors for $\lambda = 4$ are the non-zero multiples of

$$\mathbf{v}_1 = \begin{pmatrix} -1 \\ 1 \\ 0 \end{pmatrix}.$$

Activity 8.6 Determine the eigenvectors for 0 and 12. Check your answers: verify that $A\mathbf{v} = \lambda\mathbf{v}$ for each eigenvalue and one corresponding eigenvector.

Example 8.7 Let

$$A = \begin{pmatrix} -3 & -1 & -2 \\ 1 & -1 & 1 \\ 1 & 1 & 0 \end{pmatrix}.$$

Given that -1 is an eigenvalue of A, find all the eigenvalues of A.

We calculate the characteristic polynomial of A:

$$|A - \lambda I| = \begin{vmatrix} -3 - \lambda & -1 & -2 \\ 1 & -1 - \lambda & 1 \\ 1 & 1 & -\lambda \end{vmatrix}$$

$$= (-3 - \lambda) \begin{vmatrix} -1 - \lambda & 1 \\ 1 & -\lambda \end{vmatrix} - (-1) \begin{vmatrix} 1 & 1 \\ 1 & -\lambda \end{vmatrix} - 2 \begin{vmatrix} 1 & -1 - \lambda \\ 1 & 1 \end{vmatrix}$$

$$= (-3 - \lambda)(\lambda^2 + \lambda - 1) + (-\lambda - 1) - 2(2 + \lambda)$$

$$= -\lambda^3 - 4\lambda^2 - 5\lambda - 2 = -(\lambda^3 + 4\lambda^2 + 5\lambda + 2).$$

Now, the fact that -1 is an eigenvalue means that -1 is a solution of the equation $|A - \lambda I| = 0$, which means that $\lambda - (-1)$ (that is, $\lambda + 1$) is a factor of the characteristic polynomial $|A - \lambda I|$. So this characteristic polynomial can be written in the form

$$-(\lambda + 1)(a\lambda^2 + b\lambda + c).$$

Clearly, we must have $a = 1$ and $c = 2$ to obtain the correct λ^3 term and the correct constant. So the polynomial can be written as $-(\lambda + 1)(\lambda^2 + b\lambda + 2)$. Using this, and comparing the coefficients of either λ^2 or λ with the cubic polynomial, we find $b = 3$. For instance, think about the term involving λ^2. We know that the characteristic polynomial has the following term: $-4\lambda^2$. On the other hand, if we look at how the expression $-(\lambda + 1)(\lambda^2 + b\lambda + 2)$ would be expanded, it would generate the term $-\lambda^2 - b\lambda^2$. So we must have $-1 - b = -4$

and hence $b = 3$. In other words, the characteristic polynomial is

$$-(\lambda^3 + 4\lambda^2 + 5\lambda + 2) = -(\lambda + 1)(\lambda^2 + 3\lambda + 2)$$
$$= -(\lambda + 1)(\lambda + 2)(\lambda + 1).$$

Activity 8.8 Perform the calculations to check that $b = 3$ and that the characteristic polynomial factorises as stated.

We have, $|A - \lambda I| = -(\lambda + 1)^2(\lambda + 2)$. The eigenvalues are the solutions to $|A - \lambda I| = 0$, so they are $\lambda = -1$ and $\lambda = -2$.

Note that in this case there are only two distinct eigenvalues. We say that the eigenvalue -1 has occurred twice, or that $\lambda = -1$ is an eigenvalue of *multiplicity* 2. We will find the eigenvectors when we look at this example again in Section 8.3.

8.1.3 Eigenspaces

If A is an $n \times n$ matrix and λ is an eigenvalue of A, then the set of eigenvectors corresponding to the eigenvalue λ together with the zero vector, $\mathbf{0}$, is a subspace of $\mathbb{R}^n$. Why?

We have already seen that the null space of any $m \times n$ matrix is a subspace of $\mathbb{R}^n$. The null space of the $n \times n$ matrix $A - \lambda I$, consists of all solutions to the matrix equation $(A - \lambda I)\mathbf{x} = \mathbf{0}$, which is precisely the set of all eigenvectors corresponding to λ, together with the vector $\mathbf{0}$. We give this a special name.

Definition 8.9 (Eigenspace) If A is an $n \times n$ matrix and λ is an eigenvalue of A, then the *eigenspace* of the eigenvalue λ is the subspace $N(A - \lambda I)$ of $\mathbb{R}^n$.

The eigenspace of an eigenvalue λ can also be described as the set S, where

$$S = \{\mathbf{x} \mid A\mathbf{x} = \lambda\mathbf{x}\}.$$

Activity 8.10 Show this.

In Exercise 5.2, you showed that the set $S = \{\mathbf{x} \mid A\mathbf{x} = \lambda\mathbf{x}\}$ is a subspace of $\mathbb{R}^n$ for any $\lambda \in \mathbb{R}$. If λ *is not* an eigenvalue of A, then S contains only the zero vector, $S = \{\mathbf{0}\}$, and $\dim(S) = 0$. When, and only when, λ *is* an eigenvalue of A do we know that there is a non-zero vector in S, and hence $\dim(S) \geq 1$. In this case, S is the eigenspace of the eigenvalue λ.

8.1.4 Eigenvalues and the matrix

We now explore how the eigenvalues of a matrix are related to other quantities associated with it, specifically the determinant (with which we are already familiar) and the trace.

There is a straightforward relationship between the eigenvalues of a matrix A and its determinant. Suppose A is an $n \times n$ matrix. Then the characteristic polynomial of A is a polynomial of degree n in λ:

$$p(\lambda) = |A - \lambda I| = (-1)^n (\lambda^n + a_{n-1} \lambda^{n-1} + \cdots + a_0).$$

Let $\lambda_1, \lambda_2, \ldots, \lambda_n$ be the eigenvalues of A, with multiple roots listed each time they occur. In terms of the eigenvalues, the characteristic polynomial factors as

$$p(\lambda) = |A - \lambda I| = (-1)^n (\lambda - \lambda_1)(\lambda - \lambda_2) \cdots (\lambda - \lambda_n).$$

For instance, let's look at the matrix in Example 8.7,

$$A = \begin{pmatrix} -3 & -1 & -2 \\ 1 & -1 & 1 \\ 1 & 1 & 0 \end{pmatrix}.$$

The eigenvalues of A are $\lambda = -1$, of multiplicity 2, and $\lambda = -2$. So we may list the eigenvalues as $\lambda_1 = \lambda_2 = -1$ and $\lambda_3 = -2$. Then,

$$p(\lambda) = (-1)^3 (\lambda - \lambda_1)(\lambda - \lambda_2)(\lambda - \lambda_3) = -(\lambda + 1)(\lambda + 1)(\lambda + 2),$$

as we saw earlier.

If we let $\lambda = 0$ in the equation

$$p(\lambda) = |A - \lambda I| = (-1)^n (\lambda - \lambda_1)(\lambda - \lambda_2) \cdots (\lambda - \lambda_n),$$

then we obtain the constant term of the polynomial,

$$p(0) = |A| = (-1)^n a_0 = (-1)^n (-1)^n (\lambda_1 \lambda_2 \ldots \lambda_n) = \lambda_1 \lambda_2 \ldots \lambda_n.$$

Therefore, we have proved the following.

Theorem 8.11 *The determinant of an $n \times n$ matrix A is equal to the product of its eigenvalues.*

Example 8.12 Let's look again at the matrix in Example 8.7,

$$A = \begin{pmatrix} -3 & -1 & -2 \\ 1 & -1 & 1 \\ 1 & 1 & 0 \end{pmatrix}.$$

The eigenvalues of A are $\lambda_1 = \lambda_2 = -1$ and $\lambda_3 = -2$. Calculating the determinant of the matrix, we find $|A| = -2$, which is indeed the product of the three eigenvalues.

Activity 8.13 Check this. Calculate $|A|$.

Now look at the sum of the diagonal entries of the matrix in the example just given, and notice that it is equal to the sum of the three eigenvalues. This is true in general. The sum of the diagonal entries of a matrix is known as the *trace* of the matrix.

Definition 8.14 (Trace) The *trace* of a square matrix A is the sum of the entries on its main diagonal.

Theorem 8.15 *The trace of an $n \times n$ matrix A is equal to the sum of its eigenvalues.*

Proof: We can obtain this result by examining the equations

$$|A - \lambda I| = (-1)^n(\lambda^n + a_{n-1}\lambda^{n-1} + \cdots + a_0)$$
$$= (-1)^n(\lambda - \lambda_1)(\lambda - \lambda_2)\cdots(\lambda - \lambda_n)$$

again, this time looking at the coefficient of λ^{n-1}. You can consider the proof optional and safely omit it, but if you wish to see how it works, read on.

The coefficient of λ^{n-1} is $(-1)^n a_{n-1}$ in the middle expression, but what we are actually interested in is how the coefficient of λ^{n-1} is obtained from the other two expressions. First, think about how it is obtained from the factorised polynomial

$$(-1)^n(\lambda - \lambda_1)(\lambda - \lambda_2)\cdots(\lambda - \lambda_n)$$

when the factors are multiplied together. Ignoring the $(-1)^n$ for the moment, if we multiply all the λs together, one from each factor, we obtain the term λ^n. So to obtain the terms with λ^{n-1}, we need to multiply first $-\lambda_1$ times the λs in all the remaining factors, then $-\lambda_2$ times the λs in all the other factors and so on. Putting back the factor $(-1)^n$, the term involving λ^{n-1} is

$$(-1)^n(-\lambda_1 - \lambda_2 - \cdots - \lambda_n)\lambda^{n-1} = (-1)^{n-1}(\lambda_1 + \lambda_2 + \cdots + \lambda_n)\lambda^{n-1}. \tag{1}$$

Now let's look at the coefficient of λ^{n-1} in the expansion of the determinant, $|A - \lambda I|$. This is far more complicated, and we will use an inductive argument.

If A is a 2×2 matrix,

$$A = \begin{pmatrix} a_{11} & a_{12} \\ a_{21} & a_{22} \end{pmatrix},$$

then

$$|A - \lambda I| = \begin{vmatrix} a_{11} - \lambda & a_{12} \\ a_{21} & a_{22} - \lambda \end{vmatrix} = \lambda^2 - (a_{11} + a_{22})\lambda + |A|.$$

We see that the coefficient of λ is (-1) times the trace of A.

Now consider a 3×3 matrix A. We have

$$|A - \lambda I| = \begin{vmatrix} a_{11} - \lambda & a_{12} & a_{13} \\ a_{21} & a_{22} - \lambda & a_{23} \\ a_{31} & a_{32} & a_{33} - \lambda \end{vmatrix}.$$

Expanding by the first row, we see that the only term which contains powers of λ higher than 1 comes from the $(1, 1)$ entry times the $(1, 1)$ cofactor; that is, $(a_{11} - \lambda)C_{11}$. But C_{11} is the determinant of a 2×2 matrix, so we are looking at the λ^2 terms of

$$(a_{11} - \lambda)(\lambda^2 - (a_{22} + a_{33})\lambda + (a_{22}a_{33} - a_{23}a_{32})).$$

The λ^2 terms are

$$a_{11}\lambda^2 + (a_{22} + a_{33})\lambda^2 = (a_{11} + a_{22} + a_{33})\lambda^2,$$

so the coefficient of λ^2 is $(-1)^2$ times the trace of A.

What we have seen so far makes us fairly certain that the coefficient of the term λ^{n-1} in the expansion of the determinant $|A - \lambda I|$ for an $n \times n$ matrix A is equal to $(-1)^{n-1}$ times the trace of A. We have shown that this is true for any 2×2 and any 3×3 matrix. We now assume it is true for any $(n - 1) \times (n - 1)$ matrix, and then show that this implies it is also true for any $n \times n$ matrix. In this way, starting with $n = 2$, we will know it is true for all $n \times n$ matrices.

So suppose $A = (a_{ij})$ is an $n \times n$ matrix and look at the coefficient of λ^{n-1} in the cofactor expansion of $|A - \lambda I|$ by row 1:

$$|A - \lambda I| = \begin{vmatrix} a_{11} - \lambda & a_{12} & \cdots & a_{1n} \\ a_{21} & a_{22} - \lambda & \cdots & a_{2n} \\ \vdots & \vdots & \ddots & \vdots \\ a_{n1} & a_{n2} & \cdots & a_{nn} - \lambda \end{vmatrix}$$
$$= (a_{11} - \lambda)C_{11} + a_{12}C_{12} + \cdots.$$

Only the first term of the cofactor expansion, $(a_{11} - \lambda)C_{11}$, contains higher powers of λ than λ^{n-2}.

Activity 8.16 Look at the other terms to see why this is true.

Now C_{11} is the determinant of the $(n - 1) \times (n - 1)$ matrix obtained from the matrix $(A - \lambda I)$ by crossing out the first row and first column, so it is of the form $|C - \lambda I|$, where C is the $(n - 1) \times (n - 1)$

matrix obtained from A by crossing out the first row and first column. Therefore, by our assumption,

$$|A - \lambda I|$$
$$= (a_{11} - \lambda)C_{11}$$
$$= (a_{11} - \lambda)((-1)^{n-1}\lambda^{n-1} + (-1)^{n-2}(a_{22} + \cdots + a_{nn})\lambda^{n-2} + \cdots)$$
$$= (-1)^n\lambda^n + (-1)^{n-1}(a_{11} + a_{22} + \cdots + a_{nn})\lambda^{n-1} + \cdots.$$

We can now conclude that the term involving λ^{n-1} in the expansion of $|A - \lambda I|$ for any $n \times n$ matrix A is equal to

$$(-1)^{n-1}(a_{11} + a_{22} + a_{33} + \cdots + a_{nn})\lambda^{n-1}. \tag{2}$$

Comparing the coefficients of λ^{n-1} in the two expressions (1) and (2), we see that

$$a_{11} + a_{22} + a_{33} + \cdots + a_{nn} = \lambda_1 + \lambda_2 + \cdots + \lambda_n;$$

that is, the trace of A is equal to the sum of the eigenvalues, $\qquad\square$

8.2 Diagonalisation of a square matrix

8.2.1 Diagonalisation

Recall that square matrices A and M are *similar* if there is an invertible matrix P such that $P^{-1}AP = M$. We met this idea earlier when we looked at how a matrix representing a linear transformation changes when the basis is changed. We now begin to explore why this is such an important and useful concept.

Definition 8.17 (Diagonalisable matrix) The matrix A is *diagonalisable* if it is similar to a diagonal matrix; in other words, if there is a diagonal matrix D and an invertible matrix P such that $P^{-1}AP = D$.

When we find suitable P and D such that $P^{-1}AP = D$, we say that we are *diagonalising* A.

Example 8.18 The matrix

$$A = \begin{pmatrix} 7 & -15 \\ 2 & -4 \end{pmatrix}$$

from Example 8.3 is diagonalisable, because if we take P to be

$$P = \begin{pmatrix} 5 & 3 \\ 2 & 1 \end{pmatrix},$$

then P is invertible, with

$$P^{-1} = \begin{pmatrix} -1 & 3 \\ 2 & -5 \end{pmatrix}$$

and, as you can check,

$$P^{-1}AP = D = \begin{pmatrix} 1 & 0 \\ 0 & 2 \end{pmatrix},$$

which is a diagonal matrix.

Activity 8.19 Check this! Obtain the product $P^{-1}AP$ by first multiplying AP and then multiplying on the left by P^{-1}. What do you notice about AP?

The example just given probably raises a number of questions in your mind. Prominent among those will be: 'How was such a matrix P found?' (Have a look back at Example 8.4. What do you notice?) A more general question is: 'When will a matrix be diagonalisable?' To answer both of these questions, we start by outlining a general method for diagonalising a matrix (when it is possible).

8.2.2   General method

Let's first suppose that the $n \times n$ matrix A is diagonalisable. So, assume that $P^{-1}AP = D$, where D is a diagonal matrix

$$D = \text{diag}(\lambda_1, \lambda_2, \ldots, \lambda_n) = \begin{pmatrix} \lambda_1 & 0 & \cdots & 0 \\ 0 & \lambda_2 & \cdots & 0 \\ 0 & 0 & \ddots & 0 \\ 0 & 0 & \cdots & \lambda_n \end{pmatrix}.$$

(Note the useful notation for describing the diagonal matrix D.) Then, since $P^{-1}AP = D$, we have $AP = PD$. Suppose the columns of P are the vectors $\mathbf{v}_1, \mathbf{v}_2, \ldots, \mathbf{v}_n$. Then, thinking about how matrix multiplication works (see Activity 7.33), we can see that

$$AP = A(\mathbf{v}_1 \ \ldots \ \mathbf{v}_n) = (A\mathbf{v}_1 \ \ldots \ A\mathbf{v}_n).$$

Furthermore,

$$PD = (\mathbf{v}_1 \ldots \mathbf{v}_n) \begin{pmatrix} \lambda_1 & 0 & \cdots & 0 \\ 0 & \lambda_2 & \cdots & 0 \\ 0 & 0 & \ddots & 0 \\ 0 & 0 & \cdots & \lambda_n \end{pmatrix} = (\lambda_1\mathbf{v}_1 \ \ldots \ \lambda_n\mathbf{v}_n).$$

So this means that

$$A\mathbf{v}_1 = \lambda_1\mathbf{v}_1, \quad A\mathbf{v}_2 = \lambda_2\mathbf{v}_2, \quad \ldots, \quad A\mathbf{v}_n = \lambda_n\mathbf{v}_n.$$

The fact that P^{-1} exists means that none of the vectors $\mathbf{v}_i$ is the zero vector (because any matrix with a column of zeros would not be invertible). So this means that (for $i = 1, 2, \ldots, n$), $\mathbf{v}_i$ is a non-zero vector with the property that $A\mathbf{v}_i = \lambda_i\mathbf{v}_i$. But this means precisely that λ_i is an eigenvalue of A and that $\mathbf{v}_i$ is a corresponding eigenvector. Since P has an inverse, these eigenvectors are linearly independent. Therefore, A has n linearly independent eigenvectors.

Conversely, suppose A has n linearly independent eigenvectors, $\mathbf{v}_1, \mathbf{v}_2, \ldots, \mathbf{v}_n$, which correspond to eigenvalues $\lambda_1, \lambda_2, \ldots, \lambda_n$. Let P be the matrix whose columns are these eigenvectors: $P = (\mathbf{v}_1 \ldots \mathbf{v}_n)$. Because the columns are linearly independent, P will be invertible. Furthermore, since $A\mathbf{v}_i = \lambda_i\mathbf{v}_i$, it follows that

$$
\begin{aligned}
AP &= A(\mathbf{v}_1 \ldots \mathbf{v}_n) \\
&= (A\mathbf{v}_1 \ldots A\mathbf{v}_n) \\
&= (\lambda_1\mathbf{v}_1 \ldots \lambda_n\mathbf{v}_n) \\
&= (\mathbf{v}_1 \ldots \mathbf{v}_n)
\begin{pmatrix}
\lambda_1 & 0 & \cdots & 0 \\
0 & \lambda_2 & \cdots & 0 \\
0 & 0 & \ddots & 0 \\
0 & 0 & \cdots & \lambda_n
\end{pmatrix} \\
&= PD,
\end{aligned}
$$

where $D = \mathrm{diag}(\lambda_1, \lambda_2, \ldots, \lambda_n)$ is the diagonal matrix whose entries are the eigenvalues. The fact that P is invertible then implies that $P^{-1}AP = P^{-1}PD = D$. So it follows that A is diagonalisable and the matrix P is such that $P^{-1}AP$ is a diagonal matrix.

Example 8.20 Now it should be clear where P in Example 8.18 came from. In Examples 8.3 and 8.4, we discovered that the eigenvalues of

$$A = \begin{pmatrix} 7 & -15 \\ 2 & -4 \end{pmatrix}$$

are 1 and 2 and that corresponding eigenvectors are

$$\mathbf{v}_1 = \begin{pmatrix} 5 \\ 2 \end{pmatrix}, \quad \mathbf{v}_2 = \begin{pmatrix} 3 \\ 1 \end{pmatrix}.$$

This is why, if we take

$$P = (\mathbf{v}_1 \ \mathbf{v}_2) = \begin{pmatrix} 5 & 3 \\ 2 & 1 \end{pmatrix},$$

then P is invertible and $P^{-1}AP$ is the diagonal matrix

$$D = \begin{pmatrix} 1 & 0 \\ 0 & 2 \end{pmatrix}.$$

Moreover, the general discussion we have given establishes the following important result:

Theorem 8.21 *An $n \times n$ matrix A is diagonalisable if and only if it has n linearly independent eigenvectors.*

Since n linearly independent vectors in $\mathbb{R}^n$ form a basis of $\mathbb{R}^n$, another way to state this theorem is:

Theorem 8.22 *An $n \times n$ matrix A is diagonalisable if and only if there is a basis of $\mathbb{R}^n$ consisting of eigenvectors of A.*

Example 8.23 In Example 8.5 (and Activity 8.6), we found the eigenvalues and eigenvectors of the matrix

$$A = \begin{pmatrix} 4 & 0 & 4 \\ 0 & 4 & 4 \\ 4 & 4 & 8 \end{pmatrix}.$$

We will now diagonalise A. We have seen that it has three distinct eigenvalues $0, 4, 12$. From the eigenvectors we found, we take one eigenvector corresponding to each of the eigenvalues $\lambda_1 = 4, \lambda_2 = 0, \lambda_3 = 12$, in that order,

$$\mathbf{v}_1 = \begin{pmatrix} -1 \\ 1 \\ 0 \end{pmatrix}, \quad \mathbf{v}_2 = \begin{pmatrix} -1 \\ -1 \\ 1 \end{pmatrix}, \quad \mathbf{v}_3 = \begin{pmatrix} 1 \\ 1 \\ 2 \end{pmatrix}.$$

We now form the matrix P whose columns are these eigenvectors:

$$P = \begin{pmatrix} -1 & -1 & 1 \\ 1 & -1 & 1 \\ 0 & 1 & 2 \end{pmatrix}.$$

Then we know that D will be the matrix

$$D = \begin{pmatrix} 4 & 0 & 0 \\ 0 & 0 & 0 \\ 0 & 0 & 12 \end{pmatrix}.$$

You can choose any order for listing the eigenvectors as the columns of the matrix P, as long as you write the corresponding eigenvalues in the corresponding columns of D; that is, as long as the column orders in P and D match. (If, for example, we had instead chosen $P = (\mathbf{v}_2 \ \mathbf{v}_1 \ \mathbf{v}_3)$, then D would instead be diag$(0, 4, 12)$.)

As soon as you have written down the matrices P and D, you should check that your eigenvectors are correct. That is, check that

$$AP = (A\mathbf{v}_1 \ A\mathbf{v}_2 \ A\mathbf{v}_3) = (\lambda_1\mathbf{v}_1 \ \lambda_2\mathbf{v}_2 \ \lambda_3\mathbf{v}_3) = PD.$$

Activity 8.24 Carry out this calculation to check that the eigenvectors are correct; that is, check that the columns of P are eigenvectors of A corresponding to the eigenvalues 4, 0, 12.

Then, according to the theory, if P has an inverse – that is, if the eigenvectors are linearly independent – then $P^{-1}AP = D = \text{diag}(4, 0, 12)$.

Activity 8.25 Check that P is invertible. Then find P^{-1} (which may be calculated using either elementary row operations or the cofactor method) and verify that $P^{-1}AP = D$.

Note how important it is to have checked P first. Calculating the inverse of an incorrect matrix P would have been a huge wasted effort.

8.2.3   Geometrical interpretation

There is a more sophisticated way to think about diagonalisation in terms of change of basis and matrix representations of linear transformations. Suppose that $T = T_A$ is the linear transformation corresponding to A, so that $T(\mathbf{x}) = A\mathbf{x}$ for all $\mathbf{x}$. Then A is the matrix representing the linear transformation T in standard coordinates.

Suppose that A has a set of n linearly independent eigenvectors $B = \{\mathbf{v}_1, \mathbf{v}_2, \ldots, \mathbf{v}_n\}$, corresponding (respectively) to the eigenvalues $\lambda_1, \ldots, \lambda_n$. Then B is a basis of $\mathbb{R}^n$. What is the matrix representing T with respect to this basis?

By Theorem 7.37, the matrix representing T in the basis B is

$$A_{[B,B]} = P^{-1}AP,$$

where the columns of P are the basis vectors of B, so that

$$P = (\mathbf{v}_1 \ldots \mathbf{v}_n).$$

In other words, the matrices A and $A_{[B,B]}$ are similar. They represent the *same* linear transformation, but A does so with respect to the standard basis and $A_{[B,B]}$ represents T in the basis B of eigenvectors of A.

But what *is* $A_{[B,B]}$? According to Theorem 7.36, the ith column of M should be the coordinate vector of $T(\mathbf{v}_i)$ with respect to the basis B.

Now, $T(\mathbf{v}_i) = A\mathbf{v}_i = \lambda_i \mathbf{v}_i$, so the coordinate vector $[T(\mathbf{v}_i)]_B$ is just the vector with λ_i in position i and all other entries zero.

Activity 8.26 Why is this true?

It follows that $A_{[B,B]}$ must be the diagonal matrix

$$D = \begin{pmatrix} \lambda_1 & 0 & \cdots & 0 \\ 0 & \lambda_2 & \cdots & 0 \\ 0 & 0 & \ddots & 0 \\ 0 & 0 & \cdots & \lambda_n \end{pmatrix}.$$

We see, therefore, that

$$P^{-1}AP = A_{[B,B]} = D.$$

Let's explore this a little further to see what it reveals, geometrically, about the linear transformation $T = T_A$. If $\mathbf{x} \in \mathbb{R}^n$ is any vector, then its image under the linear transformation T is particularly easy to calculate in B coordinates. For example, suppose the B coordinates of $\mathbf{x}$ are given by

$$[\mathbf{x}]_B = \begin{bmatrix} b_1 \\ b_2 \\ \vdots \\ b_n \end{bmatrix}_B.$$

Then, since $[T(\mathbf{x})]_B = A_{[B,B]}[\mathbf{x}]_B = D[\mathbf{x}]_B$, we have

$$[T(\mathbf{x})]_B = \begin{pmatrix} \lambda_1 & 0 & \cdots & 0 \\ 0 & \lambda_2 & \cdots & 0 \\ 0 & 0 & \ddots & 0 \\ 0 & 0 & \cdots & \lambda_n \end{pmatrix} \begin{bmatrix} b_1 \\ b_2 \\ \vdots \\ b_n \end{bmatrix}_B = \begin{bmatrix} \lambda_1 b_1 \\ \lambda_2 b_2 \\ \vdots \\ \lambda_n b_n \end{bmatrix}_B.$$

So the effect is simply to multiply each coordinate by the corresponding eigenvalue.

This gives an interesting geometrical interpretation. We can describe the linear transformation T as a *stretch* in the direction of the eigenvector $\mathbf{v}_i$ by a factor λ_i (in the same direction if $\lambda > 0$ and in the opposite direction if $\lambda < 0$). We say that the line $\mathbf{x} = t\mathbf{v}_i$, $t \in \mathbb{R}$, is *fixed* by the linear transformation T in the sense that every point on the line is mapped to a point on the same line. Indeed, this can be seen directly. Since $A\mathbf{v}_i = \lambda_i \mathbf{v}_i$, each vector on the line $t\mathbf{v}_i$, is mapped into the scalar multiple $\lambda_i t\mathbf{v}_i$ by the linear transformation A. If $\lambda_i = 0$, the line $t\mathbf{v}_i$ is mapped to $\mathbf{0}$.

Activity 8.27 Geometrically, how would you describe the linear transformation $T_A(\mathbf{x}) = A\mathbf{x}$ for Example 8.23?

Activity 8.28 Have another look at Example 7.42.

8.2.4 Similar matrices

Now let's consider any two similar matrices A and B with $B = P^{-1}AP$. We will show that A and B have the same eigenvalues, and that they have the same corresponding eigenvectors expressed in different coordinate systems.

First, let's look at this geometrically.

If $T = T_A$, then A is the matrix of the linear transformation T in standard coordinates, and $B = P^{-1}AP$ is the matrix of the same linear transformation T in coordinates with respect to the basis given by the columns of the matrix P (see Section 7.4.2). As we have just seen, the effect of T as a mapping $T : \mathbb{R}^n \to \mathbb{R}^n$ can be described in terms of the eigenvalues and eigenvectors of A. But this description (involving fixed lines and stretches) is intrinsic to the linear transformation, and does not depend on the coordinate system being used to express the vectors. Therefore, the eigenvalues of B must be the same as those of A, and the corresponding eigenvectors must be the same vectors, only given in a different basis.

To establish these facts algebraically, we begin with the following result:

Theorem 8.29 *Similar matrices have the same characteristic polynomial.*

Proof: Let A and B be similar matrices with $B = P^{-1}AP$. The characteristic polynomial of A is given by the determinant $|A - \lambda I|$. The characteristic polynomial of B is

$$|B - \lambda I| = |P^{-1}AP - \lambda I| = |P^{-1}AP - \lambda P^{-1}IP|$$
$$= |P^{-1}AP - P^{-1}\lambda IP|,$$

since $P^{-1}IP = I$. We now factor out P^{-1} on the left and P on the right to obtain

$$|B - \lambda I| = |P^{-1}(A - \lambda I)P| = |P^{-1}||A - \lambda I||P| = |A - \lambda I|,$$

since the determinant of a product is the product of the determinants, and since $|P^{-1}| = 1/|P|$. $\qquad\square$

We can now prove the following theorem:

Theorem 8.30 *Similar matrices have the same eigenvalues, and the same corresponding eigenvectors expressed in coordinates with respect to different bases.*

Proof: That similar matrices have the same eigenvalues is a direct consequence of the previous theorem, since the matrices have the same characteristic polynomials and the eigenvalues are the solutions of the characteristic equations, $|A - \lambda I| = |B - \lambda I| = 0$.

Now for the eigenvectors. Let A and B be similar matrices, with $B = P^{-1}AP$. We consider the invertible matrix P as the transition matrix from standard coordinates to coordinates in the basis, S, consisting of the column vectors of P, so that

$$\mathbf{v} = P[\mathbf{v}]_S \quad \text{and} \quad [\mathbf{v}]_S = P^{-1}\mathbf{v}.$$

If λ is any eigenvalue of A and $\mathbf{v}$ is a corresponding eigenvector, then

$$A\mathbf{v} = \lambda\mathbf{v}.$$

Using these facts, let's see what happens if we multiply the matrix B with the same eigenvector given in the S coordinates:

$$\begin{aligned}
B[\mathbf{v}]_S &= P^{-1}AP[\mathbf{v}]_S \\
&= P^{-1}A\mathbf{v} \\
&= P^{-1}\lambda\mathbf{v} \\
&= \lambda P^{-1}\mathbf{v} \\
&= \lambda[\mathbf{v}]_S.
\end{aligned}$$

Therefore, $[\mathbf{v}]_S$ is an eigenvector of B corresponding to eigenvalue λ. $\square$

8.3   When is diagonalisation possible?

By Theorem 8.21, an $n \times n$ matrix is diagonalisable if and only if it has n linearly independent eigenvectors. However, not all $n \times n$ matrices have this property, and we now explore further the conditions under which a matrix can be diagonalised.

First, we give two examples to show that not all matrices can be diagonalised.

8.3.1　Examples of non-diagonalisable matrices

Example 8.31　The 2×2 matrix

$$A = \begin{pmatrix} 4 & 1 \\ -1 & 2 \end{pmatrix}$$

has characteristic polynomial $\lambda^2 - 6\lambda + 9 = (\lambda - 3)^2$, so there is only one eigenvalue, $\lambda = 3$. The eigenvectors are the non-zero solutions to $(A - 3I)\mathbf{x} = \mathbf{0}$: that is,

$$\begin{pmatrix} 1 & 1 \\ -1 & -1 \end{pmatrix} \begin{pmatrix} x_1 \\ x_2 \end{pmatrix} = \begin{pmatrix} 0 \\ 0 \end{pmatrix}.$$

This is equivalent to the single equation $x_1 + x_2 = 0$, with general solution $x_1 = -x_2$. Setting $x_2 = t$, we see that the solution set of the system consists of all vectors of the form $\mathbf{v} = (-t, t)^{\mathrm{T}}$ as t runs through all real numbers. So the eigenvectors are precisely the non-zero scalar multiples of the vector $\mathbf{v} = (-1, 1)^{\mathrm{T}}$. Any two eigenvectors are therefore scalar multiples of each other and hence form a linearly dependent set. In other words, there are *not* two linearly independent eigenvectors, and the matrix A is not diagonalisable.

There is another reason why a matrix A may not be diagonalisable over the real numbers. Consider the following example:

Example 8.32　If A is the matrix

$$A = \begin{pmatrix} 0 & -1 \\ 1 & 0 \end{pmatrix},$$

then the characteristic equation

$$|A - \lambda I| = \begin{vmatrix} -\lambda & -1 \\ 1 & -\lambda \end{vmatrix} = \lambda^2 + 1 = 0$$

has no real solutions.

　　This matrix A can be diagonalised over the complex numbers, but not over the real numbers. (We will look at complex numbers and matrices in Chapter 13.)

So far, and until Chapter 13, we are dealing with matrices A with real number entries. If A is diagonalisable, so that there is an invertible P (with real number entries) with $P^{-1}AP = \mathrm{diag}(\lambda_1, \ldots, \lambda_n)$, then, as we have seen, the λ_i are the eigenvalues of A. So, it follows that all the eigenvalues must be real numbers. Example 8.32 is an example of a matrix that fails to be diagonalisable because it does not have this property. On the other hand, the matrix in Example 8.31 does have only real eigenvalues, yet fails to be diagonalisable. We will return shortly to the general question of when a matrix can be diagonalised. But for now we consider the special case in which an $n \times n$ matrix has n different (real) eigenvalues.

8.3.2   Matrices with distinct eigenvalues

We now show that if a matrix has n different eigenvalues (that is, if it *has distinct eigenvalues*), then it will be diagonalisable. This is a consequence of the following useful result. The proof we give here is a *proof by contradiction*.

Theorem 8.33 *Eigenvectors corresponding to different eigenvalues are linearly independent.*

Proof: Suppose the result is false for the $n \times n$ matrix A. Let's take any smallest possible set S of eigenvectors corresponding to distinct eigenvalues of A with the property that the set is linearly dependent. (This set S will have at least 2 and at most n members.) So, S consists of eigenvectors of A, each corresponding to different eigenvalues, and it is a linearly dependent set; and, furthermore, any proper subset of S is *not* a linearly dependent set. Call the vectors in this set $\mathbf{v}_1, \mathbf{v}_2, \ldots, \mathbf{v}_k$. Then, because S is linearly dependent, there are non-zero numbers $c_1, c_2, \ldots, c_k$ such that

$$c_1 \mathbf{v}_1 + c_2 \mathbf{v}_2 + \cdots + c_k \mathbf{v}_k = \mathbf{0}.$$

(You might wonder why we assert that all the c_i are non-zero, rather than just that not all of them are zero. But remember that no proper subset of S is linearly dependent. If c_i was 0, we could delete $\mathbf{v}_i$ from S and have a proper subset of S that is linearly dependent, which can't be the case.)

Multiplying this equation by A, we have:

$$A(c_1 \mathbf{v}_1 + c_2 \mathbf{v}_2 + \cdots + c_k \mathbf{v}_k) = c_1 A \mathbf{v}_1 + c_2 A \mathbf{v}_2 + \cdots + c_k A \mathbf{v}_k$$
$$= \lambda_1 c_1 \mathbf{v}_1 + \lambda_2 c_2 \mathbf{v}_2 + \cdots + \lambda_k c_k \mathbf{v}_k.$$

But this must be equal to $A\mathbf{0} = \mathbf{0}$, since $c_1 \mathbf{v}_1 + c_2 \mathbf{v}_2 + \cdots + c_k \mathbf{v}_k = \mathbf{0}$. Hence we have:

$$L_1 = \lambda_1 c_1 \mathbf{v}_1 + \lambda_2 c_2 \mathbf{v}_2 + \cdots + \lambda_k c_k \mathbf{v}_k = \mathbf{0}.$$

Furthermore, if we simply multiply both sides of the equation

$$c_1 \mathbf{v}_1 + c_2 \mathbf{v}_2 + \cdots + c_k \mathbf{v}_k = \mathbf{0}$$

by λ_1, we obtain

$$L_2 = \lambda_1 c_1 \mathbf{v}_1 + \lambda_1 c_2 \mathbf{v}_2 + \cdots + \lambda_1 c_k \mathbf{v}_k = \mathbf{0}.$$

It follows that

$$\begin{aligned}
L_1 - L_2 &= (\lambda_1 c_1 \mathbf{v}_1 + \cdots + \lambda_k c_k \mathbf{v}_k) - (\lambda_1 c_1 \mathbf{v}_1 + \cdots + \lambda_1 c_k \mathbf{v}_k) \\
&= \mathbf{0} - \mathbf{0} \\
&= \mathbf{0},
\end{aligned}$$

which means

$$(\lambda_2 - \lambda_1) c_2 \mathbf{v}_2 + \cdots + (\lambda_k - \lambda_1) c_k \mathbf{v}_k = \mathbf{0}.$$

Since the λ_i are distinct and the c_i are non-zero, this says that the vectors $\mathbf{v}_2, \ldots, \mathbf{v}_k$ are linearly dependent, which contradicts the original assumption that no proper subset of S is linearly dependent. So we must conclude (for otherwise, there is a contradiction) that there is no such set S. That means that any set of eigenvectors corresponding to distinct eigenvalues is linearly independent. □

It follows that if an $n \times n$ matrix has n different eigenvalues, then a set consisting of one eigenvector for each eigenvalue will be a linearly independent set of size n and hence, by Theorem 8.21, the matrix will be diagonalisable. That is, we have the following theorem.

Theorem 8.34 *If an $n \times n$ matrix has n different eigenvalues, then it has a set of n linearly independent eigenvectors and is therefore diagonalisable.*

8.3.3 The general case

Theorem 8.34 provides a *sufficient* condition for an $n \times n$ matrix to be diagonalisable: *if* it has n different (real) eigenvalues, *then* it is diagonalisable. It is not, however, *necessary* for the eigenvalues to be distinct in order for the matrix to be diagonalisable. What is needed for diagonalisation is a set of n linearly independent eigenvectors, and this can happen even when there is a 'repeated' eigenvalue (that is, when there are fewer than n different eigenvalues). The following example illustrates this.

Example 8.35 Consider the matrix

$$A = \begin{pmatrix} 3 & -1 & 1 \\ 0 & 2 & 0 \\ 1 & -1 & 3 \end{pmatrix}.$$

The eigenvalues are given by the solutions of the characteristic equation $|A - \lambda I| = 0$. Expanding the determinant by the second row,

$$|A - \lambda I| = \begin{vmatrix} 3 - \lambda & -1 & 1 \\ 0 & 2 - \lambda & 0 \\ 1 & -1 & 3 - \lambda \end{vmatrix}$$

$$= (2 - \lambda) \begin{vmatrix} 3 - \lambda & 1 \\ 1 & 3 - \lambda \end{vmatrix}$$

$$= (2 - \lambda)(\lambda^2 - 6\lambda + 9 - 1)$$

$$= (2 - \lambda)(\lambda^2 - 6\lambda + 8)$$

$$= (2 - \lambda)(\lambda - 4)(\lambda - 2) = -(\lambda - 2)^2(\lambda - 4).$$

The matrix A has only two eigenvalues: $\lambda = 4$ and $\lambda = 2$. Because $(\lambda - 2)^2$ is a factor of the characteristic polynomial (or, equivalently, $\lambda = 2$ is a double root of the polynomial), we say that $\lambda = 2$ is an eigenvalue of *multiplicity* 2. If we want to diagonalise the matrix, we need to find three linearly independent eigenvectors. Any eigenvector corresponding to $\lambda = 4$ will be linearly independent of any eigenvectors corresponding to the eigenvalue 2. What we therefore need to do is to find two linearly independent eigenvectors corresponding to the eigenvalue 2 of multiplicity 2. (Then these two vectors taken together with an eigenvector corresponding to $\lambda = 4$ will give a linearly independent set.) So let's first find the eigenvectors for $\lambda = 2$. We row reduce the matrix $(A - 2I)$:

$$(A - 2I) = \begin{pmatrix} 1 & -1 & 1 \\ 0 & 0 & 0 \\ 1 & -1 & 1 \end{pmatrix} \longrightarrow \begin{pmatrix} 1 & -1 & 1 \\ 0 & 0 & 0 \\ 0 & 0 & 0 \end{pmatrix}.$$

We see immediately that this matrix has rank 1, so its null space (the eigenspace for $\lambda = 2$) will have dimension 2, and we can find a basis of this space consisting of two linearly independent eigenvectors. Setting the non-leading variables equal to arbitrary parameters s and t, we find that the solutions of $(A - 2I)\mathbf{x} = \mathbf{0}$ are

$$\mathbf{x} = s \begin{pmatrix} 1 \\ 1 \\ 0 \end{pmatrix} + t \begin{pmatrix} -1 \\ 0 \\ 1 \end{pmatrix} = s\mathbf{v}_1 + t\mathbf{v}_2, \quad s, t \in \mathbb{R},$$

where $\mathbf{v}_1$ and $\mathbf{v}_2$ are two linearly independent eigenvectors for $\lambda = 2$.

Activity 8.36 How do you know that $\mathbf{v}_1$ and $\mathbf{v}_2$ are linearly independent?

Since $\{\mathbf{v}_1, \mathbf{v}_2\}$ is a linearly independent set, and since eigenvectors corresponding to distinct eigenvalues are linearly independent, it follows that if $\mathbf{v}_3$ is any eigenvector corresponding to $\lambda = 4$, then $\{\mathbf{v}_1, \mathbf{v}_2, \mathbf{v}_3\}$ will be a linearly independent set.

We find an eigenvector for $\lambda = 4$ by reducing $(A - 4I)$.

$$(A - 4I) = \begin{pmatrix} -1 & -1 & 1 \\ 0 & -2 & 0 \\ 1 & -1 & -1 \end{pmatrix} \longrightarrow \cdots \longrightarrow \begin{pmatrix} 1 & 0 & -1 \\ 0 & 1 & 0 \\ 0 & 0 & 0 \end{pmatrix}$$

with solutions

$$\mathbf{x} = t \begin{pmatrix} 1 \\ 0 \\ 1 \end{pmatrix}, \quad t \in \mathbb{R}.$$

Let

$$\mathbf{v}_3 = \begin{pmatrix} 1 \\ 0 \\ 1 \end{pmatrix}.$$

Then $\mathbf{v}_1, \mathbf{v}_2, \mathbf{v}_3$ form a linearly independent set of eigenvectors. If we take

$$P = \begin{pmatrix} 1 & 1 & -1 \\ 0 & 1 & 0 \\ 1 & 0 & 1 \end{pmatrix},$$

then

$$P^{-1}AP = D = \begin{pmatrix} 4 & 0 & 0 \\ 0 & 2 & 0 \\ 0 & 0 & 2 \end{pmatrix}.$$

Activity 8.37 Check this! Check that $AP = PD$ and that $|P| \neq 0$. Why do these two checks enable you to find any errors?

Here is another example where, this time, diagonalisation is not possible.

Example 8.38 We found in Example 8.7 that the matrix,

$$A = \begin{pmatrix} -3 & -1 & -2 \\ 1 & -1 & 1 \\ 1 & 1 & 0 \end{pmatrix}$$

has an eigenvalue $\lambda_1 = -1$ of multiplicity 2, and a second eigenvalue, $\lambda_2 = -2$. In order to diagonalise this matrix, we need two linearly independent eigenvectors for $\lambda = -1$. To see if this is possible, we row reduce the matrix $(A + I)$:

$$(A + I) = \begin{pmatrix} -2 & -1 & -2 \\ 1 & 0 & 1 \\ 1 & 1 & 1 \end{pmatrix} \longrightarrow \cdots \longrightarrow \begin{pmatrix} 1 & 0 & 1 \\ 0 & 1 & 0 \\ 0 & 0 & 0 \end{pmatrix}.$$

This matrix has rank 2 and the null space (the eigenspace for $\lambda = -1$) therefore (by the rank–nullity theorem) has dimension 1. We can only find one linearly independent eigenvector for $\lambda = -1$. All solutions of $(A + I)\mathbf{x} = \mathbf{0}$ are of the form

$$\mathbf{x} = t \begin{pmatrix} -1 \\ 0 \\ 1 \end{pmatrix}, \qquad t \in \mathbb{R}.$$

We conclude that this matrix *cannot be diagonalised* as it is not possible to find three linearly independent eigenvectors to form the matrix P.

8.3.4 Algebraic and geometric multiplicity

To describe in more detail what it is that makes a matrix diagonalisable (and what it is that distinguishes the matrices in Example 8.35 and Example 8.38), we introduce the concepts of *algebraic* and *geometric* multiplicity of eigenvalues.

Definition 8.39 (Algebraic multiplicity) An eigenvalue λ_0 of a matrix A has *algebraic multiplicity* k if k is the largest integer such that $(\lambda - \lambda_0)^k$ is a factor of the characteristic polynomial of A.

Definition 8.40 (Geometric multiplicity) The *geometric multiplicity* of an eigenvalue λ_0 of a matrix A is the dimension of the eigenspace of λ_0 (that is, the dimension of the null space, $N(A - \lambda_0 I)$, of $A - \lambda_0 I$).

If A is an $n \times n$ matrix with an eigenvalue λ, then we know that there is at least one eigenvector corresponding to λ. Why? Since we know that $|A - \lambda I| = 0$, we know that $(A - \lambda I)\mathbf{v} = \mathbf{0}$ has a non-trivial solution $\mathbf{v}$, which is an eigenvector corresponding to λ. So the eigenspace of any eigenvalue has dimension at least 1, and hence the geometric multiplicity, $\dim(N(A - \lambda I))$, is at least 1.

In Example 8.31 we have an eigenvalue (namely, $\lambda = -1$) of algebraic multiplicity 2, but because the eigenspace only has dimension one, there does not exist two linearly independent eigenvectors. Here, the fact that the geometric multiplicity is less than the algebraic multiplicity means that the matrix cannot be diagonalised. For, it turns out that if we are to find enough linearly independent eigenvectors to diagonalise a matrix, then, for each eigenvalue, the algebraic and geometric multiplicities must be equal. We will prove this. First, though, we have a straightforward relationship between algebraic and geometric multiplicity, which has been alluded to in the above examples.

Theorem 8.41 *For any eigenvalue of a square matrix, the geometric multiplicity is no more than the algebraic multiplicity.*

Proof: Let's suppose that μ is an eigenvalue of the $n \times n$ matrix A and that μ has geometric multiplicity k. Then there is a linearly independent set $\{v_1, v_2, \ldots, v_k\}$ of eigenvectors of A corresponding to μ. By Theorem 6.45, we can extend this to a basis $B = \{v_1, v_2, \ldots, v_k, v_{k+1}, \ldots, v_n\}$ of $\mathbb{R}^n$.

Let T be the linear transformation given by multiplication by A; that is, $T(\mathbf{x}) = A\mathbf{x}$. We now apply Theorem 7.37. According to this theorem, the matrix M representing T with respect to the basis B is $P^{-1}AP$, where the columns of P are the vectors of the basis B. But, by Theorem 7.36, column i of M is equal to $[Av_i]_B$. So, since $Av_i = \mu v_i$ for $i = 1, 2, \ldots, k$, we must have

$$M = P^{-1}AP = \begin{pmatrix} \mu & 0 & 0 & \cdots & 0 & \cdots \\ 0 & \mu & 0 & \cdots & 0 & \cdots \\ 0 & 0 & \mu & \cdots & 0 & \cdots \\ \vdots & \vdots & \vdots & \ddots & \vdots & \cdots \\ 0 & 0 & 0 & \cdots & \mu & \cdots \\ \vdots & \vdots & \vdots & \vdots & \vdots & \ddots \end{pmatrix},$$

a matrix in which, for $i = 1, 2, \ldots, k$, column i has μ in position i and 0 elsewhere. (So, the top-left $k \times k$ submatrix is μ times the $k \times k$ identity matrix.) Now, it follows that the characteristic polynomial of M will be

$$|M - \lambda I| = \begin{vmatrix} \mu - \lambda & 0 & 0 & \cdots & 0 & \cdots \\ 0 & \mu - \lambda & 0 & \cdots & 0 & \cdots \\ 0 & 0 & \mu - \lambda & \cdots & 0 & \cdots \\ \vdots & \vdots & \vdots & \ddots & \vdots & \cdots \\ 0 & 0 & 0 & \cdots & \mu - \lambda & \cdots \\ \vdots & \vdots & \vdots & \vdots & \vdots & \ddots \end{vmatrix}$$

$$= (\mu - \lambda)^k q(\lambda),$$

where $q(\lambda)$ is the determinant of the bottom-right $(n-k) \times (n-k)$ submatrix of $M - \lambda I$. So $(\lambda - \mu)^k$ divides the characteristic polynomial of M, which, as we saw earlier (Theorem 8.29), is the same as the characteristic polynomial of A. So the algebraic multiplicity of μ is at least k, the geometric multiplicity. $\square$

The following theorem provides a characterisation of diagonalisable matrices in terms of algebraic and geometric multiplicities. The proof might look daunting, but its key ideas are not so hard.

Theorem 8.42 *A matrix is diagonalisable if and only if all its eigenvalues are real numbers and, for each eigenvalue, the geometric multiplicity equals the algebraic multiplicity.*

Proof: We have already noted earlier that if a matrix is to be diagonalisable, then all its eigenvalues must be real numbers. Suppose A is an $n \times n$ matrix with real eigenvalues, and denote the distinct eigenvalues by $\lambda_1, \ldots, \lambda_r$. Then $r \leq n$ and the characteristic polynomial of A takes the form

$$p(\lambda) = |A - \lambda I| = (-1)^n (\lambda - \lambda_1)^{k_1} (\lambda - \lambda_2)^{k_2} \cdots (\lambda - \lambda_r)^{k_r},$$

where k_i is the algebraic multiplicity of λ_i. But $p(\lambda)$ is of degree n and hence $n = k_1 + k_2 + \cdots + k_r$.

To be diagonalisable, there must be a basis consisting of n eigenvectors of A. We know that if m_i is the geometric multiplicity of λ_i, then $m_i \leq k_i$. Suppose that $m_j < k_j$ for some j. Then there will not be a linearly independent set of k_j eigenvectors corresponding to λ_j. But that means there cannot be a set of n linearly independent eigenvectors of A. To see why, we note that in any set S of linearly independent eigenvectors, each eigenvector must correspond to some eigenvalue λ_i and, by the definition of geometric multiplicity, no more than m_i of these can correspond to λ_i, for each i. So the maximum number of vectors in the set S is

$$m_1 + m_2 + \cdots + m_j + \cdots + m_r.$$

But since $m_i \leq k_i$ for all i, and $m_j < k_j$, we have

$$m_1 + m_2 + \cdots + m_j + \cdots + m_r < k_1 + k_2 + \cdots + k_r = n.$$

So, S contains fewer than n vectors, and A will not be diagonalisable.

The argument so far shows that A will be diagonalisable only if all its eigenvalues are real numbers and, for each eigenvalue, the geometric multiplicity equals the algebraic multiplicity. We now need to show the converse.

Suppose, then, that A has only real eigenvalues and that, for each, the algebraic and geometric multiplicities are equal. Suppose the eigenvalues are $\lambda_1, \lambda_2, \ldots, \lambda_r$ and that, for each i, the multiplicity (algebraic and geometric) of λ_i is m_i. Let $S_i = \{\mathbf{v}_1^{(i)}, \mathbf{v}_2^{(i)}, \ldots, \mathbf{v}_{m_i}^{(i)}\}$ be a linearly independent set of eigenvectors for λ_i. We know such a set exists because the geometric multiplicity is m_i. Then the set $S = S_1 \cup S_2 \cup \cdots \cup S_r$ (the union of the sets S_i) is a set of eigenvectors for A and we will show that it is linearly independent, which will imply that A is diagonalisable. So, suppose some linear combination of the vectors in S is $\mathbf{0}$. We can write this as

$$\alpha_1^{(1)}\mathbf{v}_1^{(1)} + \cdots + \alpha_{m_1}^{(1)}\mathbf{v}_{m_1}^{(1)} + \cdots + \alpha_1^{(r)}\mathbf{v}_1^{(r)} + \cdots + \alpha_{m_r}^{(r)}\mathbf{v}_{m_r}^{(r)} = \mathbf{0}.$$

For each i, let

$$\mathbf{w}^{(i)} = \alpha_1^{(i)}\mathbf{v}_1^{(i)} + \alpha_2^{(i)}\mathbf{v}_2^{(i)} + \cdots + \alpha_{m_i}^{(i)}\mathbf{v}_{m_i}^{(i)}.$$

Then this equation can be written as

$$\mathbf{w}^{(1)} + \mathbf{w}^{(2)} + \cdots + \mathbf{w}^{(r)} = \mathbf{0}. \qquad\qquad (*)$$

Now, for any i, $\mathbf{w}^{(i)}$ is a linear combination of eigenvectors corresponding to λ_i, so it is either $\mathbf{0}$ or is itself an eigenvector (since it belongs to the eigenspace). However, if any of the $\mathbf{w}^{(i)}$ is not $\mathbf{0}$, then equation $(*)$ shows that a non-trivial linear combination of eigenvectors corresponding to distinct eigenvalues is $\mathbf{0}$, and this is not possible since, by Theorem 8.33, eigenvectors for distinct eigenvalues are linearly independent. It follows that, for all i, $\mathbf{w}^{(i)} = \mathbf{0}$. Therefore,

$$\alpha_1^{(i)}\mathbf{v}_1^{(i)} + \alpha_2^{(i)}\mathbf{v}_2^{(i)} + \cdots + \alpha_{m_i}^{(i)}\mathbf{v}_{m_i}^{(i)} = \mathbf{0}.$$

But the set $S_i = \{\mathbf{v}_1^{(i)}, \mathbf{v}_2^{(i)}, \ldots, \mathbf{v}_{m_i}^{(i)}\}$ is linearly independent, so it follows that

$$\alpha_1^{(i)} = \alpha_2^{(i)} = \cdots = \alpha_{m_i}^{(i)} = 0.$$

So all the coefficients $\alpha_j^{(i)}$ are 0. This shows that the set S is linearly independent. $\qquad\qquad\qquad\qquad\qquad\qquad\qquad\qquad\qquad\square$

8.4   Learning outcomes

You should now be able to:

- state what is meant by the characteristic polynomial and the characteristic equation of a matrix
- state carefully what is meant by eigenvectors and eigenvalues, and by diagonalisation
- find eigenvalues and corresponding eigenvectors for a square matrix
- state what is meant by the eigenspace of an eigenvector
- know how the eigenvalues are related to the determinant and trace of a matrix
- diagonalise a diagonalisable matrix
- determine whether or not a matrix can be diagonalised

- recognise what diagonalisation does in terms of change of basis and matrix representation of linear transformations (similarity)
- use diagonalisation to describe the geometric effect of a linear transformation
- know how to characterise diagonalisability in terms of the algebraic and geometric multiplicities of eigenvalues.

8.5   Comments on activities

Activity 8.6 The eigenvectors for $\lambda = 0$ are the non-zero solutions of $A\mathbf{x} = \mathbf{0}$. To find these, row reduce the coefficient matrix A.

$$\begin{pmatrix} 4 & 0 & 4 \\ 0 & 4 & 4 \\ 4 & 4 & 8 \end{pmatrix} \longrightarrow \cdots \longrightarrow \begin{pmatrix} 1 & 0 & 1 \\ 0 & 1 & 1 \\ 0 & 0 & 0 \end{pmatrix}.$$

The solutions are

$$\mathbf{x} = t \begin{pmatrix} -1 \\ -1 \\ 1 \end{pmatrix}, \qquad t \in \mathbb{R},$$

so that the eigenvectors are non-zero multiples of $\mathbf{v}_2 = (-1, -1, 1)^{\mathrm{T}}$. The eigenspace of $\lambda = 0$ is the null space of the matrix A. Note that $A\mathbf{v}_2 = 0\mathbf{v}_2 = \mathbf{0}$.

Similarly, you should find that for $\lambda = 12$, the eigenvectors are non-zero multiples of

$$\mathbf{v}_3 = \begin{pmatrix} 1 \\ 1 \\ 2 \end{pmatrix}.$$

Activity 8.10 Since $A\mathbf{x} = \lambda\mathbf{x} \iff A\mathbf{x} - \lambda\mathbf{x} = (A - \lambda I)\mathbf{x} = \mathbf{0}$, the two sets contain precisely the same vectors.

Activity 8.19 You should notice that the columns of AP are $\mathbf{v}_1$ and $2\mathbf{v}_2$, where $\mathbf{v}_1, \mathbf{v}_2$ are the columns of P.

Activity 8.24 Perform the matrix multiplication to show that

$$AP = (4\mathbf{v}_1 \quad 0\mathbf{v}_2 \quad 12\mathbf{v}_3) = PD.$$

Activity 8.25 Since $|P| = 6 \neq 0$, P is invertible. Using the adjoint method (or row reduction), obtain

$$P^{-1} = \frac{1}{6} \begin{pmatrix} -3 & 3 & 0 \\ -2 & -2 & 2 \\ 1 & 1 & 2 \end{pmatrix}.$$

Check that $PP^{-1} = I$. You have calculated AP in the previous activity, so now just multiply $P^{-1}AP$ to obtain D.

Activity 8.26 Since v_i is the ith vector in the basis B, writing $T(v_i) = \lambda_i v_i$ expresses it as a linear combination of the basis vectors of B, so the B coordinates are precisely as stated: λ_i in the ith position and 0 elsewhere.

Activity 8.27 T_A is a stretch by a factor 4 in the direction of the vector $v_1 = (-1, 1, 0)^T$, a stretch by a factor of 12 in the direction of $v_3 = (1, 1, 2)^T$ and it maps the line $x = tv_2$ to 0.

Activity 8.36 This is immediately obvious since setting $sv_1 + tv_2 = 0$, the second components tell us $s = 0$ and the third components that $t = 0$. However, this was a good time to recall that the method of solution ensures that the vectors will be linearly independent; see the discussion at the end of Section 6.5.2.

Activity 8.37 If you know that $AP = PD$, then you know that the eigenvectors are correct and the eigenvalues are in the correct positions in D. If you also check that $|P| \neq 0$, then you know that you have chosen three linearly independent eigenvectors, so P^{-1} exists and then $P^{-1}AP = D$. If any of the checks fail, then you should be able to find any errors in your choice of eigenvectors and eigenvalues.

8.6   Exercises

Exercise 8.1 Diagonalise the matrix

$$A = \begin{pmatrix} 4 & 5 \\ -1 & -2 \end{pmatrix};$$

that is, find an invertible matrix P and a diagonal matrix D such that $P^{-1}AP = D$. Check your answer.

Exercise 8.2 Find the eigenvalues of the matrix

$$B = \begin{pmatrix} 0 & 2 & 1 \\ 16 & 4 & -6 \\ -16 & 4 & 10 \end{pmatrix}$$

and find an eigenvector for each eigenvalue. Hence find an invertible matrix P and a diagonal matrix D such that $P^{-1}BP = D$. Check your work.

Exercise 8.3 Determine if either of the following matrices can be diagonalised:

$$A = \begin{pmatrix} 1 & 1 \\ 0 & 1 \end{pmatrix}, \qquad B = \begin{pmatrix} 1 & 1 \\ 1 & 1 \end{pmatrix}.$$

Exercise 8.4 Let M be an $n \times n$ matrix. State precisely what is meant by the statement

'λ is an eigenvalue of M with corresponding eigenvector $\mathbf{v}$.'

Exercise 8.5 Let

$$A = \begin{pmatrix} 6 & 13 & -8 \\ 2 & 5 & -2 \\ 7 & 17 & -9 \end{pmatrix}, \qquad \mathbf{v} = \begin{pmatrix} 1 \\ 0 \\ 1 \end{pmatrix}.$$

Using the definition of eigenvector, show that $\mathbf{v}$ is an eigenvector of A and find its corresponding eigenvalue.

The matrix A defines a linear transformation $T : \mathbb{R}^3 \to \mathbb{R}^3$ by $T(\mathbf{x}) = A\mathbf{x}$. It is known that T fixes a non-zero vector $\mathbf{x}$, $T(\mathbf{x}) = \mathbf{x}$. Use this information to determine another eigenvector and eigenvalue of A. Check your result.

Diagonalise the matrix A: write down an invertible matrix P and a diagonal matrix D such that $P^{-1}AP = D$.

Describe the linear transformation T.

Exercise 8.6 Show that the vector $\mathbf{x}$ is an eigenvector of A, where:

$$A = \begin{pmatrix} -1 & 1 & 2 \\ -6 & 2 & 6 \\ 0 & 1 & 1 \end{pmatrix}, \qquad \mathbf{x} = \begin{pmatrix} 1 \\ 1 \\ 1 \end{pmatrix}.$$

What is the corresponding eigenvalue?

Find the other eigenvalues of A, and an eigenvector for each of them. Find an invertible matrix P and a diagonal matrix D such that $P^{-1}AP = D$. Check that $AP = PD$.

Exercise 8.7 Diagonalise the matrix A:

$$A = \begin{pmatrix} 0 & 0 & -2 \\ 1 & 2 & 1 \\ 1 & 0 & 3 \end{pmatrix}.$$

Describe the eigenspace of each eigenvalue.

Exercise 8.8 Prove the following statement:

0 is an eigenvalue of A if and only if $A\mathbf{x} = \mathbf{0}$ has a non-trivial solution.

Exercise 8.9 Look again at Exercise 6.7. Repeating what you did there, show that two eigenvectors corresponding to distinct eigenvalues are linearly independent.

Using an inductive argument, prove that eigenvectors corresponding to distinct eigenvalues are linearly independent; that is, give another proof of Theorem 8.33.

Exercise 8.10 Suppose that A is a real diagonalisable matrix and that all the eigenvalues of A are non-negative. Prove that there is a matrix B such that $B^2 = A$.

8.7 Problems

Problem 8.1 Determine which, if any, of the following vectors are eigenvectors for the given matrix A:

$$\mathbf{x} = \begin{pmatrix} 1 \\ -1 \\ 3 \end{pmatrix}, \quad \mathbf{y} = \begin{pmatrix} 1 \\ 5 \\ 3 \end{pmatrix}, \quad \mathbf{z} = \begin{pmatrix} 5 \\ 0 \\ 1 \end{pmatrix}; \qquad A = \begin{pmatrix} 1 & 1 & 0 \\ 1 & 4 & 3 \\ 0 & 3 & 1 \end{pmatrix}.$$

Problem 8.2 Find the eigenvalues and corresponding eigenvectors for the matrix

$$A = \begin{pmatrix} 1 & 4 \\ 3 & 2 \end{pmatrix}.$$

Hence, find an invertible matrix P such that $P^{-1}AP$ is diagonal. Calculate $P^{-1}AP$ to check your answer.

Problem 8.3 Diagonalise the matrix

$$A = \begin{pmatrix} 7 & -2 \\ -1 & 8 \end{pmatrix}.$$

Describe (geometrically) the linear transformation $T : \mathbb{R}^2 \to \mathbb{R}^2$ given by $T(\mathbf{x}) = A\mathbf{x}$.

Problem 8.4 Find the characteristic equation of the matrix B,

$$B = \begin{pmatrix} 3 & -1 & 2 \\ 5 & -3 & 5 \\ 1 & -1 & 2 \end{pmatrix}.$$

Find the eigenvalues (which are integers) and corresponding eigenvectors for B.

Find a basis of $\mathbb{R}^3$ consisting of eigenvectors of the matrix B.

Find an invertible matrix P and a diagonal matrix D such that $P^{-1}BP = D$, Check your answer for P by showing that $BP = PD$. Then calculate P^{-1} and check that $P^{-1}BP = D$.

Problem 8.5 Diagonalise the matrix

$$A = \begin{pmatrix} 5 & 0 & 4 \\ 1 & -1 & 2 \\ 2 & 0 & 3 \end{pmatrix}.$$

Problem 8.6 Explain why the matrix

$$C = \begin{pmatrix} 5 & 0 & 4 \\ a & -1 & b \\ 2 & 0 & 3 \end{pmatrix}.$$

can be diagonalised for any values of $a, b \in \mathbb{R}$.

Problem 8.7 Find the eigenvalues of the matrices

$$A = \begin{pmatrix} 1 & 1 & 1 \\ 0 & 1 & -1 \\ 1 & 0 & 2 \end{pmatrix} \quad \text{and} \quad B = \begin{pmatrix} -2 & 1 & -2 \\ -1 & 0 & 1 \\ 2 & 1 & 2 \end{pmatrix}$$

and show that neither matrix can be diagonalised over the real numbers.

Problem 8.8 Consider the matrix A and the vector $\mathbf{v}$:

$$A = \begin{pmatrix} -5 & 8 & 32 \\ 2 & 1 & -8 \\ -2 & 2 & 11 \end{pmatrix}, \qquad \mathbf{v} = \begin{pmatrix} 2 \\ -2 \\ 1 \end{pmatrix}.$$

Show that $\mathbf{v}$ is an eigenvector of A and find the corresponding eigenvalue. Find all the eigenvectors corresponding to this eigenvalue, and hence describe (geometrically) the eigenspace.

Diagonalise the matrix A.

Problem 8.9 Let the matrix A and the vector $\mathbf{v}_1$ be as follows:

$$A = \begin{pmatrix} 4 & 3 & -7 \\ 1 & 2 & 1 \\ 2 & 2 & -3 \end{pmatrix}, \qquad \mathbf{v}_1 = \begin{pmatrix} 1 \\ 2 \\ 1 \end{pmatrix}.$$

(a) Show that $\mathbf{v}_1$ is an eigenvector of A and find its corresponding eigenvalue.

Diagonalise the matrix A; that is, find an invertible matrix P and a diagonal matrix D such that $P^{-1}AP = D$. Check your answer without finding P^{-1}.

(b) Deduce the value of $|A|$ from the eigenvalues, and show that A is invertible.

Indicate how to diagonalise A^{-1} without any further calculations. (Find its matrix of eigenvectors and corresponding diagonal matrix.)

(c) Find the missing entries s_{12} and s_{31} of A^{-1}:

$$A^{-1} = \frac{1}{|A|} \begin{pmatrix} -8 & s_{12} & 17 \\ 5 & 2 & -11 \\ s_{31} & -2 & 5 \end{pmatrix}.$$

Then verify that $A^{-1}\mathbf{v} = \lambda\mathbf{v}$ for each of the eigenvalues and eigenvectors of A^{-1} found in part (b).

Problem 8.10 Show that one of the following two matrices can be diagonalised and the other cannot:

$$A = \begin{pmatrix} 2 & 3 & 0 \\ 3 & 2 & 0 \\ 1 & 1 & 5 \end{pmatrix}, \qquad B = \begin{pmatrix} 2 & 3 & 0 \\ 3 & 2 & 0 \\ 1 & -1 & 5 \end{pmatrix}.$$

Diagonalise the appropriate matrix.

Problem 8.11 Suppose that you would like to find a linear transformation $T : \mathbb{R}^3 \to \mathbb{R}^3$ which is a stretch by factor of two in the direction $\mathbf{v}_1 = (1, 0, 1)^T$, which fixes every point on the line $\mathbf{x} = t\mathbf{v}_2$, where $\mathbf{v}_2 = (1, 1, 0)^T$, and which maps the line $\mathbf{x} = t\mathbf{v}_3$, where $\mathbf{v}_3 = (2, 1, 1)^T$, to $\mathbf{0}$.

Show that no such linear transformation can exist.

Problem 8.12 Suppose that A and B are diagonalisable $n \times n$ matrices with the same eigenvalues. Prove that A and B are similar matrices.

Problem 8.13 Diagonalise each of the following matrices A and B,

$$A = \begin{pmatrix} 5 & 4 \\ -2 & -1 \end{pmatrix}, \qquad B = \begin{pmatrix} -5 & 24 \\ -2 & 9 \end{pmatrix}.$$

Show that A and B are similar by finding an invertible matrix S such that $B = S^{-1}AS$. Check your result by multiplying $S^{-1}AS$ to obtain B.

Problem 8.14 If A is an $n \times n$ matrix, show that the matrices A and A^T have the same characteristic polynomial. Deduce that A and A^T have the same eigenvalues.

9

Applications of diagonalisation

We will now look at some applications of diagonalisation. We apply diagonalisation to find powers of diagonalisable matrices. We also solve systems of simultaneous linear difference equations. In particular, we look at the important topic of Markov chains. We also look at systems of differential equations. (Do not worry if you are unfamiliar with difference or differential equations. The key ideas you'll need are discussed.) We will see that the diagonalisation process makes the solution of linear systems of difference and differential equations possible by essentially changing basis to one in which the problem is readily solvable, namely a basis of $\mathbb{R}^n$ consisting of eigenvectors of the matrix describing the system.

9.1   Powers of matrices

For a positive integer n, the nth power of a matrix A is simply

$$A^n = \underbrace{A\,A\,A\cdots A}_{n \text{ times}}.$$

Example 9.1 Consider the matrix

$$A = \begin{pmatrix} 7 & -15 \\ 2 & -4 \end{pmatrix}$$

(which we met in Example 8.3). We have

$$A^2 = AA = \begin{pmatrix} 7 & -15 \\ 2 & -4 \end{pmatrix} \begin{pmatrix} 7 & -15 \\ 2 & -4 \end{pmatrix} = \begin{pmatrix} 19 & -45 \\ 6 & -14 \end{pmatrix},$$

$$A^3 = A.A.A = A.A^2 = \begin{pmatrix} 7 & -15 \\ 2 & -4 \end{pmatrix} \begin{pmatrix} 19 & -45 \\ 6 & -14 \end{pmatrix} = \begin{pmatrix} 43 & -105 \\ 14 & -34 \end{pmatrix}.$$

It is often useful, as we shall see in this chapter, to determine A^n for a general integer n. As you can see from Example 9.1, we could calculate A^n by performing $n - 1$ matrix multiplications. But it would be more satisfying (and easier) to have a 'formula' for the nth power, a matrix expression involving n into which one could substitute any desired value of n. Diagonalisation helps here. If we can write $P^{-1}AP = D$, then $A = PDP^{-1}$ and so

$$A^n = \underbrace{A\,A\,A\cdots A}_{n \text{ times}}$$

$$= \underbrace{(PDP^{-1})(PDP^{-1})(PDP^{-1})\cdots(PDP^{-1})}_{n \text{ times}}$$

$$= PD(P^{-1}P)D(P^{-1}P)D(P^{-1}P)\cdots D(P^{-1}P)DP^{-1}$$

$$= PDIDIDI\cdots DIDP$$

$$= P\underbrace{DDD\cdots D}_{n \text{ times}}P^{-1}$$

$$= PD^nP^{-1}.$$

The product PD^nP^{-1} is easy to compute since D^n is simply the diagonal matrix with entries equal to the nth power of those of D.

Activity 9.2 Convince yourself that if

$$D = \begin{pmatrix} \lambda_1 & 0 & \cdots & 0 \\ 0 & \lambda_2 & \cdots & 0 \\ \vdots & \vdots & \ddots & \vdots \\ 0 & 0 & \cdots & \lambda_k \end{pmatrix}, \quad \text{then} \quad D^n = \begin{pmatrix} \lambda_1^n & 0 & \cdots & 0 \\ 0 & \lambda_2^n & \cdots & 0 \\ \vdots & \vdots & \ddots & \vdots \\ 0 & 0 & \cdots & \lambda_k^n \end{pmatrix}.$$

Let's look at an easy example that builds on some work we did in the previous chapter.

Example 9.3 As mentioned in the previous chapter, the matrix

$$A = \begin{pmatrix} 7 & -15 \\ 2 & -4 \end{pmatrix}$$

from Example 8.3 is diagonalisable: if

$$P = \begin{pmatrix} 5 & 3 \\ 2 & 1 \end{pmatrix},$$

then P is invertible and

$$P^{-1}AP = D = \begin{pmatrix} 1 & 0 \\ 0 & 2 \end{pmatrix}.$$

Suppose we want to find an expression for A^n. Well, since $P^{-1}AP = D$, we have $A = PDP^{-1}$ and so, as explained above,

$$A^n = PD^nP^{-1} = \begin{pmatrix} 5 & 3 \\ 2 & 1 \end{pmatrix} \begin{pmatrix} 1 & 0 \\ 0 & 2^n \end{pmatrix} \begin{pmatrix} -1 & 3 \\ 2 & -5 \end{pmatrix}$$

$$= \begin{pmatrix} -5 + 6(2^n) & 15 - 15(2^n) \\ -2 + 2(2^n) & 6 - 5(2^n) \end{pmatrix}.$$

You can see that the cases $n = 2, 3$ in Example 9.1 clearly comply with this general formula for the nth power.

Here is another fairly easy example, which we will fully work through.

Example 9.4 Suppose that we want a matrix expression for the nth power of the matrix

$$A = \begin{pmatrix} 1 & 4 \\ \frac{1}{2} & 0 \end{pmatrix}.$$

The characteristic polynomial $|A - \lambda I|$ is (check this!)

$$\lambda^2 - \lambda - 2 = (\lambda - 2)(\lambda + 1).$$

So the eigenvalues are -1 and 2. An eigenvector for -1 is a solution of $(A + I)\mathbf{v} = \mathbf{0}$, found by

$$A + I = \begin{pmatrix} 2 & 4 \\ \frac{1}{2} & 1 \end{pmatrix} \longrightarrow \begin{pmatrix} 1 & 2 \\ 0 & 0 \end{pmatrix},$$

so we may take $(2, -1)^{\mathrm{T}}$. Eigenvectors for 2 are given by

$$A - 2I = \begin{pmatrix} -1 & 4 \\ \frac{1}{2} & -2 \end{pmatrix} \longrightarrow \begin{pmatrix} 1 & -4 \\ 0 & 0 \end{pmatrix},$$

so we may take $(4, 1)^{\mathrm{T}}$. Let P be the matrix whose columns are these eigenvectors. Then

$$P = \begin{pmatrix} 2 & 4 \\ -1 & 1 \end{pmatrix}.$$

The inverse is

$$P^{-1} = \frac{1}{6} \begin{pmatrix} 1 & -4 \\ 1 & 2 \end{pmatrix}.$$

We have $P^{-1}AP = D = \mathrm{diag}(-1, 2)$. The nth power of the matrix A is given by

$$A^n = PD^nP^{-1}$$

$$= \frac{1}{6} \begin{pmatrix} 2 & 4 \\ -1 & 1 \end{pmatrix} \begin{pmatrix} (-1)^n & 0 \\ 0 & 2^n \end{pmatrix} \begin{pmatrix} 1 & -4 \\ 1 & 2 \end{pmatrix}$$

$$= \frac{1}{6} \begin{pmatrix} 2(-1)^n + 4(2^n) & -8(-1)^n + 8(2^n) \\ -(-1)^n + 2^n & 4(-1)^n + 2(2^n) \end{pmatrix}.$$

Activity 9.5 Check the calculations in the examples just given.

9.2   Systems of difference equations

9.2.1   Introduction to difference equations

A *difference equation* is an equation linking the terms of a sequence to previous terms. For example, $x_{t+1} = 5x_t - 1$ is a first-order difference equation for the sequence x_t. (It is said to be first-order because the relationship expressing x_{t+1} involves only the previous term.) If you have a first-order difference equation, once you know the first term of the sequence, the relationship determines all the other terms of the sequence. Difference equations are also often referred to as *recurrence equations*. Here t is always a non-negative *integer*: $t \in \mathbb{Z}$, $t \geq 0$.

By a *solution* of a difference equation, we mean an expression for the term $\mathbf{x}_t$ which involves t and the first term of the sequence (the initial condition). One very simple result we will need is that the solution to the difference equation

$$x_{t+1} = ax_t$$

is simply

$$x_t = a^t x_0,$$

where x_0 is the first term of the sequence. (We assume that the members of the sequence are labeled as $x_0, x_1, x_2, \ldots$, rather than $x_1, x_2, \ldots$.) You might recognise these as the terms of a geometric progression, if you have studied those before.

This result is easily established. If $x_{t+1} = ax_t$, we have

$$x_1 = ax_0$$
$$x_2 = ax_1 = a(ax_0) = a^2 x_0$$
$$x_3 = a\mathbf{x}_2 = a(a^2 x_0) = a^3 x_0$$
$$\vdots$$
$$x_t = a^t \mathbf{x}_0.$$

9.2.2   Systems of difference equations

We shall now see how we can use diagonalisation to solve (linear) systems of difference equations. This is a powerful and important application of diagonalisation. We introduce the ideas with a fairly manageable example.

Example 9.6 Suppose the sequences x_t and y_t are related as follows:
$x_0 = 1$, $y_0 = 1$, and, for $t \geq 0$,

$$x_{t+1} = 7x_t - 15y_t, \tag{9.1}$$

$$y_{t+1} = 2x_t - 4y_t. \tag{9.2}$$

This is an example of a *coupled* system of difference equations. We cannot directly solve equation (9.1) for x_t since we would need to know y_t. On the other hand, we can't work out y_t directly from equation (9.2) because to do so we would need to know x_t! You might think that it therefore seems impossible. However, there is a way to solve the problem, and it uses diagonalisation.

Example 9.6 continued Let us notice that the system we're considering can be expressed in matrix form. If we let

$$\mathbf{x}_t = \begin{pmatrix} x_t \\ y_t \end{pmatrix},$$

then the problem is to find $\mathbf{x}_t$ given that $\mathbf{x}_{t+1} = A\mathbf{x}_t$ for $t \geq 0$ and given that $\mathbf{x}_0 = (1, 1)^{\mathrm{T}}$, where A (our old friend from the previous chapter and Example 9.1) is

$$A = \begin{pmatrix} 7 & -15 \\ 2 & -4 \end{pmatrix}.$$

We're very familiar with the matrix A. We know how to diagonalise it and we know its nth power from Example 9.3. The expression we have for the nth power is immediately useful here. We have $\mathbf{x}_{t+1} = A\mathbf{x}_t$ for $t \geq 0$. So,

$$\mathbf{x}_1 = A\mathbf{x}_0,$$
$$\mathbf{x}_2 = A\mathbf{x}_1 = A(A\mathbf{x}_0) = A^2\mathbf{x}_0,$$
$$\mathbf{x}_3 = A\mathbf{x}_2 = A(A^2\mathbf{x}_0) = A^3\mathbf{x}_0$$

and, in general,

$$\mathbf{x}_t = A^t\mathbf{x}_0.$$

But we know (from Example 9.3) an expression for A^t. So we can see that

$$\begin{pmatrix} x_t \\ y_t \end{pmatrix} = \mathbf{x}_t = A^t\mathbf{x}_0 = \begin{pmatrix} -5 + 6(2^n) & 15 - 15(2^n) \\ -2 + 2(2^n) & 6 - 5(2^n) \end{pmatrix} \begin{pmatrix} 1 \\ 1 \end{pmatrix}$$

$$= \begin{pmatrix} 10 - 9(2^t) \\ 4 - 3(2^t) \end{pmatrix}.$$

Therefore, the sequences are

$$x_t = 10 - 9(2^t), \qquad y_t = 4 - 3(2^t).$$

This example demonstrates a very general approach to solving systems of difference equations where the underlying matrix A is diagonalisable.

Note that we could equally well express the system $\mathbf{x}_{t+1} = A\mathbf{x}_t$ for $t \geq 0$, as $\mathbf{x}_t = A\mathbf{x}_{t-1}$ for $t \geq 1$. The systems are exactly the same.

9.2.3 Solving using matrix powers

Suppose we want to solve a system $\mathbf{x}_{t+1} = A\mathbf{x}_t$, in which A is diagonalisable. As we have seen, we can use diagonalisation to determine the powers of the matrix and, as indicated above, this can help us to solve the system. The key ideas are encapsulated in Example 9.6. We now illustrate further with an example involving three sequences, in which the underlying matrix is therefore a 3×3 matrix.

Example 9.7 The system we consider is as follows. We want to find the sequences x_t, y_t, z_t, which satisfy the difference equations

$$x_{t+1} = 6x_t + 13y_t - 8z_t$$
$$y_{t+1} = 2x_t + 5y_t - 2z_t$$
$$z_{t+1} = 7x_t + 17y_t - 9z_t$$

and the initial conditions $x_0 = 1$, $y_0 = 1$, $z_0 = 0$.

In matrix form, this system is $\mathbf{x}_{t+1} = A\mathbf{x}_t$, where

$$A = \begin{pmatrix} 6 & 13 & -8 \\ 2 & 5 & -2 \\ 7 & 17 & -9 \end{pmatrix}, \qquad \mathbf{x}_t = \begin{pmatrix} x_t \\ y_t \\ z_t \end{pmatrix}.$$

We need to diagonalise A. You will probably have worked through this diagonalisation yourself in Exercise 8.5. If we take

$$P = \begin{pmatrix} 1 & -1 & 1 \\ 0 & 1 & 1 \\ 1 & 1 & 2 \end{pmatrix} \quad \text{and} \quad D = \begin{pmatrix} -2 & 0 & 0 \\ 0 & 1 & 0 \\ 0 & 0 & 3 \end{pmatrix},$$

then $P^{-1}AP = D$. Now, as you can calculate,

$$P^{-1} = \begin{pmatrix} -1 & -3 & 2 \\ -1 & -1 & 1 \\ 1 & 2 & -1 \end{pmatrix}.$$

It follows that $A^t = PD^tP^{-1}$, so that

$$\mathbf{x}_t = A^t\mathbf{x}_0 = PD^tP^{-1}\mathbf{x}_0.$$

Therefore, the solution is given by

$$\mathbf{x}_t = \begin{pmatrix} 1 & -1 & 1 \\ 0 & 1 & 1 \\ 1 & 1 & 2 \end{pmatrix} \begin{pmatrix} (-2)^t & 0 & 0 \\ 0 & 1^t & 0 \\ 0 & 0 & 3^t \end{pmatrix} \begin{pmatrix} -1 & -3 & 2 \\ -1 & -1 & 1 \\ 1 & 2 & -1 \end{pmatrix} \begin{pmatrix} 1 \\ 1 \\ 0 \end{pmatrix}.$$

You can multiply these matrices in any order, but the simplest way is to begin at the right with

$$P^{-1}\mathbf{x}_0 = \begin{pmatrix} -1 & -3 & 2 \\ -1 & -1 & 1 \\ 1 & 2 & -1 \end{pmatrix} \begin{pmatrix} 1 \\ 1 \\ 0 \end{pmatrix} = \begin{pmatrix} -4 \\ -2 \\ 3 \end{pmatrix}$$

so that

$$\begin{pmatrix} x_t \\ y_t \\ z_t \end{pmatrix} = \begin{pmatrix} 1 & -1 & 1 \\ 0 & 1 & 1 \\ 1 & 1 & 2 \end{pmatrix} \begin{pmatrix} -4(-2)^t \\ -2 \\ 3(3^t) \end{pmatrix} = \begin{pmatrix} -4(-2)^t + 2 + 3(3^t) \\ -2 + 3(3^t) \\ -4(-2)^t - 2 + 6(3^t) \end{pmatrix}.$$

The sequence are

$$x_t = -4(-2)^t + 2 + 3(3^t)$$
$$y_t = -2 + 3(3^t)$$
$$z_t = -4(-2)^t - 2 + 6(3^t).$$

How can we check that this solution is correct? We should at least check that it gives us the correct initial conditions by substituting $t = 0$ into the solution. This is easily done.

Activity 9.8 Do this!

We can also find $\mathbf{x}_1$ in two different ways. The original equations will give us

$$\mathbf{x}_1 = A\mathbf{x}_0 = \begin{pmatrix} 6 & 13 & -8 \\ 2 & 5 & -2 \\ 7 & 17 & -9 \end{pmatrix} \begin{pmatrix} 1 \\ 1 \\ 0 \end{pmatrix} = \begin{pmatrix} 19 \\ 7 \\ 24 \end{pmatrix}.$$

If we get the same result from our solution, then we can be fairly certain that our solution is correct. According to our solution,

$$\begin{pmatrix} x_1 \\ y_1 \\ z_1 \end{pmatrix} = \begin{pmatrix} -4(-2) + 2 + 3(3) \\ -2 + 3(3) \\ -4(-2) - 2 + 6(3) \end{pmatrix} = \begin{pmatrix} 19 \\ 7 \\ 24 \end{pmatrix}.$$

So we do indeed obtain the same answer.

Activity 9.9 Carry out any omitted calculations in this example.

9.2.4   Solving by change of variable

We can use diagonalisation as the key to another general method for solving systems of difference equations. Given a system $\mathbf{x}_{t+1} = A\mathbf{x}_t$, in which A is diagonalisable, we perform a *change of variable* or *change of coordinates*, as follows. Suppose that $P^{-1}AP = D$ (where D is diagonal) and let

$$\mathbf{x}_t = P\mathbf{u}_t.$$

Equivalently, the new variable vector $\mathbf{u}_t$ is $\mathbf{u}_t = P^{-1}\mathbf{x}_t$. One way of thinking about this is that the vector $\mathbf{x}_t$ is in standard coordinates and $\mathbf{u}_t$ is in coordinates in the basis of eigenvectors. Then substituting $\mathbf{x}_t = P\mathbf{u}_t$ into the equation $\mathbf{x}_{t+1} = A\mathbf{x}_t$, and noting that $\mathbf{x}_{t+1} = P\mathbf{u}_{t+1}$, the equation becomes

$$P\mathbf{u}_{t+1} = AP\mathbf{u}_t,$$

which means that

$$\mathbf{u}_{t+1} = P^{-1}AP\mathbf{u}_t = D\mathbf{u}_t.$$

Since D is diagonal, this is very easy to solve for $\mathbf{u}_t$. To find $\mathbf{x}_t$, we then use the fact that $\mathbf{x}_t = P\mathbf{u}_t$.

We will illustrate the method using the system in Example 9.7

Example 9.10 We find the sequences x_t, y_t, z_t which satisfy the difference equations

$$x_{t+1} = 6x_t + 13y_t - 8z_t$$
$$y_{t+1} = 2x_t + 5y_t - 2z_t$$
$$z_{t+1} = 7x_t + 17y_t - 9z_t$$

and the initial conditions $x_0 = 1$, $y_0 = 1$, $z_0 = 0$.

Using the matrices A, P and D given in Example 9.7, we let

$$\mathbf{u}_t = \begin{pmatrix} u_t \\ v_t \\ w_t \end{pmatrix}$$

be given by $\mathbf{x}_t = P\mathbf{u}_t$. Then the equation $\mathbf{x}_{t+1} = A\mathbf{x}_t$ gives rise (as explained above) to $\mathbf{u}_{t+1} = D\mathbf{u}_t$. That is,

$$\begin{pmatrix} u_{t+1} \\ v_{t+1} \\ w_{t+1} \end{pmatrix} = \begin{pmatrix} -2 & 0 & 0 \\ 0 & 1 & 0 \\ 0 & 0 & 3 \end{pmatrix} \begin{pmatrix} u_t \\ v_t \\ w_t \end{pmatrix},$$

so we have the following system for the new sequences u_t, v_t, w_t:

$$u_{t+1} = -2u_t$$
$$v_{t+1} = v_t$$
$$w_{t+1} = 3w_t.$$

This is *very* easy to solve: each equation involves only one sequence, so we have *uncoupled* the equations. We have, for all t,

$$u_t = (-2)^t u_0, \qquad v_t = v_0, \qquad w_t = 3^t w_0.$$

We have *not yet* solved the original problem, however, since we need to find x_t, y_t, z_t. We have

$$\mathbf{x}_t = \begin{pmatrix} x_t \\ y_t \\ x_t \end{pmatrix} = P\mathbf{u}_t = \begin{pmatrix} 1 & -1 & 1 \\ 0 & 1 & 1 \\ 1 & 1 & 2 \end{pmatrix} \begin{pmatrix} u_t \\ v_t \\ w_t \end{pmatrix}$$

$$= \begin{pmatrix} 1 & -1 & 1 \\ 0 & 1 & 1 \\ 1 & 1 & 2 \end{pmatrix} \begin{pmatrix} (-2)^t u_0 \\ v_0 \\ 3^t w_0 \end{pmatrix}.$$

But we have also to find out what u_0, v_0, w_0 are. These are not given in the problem, but x_0, y_0, z_0 are, and we know that

$$\begin{pmatrix} x_0 \\ y_0 \\ z_0 \end{pmatrix} = P \begin{pmatrix} u_0 \\ v_0 \\ w_0 \end{pmatrix} = \begin{pmatrix} 1 & -1 & 1 \\ 0 & 1 & 1 \\ 1 & 1 & 2 \end{pmatrix} \begin{pmatrix} u_0 \\ v_0 \\ w_0 \end{pmatrix}.$$

To find u_0, v_0, w_0, we can either solve the linear system

$$P \begin{pmatrix} u_0 \\ v_0 \\ w_0 \end{pmatrix} = \begin{pmatrix} x_0 \\ y_0 \\ z_0 \end{pmatrix} = \begin{pmatrix} 1 \\ 1 \\ 0 \end{pmatrix}$$

using row operations, or we can (though it may involve more work) find out what P^{-1} is and use the fact that $\mathbf{u}_0 = P^{-1}\mathbf{x}_0$,

$$\begin{pmatrix} u_0 \\ v_0 \\ w_0 \end{pmatrix} = P^{-1} \begin{pmatrix} x_0 \\ y_0 \\ z_0 \end{pmatrix} = P^{-1} \begin{pmatrix} 1 \\ 1 \\ 0 \end{pmatrix}.$$

Either way (and the working is omitted here, but you should check it), we find

$$\begin{pmatrix} u_0 \\ v_0 \\ w_0 \end{pmatrix} = \begin{pmatrix} -4 \\ -2 \\ 3 \end{pmatrix}.$$

Returning then to the general solution to the system, we obtain

$$\begin{pmatrix} x_t \\ y_t \\ x_t \end{pmatrix} = \begin{pmatrix} 1 & -1 & 1 \\ 0 & 1 & 1 \\ 1 & 1 & 2 \end{pmatrix} \begin{pmatrix} (-2)^t u_0 \\ v_0 \\ 3^t w_0 \end{pmatrix}$$

$$= \begin{pmatrix} 1 & -1 & 1 \\ 0 & 1 & 1 \\ 1 & 1 & 2 \end{pmatrix} \begin{pmatrix} -4(-2)^t \\ -2 \\ 3(3^t) \end{pmatrix},$$

so we have the solution

$$\begin{pmatrix} x_t \\ y_t \\ x_t \end{pmatrix} = \begin{pmatrix} -4(-2)^t + 2 + 3(3^t) \\ -2 + 3(3^t) \\ -4(-2)^t - 2 + 6(3^t) \end{pmatrix}.$$

And, of course, this is in agreement with the answer obtained earlier using matrix powers.

Activity 9.11 Perform all the omitted calculations for this example.

9.2.5   Another example

Let's find the sequences x_t, y_t, z_t which satisfy the following system of linear difference equations

$$x_{t+1} = 4x_t + 4z_t$$
$$y_{t+1} = 4y_t + 4z_t$$
$$z_{t+1} = 4x_t + 4y_t + 8z_t$$

and $x_0 = 6$, $y_0 = 12$, $z_0 = 12$.

We will do this by both methods described above (although, of course, you would only need to choose one of the methods and solve it that way). In matrix form, this system is $\mathbf{x}_{t+1} = A\mathbf{x}_t$, where

$$A = \begin{pmatrix} 4 & 0 & 4 \\ 0 & 4 & 4 \\ 4 & 4 & 8 \end{pmatrix}, \qquad \mathbf{x}_t = \begin{pmatrix} x_t \\ y_t \\ z_t \end{pmatrix}.$$

This is the matrix we diagonalised in Example 8.23. There, we found that $P^{-1}AP = D$, where

$$P = \begin{pmatrix} -1 & -1 & 1 \\ 1 & -1 & 1 \\ 0 & 1 & 2 \end{pmatrix}, \qquad D = \begin{pmatrix} 4 & 0 & 0 \\ 0 & 0 & 0 \\ 0 & 0 & 12 \end{pmatrix}.$$

First, we use matrix powers. Since $\mathbf{x}_{t+1} = A\mathbf{x}_t$, we therefore have

$$\mathbf{x}_t = A^t \mathbf{x}_0.$$

Because $A = PDP^{-1}$, we have $A^t = PD^tP^{-1}$, so

$$\mathbf{x}_t = PD^tP^{-1}\mathbf{x}_0.$$

Now, as you can calculate,

$$P^{-1} = \frac{1}{6}\begin{pmatrix} -3 & 3 & 0 \\ -2 & -2 & 2 \\ 1 & 1 & 2 \end{pmatrix}.$$

Therefore, the solution is given by

$$\mathbf{x}_t = \begin{pmatrix} -1 & -1 & 1 \\ 1 & -1 & 1 \\ 0 & 1 & 2 \end{pmatrix}\begin{pmatrix} 4^t & 0 & 0 \\ 0 & 0^t & 0 \\ 0 & 0 & 12^t \end{pmatrix}\frac{1}{6}\begin{pmatrix} -3 & 3 & 0 \\ -2 & -2 & 2 \\ 1 & 1 & 2 \end{pmatrix}\begin{pmatrix} 6 \\ 12 \\ 12 \end{pmatrix}.$$

Now (and this is a fact you may not have seen before) $0^t = 0$ for all $t \geq 1$, but $0^0 = 1$ (since by definition $x^0 = 1$ for all real numbers x). So for all $t \geq 1$, we have

$$\mathbf{x}_t = \begin{pmatrix} x_t \\ y_t \\ z_t \end{pmatrix} = \begin{pmatrix} -1 & -1 & 1 \\ 1 & -1 & 1 \\ 0 & 1 & 2 \end{pmatrix}\begin{pmatrix} 4^t & 0 & 0 \\ 0 & 0 & 0 \\ 0 & 0 & 12^t \end{pmatrix}\begin{pmatrix} 3 \\ -2 \\ 7 \end{pmatrix}.$$

That is, for $t \geq 1$,

$$\begin{pmatrix} x_t \\ y_t \\ z_t \end{pmatrix} = \begin{pmatrix} -3(4^t) + 7(12^t) \\ 3(4^t) + 7(12^t) \\ 14(12^t) \end{pmatrix},$$

and, of course (since this is given),

$$\mathbf{x}_0 = \begin{pmatrix} 6 \\ 12 \\ 12 \end{pmatrix}.$$

(You might observe that if you take the expression for $\mathbf{x}_t$ given for $t \geq 1$, and if you substitute $t = 0$, you don't get the right value for $\mathbf{x}_0$. That isn't because the solution is wrong; it's simply because that expression only works for $t \geq 1$. Look at how it is obtained: we set 0^t equal to 0, something that is true for $t \geq 1$ but not for $t = 0$.)

For the second method (in which we change the variable), let

$$\mathbf{u}_t = \begin{pmatrix} u_t \\ v_t \\ w_t \end{pmatrix}$$

be given by $\mathbf{x}_t = P\mathbf{u}_t$. Then the equation $\mathbf{x}_{t+1} = A\mathbf{x}_t$ becomes $P\mathbf{u}_{t+1} = AP\mathbf{u}_t$, or $\mathbf{u}_{t+1} = P^{-1}AP\mathbf{u}_t$; that is, $\mathbf{u}_{t+1} = D\mathbf{u}_t$,

$$\begin{pmatrix} u_{t+1} \\ v_{t+1} \\ w_{t+1} \end{pmatrix} = \begin{pmatrix} 4 & 0 & 0 \\ 0 & 0 & 0 \\ 0 & 0 & 12 \end{pmatrix}\begin{pmatrix} u_t \\ v_t \\ w_t \end{pmatrix}.$$

So we have the following system for the new sequences u_t, v_t, w_t:

$$u_{t+1} = 4\,u_t$$
$$v_{t+1} = 0\,v_t$$
$$w_{t+1} = 12\,w_t,$$

with solutions

$$u_t = 4^t u_0, \quad v_t = 0 \text{ for } t \geq 1, \quad w_t = 12^t\,w_0.$$

To find u_0, v_0, w_0, we use $\mathbf{u}_0 = P^{-1}\mathbf{x}_0$. As calculated earlier,

$$\begin{pmatrix} u_0 \\ v_0 \\ w_0 \end{pmatrix} = \frac{1}{6}\begin{pmatrix} -3 & 3 & 0 \\ -2 & -2 & 2 \\ 1 & 1 & 2 \end{pmatrix}\begin{pmatrix} 6 \\ 12 \\ 12 \end{pmatrix} = \begin{pmatrix} 3 \\ -2 \\ 7 \end{pmatrix}.$$

Then the solution is,

$$\begin{pmatrix} x_t \\ y_t \\ x_t \end{pmatrix} = \begin{pmatrix} -1 & -1 & 1 \\ 1 & -1 & 1 \\ 0 & 1 & 2 \end{pmatrix}\begin{pmatrix} 4^t u_0 \\ 0 \\ 12^t w_0 \end{pmatrix}$$

$$= \begin{pmatrix} -1 & -1 & 1 \\ 1 & -1 & 1 \\ 0 & 1 & 2 \end{pmatrix}\begin{pmatrix} 3(4^t) \\ 0 \\ 7(12^t) \end{pmatrix}$$

$$= \begin{pmatrix} -3(4^t) + 7(12^t) \\ 3(4^t) + 7(12^t) \\ 14(12^t) \end{pmatrix}, \qquad \text{for } t \geq 1$$

and $\mathbf{x}_0 = (6, 12, 12)^{\mathrm{T}}$. This is in agreement with the answer obtained earlier using matrix powers.

Activity 9.12 Check all the calculations in this section.

Activity 9.13 Check the solutions by finding $\mathbf{x}_1$. See what happens if you keep 0^t as part of your solution in either method; will the solution then work for all $t \geq 0$?

9.2.6   Markov Chains

To illustrate just what a Markov chain is, let's begin by looking at an example.

Example 9.14 Suppose two supermarkets compete for customers in a region with 20 000 shoppers. Assume that no shopper goes to both supermarkets in any week, and that the table below gives the probabilities that a shopper will change from one supermarket (or none) to another (or none) during the week.

	From A	From B	From none
To A	0.70	0.15	0.30
To B	0.20	0.80	0.20
To none	0.10	0.05	0.50

For example, the second column tells us that during any given week supermarket B will keep 80% of its customers while losing 15% to supermarket A and 5% to no supermarket. Notice that the probabilities in the column add up to 1, since every shopper has to end up somewhere in the following week.

Suppose that at the end of a certain week (call it week zero), it is known that the total population of $T = 20\,000$ shoppers was distributed as follows: $10\,000$ (that is, $0.5\,T$) went to supermarket A, $8\,000$ $(0.4\,T)$ went to supermarket B and $2\,000$ $(0.1\,T)$ did not go to a supermarket.

Given this information, the questions we wish to answer are: 'Can we predict the number of shoppers at each supermarket in any future week t?', and 'Can we predict a long-term distribution of shoppers?'

In order to answer these questions, we formulate the problem as a system of linear difference equations. Let x_t denote the (decimal) percentage of total shoppers going to supermarket A in week t, y_t the percentage going to supermarket B and z_t the percentage who do not go to any supermarket. The numbers of shoppers in week t can be predicted by this model from the numbers in the previous week; that is,

$$\mathbf{x}_t = A\mathbf{x}_{t-1},$$

where

$$A = \begin{pmatrix} 0.70 & 0.15 & 0.30 \\ 0.20 & 0.80 & 0.20 \\ 0.10 & 0.05 & 0.50 \end{pmatrix}, \quad \mathbf{x}_t = \begin{pmatrix} x_t \\ y_t \\ z_t \end{pmatrix}$$

and $x_0 = 0.5$, $y_0 = 0.4$, $z_0 = 0.1$.

What features of this problem make it a Markov chain? In general, a *Markov chain* or a *Markov process* is a closed system consisting of a fixed total population which is distributed into n different states, and which changes during specific time intervals from one distribution to another. We assume that we know the probability that a given member will change from one state into another, depending on the state the member occupied during the previous time interval.

These probabilities are listed in an $n \times n$ matrix A, where the (i, j) entry is the probability that a member of the population will change from state j to state i. Such a matrix is called a *transition matrix* of a Markov chain.

Definition 9.15 The $n \times n$ matrix $A = (a_{ij})$ is a *transition matrix of a Markov chain* if it satisfies the following two properties:

(1)   The entries of A are all non-negative.
(2)   The sum of the entries in each column of A is equal to 1:

$$a_{1j} + a_{2j} + \cdots + a_{nj} = 1.$$

Property (2) follows from the assumption that all members of the population must be in one of the n states at any given time. (Informally, all those at state j have to be somewhere at the next observation time, so the sum (over all i) of the 'transition probabilities' of going from state j to state i, must equal 1.)

The *distribution vector* (or *state vector*) for the time period t is the vector $\mathbf{x}_t$, whose ith entry is the percentage of the population in state i at time t. The entries of $\mathbf{x}_t$ sum to 1 because all members of the population must be in one of the states at any time. Our first goal is to find the state vector for any t, and to do this we need to solve the difference equation

$$\mathbf{x}_t = A\mathbf{x}_{t-1}, \quad t \geq 1.$$

A solution of the difference equation is an expression for the distribution vector $\mathbf{x}_t$ in terms of A and $\mathbf{x}_0$, and, as we have seen earlier, the solution is $\mathbf{x}_t = A^t\mathbf{x}_0$.

Now assume that A can be diagonalised. If A has eigenvalues $\lambda_1, \lambda_2, \ldots, \lambda_n$ with corresponding eigenvectors $\mathbf{v}_1, \mathbf{v}_2, \ldots, \mathbf{v}_n$, then $P^{-1}AP = D$ where P is the matrix of eigenvectors of A and D is the corresponding diagonal matrix of eigenvalues.

The solution of the difference equation is

$$\mathbf{x}_t = A^t\mathbf{x}_0 = (PD^tP^{-1})\mathbf{x}_0.$$

Let's examine this solution to see what it tells us. If we set $\mathbf{x} = P\mathbf{z}$, so that $\mathbf{z}_0 = P^{-1}\mathbf{x}_0 = (b_1, b_2, \ldots, b_n)^{\mathrm{T}}$ represents the coordinates of $\mathbf{x}_0$ in the basis of eigenvectors, then this solution can be written in vector form as

$$\mathbf{x}_t = PD^t(P^{-1}\mathbf{x}_0)$$

$$= \begin{pmatrix} | & | & & | \\ \mathbf{v}_1 & \mathbf{v}_2 & \cdots & \mathbf{v}_n \\ | & | & & | \end{pmatrix} \begin{pmatrix} \lambda_1^t & 0 & \cdots & 0 \\ 0 & \lambda_2^t & \cdots & 0 \\ \vdots & \vdots & \ddots & \vdots \\ 0 & 0 & \cdots & \lambda_n^t \end{pmatrix} \begin{pmatrix} b_1 \\ b_2 \\ \vdots \\ b_n \end{pmatrix}$$

$$= b_1\lambda_1^t\mathbf{v}_1 + b_2\lambda_2^t\mathbf{v}_2 + \cdots + b_n\lambda_n^t\mathbf{v}_n.$$

Activity 9.16 Make sure you understand how the final equality above follows from matrix multiplication properties.

Now let's return to our example.

Example 9.14 continued We will use this solution to find the number of shoppers using each of the supermarkets at the end of week t, and see if we can use this information to predict the long-term distribution of shoppers.

First, we diagonalise the matrix A. The characteristic equation of A is

$$|A - \lambda I| = \begin{vmatrix} 0.70 - \lambda & 0.15 & 0.30 \\ 0.20 & 0.80 - \lambda & 0.20 \\ 0.10 & 0.05 & 0.50 - \lambda \end{vmatrix}$$

$$= -\lambda^3 + 2\lambda^2 - 1.24\lambda + 0.24 = 0.$$

This equation is satisfied by $\lambda = 1$, and hence 1 is an eigenvalue. Using the fact that $(\lambda - 1)$ is a factor of the polynomial, we find

$$(\lambda - 1)(\lambda^2 - \lambda + 0.24) = (\lambda - 1)(\lambda - 0.6)(\lambda - 0.4) = 0,$$

so the eigenvalues are $\lambda_1 = 1$, $\lambda_2 = 0.6$, and $\lambda_3 = 0.4$. The corresponding eigenvectors $\mathbf{v}_i$ are found by solving the homogeneous systems $(A - \lambda_i I)\mathbf{v} = \mathbf{0}$. (We omit the calculations.) Writing them as the columns of a matrix P, we find that $P^{-1}AP = D$, where

$$P = \begin{pmatrix} 3 & 3 & -1 \\ 4 & -4 & 0 \\ 1 & 1 & 1 \end{pmatrix}, \quad D = \begin{pmatrix} 1 & 0 & 0 \\ 0 & 0.6 & 0 \\ 0 & 0 & 0.4 \end{pmatrix}.$$

Activity 9.17 Carry out the omitted calculations for the diagonalisation above.

The distribution vector $\mathbf{x}_t$ at any time t is then given by

$$\mathbf{x}_t = b_1(1)^t \mathbf{v}_1 + b_2(0.6)^t \mathbf{v}_2 + b_3(0.4)^t \mathbf{v}_3,$$

where it only remains to find the coordinates, b_1, b_2, b_3 of $\mathbf{x}_0$ in the basis of eigenvectors.

Before we do this, let's see what the solution tells us about a *long-term distribution* of shoppers. We want to know what happens to $\mathbf{x}_t$ for very large values of t; that is, as $t \to \infty$. Note – and this is very important – that $1^t = 1$, and that as $t \to \infty$, $(0.6)^t \to 0$ and $(0.4)^t \to 0$. So there is a long-term distribution: the limit of $\mathbf{x}_t$ as $t \to \infty$ is a scalar multiple, $\mathbf{q} = b_1 \mathbf{v}_1$, of the eigenvector $\mathbf{v}_1$ whose eigenvalue is 1.

Now we'll complete the solution by finding b_1, b_2, b_3. The coordinates of $\mathbf{x}_0$ in the basis of eigenvectors are given by

$$P^{-1}\mathbf{x}_0 = \frac{1}{8}\begin{pmatrix} 1 & 1 & 1 \\ 1 & -1 & 1 \\ -2 & 0 & 6 \end{pmatrix}\begin{pmatrix} 0.5 \\ 0.4 \\ 0.1 \end{pmatrix} = \begin{pmatrix} 0.125 \\ 0.025 \\ -0.05 \end{pmatrix} = \begin{pmatrix} b_1 \\ b_2 \\ b_3 \end{pmatrix}.$$

Hence,

$$\mathbf{x}_t = 0.125\begin{pmatrix} 3 \\ 4 \\ 1 \end{pmatrix} + 0.025(0.6)^t\begin{pmatrix} 3 \\ -4 \\ 1 \end{pmatrix} - 0.05(0.4)^t\begin{pmatrix} -1 \\ 0 \\ 1 \end{pmatrix},$$

and the long-term distribution is

$$\mathbf{q} = \lim_{t \to \infty}\mathbf{x}_t = \begin{pmatrix} 0.375 \\ 0.500 \\ 0.125 \end{pmatrix}.$$

Relating this to numbers of shoppers, and remembering that the total number of shoppers is 20000, the long-term distribution is predicted to be 20000$\mathbf{q}$: 7500 to supermarket A, 10000 to B and 2500 to no supermarket.

Activity 9.18 Verify that P^{-1} is as stated.

You will have noticed that an essential part of the solution of predicting a long-term distribution for this example is the fact that the transition matrix A has an eigenvalue $\lambda = 1$ (of multiplicity 1), and that the other eigenvalues satisfy $|\lambda_i| < 1$. In this case, as t increases, the distribution vector $\mathbf{x}_t$ will approach the unique eigenvector $\mathbf{q}$ for $\lambda = 1$ which is also a distribution vector. The fact that the entries sum to 1 makes $\mathbf{q}$ unique among the vectors satisfying $A\mathbf{q} = \mathbf{q}$.

We would like to be able to know that this is the case for any Markov chain, but there are some exceptions to this rule. A Markov chain is said to be *regular* if some power of the transition matrix A has strictly positive entries (so it has no zero entries). In this case, there will be a long-term distribution, as the following theorem implies.

Theorem 9.19 *If A is the transition matrix of a regular Markov chain, then $\lambda = 1$ is an eigenvalue of multiplicity 1, and all other eigenvalues satisfy $|\lambda_i| < 1$.*

We will not prove this theorem here. However, we will prove a similar, but weaker result, which makes it clear that the only thing that can go wrong is for the eigenvalue $\lambda = 1$ to have multiplicity greater than 1. First, we need a definition.

Definition 9.20 A matrix C is called a *stochastic matrix* if it has the following two properties:

(1)   The entries of C are all non-negative.
(2)   The sum of the entries in each *row* of C is equal to 1:

$$c_{i1} + c_{i2} + \cdots + c_{in} = 1.$$

Note that if A is a transition matrix for a Markov process, then A^T is a stochastic matrix (because all of its entries are non-negative and the sum of the entries of each row of A^T is equal to 1).

Matrices A and A^T have the same eigenvalues, because by properties of transpose and determinant, they have the same characteristic polynomials (the roots of which are the eigenvalues):

$$|A - \lambda I| = |(A - \lambda I)^T| = |A^T - \lambda I|.$$

We will prove the following theorem for stochastic matrices, and then apply it to transition matrices.

Theorem 9.21 *If C is a stochastic matrix, then:*

- $\mathbf{v} = (1, 1, \ldots, 1)^T$ *is an eigenvector of C with eigenvalue $\lambda = 1$.*
- *If λ is an eigenvalue of C, then $|\lambda| \leq 1$.*

Proof: Let $C = (c_{ij})$. That $C\mathbf{v} = \mathbf{v}$ follows immediately from property (2) of the definition of a stochastic matrix, since the ith entry of $C\mathbf{v}$ is $c_{i1}(1) + c_{i2}(1) + \cdots + c_{in}(1) = 1$.

To prove the second statement, let λ be an eigenvalue of C, let $\mathbf{u} \neq \mathbf{0}$ be any vector satisfying $C\mathbf{u} = \lambda\mathbf{u}$, and let u_i denote the largest component (in absolute value) of $\mathbf{u}$. To show that $|\lambda| \leq 1$, set

$$\mathbf{w} = \frac{1}{u_i}\mathbf{u}.$$

Then $C\mathbf{w} = \lambda\mathbf{w}$, $w_i = 1$, and $|w_k| \leq 1$ for $1 \leq k \leq n$. Consider what the ith row of the matrix equation $C\mathbf{w} = \lambda\mathbf{w}$ tells us. It says that

$$\lambda w_i = c_{i1}w_1 + c_{i2}w_2 + \cdots + c_{in}w_n,$$

and hence

$$
\begin{aligned}
|\lambda| = |\lambda w_i| \quad &\text{(since } w_i = 1\text{)} \\
= |c_{i1}w_1 + c_{i2}w_2 + \cdots + c_{in}w_n| & \\
\leq c_{i1}|w_1| + c_{i2}|w_2| + \cdots + c_{in}|w_n| & \\
\leq c_{i1} + c_{i2} + \cdots + c_{in} = 1 \quad &\text{(because } w_k \leq 1\text{)}.
\end{aligned}
$$

So we've shown that $\lambda = 1$ is an eigenvalue and that all eigenvalues λ_i satisfy $|\lambda_i| \leq 1$. □

What does this theorem imply about Markov chains? We saw earlier that if A is the transition matrix of a Markov chain, then A^T is a stochastic matrix and also that A and A^T have the same eigenvalues. Therefore, you can deduce from Theorem 9.21 that:

- $\lambda = 1$ is an eigenvalue of A, and
- if λ_i is an eigenvalue of A then $|\lambda_i| \leq 1$.

The theorem tells us that $\lambda = 1$ is an eigenvalue, but it might have multiplicity greater than 1, in which case either there would be more than one (linearly independent) eigenvector corresponding to $\lambda = 1$, or the matrix might not be diagonalisable.

In order to obtain a long-term distribution, we need to know that there is only one (linearly independent) eigenvector for the eigenvalue $\lambda = 1$. So *if* the eigenvalue $\lambda = 1$ of a transition matrix A of a Markov chain does have multiplicity 1, then Theorem 9.21 implies all the other eigenvalues λ_i satisfy $|\lambda_i| < 1$. There will be one corresponding eigenvector which is also a distribution vector, and provided A can be diagonalised, we will know that there is a long-term distribution. This is all we will need in practice.

9.3   Linear systems of differential equations

This section is aimed at those who will have studied calculus before, as many of you have (or will be doing so concurrently with your linear algebra studies). But if you have not yet studied calculus, you can simply omit this section.

A *differential equation* is, broadly speaking, an equation that involves a function and its derivatives. We are interested here only in very simple types of differential equation and it is quite easy to summarise what you need to know so that we do not need a lengthy digression into calculus (which would detract from the whole point of the exercise, which is to demonstrate the power of diagonalisation).

For a function $y = y(t)$, the derivative of y will be denoted by $y' = y'(t)$ or dy/dt. The result we will need is the following: if $y(t)$ satisfies the 'linear' differential equation

$$y' = ay,$$

then the general solution is

$$y(t) = \beta e^{at} \quad \text{for} \quad \beta \in \mathbb{R}.$$

If an *initial condition*, $y(0)$ is given, then since $y(0) = \beta e^0 = \beta$, we have a particular (unique) solution $y(t) = y(0)e^{at}$ to the differential equation.

Activity 9.22 Check that $y = 3e^{2t}$ is a solution of the differential equation $y' = 2y$ which satisfies the initial condition $y(0) = 3$.

We will look at systems consisting of these types of differential equations. In Section 9.2.4, we used a change of variable technique based on diagonalisation to solve systems of difference equations. We can apply an analogous technique to solve systems of linear differential equations.

In general, a (square) *linear system* of differential equations for the functions $y_1(t)$, $y_2(t)$, $\ldots$, $y_n(t)$ is of the form

$$y_1' = a_{11}y_1 + a_{12}y_2 + \cdots + a_{1n}y_n$$
$$y_2' = a_{21}y_1 + a_{22}y_2 + \cdots + a_{2n}y_n$$
$$\vdots$$
$$y_n' = a_{n1}y_1 + a_{n2}y_2 + \cdots + a_{nn}y_n,$$

for constants $a_{ij} \in \mathbb{R}$. So such a system takes the form

$$\mathbf{y}' = A\mathbf{y},$$

where $A = (a_{ij})$ is an $n \times n$ matrix whose entries are constants (that is, fixed numbers), and $\mathbf{y} = (y_1, y_2, \ldots, y_n)^{\mathrm{T}}, \mathbf{y}' = (y_1', y_2', \ldots, y_n')^{\mathrm{T}}$ are vectors of functions.

If A is diagonal, the system $\mathbf{y}' = A\mathbf{y}$ is easy to solve. For instance, suppose

$$A = \mathrm{diag}(\lambda_1, \lambda_2, \ldots, \lambda_n).$$

Then the system is precisely

$$y_1' = \lambda_1 y_1, \quad y_2' = \lambda_2 y_2, \quad \ldots, \quad y_n' = \lambda_n y_n,$$

and so

$$y_1 = y_1(0)e^{\lambda_1 t}, \quad y_2 = y_2(0)e^{\lambda_2 t}, \quad \ldots, \quad y_n = y_n(0)e^{\lambda_n t}.$$

Since a diagonal system is so easy to solve, it would be very helpful if we could reduce our given system to a diagonal one, and this is precisely what the method will do in the case where A is diagonalisable. We will come back to the general discussion shortly, but for now we explore with a simple example.

Example 9.23 Suppose the functions $y_1(t)$ and $y_2(t)$ are related as follows:

$$y_1' = 7y_1 - 15y_2$$
$$y_2' = 2y_1 - 4y_2.$$

In matrix form, this is $\mathbf{y}' = A\mathbf{y}$, where A is the 2×2 matrix we considered earlier:

$$A = \begin{pmatrix} 7 & -15 \\ 2 & -4 \end{pmatrix}.$$

We've seen this matrix is diagonalisable; if

$$P = \begin{pmatrix} 5 & 3 \\ 2 & 1 \end{pmatrix},$$

then P is invertible and

$$P^{-1}AP = D = \begin{pmatrix} 1 & 0 \\ 0 & 2 \end{pmatrix}.$$

We now use the matrix P to define new functions $z_1(t)$, $z_2(t)$ by setting $\mathbf{y} = P\mathbf{z}$ (or equivalently, $\mathbf{z} = P^{-1}\mathbf{y}$); that is,

$$\mathbf{y} = \begin{pmatrix} y_1 \\ y_2 \end{pmatrix} = \begin{pmatrix} 5 & 3 \\ 2 & 1 \end{pmatrix} \begin{pmatrix} z_1 \\ z_2 \end{pmatrix} = P\mathbf{z},$$

so that,

$$y_1 = 5z_1 + 3z_2$$
$$y_2 = 2z_1 + z_2.$$

By differentiating these equations, we can express y_1' and y_2' in terms of z_1' and z_2',

$$y_1' = 5z_1' + 3z_2'$$
$$y_2' = 2z_1' + z_2',$$

so that $\mathbf{y}' = (P\mathbf{z})' = P\mathbf{z}'$. Then we have,

$$P\mathbf{z}' = \mathbf{y}' = A\mathbf{y} = A(P\mathbf{z}) = AP\mathbf{z}$$

and hence

$$\mathbf{z}' = P^{-1}AP\mathbf{z} = D\mathbf{z}.$$

In other words,

$$\begin{pmatrix} z_1' \\ z_2' \end{pmatrix} = \begin{pmatrix} 1 & 0 \\ 0 & 2 \end{pmatrix} \begin{pmatrix} z_1 \\ z_2 \end{pmatrix} = \begin{pmatrix} z_1 \\ 2z_2 \end{pmatrix}.$$

So the system for the functions z_1, z_2 is diagonal and hence is easily solved. Having found z_1, z_2, we can then find y_1 and y_2 through the explicit connection between the two sets of functions, namely $\mathbf{y} = P\mathbf{z}$.

Let us now return to the general technique. Suppose we have the system $\mathbf{y}' = A\mathbf{y}$, and that A can indeed be diagonalised. Then there is an invertible matrix P and a diagonal matrix D such that $P^{-1}AP = D$. Here

$$P = (\mathbf{v}_1 \dots \mathbf{v}_n), \qquad D = \text{diag}(\lambda_1, \lambda_2, \dots, \lambda_n),$$

where λ_i are the eigenvalues and $\mathbf{v}_i$ corresponding eigenvectors. Let $\mathbf{z} = P^{-1}\mathbf{y}$ (or, equivalently, let $\mathbf{y} = P\mathbf{z}$). Then

$$\mathbf{y}' = (P\mathbf{z})' = P\mathbf{z}',$$

since P has constant entries.

Activity 9.24 Prove that $(P\mathbf{z})' = P\mathbf{z}'$.

Therefore,

$$P\mathbf{z}' = A\mathbf{y} = AP\mathbf{z},$$

and

$$\mathbf{z}' = P^{-1}AP\mathbf{z} = D\mathbf{z}.$$

We may now easily solve for $\mathbf{z}$, and hence $\mathbf{y}$.

We illustrate with an example of a 3 by 3 system of differential equations, solved using this method. Note carefully how we use the initial values $y_1(0)$, $y_2(0)$ and $y_3(0)$.

Example 9.25 We find functions $y_1(t)$, $y_2(t)$, $y_3(t)$ such that $y_1(0) = 2$, $y_2(0) = 1$ and $y_3(0) = 1$ and such that they are related by the linear system of differential equations,

$$\frac{dy_1}{dt} = 6y_1 + 13y_2 - 8y_3$$

$$\frac{dy_2}{dt} = 2y_1 + 5y_2 - 2y_3$$

$$\frac{dy_3}{dt} = 7y_1 + 17y_2 - 9y_3.$$

We can express this system in matrix form as $\mathbf{y}' = A\mathbf{y}$, where

$$A = \begin{pmatrix} 6 & 13 & -8 \\ 2 & 5 & -2 \\ 7 & 17 & -9 \end{pmatrix}.$$

As we saw earlier (in Exercise 8.5 and in Example 9.10), $P^{-1}AP = D$, where

$$P = \begin{pmatrix} 1 & -1 & 1 \\ 0 & 1 & 1 \\ 1 & 1 & 2 \end{pmatrix}, \qquad D = \begin{pmatrix} -2 & 0 & 0 \\ 0 & 1 & 0 \\ 0 & 0 & 3 \end{pmatrix}.$$

We set $\mathbf{y} = P\mathbf{z}$, and substitute into the equation, $\mathbf{y}' = A\mathbf{y}$ to obtain $(P\mathbf{z})' = A(P\mathbf{z})$. That is, $P\mathbf{z}' = AP\mathbf{z}$ and so $\mathbf{z}' = P^{-1}AP\mathbf{z} = D\mathbf{z}$. In other words, if

$$\mathbf{z} = \begin{pmatrix} z_1 \\ z_2 \\ z_3 \end{pmatrix},$$

then

$$\begin{pmatrix} z_1' \\ z_2' \\ z_3' \end{pmatrix} = \begin{pmatrix} -2 & 0 & 0 \\ 0 & 1 & 0 \\ 0 & 0 & 3 \end{pmatrix} \begin{pmatrix} z_1 \\ z_2 \\ z_3 \end{pmatrix}.$$

So,

$$z_1' = -2z_1, \quad z_2' = z_2, \quad z_3' = 3z_3.$$

Therefore,

$$z_1 = z_1(0)e^{-2t}, \quad z_2 = z_2(0)e^t, \quad z_3 = z_3(0)e^{3t}.$$

Then, using $\mathbf{y} = P\mathbf{z}$, we have

$$\begin{pmatrix} y_1 \\ y_2 \\ y_3 \end{pmatrix} = \begin{pmatrix} 1 & -1 & 1 \\ 0 & 1 & 1 \\ 1 & 1 & 2 \end{pmatrix} \begin{pmatrix} z_1(0)e^{-2t} \\ z_2(0)e^t \\ z_3(0)e^{3t} \end{pmatrix}.$$

It remains to find $z_1(0)$, $z_2(0)$, $z_3(0)$. To do so, we use the given initial values $y_1(0) = 2$, $y_2(0) = 1$, $y_3(0) = 1$. Since $\mathbf{y} = P\mathbf{z}$, we can see that $\mathbf{y}(0) = P\mathbf{z}(0)$. We could use row operations to solve this system to determine $\mathbf{z}(0)$. Alternatively, we could use $\mathbf{z}(0) = P^{-1}\mathbf{y}(0)$. Perhaps the first way is generally easier, but in this particular case we already know the inverse of P from earlier:

$$P^{-1} = \begin{pmatrix} -1 & -3 & 2 \\ -1 & -1 & 1 \\ 1 & 2 & -1 \end{pmatrix}.$$

Therefore,

$$\mathbf{z}_0 = \begin{pmatrix} z_1(0) \\ z_2(0) \\ z_3(0) \end{pmatrix} = P^{-1}\mathbf{y}(0) = \begin{pmatrix} -1 & -3 & 2 \\ -1 & -1 & 1 \\ 1 & 2 & -1 \end{pmatrix} \begin{pmatrix} 2 \\ 1 \\ 1 \end{pmatrix} = \begin{pmatrix} -3 \\ -2 \\ 3 \end{pmatrix}.$$

Therefore, finally,

$$
\begin{pmatrix} y_1 \\ y_2 \\ y_3 \end{pmatrix} = \begin{pmatrix} 1 & -1 & 1 \\ 0 & 1 & 1 \\ 1 & 1 & 2 \end{pmatrix} \begin{pmatrix} z_1(0)e^{-2t} \\ z_2(0)e^{t} \\ z_3(0)e^{3t} \end{pmatrix}
$$

$$
= \begin{pmatrix} 1 & -1 & 1 \\ 0 & 1 & 1 \\ 1 & 1 & 2 \end{pmatrix} \begin{pmatrix} -3e^{-2t} \\ -2e^{t} \\ 3e^{3t} \end{pmatrix}
$$

$$
= \begin{pmatrix} -3e^{-2t} + 2e^{t} + 3e^{3t} \\ -2e^{t} + 3e^{3t} \\ -3e^{-2t} - 2e^{t} + 6e^{3t} \end{pmatrix} .
$$

The functions are

$$
y_1(t) = -3e^{-2t} + 2e^{t} + 3e^{3t}
$$
$$
y_2(t) = -2e^{t} + 3e^{3t}
$$
$$
y_3(t) = -3e^{-2t} - 2e^{t} + 6e^{3t} .
$$

How can we check our solution? First of all, it should satisfy the initial conditions. If we substitute $t = 0$ into the equations, we should obtain the given initial conditions.

Activity 9.26 Check this!

The real check is to look at the derivatives at $t = 0$. We can take the original system, $\mathbf{y}' = A\mathbf{y}$ and use it to find $\mathbf{y}'(0)$,

$$
\begin{pmatrix} y_1'(0) \\ y_2'(0) \\ y_3'(0) \end{pmatrix} = \begin{pmatrix} 6 & 13 & -8 \\ 2 & 5 & -2 \\ 7 & 17 & -9 \end{pmatrix} \begin{pmatrix} y_1(0) \\ y_2(0) \\ y_3(0) \end{pmatrix}
$$

$$
= \begin{pmatrix} 6 & 13 & -8 \\ 2 & 5 & -2 \\ 7 & 17 & -9 \end{pmatrix} \begin{pmatrix} 2 \\ 1 \\ 1 \end{pmatrix} = \begin{pmatrix} 17 \\ 7 \\ 22 \end{pmatrix} .
$$

And we can differentiate our solution to find $\mathbf{y}'$, and then substitute $t = 0$:

$$
\begin{pmatrix} y_1'(t) \\ y_2'(t) \\ y_3'(t) \end{pmatrix} = \begin{pmatrix} 6e^{-2t} + 2e^{t} + 9e^{3t} \\ -2e^{t} + 9e^{3t} \\ 6e^{-2t} - 2e^{t} + 18e^{3t} \end{pmatrix} .
$$

Activity 9.27 Substitute $t = 0$ to obtain $\mathbf{y}'(0)$ and check that it gives the same answer.

Often it is desirable to find a general solution to a system of differential equations, where no initial conditions are given. A general solution will have n arbitrary constants, essentially one for each function, so that given different initial conditions later, different particular solutions can be easily obtained. We will show how this works using the system in Example 9.25.

Example 9.28 Let $y_1(t)$, $y_2(t)$, $y_3(t)$ be functions related by the system of differential equations

$$\frac{dy_1}{dt} = 6y_1 + 13y_2 - 8y_3$$

$$\frac{dy_2}{dt} = 2y_1 + 5y_2 - 2y_3$$

$$\frac{dy_3}{dt} = 7y_1 + 17y_2 - 9y_3.$$

Let the matrices A, P and D be exactly as before in Example 9.25, so that we still have $P^{-1}AP = D$, and setting $\mathbf{y} = P\mathbf{z}$, to define new functions $z_1(t)$, $z_2(t)$, $z_3(t)$, we have

$$\mathbf{y}' = A\mathbf{y} \iff P\mathbf{z}' = AP\mathbf{z} \iff \mathbf{z}' = P^{-1}AP\mathbf{z} = D\mathbf{z}.$$

So we need to solve the equations

$$z_1' = -2z_1, \quad z_2' = z_2, \quad z_3' = 3z_3$$

in the absence of specific initial conditions. The general solutions are

$$z_1 = \alpha e^{-2t}, \quad z_2 = \beta e^t, \quad z_3 = \gamma e^{3t},$$

for arbitrary constants $\alpha, \beta, \gamma \in \mathbb{R}$.

Therefore, the general solution of the original system is

$$\mathbf{y} = \begin{pmatrix} y_1 \\ y_2 \\ y_3 \end{pmatrix} = P\mathbf{z} = \begin{pmatrix} 1 & -1 & 1 \\ 0 & 1 & 1 \\ 1 & 1 & 2 \end{pmatrix} \begin{pmatrix} \alpha e^{-2t} \\ \beta e^t \\ \gamma e^{3t} \end{pmatrix};$$

that is,

$$y_1(t) = \alpha e^{-2t} - \beta e^t + \gamma e^{3t}$$

$$y_2(t) = \beta e^t + \gamma e^{3t} \qquad \text{for } \alpha, \beta, \gamma \in \mathbb{R}.$$

$$y_3(t) = \alpha e^{-2t} + \beta e^t + 2\gamma e^{3t}$$

Using the general solution, you can find particular solutions for any given initial conditions. For example, using the same initial conditions $y_1(0) = 2$, $y_2(0) = 1$ and $y_3(0) = 1$ as in Example 9.25, we can substitute $t = 0$ into the general solution to obtain,

$$y_1(0) = 2 = \alpha - \beta + \gamma$$

$$y_2(0) = 1 = \beta + \gamma$$

$$y_3(0) = 1 = \alpha + \beta + 2\gamma$$

and solve this linear system of equations for α, β, γ. Of course, this is precisely the same system $\mathbf{y}(0) = P\mathbf{z}(0)$ as before, with solution $P^{-1}\mathbf{y}(0)$,

$$\begin{pmatrix} \alpha \\ \beta \\ \gamma \end{pmatrix} = \begin{pmatrix} -1 & -3 & 2 \\ -1 & -1 & 1 \\ 1 & 2 & -1 \end{pmatrix} \begin{pmatrix} 2 \\ 1 \\ 1 \end{pmatrix} = \begin{pmatrix} -3 \\ -2 \\ 3 \end{pmatrix}.$$

Activity 9.29 Find the particular solution of the system of differential equations in Example 9.28 which satisfies the initial conditions $y_1(0) = 1$, $y_2(0) = 1$ and $y_3(0) = 0$. Compare your result with the solution of difference equations in Example 9.10 which uses the same initial conditions for sequences. What do you notice? Why does this happen?

9.4 Learning outcomes

You should now be able to:

- calculate the general nth power of a diagonalisable matrix using diagonalisation
- solve systems of difference equations in which the underlying matrix is diagonalisable, by using both the matrix powers method and the change of variable method
- know what is meant by a Markov chain and its properties, and be able to find the long-term distribution
- solve systems of differential equations in which the underlying matrix is diagonalisable, by using the change of variable method.

9.5 Comments on activities

Activity 9.2 Take any 2×2 diagonal matrix D. Calculate D^2 and D^3, and observe what happens. Then see how this generalises.

Activity 9.13 We have

$$\mathbf{x}_1 = A\mathbf{x}_0 = \begin{pmatrix} 4 & 0 & 4 \\ 0 & 4 & 4 \\ 4 & 4 & 8 \end{pmatrix} \begin{pmatrix} 6 \\ 12 \\ 12 \end{pmatrix} = \begin{pmatrix} 72 \\ 96 \\ 168 \end{pmatrix} = \begin{pmatrix} -3(4) + 7(12) \\ 3(4) + 7(12) \\ 14(12) \end{pmatrix}.$$

If you keep 0^t as part of the solution, the solution is

$$\begin{pmatrix} x_t \\ y_t \\ x_t \end{pmatrix} = \begin{pmatrix} -3(4^t) + 2(0^t) + 7(12^t) \\ 3(4^t) + 2(0^t) + 7(12^t) \\ -2(0^t) + 14(12^t) \end{pmatrix}.$$

Using $0^t = 0$ for $t \geq 1$ and $0^0 = 1$, this gives the correct $\mathbf{x}_0$ and the same solution for $t \geq 1$.

Activity 9.16 First multiply the two matrices on the right to obtain

$$
D^t(P^{-1}\mathbf{x}_0) = \begin{pmatrix} \lambda_1^t & 0 & \cdots & 0 \\ 0 & \lambda_2^t & \cdots & 0 \\ \vdots & \vdots & \ddots & \vdots \\ 0 & 0 & \cdots & \lambda_n^t \end{pmatrix} \begin{pmatrix} b_1 \\ b_2 \\ \vdots \\ b_n \end{pmatrix} = \begin{pmatrix} b_1\lambda_1^t \\ b_2\lambda_2^t \\ \vdots \\ b_n\lambda_n^t \end{pmatrix}.
$$

Then express the product $P(D^t(P^{-1}\mathbf{x}_0))$ as a linear combination of the columns of P (see Activity 4.14),

$$
P(D^t(P^{-1}\mathbf{x}_0)) = \begin{pmatrix} | & | & & | \\ \mathbf{v}_1 & \mathbf{v}_2 & \cdots & \mathbf{v}_n \\ | & | & & | \end{pmatrix} \begin{pmatrix} b_1\lambda_1^t \\ b_2\lambda_2^t \\ \vdots \\ b_n\lambda_n^t \end{pmatrix}
$$

$$
= b_1\lambda_1^t\mathbf{v}_1 + b_2\lambda_2^t\mathbf{v}_2 + \cdots + b_n\lambda_n^t\mathbf{v}_n.
$$

Activity 9.22 It is clear that $y(0) = 3e^0 = 3$. Furthermore,

$$
y' = 6e^{2t} = 2(3e^{2t}) = 2y.
$$

Activity 9.24 Each row of the $n \times 1$ matrix $P\mathbf{z}$ is a linear combination of the functions $z_1(t), z_2(t), \ldots, z_n(t)$. For example, row i of $P\mathbf{z}$ is

$$
p_{i1}z_1(t) + p_{i2}z_2(t) + \cdots + p_{in}z_n(t).
$$

The rows of the matrix $(P\mathbf{z})'$ are the derivatives of these linear combinations of functions, so the ith row is

$$
(p_{i1}z_1(t) + p_{i2}z_2(t) + \cdots + p_{in}z_n(t))'
$$
$$
= p_{i1}z_1'(t) + p_{i2}z_2'(t) + \cdots + p_{in}z_n'(t),
$$

using the properties of differentiation, since the entries p_{ij} of P are constants. But

$$
p_{i1}z_1'(t) + p_{i2}z_2'(t) + \cdots + p_{in}z_n'(t)
$$

is just the ith row of the $n \times 1$ matrix $P\mathbf{z}'$, so these matrices are equal.

Activity 9.29 For the initial conditions $y_1(0) = 1$, $y_2(0) = 1$, $y_3(0) = 0$ the constants α, β, γ are

$$
\begin{pmatrix} \alpha \\ \beta \\ \gamma \end{pmatrix} = \begin{pmatrix} -1 & -3 & 2 \\ -1 & -1 & 1 \\ 1 & 2 & -1 \end{pmatrix} \begin{pmatrix} 1 \\ 1 \\ 0 \end{pmatrix} = \begin{pmatrix} -4 \\ -2 \\ 3 \end{pmatrix},
$$

so the solution is

$$y_1(t) = -4e^{-2t} + 2e^t + 3e^{3t}$$
$$y_2(t) = -2e^t + 3e^{3t}$$
$$y_3(t) = -4e^{-2t} - 2e^t + 6e^{3t}.$$

Compare this with the solution of the difference equations in Example 9.10 on page 288

$$\begin{pmatrix} x_t \\ y_t \\ x_t \end{pmatrix} = \begin{pmatrix} -4(-2)^t + 2 + 3(3^t) \\ -2 + 3(3^t) \\ -4(-2)^t - 2 + 6(3^t) \end{pmatrix}.$$

The two solutions are essentially the 'same', with the functions $e^{\lambda t}$ replaced by λ^t. Why? The coefficient matrix A is the same for both systems, and so are the matrices P and D. So the change of basis used to solve the systems is the same. We are changing from a system formulated in standard coordinates to one in coordinates of the basis of eigenvectors of A in order to find the solution.

9.6 Exercises

Exercise 9.1 Given the matrix

$$A = \begin{pmatrix} 4 & 5 \\ -1 & -2 \end{pmatrix},$$

find A^n for any positive integer n.

Deduce from your result that the expression $(-1)^k - 3^k$ is divisible by 4 for all $k \geq 1$.

Exercise 9.2 Solve the following system of difference equations.

$$x_{t+1} = x_t + 4y_t$$
$$y_{t+1} = \tfrac{1}{2}x_t,$$

given that $x_0 = y_0 = 1000$.

Exercise 9.3 Sequences x_t, y_t, z_t are defined by $x_0 = -1, y_0 = 2, z_0 = 1$ and

$$x_{t+1} = 7x_t - 3z_t$$
$$y_{t+1} = x_t + 6y_t + 5z_t$$
$$z_{t+1} = 5x_t - z_t.$$

Find formulae for x_t, y_t, and z_t.

Check your solution.

Exercise 9.4 Given that

$$\mathbf{v}_1 = \begin{pmatrix} 1 \\ -1 \\ 1 \end{pmatrix}, \quad \mathbf{v}_2 = \begin{pmatrix} -3 \\ 0 \\ 1 \end{pmatrix}, \quad \mathbf{v}_3 = \begin{pmatrix} -1 \\ 1 \\ 0 \end{pmatrix}$$

are eigenvectors of the matrix

$$A = \begin{pmatrix} 1 & -2 & -6 \\ 2 & 5 & 6 \\ -2 & -2 & -3 \end{pmatrix},$$

find an invertible matrix P such that $P^{-1}AP$ is diagonal. Using the method of changing variables, find sequences x_t, y_t, z_t satisfying the equations

$$x_{t+1} = x_t - 2y_t - 6z_t$$
$$y_{t+1} = 2x_t + 5y_t + 6z_t$$
$$z_{t+1} = -2x_t - 2y_t - 3z_t,$$

and with the property that $x_0 = y_0 = 1$ and $z_0 = 0$.
Find the term x_5.

Exercise 9.5 At any time t, the total population of 210 people of Desert Island is divided into those living by the sea (x_t) and those living in the oasis (y_t). Initially, half the population is living by the sea, and half in the oasis. Yearly population movements are given by

$$\mathbf{x}_t = A\mathbf{x}_{t-1} \text{ where } A = \begin{pmatrix} 0.6 & 0.2 \\ 0.4 & 0.8 \end{pmatrix}, \quad \mathbf{x}_t = \begin{pmatrix} x_t \\ y_t \end{pmatrix}.$$

Show this is a Markov process and interpret the yearly population movements from the matrix A.

Find expressions for x_t and y_t at any future time t.

Determine the 'long-term' population distribution; that is, find what happens to $\mathbf{x}_t$ as $t \to \infty$.

Exercise 9.6 Consider the matrices

$$A = \begin{pmatrix} 0.7 & 0.2 & 0.2 \\ 0 & 0.2 & 0.4 \\ 0.3 & 0.6 & 0.4 \end{pmatrix}, \quad B = \begin{pmatrix} 7 & 2 & 2 \\ 0 & 2 & 4 \\ 3 & 6 & 4 \end{pmatrix}, \quad \mathbf{x}_t = \begin{pmatrix} x_t \\ y_t \\ z_t \end{pmatrix}.$$

(a) What is the relationship between the matrices A and B?

Show that A and B have the same eigenvectors. What is the relationship between the corresponding eigenvalues?

Show that the system $\mathbf{x}_t = A\mathbf{x}_{t-1}$ is a Markov chain by showing that the matrix A satisfies the two conditions to be the transition matrix of a Markov chain.

Deduce that $\lambda = 10$ is an eigenvalue of B.

(b) Find an eigenvector of B corresponding to the eigenvalue $\lambda = 10$.

Diagonalise the matrix B: Find an invertible matrix P and a diagonal matrix D such that $P^{-1}BP = D$. Check that $BP = PD$.

Write down the eigenvalues and corresponding eigenvectors of A.

(c) An economic model of employment of a fixed group of 1000 workers assumes that in any year t, an individual is either employed full-time, employed part-time or unemployed. Let x_t denote the percentage (as a decimal) of full-time workers in year t, y_t the percentage working part-time and z_t the percentage who are unemployed. Then according to this model, the probabilities that a worker will change from one state to another in year t are given by the matrix A above, so that $\mathbf{x}_t = A\mathbf{x}_{t-1}$. Initially, 200 are employed full-time and 300 are employed part-time.

Find the long-term population distribution of this system. Eventually, what number of workers are employed, either full or part-time?

Exercise 9.7 Suppose functions $y_1(t)$, $y_2(t)$ are related by the following system of differential equations:

$$y_1' = 4y_1 + 5y_2$$
$$y_2' = -y_1 - 2y_2.$$

Find the solutions to these equations that satisfy $y_1(0) = 2$, $y_2(0) = 6$. Check your answer.

Exercise 9.8 Find the general solution of the following system of differential equations:

$$\frac{dy_1}{dt} = -y_1 + y_2 + 2y_3$$
$$\frac{dy_2}{dt} = -6y_1 + 2y_2 + 6y_3$$
$$\frac{dy_3}{dt} = y_2 + y_3$$

for functions $y_1(t)$, $y_2(t)$, $y_3(t)$, $t \in \mathbb{R}$.

Exercise 9.9 Find functions $y_1(t)$, $y_2(t)$, $y_3(t)$ satisfying the system of differential equations:

$$y_1' = 4y_1 + 4y_3$$
$$y_2' = 4y_2 + 4y_3$$
$$y_3' = 4y_1 + 4y_2 + 8y_3$$

and with $y_1(0) = 6$, $y_2(0) = 12$, $y_3(0) = 12$.

Exercise 9.10 Consider

$$A = \begin{pmatrix} 5 & -8 & -4 \\ 3 & -5 & -3 \\ -1 & 2 & 2 \end{pmatrix}, \qquad \mathbf{v}_1 = \begin{pmatrix} 2 \\ 1 \\ 0 \end{pmatrix}.$$

Find a basis for the null space of A, $N(A)$.

Show that the vector $\mathbf{v}_1$ is an eigenvector of A and find the corresponding eigenvalue. Find all the eigenvectors of A which correspond to this eigenvalue. Hence find an invertible matrix P and a diagonal matrix D such that $P^{-1}AP = D$.

Find A^n. What do you notice about the matrix A and its powers A^n?

Find the solution to the system of difference equations given by $\mathbf{x}_{t+1} = A\mathbf{x}_t$ for sequences $\mathbf{x}_t = (x_t, y_t, z_t)^T$, $t \in \mathbb{Z}$, $t \geq 0$, with initial conditions $x_0 = 1$, $y_0 = 1$, $z_0 = 1$. Write down the first four terms of each sequence.

9.7 Problems

Problem 9.1 Find A^5 if $A = \begin{pmatrix} 1 & 4 \\ 3 & 2 \end{pmatrix}$.

Problem 9.2 Find sequences x_t and y_t which satisfy the following system of difference equations

$$x_{t+1} = x_t + 4y_t$$
$$y_{t+1} = 3x_t + 2y_t ,$$

and the initial conditions $x_0 = 1$, $y_0 = 0$.

Find the values of x_5 and y_5.

Problem 9.3 Find sequences x_t, y_t, z_t satisfying the equations

$$x_{t+1} = 4x_t + 3y_t - 7z_t$$
$$y_{t+1} = x_t + 2y_t + z_t$$
$$z_{t+1} = 2x_t + 2y_t - 3z_t,$$

and with the property that $x_0 = 4$, $y_0 = 5$ and $z_0 = 1$. (See Problem 8.9.)

Check your answer by calculating x_1, y_1, z_1 from both the solution and the original system.

Problem 9.4 A Markov process satisfies the difference equation $\mathbf{x}_k = A\mathbf{x}_{k-1}$ where

$$A = \begin{pmatrix} 0.7 & 0.6 \\ 0.3 & 0.4 \end{pmatrix}, \qquad \mathbf{x}_0 = \begin{pmatrix} 0.6 \\ 0.4 \end{pmatrix}.$$

Solve the equation and find an expression for x_k as a linear combination of the eigenvectors of A.

Use this to predict the 'long term' distribution; that is, find what happens to x_k as $k \to \infty$.

Problem 9.5 Consider the system of equations $x_t = A x_{t-1}$ where

$$A = \begin{pmatrix} 0 & \frac{1}{2} & \frac{1}{2} \\ \frac{1}{2} & \frac{1}{2} & 0 \\ \frac{1}{2} & 0 & \frac{1}{2} \end{pmatrix}, \qquad x_t = \begin{pmatrix} x_t \\ y_t \\ z_t \end{pmatrix}.$$

State the two conditions satisfied by the matrix A so that it is the transition matrix of a Markov process. What can you conclude about the eigenvalues of the matrix A?

Find an invertible matrix P and a diagonal matrix D such that $P^{-1}AP = D$.

Assume that the system represents a total population of 6000 members distributed into three states, where x_t is the number of members in state one at time t, y_t is the number in state two, and z_t is the number in state three. Initially $1/6$ of the total population is in state one, $1/3$ is in state two, and $1/2$ is in state three. Find the long term population distribution of this system. State clearly the expected number of members which will eventually be in each of the three states.

Problem 9.6 The population of osprey eagles at a certain lake is dying out. Each year the new population is only 60% of the previous year's population.

(a) Conservationists introduce a new species of trout into the lake and find that the populations satisfy the following system of difference equations, where x_t is the number of osprey in year t and y_t is the number of trout in year t: $x_t = A x_{t-1}$, where

$$x_t = \begin{pmatrix} x_t \\ y_t \end{pmatrix}, \qquad A = \begin{pmatrix} 0.6 & 0.2 \\ -0.25 & 1.2 \end{pmatrix}, \qquad x_0 = \begin{pmatrix} 20 \\ 100 \end{pmatrix}.$$

Give a reason why this system of difference equations is *not* a Markov process. Describe in words how each of the populations depends on the previous year's populations.

Solve the system of difference equations. Show that the situation is not stable: that according to this model both osprey and trout will increase without bound as $t \to \infty$. What will be the eventual ratio of osprey to trout?

(b) In order to have the populations of osprey and trout achieve a steady state, they decide to allow an amount of fishing each year, based on the number of osprey in the previous year. The new equations are

$\mathbf{x}_t = B\mathbf{x}_{t-1}$, where

$$\mathbf{x}_t = \begin{pmatrix} x_t \\ y_t \end{pmatrix}, \quad B = \begin{pmatrix} 0.6 & 0.2 \\ -\alpha & 1.2 \end{pmatrix}, \quad \mathbf{x}_0 = \begin{pmatrix} 20 \\ 100 \end{pmatrix}$$

and $\alpha > 0$ is a constant to be determined.

What property of the transition matrix of a Markov process determines that there is a (finite, non-zero) long-term distribution? Deduce a condition on the eigenvalues of the matrix B to produce the same effect. Then find the value of α which satisfies this condition.

Show that for this value of α the population now reaches a steady state as $t \to \infty$ and determine what this stable population of osprey and trout will be.

Problem 9.7 Find the general solution of the following system of linear differential equations:

$$\begin{cases} y_1'(t) = y_1(t) + 4y_2(t) \\ y_2'(t) = 3y_1(t) + 2y_2(t) . \end{cases}$$

Then find the unique solution satisfying the initial conditions $y_1(0) = 1$ and $y_2(0) = 0$.

Check your solution by finding the values of $y_1'(0)$ and $y_2'(0)$.

Problem 9.8 Write the system of differential equations

$$\begin{cases} y_1'(t) = 3y_1(t) + 2y_2(t) \\ y_2'(t) = 2y_1(t) + 6y_2(t) \end{cases}, \quad \mathbf{y} = \begin{pmatrix} y_1 \\ y_2 \end{pmatrix},$$

in matrix form, as $\mathbf{y}' = A\mathbf{y}$. Find the solution which satisfies the initial conditions $y_1(0) = 5$, $y_2(0) = 5$.

Problem 9.9 Diagonalise the matrix

$$A = \begin{pmatrix} -1 & 3 & 0 \\ 0 & 2 & 0 \\ -3 & 3 & 2 \end{pmatrix}.$$

Write out the system of linear differential equations given by $\mathbf{y}' = A\mathbf{y}$, where $\mathbf{y} = (y_1(t), y_2(t), y_3(t))^T$, and find the general solution.

Problem 9.10 Show that the vectors

$$\mathbf{v}_1 = \begin{pmatrix} 1 \\ 0 \\ 1 \end{pmatrix}, \quad \mathbf{v}_2 = \begin{pmatrix} 1 \\ 1 \\ 2 \end{pmatrix}, \quad \mathbf{v}_3 = \begin{pmatrix} 3 \\ -2 \\ 2 \end{pmatrix}$$

form a basis $B = \{\mathbf{v}_1, \mathbf{v}_2, \mathbf{v}_3\}$ of $\mathbb{R}^3$.

(a) If $\mathbf{w} = (2, -3, 1)^{\mathsf{T}}$, find $[\mathbf{w}]_B$, the B coordinates of $\mathbf{w}$.

(b) Find a matrix A for which $\mathbf{v}_1$ is an eigenvector corresponding to the eigenvalue $\lambda_1 = 1$, $\mathbf{v}_2$ is an eigenvector with eigenvalue $\lambda_2 = 2$, and $\mathbf{v}_3$ is an eigenvector with eigenvalue $\lambda_3 = 3$. Verify that your matrix A satisfies these conditions.

(c) Find a general solution of the system of differential equations

$$\mathbf{y}' = A\mathbf{y}$$

where A is the matrix in part (b) and $\mathbf{y} = (y_1(t), y_2(t), y_3(t))^{\mathsf{T}}$, $t \in \mathbb{R}$, is a vector of functions.

Then find the unique solution which satisfies $\mathbf{y}(0) = \mathbf{w}$, where $\mathbf{w}$ is the vector in part (a).

10

Inner products and orthogonality

In this chapter, we develop further some of the key geometrical ideas about vectors, specifically the concepts of inner products and orthogonality of vectors. In Chapter 1, we saw how the inner product can be useful in thinking about the geometry of vectors. We now investigate how these concepts can be extended to a general vector space.

10.1 Inner products

10.1.1 The inner product of real n-vectors

In Chapter 1, we looked at the inner product of vectors in $\mathbb{R}^n$. Recall that, for $\mathbf{x}, \mathbf{y} \in \mathbb{R}^n$, the *inner product* (sometimes called the *dot product* or *scalar product*) is defined to be the number $\langle \mathbf{x}, \mathbf{y} \rangle$ given by

$$\langle \mathbf{x}, \mathbf{y} \rangle = \mathbf{x}^T \mathbf{y} = x_1 y_1 + x_2 y_2 + \cdots + x_n y_n.$$

This is often referred to as the *standard* or *Euclidean* inner product.

We re-iterate that it is important to realise that the inner product is just a number, not another vector or a matrix. The inner product on $\mathbb{R}^n$ satisfies certain basic properties and is, as we have seen, closely linked with the geometric concepts of *length* and *angle*. This provides the background for generalising these concepts to any vector space V, as we shall see in the next section.

It is easily verified (using Theorem 1.36) that for all $\mathbf{x}, \mathbf{y}, \mathbf{z} \in \mathbb{R}^n$ and for all $\alpha, \beta \in \mathbb{R}$, the inner product satisfies the following properties:

(i) $\langle \mathbf{x}, \mathbf{y} \rangle = \langle \mathbf{y}, \mathbf{x} \rangle$.
(ii) $\langle \alpha \mathbf{x} + \beta \mathbf{y}, \mathbf{z} \rangle = \alpha \langle \mathbf{x}, \mathbf{z} \rangle + \beta \langle \mathbf{y}, \mathbf{z} \rangle$.
(iii) $\langle \mathbf{x}, \mathbf{x} \rangle \geq 0$, and $\langle \mathbf{x}, \mathbf{x} \rangle = 0$ if and only if $\mathbf{x} = \mathbf{0}$.

We have seen that the length, $\|\mathbf{a}\|$, of a vector $\mathbf{a}$ satisfies $\|\mathbf{a}\|^2 = \langle \mathbf{a}, \mathbf{a} \rangle$. We also noted that if $\mathbf{a}$, $\mathbf{b}$ are two vectors in $\mathbb{R}^2$ and θ is the angle between them, then $\langle \mathbf{a}, \mathbf{b} \rangle = \|\mathbf{a}\| \|\mathbf{b}\| \cos \theta$. In particular, non-zero vectors $\mathbf{a}$ and $\mathbf{b}$ are orthogonal (or perpendicular) if and only if $\langle \mathbf{a}, \mathbf{b} \rangle = 0$.

10.1.2   Inner products more generally

There is a more general concept of inner product than the one we met earlier, and this is very important. It is 'more general' in two ways: first, it enables us to say what we mean by an inner product on any vector space, and not just $\mathbb{R}^n$, and, second, it allows the possibility of inner products on $\mathbb{R}^n$ that are different from the standard one.

Definition 10.1 (Inner product) Let V be a vector space (over the real numbers). An *inner product* on V is a mapping from (or operation on) pairs of vectors $\mathbf{x}$, $\mathbf{y}$ to the real numbers, the result of which is a real number denoted $\langle \mathbf{x}, \mathbf{y} \rangle$, which satisfies the following properties:

(i) $\langle \mathbf{x}, \mathbf{y} \rangle = \langle \mathbf{y}, \mathbf{x} \rangle$ for all $\mathbf{x}, \mathbf{y} \in V$.
(ii) $\langle \alpha \mathbf{x} + \beta \mathbf{y}, \mathbf{z} \rangle = \alpha \langle \mathbf{x}, \mathbf{z} \rangle + \beta \langle \mathbf{y}, \mathbf{z} \rangle$ for all $\mathbf{x}, \mathbf{y}, \mathbf{z} \in V$ and all $\alpha, \beta \in \mathbb{R}$.
(iii) $\langle \mathbf{x}, \mathbf{x} \rangle \geq 0$ for all $\mathbf{x} \in V$, and $\langle \mathbf{x}, \mathbf{x} \rangle = 0$ if and only if $\mathbf{x} = \mathbf{0}$, the zero vector of the vector space.

Some other basic facts follow immediately from this definition, for example

$$\langle \mathbf{z}, \alpha \mathbf{x} + \beta \mathbf{y} \rangle = \alpha \langle \mathbf{z}, \mathbf{x} \rangle + \beta \langle \mathbf{z}, \mathbf{y} \rangle.$$

Activity 10.2 Prove that $\langle \mathbf{z}, \alpha \mathbf{x} + \beta \mathbf{y} \rangle = \alpha \langle \mathbf{z}, \mathbf{x} \rangle + \beta \langle \mathbf{z}, \mathbf{y} \rangle$.

Of course, given what we noted above, it is clear that the standard inner product on $\mathbb{R}^n$ is indeed an inner product according to this more general definition. This new, more general, abstract definition, though, applies to more than just the vector space $\mathbb{R}^n$, and there is some advantage in developing results in terms of the general notion of inner product. If a vector space has an inner product defined on it, we refer to it as an *inner product space*.

Example 10.3 (This is a deliberately strange example. Its purpose is to illustrate how we can define inner products in non-standard ways, which is why we've chosen it.) Suppose that V is the vector space consisting of all real polynomial functions of degree at most n; that is, V consists of all functions $\mathbf{p} : x \mapsto p(x)$ of the form

$$p(x) = a_0 + a_1 x + a_2 x^2 + \cdots + a_n x^n, \qquad a_0, a_1, \ldots, a_n \in \mathbb{R}.$$

The addition and scalar multiplication are, as usual, defined pointwise. (Recall that this means that $(p + q)(x) = p(x) + q(x)$ and $(\alpha p)(x) = \alpha p(x)$.) Let $x_1, x_2, \ldots, x_{n+1}$ be $n + 1$ fixed, different, real numbers, and define, for $\mathbf{p}, \mathbf{q} \in V$,

$$\langle \mathbf{p}, \mathbf{q} \rangle = \sum_{i=1}^{n+1} p(x_i) q(x_i).$$

Then this is an inner product. To see this, we check the properties in the definition of an inner product. Property (i) is clear. For (iii), we have

$$\langle \mathbf{p}, \mathbf{p} \rangle = \sum_{i=1}^{n+1} p(x_i)^2 \geq 0.$$

Clearly, if $\mathbf{p}$ is the zero vector of the vector space (which is the identically-zero function), then $\langle \mathbf{p}, \mathbf{p} \rangle = 0$. To finish verifying (iii), we need to check that if $\langle \mathbf{p}, \mathbf{p} \rangle = 0$, then $\mathbf{p}$ must be the zero function. Now, $\langle \mathbf{p}, \mathbf{p} \rangle = 0$ must mean that $p(x_i) = 0$ for $i = 1, 2, \ldots, n + 1$. So $p(x)$ has $n + 1$ different roots. But $p(x)$ has degree no more than n, so $\mathbf{p}$ must be the identically-zero function. (A non-zero polynomial of degree at most n has no more than n distinct roots.) Part (ii) is left to you.

Activity 10.4 Prove that, for any $\alpha, \beta \in \mathbb{R}$ and any $\mathbf{p}, \mathbf{q}, \mathbf{r} \in V$,

$$\langle \alpha \mathbf{p} + \beta \mathbf{q}, \mathbf{r} \rangle = \alpha \langle \mathbf{p}, \mathbf{r} \rangle + \beta \langle \mathbf{q}, \mathbf{r} \rangle.$$

Example 10.5 Let's define, for $\mathbf{x}, \mathbf{y} \in \mathbb{R}^2$,

$$\langle \mathbf{x}, \mathbf{y} \rangle = x_1 y_1 + 2 x_2 y_2.$$

Then this is an inner product. It is very easy to see that $\langle \mathbf{x}, \mathbf{y} \rangle = \langle \mathbf{y}, \mathbf{x} \rangle$. It is straightforward to check that $\langle \alpha \mathbf{x} + \beta \mathbf{y}, \mathbf{z} \rangle = \alpha \langle \mathbf{x}, \mathbf{z} \rangle + \beta \langle \mathbf{y}, \mathbf{z} \rangle$. Finally, we have $\langle \mathbf{x}, \mathbf{x} \rangle = x_1^2 + 2 x_2^2$ so $\langle \mathbf{x}, \mathbf{x} \rangle \geq 0$ and, furthermore, $\langle \mathbf{x}, \mathbf{x} \rangle = 0$ if and only if $x_1 = x_2 = 0$, meaning $\mathbf{x} = \mathbf{0}$.

Example 10.3 shows how we may define an inner product on a vector space other than $\mathbb{R}^n$, and Example 10.5 shows how we may define an inner product on $\mathbb{R}^2$ which is different from the standard Euclidean inner product (whose value would simply be $x_1 y_1 + x_2 y_2$).

10.1.3 Norms in a vector space

For any $\mathbf{x}$ in an inner product space V, the inner product $\langle \mathbf{x}, \mathbf{x} \rangle$ is non-negative (by definition). Now, because $\langle \mathbf{x}, \mathbf{x} \rangle \geq 0$, we may take its square root (obtaining a real number). We define the *norm* or *length* $\|\mathbf{x}\|$ of a vector $\mathbf{x}$ to be this real number:

Definition 10.6 (Norm) Suppose that V is an inner product space and $\mathbf{x}$ is a vector in V. Then the *norm*, or *length*, of $\mathbf{x}$, denoted $\|\mathbf{x}\|$, is

$$\|\mathbf{x}\| = \sqrt{\langle \mathbf{x}, \mathbf{x} \rangle}.$$

For example, for the standard inner product on $\mathbb{R}^n$,

$$\langle \mathbf{x}, \mathbf{x} \rangle = x_1^2 + x_2^2 + \cdots + x_n^2$$

(which is clearly non-negative since it is a sum of squares), the norm is the standard Euclidean length of a vector:

$$\|\mathbf{x}\| = \sqrt{x_1^2 + x_2^2 + \cdots + x_n^2}.$$

We say that a vector $\mathbf{v}$ is a *unit vector* if it has norm 1. If $\mathbf{v} \neq \mathbf{0}$, then it is a simple matter to create a unit vector in the same direction as $\mathbf{v}$. This is the vector

$$\mathbf{u} = \frac{1}{\|\mathbf{v}\|}\mathbf{v}.$$

The process of constructing $\mathbf{u}$ from $\mathbf{v}$ is known as *normalising* $\mathbf{v}$.

10.1.4 The Cauchy–Schwarz inequality

This important inequality will enable us to apply the geometric intuition we have developed to a much more general, completely abstract, setting.

Theorem 10.7 (Cauchy–Schwarz inequality) *Suppose that V is an inner product space. Then*

$$|\langle \mathbf{x}, \mathbf{y} \rangle| \leq \|\mathbf{x}\|\|\mathbf{y}\|$$

for all $\mathbf{x}, \mathbf{y} \in V$.

Proof: Let $\mathbf{x}, \mathbf{y}$ be any two vectors of V. For any real number α, we consider the vector $\alpha\mathbf{x} + \mathbf{y}$. Certainly, $\|\alpha\mathbf{x} + \mathbf{y}\|^2 \geq 0$ for all α. But

$$\begin{aligned}
\|\alpha\mathbf{x} + \mathbf{y}\|^2 &= \langle \alpha\mathbf{x} + \mathbf{y}, \alpha\mathbf{x} + \mathbf{y} \rangle \\
&= \alpha^2 \langle \mathbf{x}, \mathbf{x} \rangle + \alpha \langle \mathbf{x}, \mathbf{y} \rangle + \alpha \langle \mathbf{y}, \mathbf{x} \rangle + \langle \mathbf{y}, \mathbf{y} \rangle \\
&= \alpha^2 \|\mathbf{x}\|^2 + 2\alpha \langle \mathbf{x}, \mathbf{y} \rangle + \|\mathbf{y}\|^2.
\end{aligned}$$

Now, this quadratic expression in α is non-negative for all α. Generally, we know that if a quadratic expression $at^2 + bt + c$ is non-negative for all t, then $b^2 - 4ac \leq 0$.

Activity 10.8 Why is this true?

Applying this observation to the above quadratic expression in α, we see that we must have

$$(2\langle \mathbf{x}, \mathbf{y}\rangle)^2 - 4\|\mathbf{x}\|^2\|\mathbf{y}\|^2 \leq 0,$$

or

$$(\langle \mathbf{x}, \mathbf{y}\rangle)^2 \leq \|\mathbf{x}\|^2\|\mathbf{y}\|^2.$$

Taking the square root of each side, we obtain

$$|\langle \mathbf{x}, \mathbf{y}\rangle| \leq \|\mathbf{x}\|\|\mathbf{y}\|,$$

which is what we need.   □

For example, if we take V to be $\mathbb{R}^n$ and consider the standard inner product on $\mathbb{R}^n$, then for all $\mathbf{x}, \mathbf{y} \in \mathbb{R}^n$, the Cauchy–Schwarz inequality tells us that

$$\left|\sum_{i=1}^{n} x_i y_i\right| \leq \sqrt{\sum_{i=1}^{n} x_i^2} \sqrt{\sum_{i=1}^{n} y_i^2}.$$

10.2 Orthogonality

10.2.1 Orthogonal vectors

We are now ready to extend the concept of *angle* to an abstract inner product space V. To do this, we begin with the result that in $\mathbb{R}^2$, $\langle \mathbf{x}, \mathbf{y}\rangle = \|\mathbf{x}\|\|\mathbf{y}\|\cos\theta$, where θ is the angle between the vectors. This suggests that we might, more generally (in an abstract inner product space), define the cosine of the angle between vectors $\mathbf{x}$ and $\mathbf{y}$ to be

$$\cos\theta = \frac{\langle \mathbf{x}, \mathbf{y}\rangle}{\|\mathbf{x}\|\|\mathbf{y}\|}.$$

This definition will only make sense if we can show that this number $\cos\theta$ is between -1 and 1. But this follows immediately from the Cauchy–Schwartz inequality, which can be stated as

$$\left|\frac{\langle \mathbf{x}, \mathbf{y}\rangle}{\|\mathbf{x}\|\|\mathbf{y}\|}\right| \leq 1.$$

The usefulness of this definition is in the concept of orthogonality.

Definition 10.9 (Orthogonal vectors) Suppose that V is an inner product space. Then $\mathbf{x}, \mathbf{y} \in V$ are said to be *orthogonal* if and only if $\langle \mathbf{x}, \mathbf{y} \rangle = 0$. We write $\mathbf{x} \perp \mathbf{y}$ to mean that $\mathbf{x}, \mathbf{y}$ are orthogonal.

So what we have here is a *definition* of what it means, in a general inner product space, for two vectors to be orthogonal. As a special case of this, of course, we have the familiar notion of orthogonality in $\mathbb{R}^n$ (when we use the standard inner product), but the key thing to stress is that this definition gives us a way to *extend* the notion of orthogonality to inner product spaces other than $\mathbb{R}^n$.

Example 10.10 With the usual inner product on $\mathbb{R}^4$, the vectors $\mathbf{x} = (1, -1, 2, 0)^{\mathrm{T}}$ and $\mathbf{y} = (-1, 1, 1, 4)^{\mathrm{T}}$ are orthogonal.

Activity 10.11 Check this!

10.2.2 A generalised Pythagoras theorem

We can now begin to imitate the geometry of vectors discussed in Chapter 1. We are already familiar with Pythagoras' theorem in $\mathbb{R}^3$, which states that if c is the length of the longest side of a right-angled triangle, and a and b the lengths of the other two sides, then $c^2 = a^2 + b^2$. The generalised Pythagoras theorem is:

Theorem 10.12 (Generalised Pythagoras theorem) *In an inner product space V, if $\mathbf{x}, \mathbf{y} \in V$ are orthogonal, then*

$$\|\mathbf{x} + \mathbf{y}\|^2 = \|\mathbf{x}\|^2 + \|\mathbf{y}\|^2.$$

Proof: This is fairly straightforward to prove. We know that for any $\mathbf{z}$, $\|\mathbf{z}\|^2 = \langle \mathbf{z}, \mathbf{z} \rangle$, simply from the definition of the norm. So,

$$\begin{aligned}
\|\mathbf{x} + \mathbf{y}\|^2 &= \langle \mathbf{x} + \mathbf{y}, \mathbf{x} + \mathbf{y} \rangle \\
&= \langle \mathbf{x}, \mathbf{x} + \mathbf{y} \rangle + \langle \mathbf{y}, \mathbf{x} + \mathbf{y} \rangle \\
&= \langle \mathbf{x}, \mathbf{x} \rangle + \langle \mathbf{x}, \mathbf{y} \rangle + \langle \mathbf{y}, \mathbf{x} \rangle + \langle \mathbf{y}, \mathbf{y} \rangle \\
&= \|\mathbf{x}\|^2 + 2\langle \mathbf{x}, \mathbf{y} \rangle + \|\mathbf{y}\|^2 \\
&= \|\mathbf{x}\|^2 + \|\mathbf{y}\|^2,
\end{aligned}$$

where the last line follows from the fact that, $\mathbf{x}, \mathbf{y}$ being orthogonal, $\langle \mathbf{x}, \mathbf{y} \rangle = 0$. $\qquad\square$

We also have the *triangle inequality* for norms. In the special case of the standard inner product on $\mathbb{R}^2$, this states the obvious fact that the length of one side of a triangle must be no more than the sum of the lengths of the other two sides.

Theorem 10.13 (Triangle inequality for norms) *In an inner product space V, if $\mathbf{x}, \mathbf{y} \in V$, then*

$$\|\mathbf{x} + \mathbf{y}\| \le \|\mathbf{x}\| + \|\mathbf{y}\|.$$

Proof: We have

$$
\begin{aligned}
\|\mathbf{x} + \mathbf{y}\|^2 &= \langle \mathbf{x} + \mathbf{y}, \mathbf{x} + \mathbf{y} \rangle \\
&= \langle \mathbf{x}, \mathbf{x} + \mathbf{y} \rangle + \langle \mathbf{y}, \mathbf{x} + \mathbf{y} \rangle \\
&= \langle \mathbf{x}, \mathbf{x} \rangle + \langle \mathbf{x}, \mathbf{y} \rangle + \langle \mathbf{y}, \mathbf{x} \rangle + \langle \mathbf{y}, \mathbf{y} \rangle \\
&= \|\mathbf{x}\|^2 + 2\langle \mathbf{x}, \mathbf{y} \rangle + \|\mathbf{y}\|^2 \\
&\le \|\mathbf{x}\|^2 + \|\mathbf{y}\|^2 + 2\,|\langle \mathbf{x}, \mathbf{y} \rangle| \\
&\le \|\mathbf{x}\|^2 + \|\mathbf{y}\|^2 + 2\|\mathbf{x}\|\|\mathbf{y}\| \\
&= (\|\mathbf{x}\| + \|\mathbf{y}\|)^2,
\end{aligned}
$$

where the last inequality used is the Cauchy–Schwarz inequality. Thus, $\|\mathbf{x} + \mathbf{y}\| \le \|\mathbf{x}\| + \|\mathbf{y}\|$, as required. $\qquad\square$

10.2.3   Orthogonality and linear independence

If a set of (non-zero) vectors are pairwise orthogonal (that is, if any two are orthogonal), then it turns out that the vectors are linearly independent:

Theorem 10.14 *Suppose that V is an inner product space and that vectors $\mathbf{v}_1, \mathbf{v}_2, \ldots, \mathbf{v}_k \in V$ are pairwise orthogonal (meaning $\mathbf{v}_i \perp \mathbf{v}_j$ for $i \ne j$), and none is the zero-vector. Then $\{\mathbf{v}_1, \mathbf{v}_2, \ldots, \mathbf{v}_k\}$ is a linearly independent set of vectors.*

Proof: We need to show that if

$$\alpha_1 \mathbf{v}_1 + \alpha_2 \mathbf{v}_2 + \cdots + \alpha_k \mathbf{v}_k = \mathbf{0},$$

(the zero-vector), then $\alpha_1 = \alpha_2 = \cdots = \alpha_k = 0$. Let i be any integer between 1 and k. Then taking the inner product with $\mathbf{v}_i$,

$$\langle \mathbf{v}_i, \alpha_1 \mathbf{v}_1 + \alpha_2 \mathbf{v}_2 + \cdots + \alpha_k \mathbf{v}_k \rangle = \langle \mathbf{v}_i, \mathbf{0} \rangle = 0.$$

But

$$
\begin{aligned}
\langle \mathbf{v}_i, &\alpha_1 \mathbf{v}_1 + \cdots + \alpha_k \mathbf{v}_k \rangle \\
&= \alpha_1 \langle \mathbf{v}_i, \mathbf{v}_1 \rangle + \cdots + \alpha_{i-1} \langle \mathbf{v}_i, \mathbf{v}_{i-1} \rangle + \alpha_i \langle \mathbf{v}_i, \mathbf{v}_i \rangle \\
&\quad + \alpha_{i+1} \langle \mathbf{v}_i, \mathbf{v}_{i+1} \rangle + \cdots + \alpha_k \langle \mathbf{v}_i, \mathbf{v}_k \rangle.
\end{aligned}
$$

Since $\langle \mathbf{v}_i, \mathbf{v}_j \rangle = 0$ for $j \ne i$, this equals $\alpha_i \langle \mathbf{v}_i, \mathbf{v}_i \rangle$, which is $\alpha_i \|\mathbf{v}_i\|^2$. So we have $\alpha_i \|\mathbf{v}_i\|^2 = 0$. Since $\mathbf{v}_i \ne \mathbf{0}$, $\|\mathbf{v}_i\|^2 \ne 0$ and hence $\alpha_i = 0$. But i

was any integer in the range 1 to k, so we deduce that

$$\alpha_1 = \alpha_2 = \cdots = \alpha_k = 0,$$

as required. $\qquad\qquad\qquad\qquad\qquad\qquad\qquad\qquad\qquad\square$

10.3   Orthogonal matrices

10.3.1   Definition of orthogonal matrix

There is a particularly useful property that a matrix might possess, and which has links with orthogonality of vectors. This is described in the following definition.

Definition 10.15 (Orthogonal matrix) An $n \times n$ matrix P is said to be *orthogonal* if $P^{\mathrm{T}} P = P P^{\mathrm{T}} = I$; that is, if P has inverse P^{T}.

Example 10.16 The matrix

$$P = \begin{pmatrix} 3/5 & 4/5 \\ -4/5 & 3/5 \end{pmatrix}$$

is orthogonal.

Activity 10.17 Check this!

At first it appears that the definition of an orthogonal matrix has little to do with the concept of orthogonality of vectors. But, as we shall see, it is closely related. If P is an orthogonal matrix, then $P^{\mathrm{T}} P = I$, the identity matrix. Suppose that the columns of P are $\mathbf{x}_1, \mathbf{x}_2, \ldots, \mathbf{x}_n$. Then the fact that $P^{\mathrm{T}} P = I$ means that $\mathbf{x}_i^{\mathrm{T}} \mathbf{x}_j = 0$ if $i \neq j$ and $\mathbf{x}_i^{\mathrm{T}} \mathbf{x}_i = 1$, as the following theorem shows:

Theorem 10.18 *A matrix P is orthogonal if and only if, as vectors, its columns are pairwise orthogonal, and each has length 1.*

Proof: Let $P = (\mathbf{x}_1\ \mathbf{x}_2\ \cdots\ \mathbf{x}_n)$, so that P^{T} is the matrix whose rows are the vectors $\mathbf{x}_1^{\mathrm{T}}, \mathbf{x}_2^{\mathrm{T}}, \ldots, \mathbf{x}_n^{\mathrm{T}}$. Then $P^{\mathrm{T}} P = I$ can be expressed as

$$\begin{pmatrix} \mathbf{x}_1^{\mathrm{T}} \\ \mathbf{x}_2^{\mathrm{T}} \\ \vdots \\ \mathbf{x}_n^{\mathrm{T}} \end{pmatrix} (\mathbf{x}_1\ \mathbf{x}_2\ \cdots\ \mathbf{x}_n) = \begin{pmatrix} \mathbf{x}_1^{\mathrm{T}}\mathbf{x}_1 & \mathbf{x}_1^{\mathrm{T}}\mathbf{x}_2 & \cdots & \mathbf{x}_1^{\mathrm{T}}\mathbf{x}_n \\ \mathbf{x}_2^{\mathrm{T}}\mathbf{x}_1 & \mathbf{x}_2^{\mathrm{T}}\mathbf{x}_2 & \cdots & \mathbf{x}_2^{\mathrm{T}}\mathbf{x}_n \\ \vdots & \vdots & \ddots & \vdots \\ \mathbf{x}_n^{\mathrm{T}}\mathbf{x}_1 & \mathbf{x}_n^{\mathrm{T}}\mathbf{x}_2 & \cdots & \mathbf{x}_n^{\mathrm{T}}\mathbf{x}_n \end{pmatrix}$$

$$= \begin{pmatrix} 1 & 0 & \cdots & 0 \\ 0 & 1 & \cdots & 0 \\ \vdots & \vdots & \ddots & \vdots \\ 0 & 0 & \cdots & 1 \end{pmatrix}.$$

The theorem is an 'if and only if' statement, so we must prove it both ways.

If $P^{\mathrm{T}} P = I$, then the two matrices on the right are equal, and so

$$\mathbf{x}_i^{\mathrm{T}} \mathbf{x}_i = \langle \mathbf{x}_i, \mathbf{x}_i \rangle = \|\mathbf{x}_i\|^2 = 1 \quad \text{and} \quad \mathbf{x}_i^{\mathrm{T}} \mathbf{x}_j = \langle \mathbf{x}_i, \mathbf{x}_j \rangle = 0 \text{ if } i \neq j.$$

This says that the vectors are unit vectors and that the columns $\mathbf{x}_i$, $\mathbf{x}_j$ are orthogonal.

Conversely, if the columns are pairwise orthogonal unit vectors, then

$$\|\mathbf{x}_i\|^2 = \langle \mathbf{x}_i, \mathbf{x}_i \rangle = \mathbf{x}_i^{\mathrm{T}} \mathbf{x}_i = 1 \quad \text{and} \quad \langle \mathbf{x}_i, \mathbf{x}_j \rangle = \mathbf{x}_i^{\mathrm{T}} \mathbf{x}_j = 0 \text{ for } i \neq j,$$

so the matrix $P^{\mathrm{T}} P$ is equal to the identity matrix. $\qquad\square$

10.3.2 Orthonormal sets

Theorem 10.18 characterises orthogonal matrices through an important property of their columns. This important property is given a special name.

Definition 10.19 (Orthonormal) A set of vectors $\{\mathbf{x}_1, \mathbf{x}_2, \dots, \mathbf{x}_k\}$ in an inner product space V is said to be an *orthonormal set* if any two different vectors in the set are orthogonal and each vector is a unit vector; that is,

$$\langle \mathbf{x}_i, \mathbf{x}_j \rangle = 0 \quad \text{for} \quad i \neq j \quad \text{and} \quad \|\mathbf{x}_i\| = 1.$$

An important consequence of Theorem 10.14 is that an orthonormal set of n vectors in an n-dimensional vector space is a basis. If $\{\mathbf{v}_1, \mathbf{v}_2, \dots, \mathbf{v}_n\}$ is an orthonormal basis of a vector space V, then the coordinates of any vector $\mathbf{w} \in V$ are easy to calculate, as shown in the following theorem.

Theorem 10.20 Let $B = \{\mathbf{v}_1, \mathbf{v}_2, \dots, \mathbf{v}_n\}$ be an orthonormal basis of a vector space V and let $\mathbf{w} \in V$. Then the coordinates $a_1, a_2, \dots, a_n$ of $\mathbf{w}$ in the basis B are given by

$$a_i = \langle \mathbf{w}, \mathbf{v}_i \rangle.$$

Proof: We have $\mathbf{w} = a_1 \mathbf{v}_1 + a_2 \mathbf{v}_2 + \cdots + a_n \mathbf{v}_n$. We calculate the inner product of $\mathbf{w}$ with a basis vector $\mathbf{v}_i$.

$$\begin{aligned} \langle \mathbf{w}, \mathbf{v}_i \rangle &= \langle a_1 \mathbf{v}_1 + a_2 \mathbf{v}_2 + \cdots + a_n \mathbf{v}_n, \mathbf{v}_i \rangle \\ &= a_1 \langle \mathbf{v}_1, \mathbf{v}_i \rangle + a_2 \langle \mathbf{v}_2, \mathbf{v}_i \rangle + \cdots + a_n \langle \mathbf{v}_n, \mathbf{v}_i \rangle \\ &= a_i \langle \mathbf{v}_i, \mathbf{v}_i \rangle \\ &= a_i. \end{aligned}$$

The last two equalities follow from the fact that $\{\mathbf{v}_1, \mathbf{v}_2, \ldots, \mathbf{v}_n\}$ is an orthonormal set. $\qquad\square$

If P is an orthogonal matrix, then its columns form an orthonormal basis. So we can restate Theorem 10.18 as follows.

Theorem 10.21 *An $n \times n$ matrix P is orthogonal if and only if the columns of P form an orthonormal basis of $\mathbb{R}^n$.*

If the matrix P is orthogonal, then since $P = (P^{\mathrm{T}})^{\mathrm{T}}$, the matrix P^{T} is orthogonal too.

Activity 10.22 Show that if P is orthogonal, so too is P^{T}.

It therefore follows that Theorem 10.21 remains true if *column* is replaced by *row*, with rows written as vectors, and we can make the following stronger statement: a matrix P is orthogonal if and only if the columns (or rows, written as vectors) of P form an orthonormal basis of $\mathbb{R}^n$.

10.4 Gram–Schmidt orthonormalisation process

Given a set of linearly independent vectors $\{\mathbf{v}_1, \mathbf{v}_2, \ldots, \mathbf{v}_k\}$, the *Gram–Schmidt orthonormalisation process* is a way of producing k vectors that span the same space as $\{\mathbf{v}_1, \mathbf{v}_2, \ldots, \mathbf{v}_k\}$, and that form an orthonormal set. That is, the process produces a set $\{\mathbf{u}_1, \mathbf{u}_2, \ldots, \mathbf{u}_k\}$ such that:

- $\mathrm{Lin}\{\mathbf{u}_1, \mathbf{u}_2, \ldots, \mathbf{u}_k\} = \mathrm{Lin}\{\mathbf{v}_1, \mathbf{v}_2, \ldots, \mathbf{v}_k\}$
- $\{\mathbf{u}_1, \mathbf{u}_2, \ldots, \mathbf{u}_k\}$ is an orthonormal set.

We will see in the next chapter why this is a useful process to be able to perform. It works as follows. First, we set

$$\mathbf{u}_1 = \frac{\mathbf{v}_1}{\|\mathbf{v}_1\|}$$

so that $\mathbf{u}_1$ is a unit vector and $\mathrm{Lin}\{\mathbf{u}_1\} = \mathrm{Lin}\{\mathbf{v}_1\}$.

Then we define

$$\mathbf{w}_2 = \mathbf{v}_2 - \langle \mathbf{v}_2, \mathbf{u}_1 \rangle \mathbf{u}_1,$$

and set

$$\mathbf{u}_2 = \frac{\mathbf{w}_2}{\|\mathbf{w}_2\|}.$$

Then $\{\mathbf{u}_1, \mathbf{u}_2\}$ is an orthonormal set and $\mathrm{Lin}\{\mathbf{u}_1, \mathbf{u}_2\} = \mathrm{Lin}\{\mathbf{v}_1, \mathbf{v}_2\}$.

Activity 10.23 Make sure you understand why this works. Show that $w_2 \perp u_1$ and conclude that $u_2 \perp u_1$. Why are the linear spans of $\{u_1, u_2\}$ and $\{v_1, v_2\}$ the same?

Next, we define

$$w_3 = v_3 - \langle v_3, u_1 \rangle u_1 - \langle v_3, u_2 \rangle u_2$$

and set

$$u_3 = \frac{w_3}{\|w_3\|}.$$

Then $\{u_1, u_2, u_3\}$ is an orthonormal set and $\text{Lin}\{u_1, u_2, u_3\}$ is the same as $\text{Lin}\{v_1, v_2, v_3\}$. Generally, when we have $u_1, u_2, \ldots, u_i$, we let

$$w_{i+1} = v_{i+1} - \sum_{j=1}^{i} \langle v_{i+1}, u_j \rangle u_j,$$

$$u_{i+1} = \frac{w_{i+1}}{\|w_{i+1}\|}.$$

Then the resulting set $\{u_1, u_2, \ldots, u_k\}$ has the required properties.

Example 10.24 In $\mathbb{R}^4$, let us find an orthonormal basis for the linear span of the three vectors

$$v_1 = \begin{pmatrix} 1 \\ 1 \\ 1 \\ 1 \end{pmatrix}, \quad v_2 = \begin{pmatrix} -1 \\ 4 \\ 4 \\ -1 \end{pmatrix}, \quad v_3 = \begin{pmatrix} 4 \\ -2 \\ 2 \\ 0 \end{pmatrix}.$$

First, we have

$$u_1 = \frac{v_1}{\|v_1\|} = \frac{v_1}{\sqrt{1^2 + 1^2 + 1^2 + 1^2}} = \frac{1}{2}v_1 = \begin{pmatrix} 1/2 \\ 1/2 \\ 1/2 \\ 1/2 \end{pmatrix}.$$

Next, we have

$$w_2 = v_2 - \langle v_2, u_1 \rangle u_1 = \begin{pmatrix} -1 \\ 4 \\ 4 \\ -1 \end{pmatrix} - 3 \begin{pmatrix} 1/2 \\ 1/2 \\ 1/2 \\ 1/2 \end{pmatrix} = \begin{pmatrix} -5/2 \\ 5/2 \\ 5/2 \\ -5/2 \end{pmatrix},$$

and we set

$$u_2 = \frac{w_2}{\|w_2\|} = \begin{pmatrix} -1/2 \\ 1/2 \\ 1/2 \\ -1/2 \end{pmatrix}.$$

(Note: to do this last step, we merely noted that a normalised vector in the same direction as $\mathbf{w}_2$ is also a normalised vector in the same direction as $(-1, 1, 1, -1)^T$, and this second vector is easier to work with.) At this stage, you should check that $\mathbf{u}_2 \perp \mathbf{u}_1$. Continuing, we have

$$
\mathbf{w}_3 = \mathbf{v}_3 - \langle \mathbf{v}_3, \mathbf{u}_1 \rangle \mathbf{u}_1 - \langle \mathbf{v}_3, \mathbf{u}_2 \rangle \mathbf{u}_2
$$

$$
= \begin{pmatrix} 4 \\ -2 \\ 2 \\ 0 \end{pmatrix} - 2 \begin{pmatrix} 1/2 \\ 1/2 \\ 1/2 \\ 1/2 \end{pmatrix} - (-2) \begin{pmatrix} -1/2 \\ 1/2 \\ 1/2 \\ -1/2 \end{pmatrix} = \begin{pmatrix} 2 \\ -2 \\ 2 \\ -2 \end{pmatrix}.
$$

Then,

$$
\mathbf{u}_3 = \frac{\mathbf{w}_3}{\|\mathbf{w}_3\|} = (1/2, -1/2, 1/2, -1/2)^T.
$$

So

$$
\{\mathbf{u}_1, \mathbf{u}_2, \mathbf{u}_3\} = \left\{ \begin{pmatrix} 1/2 \\ 1/2 \\ 1/2 \\ 1/2 \end{pmatrix}, \begin{pmatrix} -1/2 \\ 1/2 \\ 1/2 \\ -1/2 \end{pmatrix}, \begin{pmatrix} 1/2 \\ -1/2 \\ 1/2 \\ -1/2 \end{pmatrix} \right\}.
$$

Activity 10.25 Work through all the calculations in this example. Then verify that the set $\{\mathbf{u}_1, \mathbf{u}_2, \mathbf{u}_3\}$ is an orthonormal set.

10.5 Learning outcomes

You should now be able to:

- explain what is meant by an inner product on a vector space
- verify that a given inner product is indeed an inner product
- compute norms in inner product spaces
- state and apply the Cauchy–Schwarz inequality, the generalised Pythagoras theorem, and the triangle inequality for norms
- prove that orthogonality of a set of vectors implies linear independence
- explain what is meant by an orthonormal set of vectors
- use the Gram–Schmidt orthonormalisation process
- state what is meant by an orthogonal matrix
- explain why an $n \times n$ matrix is orthogonal if and only if its columns are an orthonormal basis of $\mathbb{R}^n$

10.6   Comments on activities

Activity 10.2 By property (i) $\langle \mathbf{z}, \alpha\mathbf{x} + \beta\mathbf{y} \rangle = \langle \alpha\mathbf{x} + \beta\mathbf{y}, \mathbf{z} \rangle$. Then applying property (ii), and then property (i) again, the result follows.

Activity 10.4 Since $\alpha\mathbf{p} + \beta\mathbf{q}$ is the polynomial function

$$x \mapsto \alpha p(x) + \beta q(x),$$

we have

$$\langle \alpha\mathbf{p} + \beta\mathbf{q}, \mathbf{r} \rangle = \sum_{i=1}^{n+1} (\alpha p(x_i) + \beta q(x_i)) r(x_i)$$

$$= \alpha \sum_{i=1}^{n+1} p(x_i) r(x_i) + \beta \sum_{i=1}^{n+1} q(x_i) r(x_i)$$

$$= \alpha \langle \mathbf{p}, \mathbf{r} \rangle + \beta \langle \mathbf{q}, \mathbf{r} \rangle,$$

as required.

Activity 10.8 By the quadratic formula, the solutions of

$$at^2 + bt + c = 0,$$

are given by

$$t = \frac{-b \pm \sqrt{b^2 - 4ac}}{2a}.$$

If $b^2 - 4ac > 0$, then this will have two real solutions, so the graph of the function $f(t) = at^2 + bt + c$ will cross the t axis twice, and so it must have both positive and negative values. Therefore it would not be true that $at^2 + bt + c \geq 0$ for all $t \in \mathbb{R}$.

Activity 10.11 Just check that $\langle \mathbf{x}, \mathbf{y} \rangle = 0$.

Activity 10.17 Multiply $P^{\mathrm{T}} P$ and show that you get the identity matrix.

Activity 10.22 The matrix P is orthogonal if and only if $PP^{\mathrm{T}} = P^{\mathrm{T}} P = I$. Since $(P^{\mathrm{T}})^{\mathrm{T}} = P$, this statement can be written as $(P^{\mathrm{T}})^{\mathrm{T}} P^{\mathrm{T}} = P^{\mathrm{T}} (P^{\mathrm{T}})^{\mathrm{T}} = I$, which says that P^{T} is orthogonal.

Activity 10.23 Using the fact that $\langle \mathbf{u}_1, \mathbf{u}_1 \rangle = 1$, we have

$$\langle \mathbf{w}_2, \mathbf{u}_1 \rangle = \langle \mathbf{v}_2 - \langle \mathbf{v}_2, \mathbf{u}_1 \rangle \mathbf{u}_1, \mathbf{u}_1 \rangle = \langle \mathbf{v}_2, \mathbf{u}_1 \rangle - \langle \mathbf{v}_2, \mathbf{u}_1 \rangle \langle \mathbf{u}_1, \mathbf{u}_1 \rangle = 0.$$

The fact that $\mathbf{w}_2 \perp \mathbf{u}_1$ if and only if $\mathbf{u}_2 \perp \mathbf{u}_1$ follows from property (ii) of the definition of inner product since $\mathbf{w}_2 = \alpha\mathbf{u}_2$ for some constant α.

The linear spans are the same because $\mathbf{u}_1$, $\mathbf{u}_2$ are linear combinations of $\mathbf{v}_1$, $\mathbf{v}_2$ and conversely.

Activity 10.25 We only need to check that each $\mathbf{u}_i$ satisfies $\|\mathbf{u}_i\| = 1$, and that $\langle \mathbf{u}_1, \mathbf{u}_2 \rangle = \langle \mathbf{u}_1, \mathbf{u}_3 \rangle = \langle \mathbf{u}_2, \mathbf{u}_3 \rangle = 0$. All of this is very easily checked. (It is much harder to *find* the $\mathbf{u}_i$ in the first place. But once you think you have found them, it is always fairly easy to check whether they form an orthonormal set, as they should.)

10.7   Exercises

Exercise 10.1 Let V be the vector space of all $m \times n$ real matrices (with matrix addition and scalar multiplication). Define, for $A = (a_{ij})$ and $B = (b_{ij}) \in V$,

$$\langle A, B \rangle = \sum_{i=1}^{m} \sum_{j=1}^{n} a_{ij} b_{ij}.$$

Prove that this is an inner product on V.

Exercise 10.2 Prove that in any inner product space V,

$$\|\mathbf{x} + \mathbf{y}\|^2 + \|\mathbf{x} - \mathbf{y}\|^2 = 2\|\mathbf{x}\|^2 + 2\|\mathbf{y}\|^2,$$

for all $\mathbf{x}, \mathbf{y} \in V$.

Exercise 10.3 Suppose that $\mathbf{v} \in \mathbb{R}^n$. Prove that $W = \{\mathbf{x} \in \mathbb{R}^n \mid \mathbf{x} \perp \mathbf{v}\}$, the set of vectors orthogonal to $\mathbf{v}$, is a subspace of $\mathbb{R}^n$. How would you describe this subspace geometrically?

More generally, suppose that S is any (not necessarily finite) set of vectors in $\mathbb{R}^n$ and let $S^\perp$ denote the set

$$S^\perp = \{\mathbf{x} \in \mathbb{R}^n \mid \mathbf{x} \perp \mathbf{v} \text{ for all } \mathbf{v} \in S\}.$$

Prove that $S^\perp$ is a subspace of $\mathbb{R}^n$.

Exercise 10.4 Show that if P is an orthogonal matrix, then $|P| = \pm 1$.

Exercise 10.5 Consider the mapping from pairs of vectors $\mathbf{x}, \mathbf{y} \in \mathbb{R}^2$ to the real numbers given by

$$\langle \mathbf{x}, \mathbf{y} \rangle = \mathbf{x}^T A \mathbf{y}, \qquad \text{with} \quad A = \begin{pmatrix} 5 & 2 \\ 2 & 1 \end{pmatrix},$$

where the 1×1 matrix $\mathbf{x}^T A \mathbf{y}$ is interpreted as the real number which is its only entry. Show that this is an inner product on $\mathbb{R}^2$.

Let

$$\mathbf{v} = \begin{pmatrix} 1 \\ 1 \end{pmatrix}, \qquad \mathbf{w} = \begin{pmatrix} -1 \\ 2 \end{pmatrix}.$$

(a) Find $\langle \mathbf{v}, \mathbf{w} \rangle$ under this inner product.

(b) Find the length of the vector $\mathbf{v}$ in the norm defined by this inner product.

(c) Find the set of all vectors which are orthogonal to the vector $\mathbf{v}$ under this inner product. That is, if $S = \text{Lin}(\mathbf{v})$, find

$$S^\perp = \{\mathbf{x} \in \mathbb{R}^2 \mid \mathbf{x} \perp \mathbf{v}\}.$$

Write down a basis of $S^\perp$.

(d) Express the vector $\mathbf{w}$ above as $\mathbf{w} = \mathbf{w}_1 + \mathbf{w}_2$ where $\mathbf{w}_1 \in S$ and $\mathbf{w}_2 \in S^\perp$.

(e) Write down an orthonormal basis of $\mathbb{R}^2$ with respect to this inner product.

Exercise 10.6 Use the Gram–Schmidt process to find an orthonormal basis for the subspace of $\mathbb{R}^4$ spanned by the vectors

$$\mathbf{v}_1 = \begin{pmatrix} 1 \\ 0 \\ 1 \\ 0 \end{pmatrix}, \quad \mathbf{v}_2 = \begin{pmatrix} 1 \\ 2 \\ 1 \\ 1 \end{pmatrix}, \quad \mathbf{v}_3 = \begin{pmatrix} 0 \\ 1 \\ 2 \\ 1 \end{pmatrix}.$$

Exercise 10.7 Let

$$W = \left\{ \mathbf{x} = \begin{pmatrix} x \\ y \\ z \end{pmatrix} \;\middle|\; x - 2y + 3z = 0 \right\}.$$

Find an orthonormal basis of W. Extend it to an orthonormal basis of $\mathbb{R}^3$.

10.8   Problems

Problem 10.1 Consider the vectors

$$\mathbf{a} = \begin{pmatrix} 1 \\ 1 \\ 2 \\ 1 \end{pmatrix}, \quad \mathbf{b} = \begin{pmatrix} 2 \\ 3 \\ 1 \\ 0 \end{pmatrix}.$$

Show that the vectors $\mathbf{a}$, $\mathbf{b}$, $\mathbf{b} - \mathbf{a}$ form an isosceles right-angled triangle. Verify that these vectors satisfy the generalised Pythagoras theorem.

Problem 10.2 Let A be an $m \times k$ matrix with full column rank, meaning that $\text{rank}(A) = k$.

(a) Show that $A^T A$ is a $k \times k$ symmetric matrix. Show also that $\mathbf{x}^T(A^T A)\mathbf{x} > 0$ for all $\mathbf{x} \neq \mathbf{0}$, $\mathbf{x} \in \mathbb{R}^k$.

(b) Using the results in part (a), show that the mapping from pairs of vectors in $\mathbb{R}^k$ to the real numbers given by the rule

$$\langle \mathbf{x}, \mathbf{y} \rangle = \mathbf{x}^{\mathrm{T}}(A^{\mathrm{T}}A)\mathbf{y}$$

defines an inner product on $\mathbb{R}^k$, where the 1×1 matrix $\mathbf{x}^{\mathrm{T}}(A^{\mathrm{T}}A)\mathbf{y}$ is identified with the scalar which is its unique entry.

Problem 10.3 If P is an orthogonal $n \times n$ matrix and $\mathbf{x} = P\mathbf{z}$, show that $\|\mathbf{x}\| = \|\mathbf{z}\|$ using the standard inner product on $\mathbb{R}^n$.

Problem 10.4 Let P be an orthogonal $n \times n$ matrix and let T be the linear transformation defined by $T(\mathbf{x}) = P\mathbf{x}$. Using the standard inner product, show that for any $\mathbf{x}, \mathbf{y} \in \mathbb{R}^n$,

$$\langle T(\mathbf{x}), T(\mathbf{y}) \rangle = \langle \mathbf{x}, \mathbf{y} \rangle.$$

Problem 10.5 Suppose T and S are linear transformations of $\mathbb{R}^2$ to $\mathbb{R}^2$ with respective matrices:

$$A_T = \begin{pmatrix} \frac{1}{\sqrt{2}} & -\frac{1}{\sqrt{2}} \\ \frac{1}{\sqrt{2}} & \frac{1}{\sqrt{2}} \end{pmatrix}, \qquad A_S = \begin{pmatrix} -1 & 0 \\ 0 & 1 \end{pmatrix}.$$

Describe the effect of T and S in words. Show that the both A_T and A_S are orthogonal matrices.

Write down the matrix A that represents the linear transformation of $\mathbb{R}^2$ which is a rotation anticlockwise about the origin by an angle θ. Show that A is an orthogonal matrix.

Problem 10.6 Find an orthonormal basis for the subspace of $\mathbb{R}^3$ given by

$$V = \left\{ \begin{pmatrix} x \\ y \\ z \end{pmatrix} \;\middle|\; 5x - y + 2z = 0 \right\}.$$

Extend this to an orthonormal basis of $\mathbb{R}^3$.

Problem 10.7 Show that $S = \{\mathbf{v}_1, \mathbf{v}_2, \mathbf{v}_3\}$ is a basis of $\mathbb{R}^3$, where

$$\mathbf{v}_1 = \begin{pmatrix} 1 \\ 0 \\ 1 \end{pmatrix}, \quad \mathbf{v}_2 = \begin{pmatrix} 2 \\ -1 \\ 1 \end{pmatrix}, \quad \mathbf{v}_3 = \begin{pmatrix} 1 \\ 1 \\ 5 \end{pmatrix}.$$

Beginning with the vector $\mathbf{v}_1$, find an orthonormal basis of the subspace $\mathrm{Lin}\{\mathbf{v}_1, \mathbf{v}_2\}$. Using any method, extend this to an orthonormal basis B of $\mathbb{R}^3$.

Find the B coordinates of the vectors $\mathbf{v}_2$ and $\mathbf{v}_3$.

Find the transition matrix P from coordinates in the basis S to coordinates in the basis B.

Check that $[\mathbf{v}_3]_B = P[\mathbf{v}_3]_S$.

Problem 10.8 Beginning with the vector $\mathbf{v}_1$, use the Gram–Schmidt orthonormalisation process to obtain an orthonormal basis for the subspace of $\mathbb{R}^4$ spanned by the following vectors:

$$\mathbf{v}_1 = \begin{pmatrix} 1 \\ 0 \\ 1 \\ 0 \end{pmatrix} \qquad \mathbf{v}_2 = \begin{pmatrix} 3 \\ 0 \\ 2 \\ 0 \end{pmatrix} \qquad \mathbf{v}_3 = \begin{pmatrix} 2 \\ 1 \\ -1 \\ 3 \end{pmatrix}.$$

Problem 10.9 Put the following matrix into reduced row echelon form:

$$A = \begin{pmatrix} 1 & 1 & -1 & 2 \\ -1 & 0 & 1 & 1 \\ 1 & 2 & -1 & 5 \end{pmatrix}.$$

Find an orthonormal basis of the null space of A. Extend this to an orthonormal basis of $\mathbb{R}^4$ using the row space of A.

Problem 10.10 Consider the planes in $\mathbb{R}^3$:

$$U = \left\{ \begin{pmatrix} x \\ y \\ z \end{pmatrix} \middle| x - y + 2z = 0 \right\}, \qquad V = \left\{ \begin{pmatrix} x \\ y \\ z \end{pmatrix} \middle| 3x + 2y + z = 0 \right\}.$$

Find the vector equation of the line of intersection of U and V.

Find vectors $\mathbf{x}, \mathbf{y}, \mathbf{z}$ in $\mathbb{R}^3$ with the following properties:

(i)   The vector $\mathbf{x}$ is on both planes, that is, $\mathbf{x} \in U \cap V$;
(ii) The set $\{\mathbf{x}, \mathbf{y}\}$ is an orthonormal basis of U;
(iii) The set $\{\mathbf{x}, \mathbf{z}\}$ is an orthonormal basis of V.

Is your set $\{\mathbf{x}, \mathbf{y}, \mathbf{z}\}$ a basis of $\mathbb{R}^3$? Is it an orthonormal basis of $\mathbb{R}^3$? Justify your answers.

11

Orthogonal diagonalisation and its applications

In this chapter, we look at orthogonal diagonalisation, a special form of diagonalisation for real symmetric matrices. This has some useful applications: to quadratic forms, in particular.

11.1 Orthogonal diagonalisation of symmetric matrices

Recall that a square matrix $A = (a_{ij})$ is symmetric if $A^T = A$. Equivalently, A is symmetric if $a_{ij} = a_{ji}$ for all i, j; that is, if the entries in opposite positions relative to the main diagonal are equal. It turns out that symmetric matrices are always diagonalisable. They are, furthermore, diagonalisable in a special way.

11.1.1 Orthogonal diagonalisation

We know what it means to diagonalise a square matrix A. It means to find an invertible matrix P and a diagonal matrix D such that $P^{-1}AP = D$. If, in addition, we can find an orthogonal matrix P which diagonalises A, so that $P^{-1}AP = P^T AP = D$, then this is *orthogonal diagonalisation*.

Definition 11.1 A matrix A is said to be *orthogonally diagonalisable* if there is an orthogonal matrix P such that $P^T AP = D$ where D is a diagonal matrix.

As P is orthogonal, $P^T = P^{-1}$, so $P^T AP = P^{-1}AP = D$. The fact that A is diagonalisable means that the columns of P are a basis of

$\mathbb{R}^n$ consisting of eigenvectors of A (Theorem 8.22). The fact that A is orthogonally diagonalisable means that the columns of P are an orthonormal basis of $\mathbb{R}^n$ consisting of an orthonormal set of eigenvectors of A (Theorem 10.21). Putting these facts together, we have the following theorem:

Theorem 11.2 *A matrix A is orthogonally diagonalisable if and only if there is an orthonormal basis of $\mathbb{R}^n$ consisting of eigenvectors of A.*

Let's look at some examples.

Example 11.3 The matrix

$$B = \begin{pmatrix} 7 & -15 \\ 2 & -4 \end{pmatrix},$$

which we have met in previous examples, cannot be orthogonally diagonalised. The eigenvalues are $\lambda_1 = 1$ and $\lambda_2 = 2$. All the eigenvectors corresponding to $\lambda = 1$ are scalar multiples of $\mathbf{v}_1 = (5, 2)^T$, and all the eigenvectors corresponding to $\lambda = 2$ are scalar multiples of $\mathbf{v}_2 = (3, 1)^T$. Since

$$\langle \mathbf{v}_1, \mathbf{v}_2 \rangle = \left\langle \begin{pmatrix} 5 \\ 2 \end{pmatrix}, \begin{pmatrix} 3 \\ 1 \end{pmatrix} \right\rangle \neq 0$$

no eigenvector in the eigenspace of λ_1 is perpendicular to any eigenvector for λ_2, so it is not possible to find an orthogonal set of eigenvectors for B.

Example 11.4 Now consider the matrix

$$A = \begin{pmatrix} 5 & -3 \\ -3 & 5 \end{pmatrix}.$$

The eigenvalues are given by

$$|A - \lambda I| = \begin{pmatrix} 5 - \lambda & -3 \\ -3 & 5 - \lambda \end{pmatrix} = \lambda^2 - 10\lambda + 16 = 0.$$

So the eigenvalues are $\lambda_1 = 2$ and $\lambda_2 = 8$. The corresponding eigenvectors are the solutions of $(A - \lambda I)\mathbf{v} = \mathbf{0}$, so

$$(A - 2I) = \begin{pmatrix} 3 & -3 \\ -3 & 3 \end{pmatrix} \longrightarrow \begin{pmatrix} 1 & -1 \\ 0 & 0 \end{pmatrix} \implies \mathbf{w}_1 = \begin{pmatrix} 1 \\ 1 \end{pmatrix}$$

$$(A - 8I) = \begin{pmatrix} -3 & -3 \\ -3 & -3 \end{pmatrix} \longrightarrow \begin{pmatrix} 1 & 1 \\ 0 & 0 \end{pmatrix} \implies \mathbf{w}_2 = \begin{pmatrix} -1 \\ 1 \end{pmatrix}.$$

Because $\langle \mathbf{w}_1, \mathbf{w}_2 \rangle = 0$, the eigenvectors $\mathbf{w}_1$ and $\mathbf{w}_2$ are orthogonal! So A can be orthogonally diagonalised. We just need to normalise the vectors by making them into unit vectors. If

$$P = \begin{pmatrix} \frac{1}{\sqrt{2}} & -\frac{1}{\sqrt{2}} \\ \frac{1}{\sqrt{2}} & \frac{1}{\sqrt{2}} \end{pmatrix} \quad \text{and} \quad D = \begin{pmatrix} 2 & 0 \\ 0 & 8 \end{pmatrix},$$

then P is orthogonal and $P^{\mathrm{T}}AP = P^{-1}AP = D$.

Note that the matrix A in this example is symmetric, whereas the matrix B in the first example is not.

11.1.2 When is orthogonal diagonalisation possible?

It's natural to ask which matrices can be orthogonally diagonalised. The answer is remarkably straightforward and is given by the following important result.

Theorem 11.5 (Spectral theorem for symmetric matrices) *The matrix A is orthogonally diagonalisable if and only if A is symmetric.*

Since this is an if and only if statement, it needs to be proved in both directions. One way is easy: if A can be orthogonally diagonalised, then it must be symmetric.

Activity 11.6 Try to prove this yourself before you continue reading. Assuming that A can be orthogonally diagonalised, write down what this means, and then show that $A^{\mathrm{T}} = A$.

The argument goes as follows. If A is orthogonally diagonalisable, then there exists an orthogonal matrix P and a diagonal matrix D such that $P^{\mathrm{T}}AP = P^{-1}AP = D$. Then solving for the matrix A,

$$A = PDP^{-1} = PDP^{\mathrm{T}}.$$

Taking the transposes of both sides of this equation (using properties of transpose), and using the fact that $D^{\mathrm{T}} = D$ since D is diagonal, we have

$$A^{\mathrm{T}} = (PDP^{\mathrm{T}})^{\mathrm{T}} = PD^{\mathrm{T}}P^{\mathrm{T}} = PDP^{\mathrm{T}} = A,$$

which shows that A is symmetric.

So only symmetric matrices can be orthogonally diagonalised. That's the 'only if' part of the proof. It is much more difficult to prove the 'if' part: if a matrix is symmetric, then it can be orthogonally diagonalised. We will first prove this for the special case in which the matrix has distinct eigenvalues, and then prove it for the more general case in Section 11.1.5.

For both of these, we will need one important fact about symmetric matrices: symmetric matrices have only real eigenvalues. (As we noted in Theorem 8.42, this is a necessary condition for diagonalisability, so we certainly need it.) We state it here as a theorem, but we will defer the proof of this fact until Chapter 13, as it is most easily established as a corollary (that is, a consequence) of a similar theorem on complex matrices.

Theorem 11.7 *If A is a symmetric matrix, then all of its eigenvalues are real numbers.*

This means that the characteristic polynomial of an $n \times n$ symmetric matrix factorises into n linear factors over the real numbers (repeating any roots with multiplicity greater than 1).

11.1.3   The case of distinct eigenvalues

Assuming Theorem 11.7, we now prove Theorem 11.5 for symmetric $n \times n$ matrices which have n different eigenvalues. To do so, we need the following result:

Theorem 11.8 *If the matrix A is symmetric, then eigenvectors corresponding to distinct eigenvalues are orthogonal.*

Proof: Suppose that λ and μ are any two different eigenvalues of A and that $\mathbf{x}, \mathbf{y}$ are corresponding eigenvectors. Then $A\mathbf{x} = \lambda\mathbf{x}$ and $A\mathbf{y} = \mu\mathbf{y}$. The trick in this proof is to find two different expressions for the product $\mathbf{x}^{\mathrm{T}}A\mathbf{y}$ (which then, of course, must be equal to each other). Note that the matrix product $\mathbf{x}^{\mathrm{T}}A\mathbf{y}$ is a 1×1 matrix or, equivalently, a number.

First, since $A\mathbf{y} = \mu\mathbf{y}$, we have

$$\mathbf{x}^{\mathrm{T}}A\mathbf{y} = \mathbf{x}^{\mathrm{T}}(A\mathbf{y}) = \mathbf{x}^{\mathrm{T}}(\mu\mathbf{y}) = \mu\mathbf{x}^{\mathrm{T}}\mathbf{y}.$$

But also, $A\mathbf{x} = \lambda\mathbf{x}$. Since A is symmetric, $A = A^{\mathrm{T}}$. Substituting and using the properties of the transpose of a matrix, we have

$$\mathbf{x}^{\mathrm{T}}A\mathbf{y} = \mathbf{x}^{\mathrm{T}}A^{\mathrm{T}}\mathbf{y} = (\mathbf{x}^{\mathrm{T}}A^{\mathrm{T}})\mathbf{y} = (A\mathbf{x})^{\mathrm{T}}\mathbf{y} = (\lambda\mathbf{x})^{\mathrm{T}}\mathbf{y} = \lambda\mathbf{x}^{\mathrm{T}}\mathbf{y}.$$

Equating these two different expressions for $\mathbf{x}^{\mathrm{T}}A\mathbf{y}$, we have $\mu\mathbf{x}^{\mathrm{T}}\mathbf{y} = \lambda\mathbf{x}^{\mathrm{T}}\mathbf{y}$, or

$$(\mu - \lambda)\mathbf{x}^{\mathrm{T}}\mathbf{y} = 0.$$

But since $\lambda \neq \mu$ (they are different eigenvalues), we have $\mu - \lambda \neq 0$. We deduce therefore that $x^T y = \langle x, y \rangle = 0$. But this says precisely that x and y are orthogonal. $\qquad \square$

Theorem 11.8 shows that if an $n \times n$ symmetric matrix has exactly n different eigenvalues and if we take a set of n eigenvectors with one eigenvector corresponding to each eigenvalue, then any two of these eigenvectors are orthogonal to one another. We may take the eigenvectors to have length 1, simply by normalising them. This shows that we have an orthonormal set of n eigenvectors, which is therefore a basis of $\mathbb{R}^n$. So by Theorem 11.2, the matrix can be orthogonally diagonalised. But let's spell it out. If P is the matrix with this set of eigenvectors as its columns, then (as usual) $P^{-1}AP = D$, the diagonal matrix of eigenvalues. Moreover, since the columns of P form an orthonormal set, by Theorem 10.18, P is an orthogonal matrix. So $P^{-1} = P^T$ and hence $P^T AP = D$. In other words, we have the following result (which outlines the method):

Theorem 11.9 *Suppose that A is symmetric and has n different eigenvalues. Take n corresponding eigenvectors, each of length* 1. *Form the matrix P which has these unit eigenvectors as its columns. Then $P^{-1} = P^T$ (that is, P is an orthogonal matrix) and $P^T AP = D$, the diagonal matrix whose entries are the eigenvalues of A.*

Here is an example of the technique.

Example 11.10 The matrix

$$A = \begin{pmatrix} 4 & 0 & 4 \\ 0 & 4 & 4 \\ 4 & 4 & 8 \end{pmatrix}$$

is symmetric. We have seen in Example 8.23 that it has three distinct eigenvalues, $\lambda_1 = 4$, $\lambda_2 = 0$, $\lambda_3 = 12$, and we found that corresponding eigenvectors are (in that order)

$$v_1 = \begin{pmatrix} 1 \\ -1 \\ 0 \end{pmatrix}, \quad v_2 = \begin{pmatrix} 1 \\ 1 \\ -1 \end{pmatrix}, \quad v_3 = \begin{pmatrix} 1 \\ 1 \\ 2 \end{pmatrix}.$$

Activity 11.11 Check that any two of these three eigenvectors are orthogonal.

These eigenvectors are mutually orthogonal, but *not* of length 1, so we normalise them. For example, the first one has length $\sqrt{2}$. If we divide each entry of it by $\sqrt{2}$, we obtain a unit eigenvector. We can similarly normalise the other two vectors, obtaining

$$\mathbf{u}_1 = \begin{pmatrix} 1/\sqrt{2} \\ -1/\sqrt{2} \\ 0 \end{pmatrix}, \quad \mathbf{u}_2 = \begin{pmatrix} 1/\sqrt{3} \\ 1/\sqrt{3} \\ -1/\sqrt{3} \end{pmatrix}, \quad \mathbf{u}_3 = \begin{pmatrix} 1/\sqrt{6} \\ 1/\sqrt{6} \\ 2/\sqrt{6} \end{pmatrix}.$$

Activity 11.12 Verify that the normalisations of the second and third vectors are as just stated.

We now form the matrix P whose columns are these unit eigenvectors:

$$P = \begin{pmatrix} 1/\sqrt{2} & 1/\sqrt{3} & 1/\sqrt{6} \\ -1/\sqrt{2} & 1/\sqrt{3} & 1/\sqrt{6} \\ 0 & -1/\sqrt{3} & 2/\sqrt{6} \end{pmatrix}.$$

Then P is orthogonal and $P^{\mathrm{T}}AP = D = \mathrm{diag}(4, 0, 12)$.

Activity 11.13 Check that P is orthogonal by calculating $P^{\mathrm{T}}P$.

11.1.4   When eigenvalues are not distinct

We have seen that if a symmetric matrix has distinct eigenvalues, then (since eigenvectors corresponding to different eigenvalues are orthogonal) it is orthogonally diagonalisable. But, as stated in Theorem 11.5, all $n \times n$ symmetric matrices are orthogonally diagonalisable, even if they do not have n distinct eigenvalues. We will prove this in the next section, but first we discuss how, in practice, we would go about orthogonally diagonalising a matrix in the case when it does not have distinct eigenvalues.

What we need for orthogonal diagonalisation is an orthonormal set of n eigenvectors. As we have seen, if it so happens that there are n different eigenvalues, then any set of n corresponding eigenvectors form a pairwise orthogonal set of vectors, and all we need to do is normalise each vector. However, if we have repeated eigenvalues, more care is required.

Suppose that λ_0 is a repeated eigenvalue of A, by which we mean that, for some $k \geq 2$, $(\lambda - \lambda_0)^k$ is a factor of the characteristic polynomial of A. As we saw in Definition 8.39, the *algebraic multiplicity* of λ_0 is the largest k for which this is the case. The eigenspace corresponding to λ_0 is (see Definition 8.9)

$$E(\lambda_0) = \{\mathbf{x} \mid (A - \lambda_0 I)\mathbf{x} = \mathbf{0}\},$$

the subspace consisting of all eigenvectors corresponding to λ_0, together with the zero-vector $\mathbf{0}$. It turns out (and, indeed, by Theorem 8.42, it must be the case if A is diagonalisable) that, for any symmetric matrix A, the dimension of $E(\lambda_0)$ (that is, the *geometric multiplicity*) is exactly the algebraic multiplicity k of λ_0. This means that there is some basis $\{v_1, v_2, \ldots, v_k\}$ of k vectors of the eigenspace $E(\lambda_0)$. So far, we are proceeding just as we would in diagonalisation, generally. But remember that we are trying to *orthogonally* diagonalise. We therefore use the Gram-Schmidt orthonormalisation process to take any such basis and produce an orthonormal basis of $E(\lambda_0)$.

Since, by Theorem 11.8, eigenvectors from different eigenspaces are orthogonal (and hence linearly independent), if we construct a set of n vectors by taking orthonormal bases for each of the eigenspaces, the resulting set is an orthonormal basis of $\mathbb{R}^n$. We can therefore orthogonally diagonalise the matrix A by means of the matrix P with these vectors as its columns. Here is an example of how we can carry out this process.

Example 11.14 We orthogonally diagonalise the symmetric matrix

$$B = \begin{pmatrix} 2 & 1 & 1 \\ 1 & 2 & 1 \\ 1 & 1 & 2 \end{pmatrix}.$$

The eigenvalues of B are given by the characteristic equation

$$|B - \lambda I| = -\lambda^3 + 6\lambda^2 - 9\lambda + 4 = -(\lambda - 1)^2(\lambda - 4) = 0.$$

The eigenvalues are 4 and 1, where 1 is an eigenvalue of multiplicity 2.

We will find the eigenvectors for $\lambda = 1$ first. Reducing $B - I$, we have

$$\begin{pmatrix} 1 & 1 & 1 \\ 1 & 1 & 1 \\ 1 & 1 & 1 \end{pmatrix} \rightarrow \cdots \rightarrow \begin{pmatrix} 1 & 1 & 1 \\ 0 & 0 & 0 \\ 0 & 0 & 0 \end{pmatrix},$$

so the eigenspace for $\lambda = 1$ does indeed have dimension 2. From the reduced row echelon form, we deduce the linearly independent eigenvectors

$$v_1 = \begin{pmatrix} -1 \\ 1 \\ 0 \end{pmatrix}, \qquad v_2 = \begin{pmatrix} -1 \\ 0 \\ 1 \end{pmatrix}.$$

For $\lambda = 4$,

$$B - 4I = \begin{pmatrix} -2 & 1 & 1 \\ 1 & -2 & 1 \\ 1 & 1 & -2 \end{pmatrix} \rightarrow \cdots \rightarrow \begin{pmatrix} 1 & 0 & -1 \\ 0 & 1 & -1 \\ 0 & 0 & 0 \end{pmatrix},$$

so we may take

$$\mathbf{v}_3 = \begin{pmatrix} 1 \\ 1 \\ 1 \end{pmatrix}.$$

The vectors $\mathbf{v}_1$ and $\mathbf{v}_2$ are *not* orthogonal. However, each of the vectors $\mathbf{v}_1$ and $\mathbf{v}_2$ is orthogonal to $\mathbf{v}_3$, the eigenvector for $\lambda = 4$. (This must be the case since they correspond to distinct eigenvalues.)

Activity 11.15 Check that $\langle \mathbf{v}_1, \mathbf{v}_3 \rangle = 0$, $\langle \mathbf{v}_2, \mathbf{v}_3 \rangle = 0$, and $\langle \mathbf{v}_1, \mathbf{v}_2 \rangle \neq 0$.

Notice that the eigenspace for $\lambda = 1$ can be described geometrically as a plane through the origin in $\mathbb{R}^3$ with normal vector $\mathbf{v}_3$. It consists of all linear combinations of $\mathbf{v}_1$ and $\mathbf{v}_2$; that is, all vectors which are perpendicular to $\mathbf{v}_3$.

Activity 11.16 Look at the reduced row echelon form of the matrix $B - I$. Could you have deduced the last eigenvector from this matrix? Why?

We still need to obtain an orthonormal basis of eigenvectors, so we now apply the Gram–Schmidt orthonormalisation process to $\text{Lin}\{\mathbf{v}_1, \mathbf{v}_2\}$. First we set

$$\mathbf{u}_1 = \begin{pmatrix} -1/\sqrt{2} \\ 1/\sqrt{2} \\ 0 \end{pmatrix}.$$

Then we define

$$\mathbf{w}_2 = \begin{pmatrix} -1 \\ 0 \\ 1 \end{pmatrix} - \left\langle \begin{pmatrix} -1 \\ 0 \\ 1 \end{pmatrix}, \begin{pmatrix} -1/\sqrt{2} \\ 1/\sqrt{2} \\ 0 \end{pmatrix} \right\rangle \begin{pmatrix} -1/\sqrt{2} \\ 1/\sqrt{2} \\ 0 \end{pmatrix} = \begin{pmatrix} -1/2 \\ -1/2 \\ 1 \end{pmatrix}.$$

This vector is parallel to $(-1, -1, 2)$ with length $\sqrt{6}$, so we have

$$\mathbf{u}_2 = \begin{pmatrix} -1/\sqrt{6} \\ -1/\sqrt{6} \\ 2/\sqrt{6} \end{pmatrix}.$$

Activity 11.17 What should you check now, before you proceed to the next step?

Normalising the vector $\mathbf{v}_3$, we can let P be the matrix

$$P = \begin{pmatrix} -1/\sqrt{2} & -1/\sqrt{6} & 1/\sqrt{3} \\ 1/\sqrt{2} & -1/\sqrt{6} & 1/\sqrt{3} \\ 0 & 2/\sqrt{6} & 1/\sqrt{3} \end{pmatrix}$$

and D the diagonal matrix

$$D = \begin{pmatrix} 1 & 0 & 0 \\ 0 & 1 & 0 \\ 0 & 0 & 4 \end{pmatrix}.$$

Then $P^T = P^{-1}$ and $P^T B P = D$.

11.1.5   The general case

Assuming for now the fact that symmetric matrices have only real eigen-values (Theorem 11.7, which will be proved in Chapter 13), we have proved Theorem 11.5 for symmetric matrices with distinct eigenvalues. We have also indicated how, in practice, to orthogonally diagonalise any symmetric matrix (in general, even if the eigenvalues are not distinct). To complete the picture, we will now prove the Spectral theorem for symmetric matrices in general. This is a fairly long and difficult proof, and it can safely be omitted without affecting your ability to carry out orthogonal diagonalisation. You can skip it and proceed on to the next section, where we begin to look at applications of orthogonal diagonalisation. However, we include the proof for two reasons: first, for completeness and, second, because it draws on many of the most important ideas we have studied so far, so trying to understand it will be a good exercise.

    We will give a proof by induction on n, the size of the matrix. That means we establish the theorem for the case $n = 1$ and we show that, for $n \geq 2$, if the theorem holds for all symmetric $(n-1) \times (n-1)$ matrices, then it will also be true for $n \times n$ matrices. (So, the $n = 2$ case then follows from the $n = 1$ case; the $n = 3$ from the $n = 2$, and so on.)

Proof of Theorem 11.5 Let's just remind ourselves what it is we are trying to prove. It is that, for any symmetric matrix A, there is an orthogonal matrix P and a diagonal matrix D such that $P^T A P = D$.

    Any 1×1 symmetric matrix is already diagonal, so we can take $P = I$, which is an orthogonal matrix. So the result is true when $n = 1$.

    Now let us consider a general value of $n \geq 2$ and assume that the theorem holds for all $(n-1) \times (n-1)$ symmetric matrices. Let A be any $n \times n$ symmetric matrix. As mentioned above, we take for granted now (and will prove in Chapter 13) the fact that A has real eigenvalues. Let λ_1 be any eigenvalue of A and let $\mathbf{v}_1$ be a corresponding eigenvector which satisfies $\|\mathbf{v}_1\| = 1$. By Theorem 6.45, we can extend the basis

$\{v_1\}$ of $\text{Lin}\{v_1\}$ to a basis $\{v_1, x_2, x_3, \ldots, x_n\}$ of $\mathbb{R}^n$. We can then use the Gram–Schmidt process to transform this into an *orthonormal* basis $B = \{v_1, v_2, \ldots, v_n\}$ of $\mathbb{R}^n$. (Remember that we chose v_1 to be a unit vector, so we can take it to be the first member of the orthonormal basis.)

Let P be the matrix whose columns are the vectors in B, with the first column being v_1. Then P is orthogonal, by Theorem 10.18, and (by Theorem 7.37) $P^T A P = P^{-1} A P$ represents the linear transformation $T : x \mapsto Ax$ in the basis B. But we know, by Theorem 7.36, that the first column of $P^T A P$ will be the coordinate vector of $T(v_1)$ with respect to the basis B. Now, $T(v_1) = Av_1 = \lambda_1 v_1$, so this coordinate vector is

$$\begin{pmatrix} \lambda_1 \\ 0 \\ \vdots \\ 0 \end{pmatrix}.$$

It follows that, for some numbers $d_1, d_2, \ldots, d_{n-1}$ and $c_{(i,j)}$ for $i, j = 1, \ldots, n - 1$, $P^T A P$ takes the form

$$P^T A P = \begin{pmatrix} \lambda_1 & d_1 & \cdots & d_{n-1} \\ 0 & c_{(1,1)} & \cdots & c_{(1,n-1)} \\ 0 & c_{(2,1)} & \cdots & c_{(2,n-1)} \\ \vdots & \vdots & \ddots & \vdots \\ 0 & c_{(n-1,1)} & \cdots & c_{(n-1,n-1)} \end{pmatrix}.$$

But A is symmetric, and so therefore is $P^T A P$, since

$$(P^T A P)^T = P^T A^T P = P^T A P.$$

The fact that this matrix is symmetric has two immediate consequences:

- $d_1 = d_2 = \cdots = d_{n-1} = 0$;
- the $(n - 1) \times (n - 1)$ matrix $C = (c_{(i,j)})$ is symmetric.

So we can write

$$P^T A P = \begin{pmatrix} \lambda_1 & \mathbf{0}^T \\ \mathbf{0} & C \end{pmatrix},$$

where $\mathbf{0}$ is the all-zero vector of length $n - 1$ and C is a symmetric $(n - 1) \times (n - 1)$ matrix.

We are assuming that the theorem holds for $(n - 1) \times (n - 1)$ symmetric matrices, so it holds for the matrix C. That means there is some orthogonal $(n-1) \times (n-1)$ matrix R such that $R^{\mathrm{T}} C R = D$, where D is a diagonal matrix. Consider the $n \times n$ matrix

$$Q = \begin{pmatrix} 1 & \mathbf{0}^{\mathrm{T}} \\ \mathbf{0} & R \end{pmatrix}.$$

This is an orthogonal matrix because the fact that R is orthogonal means columns $2, 3, \ldots, n$ of Q are mutually orthogonal and of length 1; and, furthermore, the first column is evidently orthogonal with all the other columns, and also has length 1. Let $S = PQ$. Then S is orthogonal because P and Q are: we have

$$S^{-1} = (PQ)^{-1} = Q^{-1} P^{-1} = Q^{\mathrm{T}} P^{\mathrm{T}} = (PQ)^{\mathrm{T}} = S^{\mathrm{T}}.$$

Now, let us think about $S^{\mathrm{T}} A S$. We have:

$$\begin{aligned} S^{\mathrm{T}} A S &= (PQ)^{\mathrm{T}} A (PQ) \\ &= Q^{\mathrm{T}} P^{\mathrm{T}} A P Q \\ &= Q^{\mathrm{T}} (P^{\mathrm{T}} A P) Q \\ &= \begin{pmatrix} 1 & \mathbf{0}^{\mathrm{T}} \\ \mathbf{0} & R \end{pmatrix}^{\mathrm{T}} \begin{pmatrix} \lambda_1 & \mathbf{0}^{\mathrm{T}} \\ \mathbf{0} & C \end{pmatrix} \begin{pmatrix} 1 & \mathbf{0}^{\mathrm{T}} \\ \mathbf{0} & R \end{pmatrix} \\ &= \begin{pmatrix} 1 & \mathbf{0}^{\mathrm{T}} \\ \mathbf{0} & R^{\mathrm{T}} \end{pmatrix} \begin{pmatrix} \lambda_1 & \mathbf{0}^{\mathrm{T}} \\ \mathbf{0} & C \end{pmatrix} \begin{pmatrix} 1 & \mathbf{0}^{\mathrm{T}} \\ \mathbf{0} & R \end{pmatrix} \\ &= \begin{pmatrix} \lambda_1 & \mathbf{0}^{\mathrm{T}} \\ \mathbf{0} & R^{\mathrm{T}} C R \end{pmatrix} \\ &= \begin{pmatrix} \lambda_1 & \mathbf{0}^{\mathrm{T}} \\ \mathbf{0} & D \end{pmatrix}, \end{aligned}$$

which is a diagonal matrix, because D is diagonal. So we're done! We have established that there is an orthogonal matrix, S, such that $S^{\mathrm{T}} A S$ is diagonal. $\qquad\square$

Activity 11.18 In order to understand this proof fully, work through it for a 2×2 symmetric matrix A. That is, assuming only that A has real eigenvalues, show that there is an orthogonal matrix P and a diagonal matrix D such that $P^{\mathrm{T}} A P = D$.

11.2 Quadratic forms

A very useful application of orthogonal diagonalisation is to the analysis of quadratic forms.

11.2.1 Quadratic forms

A quadratic form in two variables x and y is an expression of the form

$$q(x, y) = ax^2 + 2cxy + by^2.$$

This can be written as

$$q = \mathbf{x}^T A \mathbf{x},$$

where $\mathbf{x} = \begin{pmatrix} x \\ y \end{pmatrix}$ and A is the symmetric matrix

$$A = \begin{pmatrix} a & c \\ c & b \end{pmatrix}.$$

Activity 11.19 Check this. Perform the matrix multiplication $\mathbf{x}^T A \mathbf{x}$ to see how the expression $q(x, y)$ is obtained. Notice how the coefficients of the expression of $q(x, y)$ correspond to the entries of A.

Of course, there are other ways of writing $q(x, y)$ as a product of matrices, $\mathbf{x}^T B \mathbf{x}$, where B is not symmetric, but these are of no interest to us here; our focus is on the case where the matrix is symmetric. We say that q is written *in matrix form* when we express it as $q = \mathbf{x}^T A \mathbf{x}$, where A is symmetric.

Activity 11.20 Find an expression for $q(x, y) = \mathbf{x}^T B \mathbf{x}$, where B is not symmetric.

Here is a specific example of how a two-variable quadratic form can be expressed in matrix form.

Example 11.21 The quadratic form $q = x^2 + xy + 3y^2$ in matrix form is

$$q = (x \quad y) \begin{pmatrix} 1 & 1/2 \\ 1/2 & 3 \end{pmatrix} \begin{pmatrix} x \\ y \end{pmatrix}.$$

More generally, we consider quadratic forms in n variables.

Definition 11.22 (Quadratic form) A quadratic form in $n \geq 2$ variables is an expression of the form

$$q = \mathbf{x}^T A \mathbf{x},$$

where A is a symmetric $n \times n$ matrix and $\mathbf{x} \in \mathbb{R}^n$.

Example 11.23 The following is a quadratic form in three variables:

$$q(x_1, x_2, x_3) = 5x_1^2 + 10x_2^2 + 2x_3^2 + 4x_1x_2 + 2x_1x_3 - 6x_2x_3.$$

In matrix form, it is $\mathbf{x}^{\mathrm{T}} A \mathbf{x}$, where $\mathbf{x} = (x_1, x_2, x_3)^{\mathrm{T}}$ and A is the symmetric matrix

$$A = \begin{pmatrix} 5 & 2 & 1 \\ 2 & 10 & -3 \\ 1 & -3 & 2 \end{pmatrix}.$$

Activity 11.24 Check this.

You should be able to write down the $n \times n$ symmetric matrix A from the expression of the quadratic form, and conversely, without having to multiply out the matrices. If you don't already understand the correspondence between A and the expression $q(x_1, x_2, \ldots, x_n)$ (that is, how to obtain one from the other by inspection), take a general 3×3 symmetric matrix with (i, j) entry equal to a_{ij} and calculate explicitly what the product $\mathbf{x}^{\mathrm{T}} A \mathbf{x}$ equals, where $\mathbf{x} = (x_1, x_2, x_3)^{\mathrm{T}}$. You will find that the diagonal entries of A are the coefficients of the corresponding x_i^2 terms, and that the coefficient of the $x_i x_j$ term comes from the sum of the entries a_{ij} and a_{ji}, where $a_{ij} = a_{ji}$, since A is symmetric. Try the following activity.

Activity 11.25 As practice, write down an expression for $q(x, y, z) = \mathbf{x}^{\mathrm{T}} B \mathbf{x}$, where $\mathbf{x} = (x, y, z)^{\mathrm{T}}$ and B is the symmetric matrix

$$B = \begin{pmatrix} 3 & 2 & -1 \\ 2 & 7 & 4 \\ -1 & 4 & -5 \end{pmatrix}.$$

11.2.2   Definiteness of quadratic forms

Consider the quadratic form $q_1(x, y) = x^2 + y^2$. For any choices of x and y, $q_1(x, y) \geq 0$ and, furthermore, $q_1(x, y) = 0$ only when $x = y = 0$. On the other hand, the quadratic form $q_2(x, y) = x^2 + 3xy + y^2$ is not always non-negative: note, for example, that $q_2(1, -1) = -1 < 0$. An important general question we might ask (and one which has useful applications) is whether a quadratic form q is always positive (except when $x = y = 0$). Here, eigenvalue techniques help: specifically, we can use orthogonal diagonalisation. First, we need some terminology.

Definition 11.26 Suppose that $q(\mathbf{x})$ is a quadratic form. Then:

- $q(\mathbf{x})$ is *positive definite* if $q(\mathbf{x}) \geq 0$ for all $\mathbf{x}$, and $q(\mathbf{x}) = 0$ only when $\mathbf{x} = \mathbf{0}$, the zero-vector,
- $q(\mathbf{x})$ is *positive semi-definite* if $q(\mathbf{x}) \geq 0$ for all $\mathbf{x}$,

- $q(\mathbf{x})$ is *negative definite* if $q(\mathbf{x}) \leq 0$ for all $\mathbf{x}$, and $q(\mathbf{x}) = 0$ only when $\mathbf{x} = \mathbf{0}$, the zero-vector,
- $q(\mathbf{x})$ is *negative semi-definite* if $q(\mathbf{x}) \leq 0$ for all $\mathbf{x}$,
- $q(\mathbf{x})$ is *indefinite* if it is neither positive definite, nor positive semi-definite, nor negative definite, nor negative semi-definite; in other words, if there are $\mathbf{x}_1, \mathbf{x}_2$ such that $q(\mathbf{x}_1) < 0$ and $q(\mathbf{x}_2) > 0$.

Consider the quadratic form $q = \mathbf{x}^T A \mathbf{x}$, where A is symmetric, and suppose that we have found P that will orthogonally diagonalise A; that is, which is such that $P^T = P^{-1}$ and $P^T A P = D$, where D is a diagonal matrix. We make the (usual) change of variable as follows: define $\mathbf{z}$ by $\mathbf{x} = P\mathbf{z}$ (or, equivalently, $\mathbf{z} = P^{-1}\mathbf{x} = P^T\mathbf{x}$). P is the transition matrix from coordinates in the orthonormal basis of eigenvectors of A to standard coordinates. Then

$$q = \mathbf{x}^T A \mathbf{x} = (P\mathbf{z})^T A (P\mathbf{z}) = \mathbf{z}^T (P^T A P)\mathbf{z} = \mathbf{z}^T D \mathbf{z}.$$

Now, the entries of D must be the eigenvalues of A: let us suppose these are (in the order in which they appear in D) $\lambda_1, \lambda_2, \ldots, \lambda_n$. Let $\mathbf{z} = (z_1, z_2, \ldots, z_n)^T$ be the coordinates in the orthonormal basis of eigenvectors. Then

$$q = \mathbf{z}^T D \mathbf{z} = \lambda_1 z_1^2 + \lambda_2 z_2^2 + \cdots + \lambda_n z_n^2.$$

This is a linear combination of squares.

Now suppose that all the eigenvalues are positive. Then we can conclude that, for all $\mathbf{z}$, $q \geq 0$, and also that $q = 0$ only when $\mathbf{z}$ is the zero-vector. But because of the way in which $\mathbf{x}$ and $\mathbf{z}$ are related ($\mathbf{x} = P\mathbf{z}$ and $\mathbf{z} = P^T\mathbf{x}$), $\mathbf{x} = \mathbf{0}$ if and only if $\mathbf{z} = \mathbf{0}$. Therefore, if all the eigenvalues are positive, the quadratic form is positive definite. Conversely, assume the quadratic form is positive definite, so that $\mathbf{x}^T A \mathbf{x} > 0$ for all $\mathbf{x} \neq \mathbf{0}$. If $\mathbf{u}_i$ is a unit eigenvector corresponding to the eigenvalue λ_i, then

$$\mathbf{u}_i^T A \mathbf{u}_i = \mathbf{u}_i^T \lambda_i \mathbf{u}_i = \lambda_i \mathbf{u}_i^T \mathbf{u}_i = \lambda_i \|\mathbf{u}_i\|^2 = \lambda_i > 0.$$

So the eigenvalues of A are positive. Therefore, we have the first part of the following result. (The other parts arise from similar reasoning.)

Theorem 11.27 *Suppose that the quadratic form $q(\mathbf{x})$ has matrix representation $q(\mathbf{x}) = \mathbf{x}^T A \mathbf{x}$. Then:*

- *q is positive definite if and only if all eigenvalues of A are positive,*
- *q is positive semi-definite if and only if all eigenvalues of A are non-negative,*
- *q is negative definite if and only if all eigenvalues of A are negative,*

- *q is negative semi-definite if and only if all eigenvalues of A are non-positive,*
- *q is indefinite if and only if some eigenvalues of A are negative, and some are positive.*

Activity 11.28 Assume an $n \times n$ matrix A has 0 as one of its eigenvalues and that all other eigenvalues are non-negative. Show that $q = \mathbf{x}^\mathsf{T} A \mathbf{x}$ is positive semi-definite but *not* positive definite.

We say that a symmetric matrix A is *positive definite* if the corresponding quadratic form $q = \mathbf{x}^\mathsf{T} A \mathbf{x}$ is positive definite (and, similarly, we speak of negative definite, positive semi-definite, negative semi-definite, and indefinite matrices).

As a consequence of Theorem 11.27, in order to establish if a matrix is positive definite or negative definite, we only need to know the signs of the eigenvalues and not their values. It is possible to obtain this information directly from the matrix A.

We first examine the case where A is a symmetric 2×2 matrix,

$$A = \begin{pmatrix} a & c \\ c & b \end{pmatrix}.$$

Let λ_1 and λ_2 be the eigenvalues of a matrix A whose characteristic equation is

$$|A - \lambda I| = \begin{vmatrix} a - \lambda & c \\ c & b - \lambda \end{vmatrix} = \lambda^2 - (a + b)\lambda + (ab - c^2) = 0.$$

Since the eigenvalues λ_1 and λ_2 are the roots of the characteristic equation, we have

$$|A - \lambda I| = (\lambda - \lambda_1)(\lambda - \lambda_2) = \lambda^2 - (\lambda_1 + \lambda_2)\lambda + \lambda_1\lambda_2 = 0.$$

Comparing terms of these two polynomials in λ, we have

$$\lambda_1\lambda_2 = ab - c^2 \qquad \text{and} \qquad \lambda_1 + \lambda_2 = a + b.$$

These observations are, in fact, simply special cases of Theorem 8.11 and Theorem 8.15: namely, that the determinant of A is the product of the eigenvalues (explicitly, in this case, $ab - c^2 = \lambda_1\lambda_2$); and that the trace of A is the sum of the eigenvalues ($a + b = \lambda_1 + \lambda_2$).

If $|A| = ab - c^2 > 0$, then both eigenvalues λ_1, λ_2 have the same sign (since their product, which is equal to $|A|$, is positive). Since also a and b must have the same sign in this case (since $ab > c^2 \geq 0$), we can deduce the signs of the eigenvalues from the sign of a.

Consider the following example:

Example 11.29 Let A be the following matrix:

$$A = \begin{pmatrix} 9 & -2 \\ -2 & 6 \end{pmatrix}.$$

Then (because it is symmetric) A has real eigenvalues λ_1, λ_2 which must satisfy the following equations:

$$\lambda_1\lambda_2 = |A| = 9(6) - (-2)(-2) = 50, \quad \text{and} \quad \lambda_1 + \lambda_2 = 9 + 6 = 15.$$

Since $\lambda_1\lambda_2 > 0$, the eigenvalues are non-zero and have the same sign. Since $\lambda_1 + \lambda_2 > 0$, both must be positive. Therefore, the matrix A is positive definite. (In fact, the eigenvalues are 5 and 10, but we do not need to do the extra work of finding them explicitly if we only want to know about their signs.)

In a similar way, if $|A| = ab - c^2 < 0$, then the eigenvalues have opposite signs and the form is therefore indefinite. So we can conclude:

- If $|A| > 0$ and $a > 0$, then $\lambda_1 > 0$, $\lambda_2 > 0$ and A is positive definite.
- If $|A| > 0$ and $a < 0$, then $\lambda_1 < 0$, $\lambda_2 < 0$ and A is negative definite.
- If $|A| < 0$, then λ_1 and λ_2 have opposite signs and A is indefinite.

If $|A| = 0$, we conclude that one of the eigenvalues is 0.

This kind of test on the matrix can be generalised to an $n \times n$ symmetric matrix A. But first we need a definition.

Definition 11.30 If A is an $n \times n$ matrix, the *principal minors* of A are the n determinants formed from the first r rows and the first r columns of A, for $r = 1, 2, \ldots, n$; that is,

$$a_{11}, \quad \begin{vmatrix} a_{11} & a_{12} \\ a_{21} & a_{22} \end{vmatrix}, \quad \begin{vmatrix} a_{11} & a_{12} & a_{13} \\ a_{21} & a_{22} & a_{23} \\ a_{31} & a_{32} & a_{33} \end{vmatrix}, \quad \ldots, \quad \begin{vmatrix} a_{11} & a_{12} & \cdots & a_{1n} \\ a_{21} & a_{22} & \cdots & a_{2n} \\ \vdots & \vdots & \ddots & \vdots \\ a_{n1} & a_{n2} & \cdots & a_{nn} \end{vmatrix}.$$

Notice that the $n \times n$ principal minor is just the determinant of A. If, for example,

$$A = \begin{pmatrix} 5 & 2 & 1 \\ 2 & 10 & -3 \\ 1 & -3 & 2 \end{pmatrix},$$

then the principal minors are

$$a_{11} = 5, \quad \begin{vmatrix} 5 & 2 \\ 2 & 10 \end{vmatrix} = 46, \quad |A| = 25.$$

Activity 11.31 Check these determinant calculations.

Notice that all three principal minors are positive. In fact, this is enough to show that A is positive definite, as stated in the following result.

Theorem 11.32 *Suppose that A is an $n \times n$ symmetric matrix. Then A is positive definite if and only if all its principal minors are positive.*

We will prove this result in the next section. For now, let's assume it is true and look at some of the consequences.

Theorem 11.32 gives us a test to see if a matrix is positive definite. What about the other possibilities?

A matrix A is negative definite if and only if its negative, $-A$, is positive definite. (You can see this by noting that the quadratic form determined by $-A$ is the negative of that determined by A.) Now, if A_r is any $r \times r$ matrix, then $|-A_r| = (-1)^r |A|$.

Activity 11.33 Show this using properties of the determinant.

So if r is even, the $r \times r$ principal minor of A (the principal minor of *order r*) and that of $-A$ have the same sign, and if r is odd, they have opposite signs. If $-A$ is positive definite, Theorem 11.32 tells us that all of its principal minors are positive. So we have the following characterisation of a negative definite matrix.

Theorem 11.34 *Suppose that A is an $n \times n$ symmetric matrix. Then A is negative definite if and only if its principal minors of even order are positive and its principal minors of odd order are negative.*

Another way of stating this is: the symmetric $n \times n$ matrix A is negative definite if and only if its principal minors alternate in sign, with the first one negative.

Activity 11.35 Convince yourself that these two statements are equivalent.

If A is an $n \times n$ symmetric matrix which is neither positive nor negative definite, and if $|A| \neq 0$, then A is indefinite because we can conclude that A has both positive and negative eigenvalues. If $|A| = 0$, the only thing we can conclude is that one of the eigenvalues is 0. These statements follow from Theorem 8.11, which states that if A has eigenvalues $\lambda_1, \lambda_2, \cdots, \lambda_n$, then

$$|A| = \lambda_1 \lambda_2 \cdots \lambda_n.$$

Activity 11.36 Explain why Theorem 8.11 establishes the following: if A is an $n \times n$ symmetric matrix and $|A| = 0$, then one of the eigenvalues is 0; and if A is neither positive nor negative definite and $|A| \neq 0$, then A has both positive and negative eigenvalues and is therefore indefinite.

It should be noted that there is no test quite as simple as those of Theorem 11.32 and Theorem 11.34 to check whether a matrix is positive or negative *semi*-definite. Consider the following example.

Example 11.37 Let A be the matrix

$$A = \begin{pmatrix} 1 & 1 & 0 \\ 1 & 1 & 0 \\ 0 & 0 & t \end{pmatrix}.$$

Solving $|A - \lambda I| = 0$, we find that the eigenvalues are 0, 2, t. The principal minors of A are

$$a_{11} = 1, \qquad \begin{vmatrix} 1 & 1 \\ 1 & 1 \end{vmatrix} = 0, \qquad |A| = 0.$$

But t can be any real number, either positive or negative. So in this case the principal minors are no indication of the signs of the eigenvalues.

11.2.3   The characterisation of positive-definiteness

Before we embark on this proof, there are two general results which we will need, so we state them and prove them now. The first is really an observation:

If D is an $n \times n$ diagonal matrix with positive entries on the diagonal, then D is positive definite.

Activity 11.38 Prove this.

The second we will state as a theorem.

Theorem 11.39 *If A and B are any $n \times n$ symmetric matrices such that $E A E^T = B$ for an invertible matrix E, then A is positive definite if and only if B is positive definite.*

Proof: To see this, assume B is positive definite. Let $\mathbf{x} \in \mathbb{R}^n$ and let $\mathbf{y} = (E^T)^{-1}\mathbf{x}$ (or, equivalently, $\mathbf{x} = E^T\mathbf{y}$). Then $\mathbf{x} = \mathbf{0}$ if and only if $\mathbf{y} = \mathbf{0}$. Since B is positive definite, we have, for all $\mathbf{x} \neq \mathbf{0}$,

$$\mathbf{x}^T A \mathbf{x} = (E^T\mathbf{y})^T A (E^T\mathbf{y}) = \mathbf{y}^T E A E^T \mathbf{y} = \mathbf{y}^T B \mathbf{y} > 0,$$

so A is also positive definite. The converse follows immediately by noting that

$$A = E^{-1}B(E^T)^{-1} = E^{-1}B(E^{-1})^T = FBF^T, \quad \text{where} \quad F = E^{-1},$$

so if A is positive definite then so is B.   □

Proof of Theorem 11.32 To prove this, we (once again) need to use an inductive proof on the size n of an $n \times n$ symmetric matrix A. We have already shown that the result is true for a 2×2 matrix by looking at the trace and the determinant of the matrix. We will prove this again in a different way so that we can extend the proof to $n \times n$ matrices. You can safely omit this proof. However, we include it for completeness and because it uses ideas from earlier chapters, namely row operations and elementary matrices.

We want to show that a symmetric matrix A is positive definite if and only if all its principal minors are positive.

We will first prove the difficult part of the 'if and only if' statement: assuming that all the principal minors of the matrix A are positive, we will show that A is positive definite. First, we do this for a 2×2 matrix, and then we show how this implies the statement for a 3×3 matrix. After that, assuming that the result is true for $(n-1) \times (n-1)$ matrices, it is not difficult to show it is true for $n \times n$ matrices. The main idea in this proof is to use carefully chosen row (and column) operations to diagonalise the matrix A.

Let A be a 2×2 symmetric matrix,

$$A = \begin{pmatrix} a & c \\ c & b \end{pmatrix}, \qquad a, b, c \in \mathbb{R}$$

with positive principal minors. Then $a > 0$ and $|A| > 0$. We perform the following row operation on A by multiplying it on the left by an elementary matrix:

$$EA = \begin{pmatrix} 1 & 0 \\ -(c/a) & 1 \end{pmatrix} \begin{pmatrix} a & c \\ c & b \end{pmatrix} = \begin{pmatrix} a & c \\ 0 & b - (c^2/a) \end{pmatrix} = \begin{pmatrix} a & c \\ 0 & |A|/a \end{pmatrix}.$$

We then perform the analogous column operation on this, by multiplying on the right by the matrix E^{T}:

$$EAE^{\mathrm{T}} = \begin{pmatrix} a & c \\ 0 & |A|/a \end{pmatrix} \begin{pmatrix} 1 & -(c/a) \\ 0 & 1 \end{pmatrix} = \begin{pmatrix} a & 0 \\ 0 & |A|/a \end{pmatrix} = D.$$

It turns out that the diagonal matrix $EAE^{\mathrm{T}} = D$ has the same principal minors as the matrix A. For the first principal minor of D (which is just the $(1, 1)$ entry) is equal to a, since this was unchanged; and the 2×2 principal minor of D is $|D|$, where

$$|D| = |EAE^{\mathrm{T}}| = |E||A||E^{\mathrm{T}}| = |A|, \quad \text{since} \quad |E| = |E^{\mathrm{T}}| = 1.$$

Note that, as a consequence of our method, the diagonal entries of $D = (d_{ij})$ are

$$d_{11} = a \quad \text{and} \quad d_{22} = \frac{|A|}{a}.$$

So $D = EAE^{\mathrm{T}}$ is a diagonal matrix with all its diagonal entries positive. Therefore, D is positive definite, and by Theorem 11.39, A is positive definite.

In order to continue, we introduce some notation. If A is an $n \times n$ matrix, let $A_{r \times r}$ denote the $r \times r$ matrix consisting of the first r rows and r columns of A. Then the principal minors of A are

$$a_{11} = |A_{1 \times 1}|, \ |A_{2 \times 2}|, \ |A_{3 \times 3}|, \ \ldots, \ |A_{n \times n}| = |A|.$$

The idea of this proof is to reduce the $n \times n$ matrix A to a diagonal matrix in the same way as we did for the 2×2 matrix, using only row operations which add a multiple of one row to another (and also by using corresponding column operations). An elementary matrix which corresponds to this type of row operation is, of course, invertible, and, most importantly, it has determinant equal to 1.

Let E_{i1} denote the elementary matrix that performs the row operation: 'row $i - (a_{i1}/a_{11})$ row 1', where the size of this elementary matrix will depend on the size of the matrix on which the row operation is being performed. For example, if A is a 2×2 matrix, then E_{21} is just the matrix E we used above.

If A is a 3×3 matrix, then, for instance, $E_{21}A$ is

$$\begin{pmatrix} 1 & 0 & 0 \\ -(a_{21}/a_{11}) & 1 & 0 \\ 0 & 0 & 1 \end{pmatrix} \begin{pmatrix} a_{11} & a_{12} & a_{13} \\ a_{21} & a_{22} & a_{23} \\ a_{31} & a_{32} & a_{33} \end{pmatrix}$$

$$= \begin{pmatrix} a_{11} & a_{12} & a_{13} \\ 0 & a_{22} - (a_{21}a_{12}/a_{11}) & * \\ a_{31} & a_{32} & a_{33} \end{pmatrix},$$

where we have written $*$ to indicate the $(2, 3)$ entry in $E_{21}A$. Notice that the $(2, 2)$ entry of $E_{21}A$ is already equal to the determinant of the 2×2 principal minor of A divided by the entry a_{11}, so the 2×2 principal minors of the matrices $E_{21}A$ and A are the same.

We now show how the 2×2 result implies that the theorem is also true for a 3×3 matrix. We apply the elementary row and column operations (as indicated above) to A to reduce the 2×2 principal minor to diagonal form:

$$E_{21}AE_{21}^{\mathrm{T}} = \begin{pmatrix} a_{11} & 0 & * \\ 0 & d_{22} & * \\ * & * & * \end{pmatrix},$$

so that the first two principal minors of the matrix A and the matrix $E_{21}AE_{21}^{\mathrm{T}}$ are equal. Then $d_{22} = |A_{2 \times 2}|/a_{11}$. We now continue reducing the matrix to diagonal form using the same type of elementary row operations (adding a multiple of one row to another) and the corresponding column operations. All of the elementary matrices which perform this type of operation have determinant equal to 1. We have

$$E_{32} E_{31} E_{21} A E_{21}^T E_{31}^T E_{32}^T = \begin{pmatrix} a_{11} & 0 & 0 \\ 0 & d_{22} & 0 \\ 0 & 0 & d_{33} \end{pmatrix} = D,$$

where E_{32} is the elementary matrix that performs the (obvious) row operation needed to complete the diagonalisation. Now we already know that the first two principal minors of these matrices are equal, and since

$$|D| = |E_{32} E_{31} E_{21} A E_{21}^T E_{31}^T E_{32}^T|$$
$$= |E_{32}||E_{31}||E_{21}||A||E_{21}^T||E_{31}^T||E_{32}^T| = |A|,$$

all the principal minors are equal.

In addition, since each principal minor of D is just the product of the diagonal entries, we can deduce that the entries of D are

$$d_{11} = a_{11}, \quad d_{22} = \frac{|A_{2\times2}|}{a_{11}}, \quad d_{33} = \frac{|A_{3\times3}|}{|A_{2\times2}|}.$$

For we know from above that $d_{11} = a_{11}$ and $d_{22} = |A_{2\times2}|/a_{11}$; and the fact that d_{33} takes the value indicated then follows directly from the observation that

$$|D| = d_{11} d_{22} d_{33} = |A| = |A_{3\times3}|.$$

Since the diagonal entries of D are positive, we conclude as earlier that D is positive definite, and therefore A is positive definite.

We are now ready to consider an $n \times n$ symmetric matrix A, assuming the result is true for any $(n-1) \times (n-1)$ matrix. We apply the elementary row and column operations to A to reduce the $(n-1) \times (n-1)$ principal minor to diagonal form, and then continue to reduce the matrix A so that we obtain a matrix

$$E A E^T = \mathrm{diag}(d_{11}, d_{22}, \dots, d_{nn}),$$

where we used E to denote the product of the elementary matrices which achieve this diagonalisation. All of these elementary matrices have determinant equal to 1, and therefore so does E. The method ensures (by the underlying assumption about the $(n-1) \times (n-1)$ case) that the first $n-1$ principal minors of the matrices A and D are the same. It only remains to show that $|A| = |D|$, which follows immediately from

$$|D| = |E A E^T| = |E||A||E^T| = |A|.$$

Therefore, using the same arguments as earlier, D is a diagonal matrix with positive entries along the diagonal. Therefore D is positive definite, and so therefore is A.

To complete the proof, we show that *if* the quadratic form is positive definite, *then* all the principal minors are positive. Recall that the principal minors of A are

$$a_{11} = |A_{1\times 1}|, \ |A_{2\times 2}|, \ |A_{3\times 3}|, \ \ldots, \ |A_{n\times n}| = |A|,$$

where $A_{r\times r}$ denotes the $r \times r$ matrix consisting of the first r rows and r columns of A. We will prove that $A_{r\times r}$ is a positive definite $r \times r$ matrix for $r = 1, 2, \ldots, n-1$. We already know this to be the case for $r = n$. It will then follow, by Theorem 11.27, that the eigenvalues of $A_{r\times r}$ are all positive. Then Theorem 8.11 will tell us that $|A_{r\times r}|$ is the product of these positive eigenvalues, and is therefore positive. So, let's show that $A_{r\times r}$ is positive definite, using the fact that A is.

We know that, for all $\mathbf{y} \in \mathbb{R}^n$, $\mathbf{y}^T A \mathbf{y} > 0$, unless $\mathbf{y} = \mathbf{0}$. Fix r, a number between 1 and $n-1$. Let $\mathbf{x} = (x_1, x_2, \ldots, x_r)^T \in \mathbb{R}^r$ and let $\mathbf{x}_r = (x_1, x_2, \ldots, x_r, 0, 0, \ldots, 0)^T \in \mathbb{R}^n$ be the n-vector with first r entries the same as $\mathbf{x}$, and all other entries zero.

Suppose $\mathbf{x}$ is a non-zero vector. Then so is $\mathbf{x}_r$ and we must have $\mathbf{x}_r^T A \mathbf{x}_r > 0$. But

$$\mathbf{x}_r^T A \mathbf{x}_r = (x_1 \ \cdots \ x_r \ 0 \ \cdots \ 0) \begin{pmatrix} a_{11} & a_{12} & \cdots & a_{1r} & \cdots \\ a_{21} & a_{22} & \cdots & a_{2r} & \cdots \\ a_{31} & a_{32} & \cdots & a_{3r} & \cdots \\ \vdots & \vdots & \ddots & \vdots & \cdots \\ a_{r1} & a_{\kappa 2} & \cdots & a_{rr} & \cdots \\ \vdots & \vdots & & \vdots & \ddots \end{pmatrix} \begin{pmatrix} x_1 \\ \vdots \\ x_r \\ 0 \\ \vdots \\ 0 \end{pmatrix}.$$

Think about how this product evaluates. It is a 1×1 matrix. Because of the zero entries in $\mathbf{x}_r$, and because of the way in which matrix multiplication works, we have, for $\mathbf{x} = (x_1, x_2, \ldots, x_r)^T \in \mathbb{R}^r$,

$$\mathbf{x}_r^T A \mathbf{x}_r = \mathbf{x}^T A_{r\times r} \mathbf{x}.$$

So, since A is positive definite, for all $\mathbf{x} \in \mathbb{R}^r$, with $\mathbf{x} \neq \mathbf{0}$

$$\mathbf{x}^T A_{r\times r} \mathbf{x} = \mathbf{x}_r^T A \mathbf{x}_r > 0.$$

So, indeed, $A_{r\times r}$ is positive definite, as required.   □

Figure 11.1 The graphs of (left) $x^2 - 2y^2 = 2$ and (right) $y^2 - 2x^2 = 2$

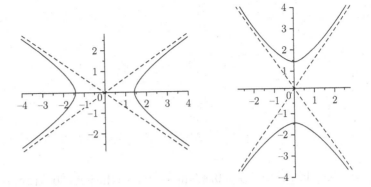

11.2.4   Quadratic forms in $\mathbb{R}^2$: conic sections

Conic sections are traditionally described as curves which can be obtained as the intersection of a plane and a double cone, such as a circle, ellipse, parabola or hyperbola. It is more common, however, to think of them as defined by certain types of equation and here there is a very useful link to quadratic forms. The technique of orthogonal diagonalisation enables us to determine what type of conic section we have, and to sketch them accurately.

If A is a 2×2 symmetric matrix, the equation $\mathbf{x}^T A\mathbf{x} = k$, where k is a constant, represents a curve whose graph in the xy-plane is a conic section. For example, the equation

$$\frac{x^2}{a^2} + \frac{y^2}{b^2} = 1$$

represents an ellipse which intersects the x axis at $(-a, 0)$ and $(a, 0)$, and intersects the y axis at $(0, b)$ and $(0, -b)$. If $a = b$, this is a circle of radius a. These curves are said to be in *standard position* relative to the coordinate axes (meaning that their axes of symmetry are the x axis and the y axis), as are the two hyperbolas whose graphs are shown in Figure 11.1.

The graphs of each of the hyperbolas $x^2 - 2y^2 = 2$ and $y^2 - 2x^2 = 2$ are shown in this figure, together with the two straight lines which are the *asymptotes* of the hyperbola. From each equation, we see that if x is large, then y must also be large, so that the difference in the squared terms remains constant. For example, for the first hyperbola, $x^2 - 2y^2 = 2$, the asymptotes can be easily found by rewriting this equation as

$$\frac{y^2}{x^2} = \frac{1}{2} - \frac{1}{x^2}.$$

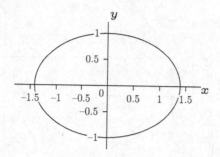

Figure 11.2 The ellipse $x^2 + 2y^2 = 2$

As x gets very large, $1/x^2 \to 0$, so the points on the hyperbola approach the lines given by $y^2 = x^2/2$. The asymptotes are therefore

$$y = \frac{1}{\sqrt{2}}x \qquad \text{and} \qquad y = -\frac{1}{\sqrt{2}}x.$$

Activity 11.40 Find the equations of the asymptotes for the hyperbola $y^2 - 2x^2 = 2$.

On the other hand, in the equation of an ellipse, such as

$$x^2 + 2y^2 = 2,$$

the values of x and y are constrained by the equation: the largest value that x can obtain is when $y = 0$.

If A is a diagonal matrix, so that the equation $\mathbf{x}^T A\mathbf{x} = k$ has no xy term, then this equation represents a conic section in standard position and it is straightforward to sketch its graph. For example, if $A = \text{diag}(1, -2)$, then the graph of $\mathbf{x}^T A\mathbf{x} = x^2 - 2y^2 = 2$ is the hyperbola shown on the left in Figure 11.1, whereas if $A = \text{diag}(1, 2)$, then the graph of $\mathbf{x}^T A\mathbf{x} = x^2 + 2y^2 = 2$ is an ellipse which intersects the x axis at $x = \pm\sqrt{2}$ and the y axis at $y = \pm 1$. This is shown in Figure 11.2.

But how do we sketch the graph if A is not diagonal? To achieve this, we can use orthogonal diagonalisation. We illustrate the method using the following example (which you have seen before as Example 7.40).

Example 11.41 Consider the curve $\mathcal{C}$ with equation

$$5x^2 - 6xy + 5y^2 = 2.$$

In matrix form, this equation is

$$\mathbf{x}^T A\mathbf{x} = (x \quad y)\begin{pmatrix} 5 & -3 \\ -3 & 5 \end{pmatrix}\begin{pmatrix} x \\ y \end{pmatrix} = 2,$$

where A is symmetric. We orthogonally diagonalised the matrix A in Example 11.4. We found that the eigenvalues are $\lambda_1 = 2$ and $\lambda_2 = 8$, with corresponding eigenvectors

$$\mathbf{v}_1 = \begin{pmatrix} 1 \\ 1 \end{pmatrix}, \qquad \mathbf{v}_2 = \begin{pmatrix} -1 \\ 1 \end{pmatrix}.$$

Now, if P is the matrix

$$P = \begin{pmatrix} \frac{1}{\sqrt{2}} & -\frac{1}{\sqrt{2}} \\ \frac{1}{\sqrt{2}} & \frac{1}{\sqrt{2}} \end{pmatrix},$$

then

$$P^{-1}AP = P^{\mathrm{T}}AP = D = \mathrm{diag}(2, 8).$$

Activity 11.42 Check these calculations.

We set $\mathbf{x} = P\mathbf{z}$, and interpret P both as a change of basis *and* as a linear transformation. The matrix P is the transition matrix from the (orthonormal) basis B of eigenvectors, $B = \{\mathbf{v}_1, \mathbf{v}_2\}$, to the (orthonormal) standard basis. If we let $\mathbf{z} = (X, Y)^{\mathrm{T}}$, then as we saw earlier (Section 11.2.2), the quadratic form – and hence the curve C – can be expressed in the coordinates of the basis of eigenvectors as

$$\mathbf{x}^{\mathrm{T}}A\mathbf{x} = \mathbf{z}^{\mathrm{T}}D\mathbf{z} = 2X^2 + 8Y^2 = 2;$$

that is, as $X^2 + 4Y^2 = 1$. This is an ellipse in standard position with respect to the X and Y axes.

But how do we sketch this? We first need to find the positions of the X and Y axes in our xy-plane. If we think of P as defining a linear transformation T which maps $\mathbb{R}^2$ onto itself, with $T(\mathbf{x}) = P\mathbf{x}$, then the X and Y axes are the images of the x and y axes under the linear transformation T. Why? The positive x axis is described as all positive multiples of the vector $\mathbf{e}_1 = (1, 0)^{\mathrm{T}}$, and the positive y axis as all positive multiples of $\mathbf{e}_2 = (0, 1)^{\mathrm{T}}$. The images of these vectors are

$$T(\mathbf{e}_1) = \mathbf{v}_1 \quad \text{and} \quad T(\mathbf{e}_2) = \mathbf{v}_2.$$

Analogous descriptions of the X and Y axes are that the positive X axis is described as all positive multiples of the vector $[1, 0]_B$ and the positive Y axis as positive multiples of $[0, 1]_B$. But these are just the coordinates in the basis B of the vectors $\mathbf{v}_1$ and $\mathbf{v}_2$, respectively.

This allows us to draw the new X and Y axes in the xy-plane as the lines in the directions of the vectors $\mathbf{v}_1$ and $\mathbf{v}_2$.

In this example, the new X, Y axes are a rotation (anticlockwise) of the old x, y axes by $\pi/4$ radians. We looked at rotations in Section 7.1.3, where we showed that the matrix representing a rotation anticlockwise by an angle θ is given by

$$\begin{pmatrix} \cos\theta & -\sin\theta \\ \sin\theta & \cos\theta \end{pmatrix}.$$

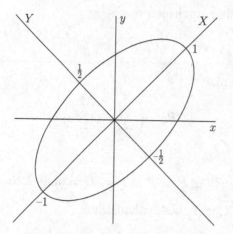

Figure 11.3 The ellipse $5x^2 - 6xy + 5y^2 = 2$

In fact, in this example we carefully chose the column positions of the eigenvectors so that P would define a rotation anticlockwise, and it is always possible to do so. Why is that? Well, an orthonormal basis of $\mathbb{R}^2$ consists of two unit vectors which are orthogonal. Suppose one of the vectors is $\mathbf{u}_1 = (u_1, u_2)^T$ with $u_1 > 0$ and $u_2 > 0$, then the other vector must be either $\mathbf{u}_2 = (-u_2, u_1)^T$ or $-\mathbf{u}_2$, and we can choose $\mathbf{u}_2$. If

$$P = \begin{pmatrix} u_1 & -u_2 \\ u_2 & u_1 \end{pmatrix},$$

then P is the matrix of a rotation anticlockwise, since it is possible to find an angle θ such that $\cos\theta = u_1$ and $\sin\theta = u_2$.

Activity 11.43 Think about why these two assertions are true: why is $\mathbf{u}_2$ (or $-\mathbf{u}_2$) the second vector in the orthonormal basis, and why is it possible to find such an an angle θ?

By choosing to write the unit eigenvectors as the columns of P in this way, it is easy to find the positions of the new axes because we can recognise the linear transformation as a rotation anticlockwise. However, *any* choice of P would still enable us to sketch this graph (see Exercise 11.8).

Continuing our example, we are now in a position to sketch the graph of $\mathcal{C}$ in the xy-plane. First draw the usual x and y axes. The positive X axis is in the direction of the vector $(1, 1)^T$ and the positive Y axis is along the direction of the vector $(-1, 1)^T$. These new X, Y axes are a rotation of $\pi/4$ radians anticlockwise of the old x, y axes. So we draw the X and Y axes along the lines $y = x$ and $y = -x$. We now sketch the ellipse $X^2 + 4Y^2 = 1$ in standard position with respect to the X and Y axes. It intersects the X axis at $X = \pm 1$ and the Y axis at $Y = \pm 1/2$. See Figure 11.3.

Activity 11.44 Where does the curve $\mathcal{C}$ intersect the x and y axes?

You should be asking another question about this method. How do we know that the linear transformation defined by P did not change the shape of the curve?

It turns out that a linear transformation given by an orthogonal matrix P, $P(\mathbf{x}) = P\mathbf{x}$, is a *rigid* motion of the plane: nothing changes shape because the linear transformation preserves both lengths and angles. Such a linear transformation is called an *isometry*. In order to prove this assertion, note that both the length of a vector $\mathbf{v}$ and the angle between two vectors $\mathbf{v}$, $\mathbf{w}$ are defined in $\mathbb{R}^n$ by the inner product,

$$\|\mathbf{v}\| = \sqrt{\langle \mathbf{v}, \mathbf{v} \rangle} \qquad \cos\theta = \frac{\langle \mathbf{v}, \mathbf{w} \rangle}{\|\mathbf{v}\|\,\|\mathbf{w}\|}.$$

So we only need to show that P preserves the inner product. We have the following general result:

Theorem 11.45 *The linear transformation defined by an orthogonal matrix P preserves the standard inner product on $\mathbb{R}^n$.*

Proof: If the linear transformation defined by P is denoted by $T : \mathbb{R}^n \to \mathbb{R}^n$, then $T(\mathbf{x}) = P\mathbf{x}$. Let $\mathbf{v}, \mathbf{w} \in \mathbb{R}^n$. Then, taking the inner product of the images, we have

$$\langle P\mathbf{v}, P\mathbf{w} \rangle = (P\mathbf{v})^{\mathrm{T}}(P\mathbf{w}) = \mathbf{v}^{\mathrm{T}} P^{\mathrm{T}} P \mathbf{w} = \mathbf{v}^{\mathrm{T}} \mathbf{w} = \langle \mathbf{v}, \mathbf{w} \rangle.$$

The inner product between two vectors is equal to the inner product between their images under P. $\qquad\qquad\square$

Therefore, length and angle are preserved by such a linear transformation, and hence so also is the shape of any curve. This validates our method.

11.3 Learning outcomes

You should now be able to:

- know what is meant by orthogonal diagonalisation
- explain why an $n \times n$ matrix can be orthogonally diagonalised if and only if it possesses an orthonormal set of n eigenvectors
- orthogonally diagonalise a symmetric matrix and know that only symmetric matrices can be orthogonally diagonalised
- know what is meant by a quadratic form and what it means to say that a quadratic form or a symmetric matrix is positive definite, positive semi-definite, negative definite, negative semi-definite and indefinite; and be able to determine which of these is the case
- use orthogonal diagonalisation to analyse conic sections.

11.4   Comments on activities

Activity 11.11 Check that the inner product of any two vectors is equal to 0.

Activity 11.13 You should obtain that $P^T P = I$.

Activity 11.16 The reduced row echelon form shows that a basis of the row space of $B - I$ is given by the vector $\mathbf{v}_3 = (1, 1, 1)^T$. Since the row space and the null space of a matrix are orthogonal, this vector $\mathbf{v}_3$ is orthogonal to every vector in the null space of $B - I$, which is the eigenspace of $\lambda = 1$. Therefore, $\mathbf{v}_3$ must be an eigenvector for the third eigenvalue, $\lambda = 4$, and this can be easily checked by finding $B\mathbf{v}_3$.

Activity 11.17 You should check that $\mathbf{u}_2 \perp \mathbf{u}_1$. You should also check that $\mathbf{u}_2 \perp \mathbf{v}_3$ to show that it is in the eigenspace for $\lambda = 1$.

Activity 11.18 Let A be a 2×2 symmetric matrix. Then by Theorem 11.7, A has real eigenvalues. Let λ_1 be an eigenvalue of A, and let $\mathbf{v}_1$ by a corresponding unit eigenvector, so $A\mathbf{v}_1 = \lambda_1 \mathbf{v}_1$ and $\|\mathbf{v}_1\| = 1$. Extend $\{\mathbf{v}_1\}$ to a basis $\{\mathbf{v}_1, \mathbf{x}_2\}$ of $\mathbb{R}^2$, then using Gram–Schmidt (starting with $\mathbf{v}_1$ which is a unit vector) make this into an orthonormal basis $B = \{\mathbf{v}_1, \mathbf{v}_2\}$ of $\mathbb{R}^2$. Let P be the matrix whose columns are the vectors in B. Then P is the transition matrix from B coordinates to standard coordinates, and P is orthogonal by Theorem 10.18. By Theorem 7.37, the matrix $P^{-1}AP = P^T AP$ represents the linear transformation $T(\mathbf{x}) = A\mathbf{x}$ in the basis B. By Theorem 7.36, the first column of $P^T AP$ will be the coordinate vector of $T(\mathbf{v}_1)$ with respect to the basis B. Now, $T(\mathbf{v}_1) = A\mathbf{v}_1 = \lambda_1 \mathbf{v}_1$, so this coordinate vector is

$$\begin{bmatrix} \lambda_1 \\ 0 \end{bmatrix}_B.$$

Then

$$P^T AP = \begin{pmatrix} \lambda_1 & a_1 \\ 0 & a_2 \end{pmatrix}.$$

But the matrix $P^T AP$ is symmetric, since

$$(P^T AP)^T = P^T A^T P = P^T AP,$$

so it must be of the form

$$P^T AP = \begin{pmatrix} \lambda_1 & 0 \\ 0 & a_2 \end{pmatrix} = D.$$

Therefore, A can be orthogonally diagonalised.

Activity 11.20 For example, let

$$B = \begin{pmatrix} a & 2c \\ 0 & b \end{pmatrix}.$$

Activity 11.25 $q(x, y, z) = 3x^2 + 7y^2 - 5z^2 + 4xy - 2xz + 8yz.$

Activity 11.28 Let the eigenvalues of A be $\lambda_1 = 0, \lambda_2, \ldots, \lambda_n, \lambda_i \geq 0$, and let $\mathbf{v}_1$ be an eigenvector corresponding to $\lambda_1 = 0$. Then $\mathbf{v}_1^T A \mathbf{v}_1 = 0$ so $\mathbf{x}^T A \mathbf{x}$ is not positive definite. But

$$\mathbf{x}^T A \mathbf{x} = \mathbf{z}^T D \mathbf{z} = \lambda_2 z_2^2 + \cdots + \lambda_r z_n^2 \geq 0,$$

since it is the sum of non-negative numbers, so $\mathbf{x}^T A \mathbf{x}$ is positive semi-definite.

Activity 11.33 If a row (or column) of a matrix A is multiplied by a constant, then the determinant of A is multiplied by that constant. Since the rth principal minor has r rows and each is multiplied by -1 to form $|-A_r|$, we have $|-A_r| = (-1)^r |A_r|$.

Activity 11.36 We have $|A| = \lambda_1 \lambda_2 \cdots \lambda_n$. If $|A| = 0$, one of these factors (that is, one of the eigenvalues) must be 0. On the other hand, if $|A| \neq 0$, then none of these factors is 0. If A is neither positive nor negative definite, then A must have some positive eigenvalues and some negative eigenvalues, so A must be indefinite.

Activity 11.38 Let $\mathbf{y} \in \mathbb{R}^n$. Then $\mathbf{y}^T D \mathbf{y}$ is a sum of squares with positive coefficients, therefore $\mathbf{y}^T D \mathbf{y} > 0$ for all $\mathbf{y} \neq \mathbf{0}$, and $\mathbf{y}^T D \mathbf{y} = 0$ if and only if $\mathbf{y} = \mathbf{0}$.

Activity 11.40 These are $y = \pm\sqrt{2}x$.

Activity 11.43 Certainly $\langle \mathbf{u}_1, \mathbf{u}_2 \rangle = 0$, and each is a unit vector, so $\{\mathbf{u}_1, \mathbf{u}_2\}$ is an orthonormal basis of $\mathbb{R}^2$. (You can show that any vector $\mathbf{a} = (a, b)^T$ such that $u_1 a + u_2 b = 0$ must be a scalar multiple of $\mathbf{u}_2$ and there are only two such unit vectors, namely $\pm \mathbf{u}_2$.)

The angle θ is defined by the two equations, $\cos\theta = u_1$ and $\sin\theta = u_2$. This works since

$$\cos^2\theta + \sin^2\theta = u_1^2 + u_2^2 = 1.$$

Activity 11.44 These points are found using the original equation, $5x^2 - 6xy + 5y^2 = 2$. If $x = 0$, $y = \pm\sqrt{2/5}$, and, similarly, if $y = 0$, $x = \pm\sqrt{2/5}$.

11.5   Exercises

Exercise 11.1 Orthogonally diagonalise the matrix

$$A = \begin{pmatrix} 7 & 0 & 9 \\ 0 & 2 & 0 \\ 9 & 0 & 7 \end{pmatrix}.$$

Why do you know this can be done before even attempting to do it?

Exercise 11.2 Let

$$A = \begin{pmatrix} 2 & 1 & -2 \\ 1 & 2 & 2 \\ -2 & 2 & -1 \end{pmatrix}, \qquad \mathbf{v}_1 = \begin{pmatrix} 1 \\ 1 \\ 0 \end{pmatrix}.$$

Show that $\mathbf{v}_1$ is an eigenvector of A and find its corresponding eigenvalue. Find a basis of the eigenspace corresponding to this eigenvalue.

Orthogonally diagonalise the matrix A.

Exercise 11.3 Prove that the following quadratic form

$$q(x, y, z) = 6xy - 4yz + 2xz - 4x^2 - 2y^2 - 4z^2$$

is neither positive definite nor negative definite. Is it indefinite?

Determine whether the quadratic form

$$f(x, y, z) = 2xy - 4yz + 6xz - 4x^2 - 2y^2 - 4z^2$$

is positive definite, negative definite or indefinite.

Exercise 11.4 Consider again the matrix A in Exercise 11.1. Express the quadratic form $f(x, y, z) = \mathbf{x}^{\mathrm{T}} A \mathbf{x}$ as a function of the variables x, y and z. Write down a matrix Q so that if $\mathbf{x} = Q\mathbf{z}$ with $\mathbf{z} = (X, Y, Z)^{\mathrm{T}}$, then

$$f(x, y, z) = \mathbf{z}^{\mathrm{T}} D \mathbf{z} = \lambda_1 X^2 + \lambda_2 Y^2 + \lambda_3 Z^2, \quad \text{where} \quad \lambda_1 \geq \lambda_2 \geq \lambda_3.$$

Is the quadratic form $f(x, y, z)$ positive definite, negative definite or indefinite?

Find, if possible, a vector $\mathbf{a} = (a, b, c)$ such that $f(a, b, c) = -8$.

Exercise 11.5 Prove that the diagonal entries of a positive definite $n \times n$ matrix A must be positive numbers. (Do this by considering $\mathbf{e}_i^{\mathrm{T}} A \mathbf{e}_i$, where $\{\mathbf{e}_1, \mathbf{e}_2, \ldots, \mathbf{e}_n\}$ is the standard basis in $\mathbb{R}^n$.)

Give an example to show that the converse statement is not true; that is, write down a symmetric matrix which has positive numbers along the diagonal but which is not positive definite.

Exercise 11.6 Let B be an $m \times k$ matrix with full column rank (meaning rank$(B) = k$). Show that the matrix $B^{\mathrm{T}}B$ is a positive definite symmetric matrix.

Show also $B^{\mathrm{T}}B$ is invertible, by proving that any positive definite matrix is invertible.

Exercise 11.7 Consider the quadratic form

$$f(x, y, z) = x^2 - 4xy + 5y^2 - 2xz + 6yz + 2z^2.$$

Write down a symmetric matrix A such that $f(x, y, z) = \mathbf{x}^{\mathrm{T}}A\mathbf{x}$ for $\mathbf{x} = (x, y, z)^{\mathrm{T}}$. Is the matrix A negative definite, positive definite or indefinite?

Is there a vector $\mathbf{a} = (a, b, c)^{\mathrm{T}}$ such that $f(a, b, c) < 0$? Investigate carefully and justify your answer. Write down such a vector if one exists.

Exercise 11.8 Sketch the curve: $5x^2 - 6xy + 5y^2 = 2$, by reworking Example 11.41, this time choosing Q to be the orthogonal matrix

$$Q = \begin{pmatrix} -\frac{1}{\sqrt{2}} & \frac{1}{\sqrt{2}} \\ \frac{1}{\sqrt{2}} & \frac{1}{\sqrt{2}} \end{pmatrix}.$$

Exercise 11.9 Express the quadratic form $9x^2 + 4xy + 6y^2$ as $\mathbf{x}^{\mathrm{T}}A\mathbf{x}$, where A is a symmetric 2×2 matrix, and find the eigenvalues of A. Deduce whether the quadratic form is positive definite or otherwise, and determine what type of conic section is given by the equation.

$$9x^2 + 4xy + 6y^2 = 10.$$

Orthogonally diagonalise the matrix A and use this information to sketch the curve $9x^2 + 4xy + 6y^2 = 10$ in the xy-plane.

Exercise 11.10 Show that the vectors $\mathbf{v}_1 = (1, 1)^{\mathrm{T}}$ and $\mathbf{v}_2 = (-1, 1)^{\mathrm{T}}$ are eigenvectors of the symmetric matrix

$$A = \begin{pmatrix} a & b \\ b & a \end{pmatrix} \qquad a, b \in \mathbb{R}.$$

What are the corresponding eigenvalues?

Exercise 11.11 Sketch the curve $x^2 + y^2 + 6xy = 4$ in the xy-plane. Find the points of intersection with the old and new axes.

11.6 Problems

Problem 11.1 Find the eigenvalues and corresponding eigenvectors of the matrix

$$A = \begin{pmatrix} 3 & 5 \\ 5 & 3 \end{pmatrix}.$$

Find an orthogonal matrix P and a diagonal matrix D such that $P^{\mathrm{T}}AP = D$.

Problem 11.2 Consider the matrix B and the vector $\mathbf{v}_1$, where

$$B = \begin{pmatrix} 1 & 1 & 0 \\ 1 & 4 & 3 \\ 0 & 3 & 1 \end{pmatrix}, \qquad \mathbf{v}_1 = \begin{pmatrix} 1 \\ 5 \\ 3 \end{pmatrix}.$$

Show that $\mathbf{v}_1$ is an eigenvector of B and find its corresponding eigenvalue.

Orthogonally diagonalise the matrix B.

Problem 11.3 Let

$$A = \begin{pmatrix} 1 & -4 & 2 \\ -4 & 1 & -2 \\ 2 & -2 & -2 \end{pmatrix}.$$

Find the eigenvalues of A and for each eigenvalue find an orthonormal basis for the corresponding eigenspace.

Hence find an orthogonal matrix P such that

$$P^{\mathrm{T}}AP = P^{-1}AP = D.$$

Write down D and check that $P^{\mathrm{T}}AP = D$.

Problem 11.4 Consider the matrix A and the vector $\mathbf{v}_1$, where

$$A = \begin{pmatrix} 3 & -2 & 1 \\ -2 & 6 & -2 \\ 1 & -2 & 3 \end{pmatrix}, \qquad \mathbf{v}_1 = \begin{pmatrix} -1 \\ 0 \\ 1. \end{pmatrix}$$

(a) Show that $\mathbf{v}_1$ is an eigenvector of A and find its corresponding eigenvalue.

Given that $\lambda = 8$ is the only other eigenvalue of A, how do you know that the matrix A can be diagonalised before attempting to do so?

Diagonalise the matrix A; that is, find an invertible matrix P and a diagonal matrix D such that $P^{-1}AP = D$.

Then orthogonally diagonalise A; that is, find an orthogonal matrix Q such that $Q^{\mathrm{T}}AQ = D$.

(b) Write out an expression for the quadratic form

$$f(x, y, z) = \mathbf{x}^T A \mathbf{x}, \quad \text{where} \quad \mathbf{x} = \begin{pmatrix} x \\ y \\ z \end{pmatrix},$$

in terms of x, y and z. Is the quadratic form f positive definite, negative definite or indefinite?

Write down a basis B, and numbers $\lambda_1, \lambda_2, \lambda_3$ such that f can be expressed as

$$f(x, y, z) = \lambda_1 X^2 + \lambda_2 Y^2 + \lambda_3 Z^2,$$

where X, Y, Z are coordinates in the basis B Write down the transition matrix from coordinates in this basis B to standard coordinates.

Evaluate $f(x, y, z)$ at one unit eigenvector corresponding to each eigenvalue.

Problem 11.5 Suppose $\mathbf{x}^T A \mathbf{x}$ is a quadratic form, and λ is an eigenvalue of A. If $\mathbf{u}$ is a unit eigenvector corresponding to λ, show that

$$\mathbf{u}^T A \mathbf{u} = \lambda.$$

Problem 11.6 Consider the following matrix A and vector $\mathbf{v}$:

$$A = \begin{pmatrix} -5 & 1 & 2 \\ 1 & -5 & 2 \\ 2 & 2 & -2 \end{pmatrix}, \quad \mathbf{v} = \begin{pmatrix} -1 \\ 1 \\ 0 \end{pmatrix}.$$

Find a basis of the null space of A, $N(A)$. For this matrix A, why is the row space equal to the column space, $RS(A) = CS(A)$? Show that $CS(A)$ is a plane in $\mathbb{R}^3$, and find its Cartesian equation.

Show that $\mathbf{v}$ is an eigenvector of A and find the corresponding eigenvalue λ. Find all eigenvectors corresponding to this eigenvalue. Orthogonally diagonalise A.

What is the relationship between the column space of A and the eigenspace corresponding to the eigenvalue of multiplicity 2? Why does this happen for this particular matrix A?

Problem 11.7 Determine whether either of the following quadratic forms is positive definite:

$$F(x, y) = 3x^2 - 8xy + 3y^2, \qquad G(x, y) = 43x^2 - 48xy + 57y^2.$$

Find, if possible, points (a_i, b_i), $(i = 1, 2, 3, 4)$, such that

$$F(a_1, b_1) > 0, \quad F(a_2, b_2) < 0, \quad G(a_3, b_3) > 0, \quad G(a_4, b_4) < 0.$$

Problem 11.8

(a) Express the following quadratic form as $\mathbf{x}^T A \mathbf{x}$ where A is a symmetric matrix:

$$f(x, y, z) = 6yz - x^2 + 2xy - 4y^2 - 6z^2.$$

Determine whether the quadratic form is positive definite, negative definite or indefinite.

(b) Do the same for the quadratic form

$$g(x, y, z) = 6yz - x^2 + 2xy - 4y^2.$$

Problem 11.9 Find a symmetric matrix A such that the quadratic form

$$f(x, y, z) = 3x^2 + 4xy + 2y^2 + 5z^2 - 2xz + 2yz$$

can be expressed as $\mathbf{x}^T A \mathbf{x}$. Determine the signs of the eigenvalues of A.

Problem 11.10 Let A be a positive definite symmetric $n \times n$ matrix. Show that the mapping from pairs of vectors $\mathbf{x}, \mathbf{y} \in \mathbb{R}^n$ to the real numbers defined by

$$\langle \mathbf{x}, \mathbf{y} \rangle = \mathbf{x}^T A \mathbf{y}$$

defines an inner product on $\mathbb{R}^n$, where the 1×1 matrix $\mathbf{x}^T A \mathbf{y}$ is interpreted as the real number which is its only entry.

Problem 11.11 Let λ be a constant and let

$$B = \begin{pmatrix} 1 & -3 \\ -1 & 3 \\ 2 & \lambda \end{pmatrix}.$$

For what value(s) of λ will the matrix $B^T B$ be invertible? For what value(s) of λ will the matrix $B^T B$ be positive definite? Justify your answers.

Problem 11.12 If A is an $m \times k$ matrix, show that the matrix $A^T A$ can never be negative definite.

Problem 11.13 Orthogonally diagonalise the matrix

$$A = \begin{pmatrix} 1 & 2 \\ 2 & 1 \end{pmatrix},$$

and use this to sketch the curve $\mathbf{x}^T A \mathbf{x} = 3$ in the xy-plane.

Find the points of intersection of the curve with the old and the new axes.

Problem 11.14 Let C be the curve defined by

$$3x^2 + 2\sqrt{3}xy + 5y^2 = 6.$$

Find a symmetric matrix A such that C is given by $\mathbf{x}^{\mathrm{T}} A \mathbf{x} = 6$.

Find an orthogonal matrix P and a diagonal matrix D, such that $P^{\mathrm{T}} A P = D$, and such that the linear transformation $T : \mathbb{R}^2 \to \mathbb{R}^2$ defined by $T(\mathbf{x}) = P\mathbf{x}$ is an anti-clockwise rotation. Use this to sketch the curve in the xy-plane, showing the old and new axes on your diagram. Compare this with Problem 7.11, where the same curve was sketched using a different linear transformation.

Problem 11.15 Sketch the curve

$$3x^2 + 4xy + 6y^2 = 14$$

in the xy-plane.

12

Direct sums and projections

In this chapter, we meet several important new ideas: direct sum, orthogonal complement and projection. These are useful in the theoretical study of linear algebra, but they also lead us to a very useful practical solution to a real-life problem, namely that of finding the 'best fit' of a particular type to a set of data.

12.1 The direct sum of two subspaces

A very useful idea is the *sum* of two subspaces of a vector space. A special case of this, a *direct sum*, is of particular importance.

12.1.1 The sum of two subspaces

For subspaces U and W of a vector space V, the sum of U and W, written $U + W$, is simply the set of all vectors in V which are obtained by adding together a vector in U and a vector in W.

Definition 12.1 If U and W are subspaces of a vector space V, then the *sum* of U and W, denoted by $U + W$, is the set

$$U + W = \{\mathbf{u} + \mathbf{w} \mid \mathbf{u} \in U, \ \mathbf{w} \in W\}.$$

The sum $U + W$ is also a subspace of V.

Activity 12.2 Prove that $U + W$ is a subspace of V.

Note the difference between $U + W$ and $U \cup W$, which is the set that contains all vectors that are in U or in W. The set $U \cup W$ is *not* generally a subspace, since if, say, $\mathbf{u} \in U$, $\mathbf{u} \notin W$ and $\mathbf{w} \in W$, $\mathbf{w} \notin U$, then $\mathbf{u} + \mathbf{w} \notin U \cup W$; in which case it is not closed under addition (see

Exercise 5.6). The subspace $U + W$ contains both U and W, but is generally much larger.

In fact, $U + W$ is the 'smallest' subspace of V that contains both U and W. By this, we mean that if S is a subspace of V and we have both $U \subseteq S$ and $W \subseteq S$, then $U + W \subseteq S$. To see this, we can simply note that for any $\mathbf{u} \in U$ and any $\mathbf{w} \in W$, we will have $\mathbf{u} \in S$ and $\mathbf{w} \in S$ and so, because S is a subspace, $\mathbf{u} + \mathbf{w} \in S$. This shows that any vector of the form $\mathbf{u} + \mathbf{w}$ is in S, which means that $U + W \subseteq S$.

Activity 12.3 Suppose that $\mathbf{u}, \mathbf{w} \in \mathbb{R}^n$. Prove that

$$\text{Lin}\{\mathbf{u}\} + \text{Lin}\{\mathbf{w}\} = \text{Lin}\{\mathbf{u}, \mathbf{w}\}.$$

12.1.2 Direct sums

A sum of two subspaces is sometimes a *direct sum*.

Definition 12.4 A sum of two subspaces $U + W$ is said to be a *direct sum* if $U \cap W = \{\mathbf{0}\}$.

That is, the sum is direct if the intersection of U and W is as small as it can be. (Since both are subspaces, they will both contain $\mathbf{0}$; the sum is direct if that is all they have in common.) When a sum of subspaces is direct, we use the special notation $U \oplus W$ to mean $U + W$. So the notation means the sum of the subspaces, and the use of the special symbol signifies that the sum is direct.

It turns out that there is another, often very useful, way of characterising when a sum of subspaces is direct, as the following theorem shows:

Theorem 12.5 *Suppose U and W are subspaces of a vector space. Then the sum of U and W is direct if and only if every vector $\mathbf{z}$ in the sum can be written uniquely (that is, in one way only) as $\mathbf{z} = \mathbf{u} + \mathbf{w}$, where $\mathbf{u} \in U$ and $\mathbf{w} \in W$. Explicitly, the sum is direct if and only if whenever $\mathbf{u}, \mathbf{u}' \in U$ and $\mathbf{w}, \mathbf{w}' \in W$ and $\mathbf{u} + \mathbf{w} = \mathbf{u}' + \mathbf{w}'$, then $\mathbf{u} = \mathbf{u}'$ and $\mathbf{w} = \mathbf{w}'$.*

Proof: There are two things we need to prove to establish the theorem. First, we need to show that if the sum is direct, then any element in the sum has a unique expression as a sum of a vector in U and a vector in W. Secondly, we need to show that, conversely, if it's the case that any vector in the sum can be expressed in such a way uniquely, then it follows that the sum is direct (that is, $U \cap W = \{\mathbf{0}\}$).

Suppose first that the sum is direct, so that $U \cap W = \{\mathbf{0}\}$. Now suppose that $\mathbf{u}, \mathbf{u}' \in U$ and $\mathbf{w}, \mathbf{w}' \in W$ and that $\mathbf{u} + \mathbf{w} = \mathbf{u}' + \mathbf{w}'$. Then we may rearrange this as follows:

$$\mathbf{u} - \mathbf{u}' = \mathbf{w}' - \mathbf{w}.$$

Now, the vector on the left-hand side is in U, because $\mathbf{u}$ and $\mathbf{u}'$ are and because U is a subspace. Similarly, the vector on the right is in W. So the vector $\mathbf{v} = \mathbf{u} - \mathbf{u}' = \mathbf{w}' - \mathbf{w}$ is in both U and W. But $U \cap W = \{\mathbf{0}\}$, so $\mathbf{v} = \mathbf{0}$, which means $\mathbf{u} - \mathbf{u}' = \mathbf{0} = \mathbf{w}' - \mathbf{w}$ and so $\mathbf{u} = \mathbf{u}'$ and $\mathbf{w} = \mathbf{w}'$.

Now suppose that every vector $\mathbf{z}$ in the sum can be written *uniquely* as $\mathbf{z} = \mathbf{u} + \mathbf{w}$. We want to show that this means $U \cap W = \{\mathbf{0}\}$. Now, suppose that $\mathbf{z} \in U \cap W$. Then we can write $\mathbf{z}$ as $\mathbf{z} = \mathbf{z} + \mathbf{0}$, where $\mathbf{z} \in U$ and $\mathbf{0} \in W$ and we can also write $\mathbf{z} = \mathbf{0} + \mathbf{z}$, where $\mathbf{0} \in U$ and $\mathbf{z} \in W$. But $\mathbf{z}$ can be expressed in only one way as a sum of a vector in U and a vector in W. So it must be the case that $\mathbf{z} = \mathbf{0}$. Otherwise, these two expressions are different ways of expressing $\mathbf{z}$ as a vector in U plus a vector in W. Here, we assumed $\mathbf{z}$ was any member of $U \cap W$ and we showed that $\mathbf{z} = \mathbf{0}$. It follows that $U \cap W = \{\mathbf{0}\}$ and the sum is direct. □

This theorem shows that there are really two equivalent definitions of what it means for a sum of subspaces to be direct, and it is useful to be able to work with either one. A sum of subspaces U and W is direct if either of the following equivalent conditions holds:

- $U \cap W = \{\mathbf{0}\}$.
- Any vector in $U + W$ can be written uniquely in the form $\mathbf{u} + \mathbf{w}$ with $\mathbf{u} \in U$ and $\mathbf{w} \in W$.

Example 12.6 Suppose that $\mathbf{u}, \mathbf{w} \in \mathbb{R}^n$ and that $\mathbf{u}$ and $\mathbf{w}$ are linearly independent. Then the sum $\text{Lin}\{\mathbf{u}\} + \text{Lin}\{\mathbf{w}\}$ is direct. In other words,

$$\text{Lin}\{\mathbf{u}, \mathbf{w}\} = \text{Lin}\{\mathbf{u}\} \oplus \text{Lin}\{\mathbf{w}\}.$$

(We saw in Activity 12.3 that $\text{Lin}\{\mathbf{u}, \mathbf{w}\} = \text{Lin}\{\mathbf{u}\} + \text{Lin}\{\mathbf{w}\}$; what's new here is that the sum is direct.)

To show the sum is direct, we can verify that if $U = \text{Lin}\{\mathbf{u}\}$ and $W = \text{Lin}\{\mathbf{w}\}$, then $U \cap W = \{\mathbf{0}\}$. So suppose $\mathbf{z} \in U \cap W$. Then $\mathbf{z} \in \text{Lin}\{\mathbf{u}\}$ and $\mathbf{z} \in \text{Lin}\{\mathbf{w}\}$, so there are scalars α, β such that $\mathbf{z} = \alpha \mathbf{u}$ and $\mathbf{z} = \beta \mathbf{w}$. If $\mathbf{z} \neq \mathbf{0}$, then we must have $\alpha \neq 0$ and $\beta \neq 0$. So $\alpha \mathbf{u} = \beta \mathbf{w}$ and therefore $\mathbf{u} = (\beta/\alpha)\mathbf{w}$. But this can't be, since $\mathbf{u}$ and $\mathbf{w}$ are linearly independent. So we can only have $\mathbf{z} = \mathbf{0}$. This shows that the only vector in both U and W is the zero vector. And, clearly, $\mathbf{0} \in U \cap W$ because, since U and W are subspaces, $\mathbf{0}$ belongs to both of them.

12.2 Orthogonal complements

For a subspace S of a vector space V with an inner product, there is a particularly important direct sum involving S. This involves a subspace known as the *orthogonal complement* of S.

12.2.1 The orthogonal complement of a subspace

Suppose that V is a vector space with an inner product, denoted as $\langle \mathbf{x}, \mathbf{y} \rangle$ for $\mathbf{x}, \mathbf{y} \in V$. Given a subset S of V, we define the following set, denoted by $S^{\perp}$.

Definition 12.7 (Orthogonal complement) The *orthogonal complement* of a subset S of a vector space V with inner product $\langle \mathbf{x}, \mathbf{y} \rangle$ is

$$S^{\perp} = \{ \mathbf{v} \in V \mid \text{for all } \mathbf{s} \in S, \langle \mathbf{v}, \mathbf{s} \rangle = 0 \}$$
$$= \{ \mathbf{v} \in V \mid \text{for all } \mathbf{s} \in S, \mathbf{v} \perp \mathbf{s} \}.$$

In other words, $S^{\perp}$ is the set of vectors that are orthogonal to *every* vector in S.

It turns out that $S^{\perp}$ is a subspace (and not merely a subset) of V. (This is true for any set S: S itself need not be a subspace.)

Theorem 12.8 *For any subset S of V, $S^{\perp}$ is a subspace of V.*

Activity 12.9 Prove Theorem 12.8.

Example 12.10 Suppose that $V = \mathbb{R}^3$ with the standard inner product and suppose that $S = \text{Lin}\{\mathbf{u}\}$, where $\mathbf{u} = (1, 2, -1)^{\mathrm{T}}$. Then $S^{\perp}$ is the set of all vectors $\mathbf{v}$ such that, for all $\mathbf{s} \in S$, $\langle \mathbf{v}, \mathbf{s} \rangle = 0$. Now, any member of S is of the form $\alpha \mathbf{u}$. We have $\langle \mathbf{v}, \alpha \mathbf{u} \rangle = \alpha \langle \mathbf{v}, \mathbf{u} \rangle$, so $\mathbf{v} = (x, y, z)^{\mathrm{T}}$ is in $S^{\perp}$ precisely when, for all α, $\alpha \langle \mathbf{v}, \mathbf{u} \rangle = 0$, which means $\langle \mathbf{v}, \mathbf{u} \rangle = 0$. So we see that

$$S^{\perp} = \left\{ \begin{pmatrix} x \\ y \\ z \end{pmatrix} \;\middle|\; x + 2y - z = 0 \right\}.$$

That is, S is the line through the origin in the direction of $\mathbf{u}$ and $S^{\perp}$ is the plane through the origin perpendicular to this line; that is, with normal vector $\mathbf{u}$.

Example 12.11 Suppose again that $V = \mathbb{R}^3$ with the standard inner product and this time suppose that $S = \text{Lin}\{\mathbf{u}, \mathbf{w}\}$, where $\mathbf{u} = (1, 2, -1)^{\mathrm{T}}$ and $\mathbf{w} = (1, 0, 1)^{\mathrm{T}}$. Then what is $S^{\perp}$? Considering

Example 12.10, you might expect that since S is a plane through the origin, then $S^\perp$ is the normal line to the plane through the origin, and indeed this is the case, but let us see precisely why.

The typical element of S is of the form $\alpha\mathbf{u} + \beta\mathbf{w}$. Suppose $\mathbf{z} \in S^\perp$. Then $\mathbf{z}$ is orthogonal to every member of S. In particular, since $\mathbf{u} \in S$ and $\mathbf{w} \in S$, $\mathbf{z} \perp \mathbf{u}$ and $\mathbf{z} \perp \mathbf{w}$. Conversely, if $\mathbf{z} \perp \mathbf{u}$ and $\mathbf{z} \perp \mathbf{w}$, then for any α, β

$$\langle \mathbf{z}, \alpha\mathbf{u} + \beta\mathbf{w} \rangle = \alpha \langle \mathbf{z}, \mathbf{u} \rangle + \beta \langle \mathbf{z}, \mathbf{w} \rangle = 0 + 0 = 0,$$

so $\mathbf{z}$ is orthogonal to every vector in S. So we see that $S^\perp$ is exactly the set of vectors $\mathbf{z}$ orthogonal to both $\mathbf{u}$ and $\mathbf{w}$. Now, for $\mathbf{z} = (x, y, z)^T$, this means we must have both

$$x + 2y - z = 0 \qquad \text{and} \qquad x + z = 0.$$

Solving this system, we see that

$$S^\perp = \left\{ \begin{pmatrix} r \\ -r \\ -r \end{pmatrix} = r \begin{pmatrix} 1 \\ -1 \\ -1 \end{pmatrix} \,\middle|\, r \in \mathbb{R} \right\},$$

which is a line through the origin. So, here, S is a plane through the origin and $S^\perp$ is the line through the origin that is perpendicular to the plane S.

Activity 12.12 Can you think of another way of finding $S^\perp$?

If V is a finite-dimensional inner product space (such as $\mathbb{R}^n$), and S is a subspace of V, then an important fact is that every element of V can be written uniquely as the sum of a vector in S and a vector in $S^\perp$. In other words, we have the following theorem:

Theorem 12.13 *For any subspace S of a finite-dimensional inner product space V, $V = S \oplus S^\perp$.*

Proof: Suppose $\mathbf{z} \in S \cap S^\perp$. Then $\mathbf{z} \in S$ and $\mathbf{z} \perp \mathbf{s}$ for all $\mathbf{s} \in S$. So $\mathbf{z} \perp \mathbf{z}$, which means $\langle \mathbf{z}, \mathbf{z} \rangle = 0$. So $\|\mathbf{z}\|^2 = 0$ and hence $\mathbf{z} = \mathbf{0}$. This shows that $S \cap S^\perp \subseteq \{\mathbf{0}\}$. On the other hand, $\mathbf{0} \in S \cap S^\perp$, so $\{\mathbf{0}\} \subseteq S \cap S^\perp$. It follows that $S \cap S^\perp = \{\mathbf{0}\}$.

Next, we show $V = S + S^\perp$. The cases in which $S = \{\mathbf{0}\}$ or $S = V$ are easily dealt with. So suppose $S \neq \{\mathbf{0}\}$ and $S \neq V$.

Let $\dim(V) = n$ and let $\{\mathbf{e}_1, \mathbf{e}_2, \ldots, \mathbf{e}_r\}$ be an orthonormal basis of the subspace S. Such a basis exists by the Gram–Schmidt orthonormalisation process. We can extend this to an orthonormal basis of V,

$$\{\mathbf{e}_1, \ldots, \mathbf{e}_r, \mathbf{e}_{r+1}, \ldots, \mathbf{e}_n\},$$

again using the Gram–Schmidt process if necessary. We will show that $S^\perp = \text{Lin}\{\mathbf{e}_{r+1}, \ldots, \mathbf{e}_n\}$. If we can do this, then since any $\mathbf{v} \in V$ can be written as

$$\mathbf{v} = \sum_{i=1}^{n} \alpha_i \mathbf{e}_i = (\alpha_1 \mathbf{e}_1 + \cdots + \alpha_r \mathbf{e}_r) + (\alpha_{r+1} \mathbf{e}_{r+1} + \cdots + \alpha_n \mathbf{e}_n),$$

the quantity in the first parentheses will be in S and the quantity in the second in $S^\perp$, showing that $V = S + S^\perp$.

So suppose $\mathbf{v} \in \text{Lin}\{\mathbf{e}_{r+1}, \ldots, \mathbf{e}_n\}$. Then, for some $\alpha_{r+1}, \ldots, \alpha_n$,

$$\mathbf{v} = \alpha_{r+1} \mathbf{e}_{r+1} + \cdots + \alpha_n \mathbf{e}_n.$$

Any $\mathbf{s} \in S$ can be written in the form

$$\alpha_1 \mathbf{e}_1 + \cdots + \alpha_r \mathbf{e}_r.$$

So

$$\langle \mathbf{v}, \mathbf{s} \rangle = \langle \alpha_{r+1} \mathbf{e}_{r+1} + \cdots + \alpha_n \mathbf{e}_n , \ \alpha_1 \mathbf{e}_1 + \cdots + \alpha_r \mathbf{e}_r \rangle.$$

When you expand this inner product, all the terms are of the form $\alpha_i \alpha_j \langle \mathbf{e}_i, \mathbf{e}_j \rangle$ with $i \neq j$. But these are all 0, by orthonormality. So $\mathbf{v} \perp \mathbf{s}$ for all $\mathbf{s} \in S$ and hence $\mathbf{v} \in S^\perp$. Conversely, suppose that $\mathbf{v} \in S^\perp$. Because $\{\mathbf{e}_1, \ldots, \mathbf{e}_n\}$ is a basis of V, there are $\alpha_1, \ldots, \alpha_n$ with $\mathbf{v} = \sum_{i=1}^{n} \alpha_i \mathbf{e}_i$. Now, $\mathbf{e}_1, \mathbf{e}_2, \ldots, \mathbf{e}_r \in S$ and $\mathbf{v} \in S^\perp$, so $\langle \mathbf{v}, \mathbf{e}_i \rangle = 0$ for $i = 1, \ldots, r$. But $\langle \mathbf{v}, \mathbf{e}_i \rangle = \alpha_i$. So $\alpha_1 = \alpha_2 = \cdots = \alpha_r = 0$ and hence $\mathbf{v}$ is a linear combination of $\mathbf{e}_{r+1}, \ldots, \mathbf{e}_n$ only; that is, $\mathbf{v} \in \text{Lin}\{\mathbf{e}_{r+1}, \ldots, \mathbf{e}_n\}$.

Because $S \cap S^\perp = \{\mathbf{0}\}$, the sum $S + S^\perp$ is direct and therefore $V = S \oplus S^\perp$. $\qquad\square$

Another useful result is the following:

Theorem 12.14 *If S is a subspace of a finite-dimensional inner product space V, then $(S^\perp)^\perp = S$.*

You will be able to prove this result yourself when you have worked through the exercises at the end of this chapter.

12.2.2 Orthogonal complements of null spaces and ranges

There are four important subspaces associated with a matrix: the null space and the range of A, and the null space and range of A^T. If A is an $m \times n$ real matrix, then $N(A) = \{\mathbf{x} \in \mathbb{R}^n \mid A\mathbf{x} = \mathbf{0}\}$ is a subspace of $\mathbb{R}^n$, $R(A) = \{A\mathbf{x} \mid \mathbf{x} \in \mathbb{R}^n\}$ is a subspace of $\mathbb{R}^m$, $N(A^\mathrm{T})$ is a subspace of $\mathbb{R}^m$ and $R(A^\mathrm{T})$ is a subspace of $\mathbb{R}^n$. In fact, $R(A^\mathrm{T})$ is just the row space of A, which we met earlier in Section 5.3.1, since the rows of A are the columns of A^T.

It's quite natural, therefore, for us to ask what the orthogonal complements of these subspaces are. The following result answers these questions: the orthogonal complements of the null space and range of any matrix are the range and null space of the *transpose* matrix, respectively.

Theorem 12.15 *Suppose that A is an $m \times n$ real matrix. Then $R(A)^\perp = N(A^T)$ and $N(A)^\perp = R(A^T)$.*

Proof: We prove that $R(A)^\perp = N(A^T)$. The other result then follows by substituting A^T for A, obtaining $R(A^T)^\perp = N(A)$ and then, taking orthogonal complements,

$$R(A^T) = (R(A^T)^\perp)^\perp = N(A)^\perp.$$

The easiest way to show that $R(A)^\perp = N(A^T)$ is to show that $R(A)^\perp \subseteq N(A^T)$ and that $N(A^T) \subseteq R(A)^\perp$. The key fact we're going to use in this proof is that for any matrix A and any vectors $\mathbf{x} \in \mathbb{R}^n$, $\mathbf{y} \in \mathbb{R}^m$,

$$\langle \mathbf{y}, A\mathbf{x} \rangle = \langle A^T\mathbf{y}, \mathbf{x} \rangle,$$

even though $\mathbf{y}, A\mathbf{x} \in \mathbb{R}^m$ and $A^T\mathbf{y}, \mathbf{x} \in \mathbb{R}^n$. This is because the inner product is a scalar given by $\langle \mathbf{a}, \mathbf{b} \rangle = \mathbf{a}^T\mathbf{b}$, and so

$$\langle A^T\mathbf{y}, \mathbf{x} \rangle = (A^T\mathbf{y})^T\mathbf{x} = \mathbf{y}^T(A^T)^T\mathbf{x} = \mathbf{y}^T A\mathbf{x} = \mathbf{y}^T(A\mathbf{x}) = \langle \mathbf{y}, A\mathbf{x} \rangle.$$

Suppose $\mathbf{z} \in R(A)^\perp$. This means that for all $\mathbf{y} \in R(A)$, $\langle \mathbf{z}, \mathbf{y} \rangle = 0$. Every $\mathbf{y} \in R(A)$ is of the form $A\mathbf{x}$, by definition of $R(A)$, so if $\mathbf{z} \in R(A)^\perp$, then for all $\mathbf{x}$, $\langle \mathbf{z}, A\mathbf{x} \rangle = 0$, which means $\langle A^T\mathbf{z}, \mathbf{x} \rangle = 0$ for all $\mathbf{x}$. If we take $\mathbf{x} = A^T\mathbf{z}$, we see that $\|A^T\mathbf{z}\|^2 = \langle A^T\mathbf{z}, A^T\mathbf{z} \rangle = 0$, so we have $A^T\mathbf{z} = \mathbf{0}$. This shows that $\mathbf{z} \in N(A^T)$. Hence $R(A)^\perp \subseteq N(A^T)$.

Now suppose $\mathbf{z} \in N(A^T)$. Then $A^T\mathbf{z} = \mathbf{0}$. So, for all $\mathbf{x} \in \mathbb{R}^n$, $\langle A^T\mathbf{z}, \mathbf{x} \rangle = 0$ and hence $\langle \mathbf{z}, A\mathbf{x} \rangle = 0$, for all $\mathbf{x}$. But this means $\langle \mathbf{z}, \mathbf{y} \rangle = 0$ for all $\mathbf{y} \in R(A)$, and so $\mathbf{z} \in R(A)^\perp$. This shows that $N(A^T) \subseteq R(A)^\perp$. $\square$

Example 12.16 Suppose again that $V = \mathbb{R}^3$ with the standard inner product and suppose that $S = \text{Lin}\{\mathbf{u}, \mathbf{w}\}$, where $\mathbf{u} = (1, 2, -1)^T$ and $\mathbf{w} = (1, 0, 1)^T$. Then what is $S^\perp$? Earlier, in Example 12.11, we found that

$$S^\perp = \left\{ \begin{pmatrix} r \\ -r \\ -r \end{pmatrix} \,\middle|\, r \in \mathbb{R} \right\}$$

by obtaining the solution $\mathbf{x}$ of the homogeneous system of equations given by $\langle \mathbf{u}, \mathbf{x} \rangle = 0$ and $\langle \mathbf{w}, \mathbf{x} \rangle = 0$.

We now have another way to confirm this, by using Theorem 12.15. For $S = R(A)$ where

$$A = \begin{pmatrix} 1 & 1 \\ 2 & 0 \\ -1 & 1 \end{pmatrix}.$$

By Theorem 12.15, $S^\perp = (R(A))^\perp = N(A^T)$. Now,

$$A^T = \begin{pmatrix} 1 & 2 & -1 \\ 1 & 0 & 1 \end{pmatrix}.$$

If we determine $N(A^T)$, we find $S^\perp$. Since A^T is precisely the coefficient matrix of the homogeneous system of equations we had before, we get exactly the same answer.

The result that $R(A^T) = N(A)^\perp$ for an $m \times n$ matrix A is just another way of looking at familiar results concerning a linear system of homogeneous equations in light of what we now know about orthogonal complements. The subspace $R(A^T)$ is the linear span of the columns of A^T, so it is the linear span of the rows of A. We denoted this subspace by $RS(A)$ in Section 5.3.1. If $\mathbf{v}$ is any solution of $A\mathbf{x} = \mathbf{0}$, then since $A\mathbf{v} = \mathbf{0}$ we must have $\langle \mathbf{r}_i, \mathbf{v} \rangle = 0$ for each row $\mathbf{r}_i$, $i = 1, \ldots, m$ of A. Therefore, any and every vector $\mathbf{v}$ in $N(A)$ is orthogonal to any and every vector in $RS(A) = \text{Lin}\{\mathbf{r}_1, \ldots, \mathbf{r}_n\}$, so these subspaces are orthogonal. In particular, the only vector which is in both subspaces is the zero vector, $\mathbf{0}$, since such a vector will be orthogonal to itself. But now we have additional information. We know by Theorem 12.15 that the row space and the null space are orthogonal complements, so $\mathbb{R}^n$ is the direct sum of these two subspaces.

As a direct consequence of this observation, we can prove the following useful fact.

- If A is an $m \times n$ matrix of rank n, then $A^T A$ is invertible.

You have already proved this in Exercise 6.13 and again in Exercise 11.6, but let's look at it once again using Theorem 12.15. Since A is $m \times n$ and A^T is $n \times m$, the matrix $A^T A$ is a square $n \times n$ matrix. We will show that the only solution of the system $A^T A\mathbf{x} = \mathbf{0}$ is the trivial solution, so by Theorem 4.5 this will prove that $A^T A$ is invertible.

Let $\mathbf{v} \in \mathbb{R}^n$ be any solution of $A^T A\mathbf{x} = \mathbf{0}$, so $A^T A\mathbf{v} = \mathbf{0}$. Then the vector $A\mathbf{v} \in \mathbb{R}^m$ is in the null space of A^T since $A^T(A\mathbf{v}) = \mathbf{0}$ and, also $A\mathbf{v}$ is in the range of A. But $N(A^T) = R(A)^\perp$, so $N(A^T) \cap R(A) = \{\mathbf{0}\}$. Therefore we must have $A\mathbf{v} = \mathbf{0}$. But A has full column rank, so the columns of A are linearly independent and the only solution of this system is the trivial solution $\mathbf{v} = \mathbf{0}$. This completes the argument.

12.3   Projections

12.3.1   The definition of a projection

Suppose that a vector space V can be written as a direct sum of two subspaces, U and W, so $V = U \oplus W$. This means that for each $\mathbf{v} \in V$ there is a unique $\mathbf{u} \in U$ and a unique $\mathbf{w} \in W$ such that $\mathbf{v} = \mathbf{u} + \mathbf{w}$. We can use this fact to define two functions $P_U : V \to U$ and $P_W : V \to W$ as follows.

Definition 12.17 Suppose that the vector space V is such that $V = U \oplus W$ where U and W are subspaces of V. Define the functions $P_U : V \to U$ and $P_W : V \to W$ as follows: for each $\mathbf{v} \in V$, if $\mathbf{v} = \mathbf{u} + \mathbf{w}$ where $\mathbf{u} \in U$ and $\mathbf{w} \in W$ (these being unique, since the sum is direct), then let $P_U(\mathbf{v}) = \mathbf{u}$ and $P_W(\mathbf{v}) = \mathbf{w}$. The mapping P_U is called the *projection of V onto U, parallel to W*. The mapping P_W is called the *projection of V onto W, parallel to U*.

Activity 12.18 Why does the sum $V = U \oplus W$ have to be direct for us to be able to define P_U and P_W?

Each of the projections P_U and P_W is a linear transformation. For, suppose that $\mathbf{v}_1, \mathbf{v}_2 \in V$ and α_1, α_2 are scalars, and suppose $P_U(\mathbf{v}_1) = \mathbf{u}_1$, $P_W(\mathbf{v}_1) = \mathbf{w}_1$, $P_U(\mathbf{v}_2) = \mathbf{u}_2$, and $P_W(\mathbf{v}_2) = \mathbf{w}_2$. This means that $\mathbf{v}_1 = \mathbf{u}_1 + \mathbf{w}_1$ and that $\mathbf{v}_2 = \mathbf{u}_2 + \mathbf{w}_2$ (and also that there are no other ways of writing $\mathbf{v}_1$ and $\mathbf{v}_2$ in the form $\mathbf{u} + \mathbf{w}$, where $\mathbf{u} \in U$ and $\mathbf{w} \in W$). Now,

$$\alpha_1 \mathbf{v}_1 + \alpha_2 \mathbf{v}_2 = \alpha_1(\mathbf{u}_1 + \mathbf{w}_1) + \alpha_2(\mathbf{u}_2 + \mathbf{w}_2)$$
$$= \underbrace{(\alpha_1 \mathbf{u}_1 + \alpha_2 \mathbf{u}_2)}_{\text{in } U} + \underbrace{(\alpha_1 \mathbf{w}_1 + \alpha_2 \mathbf{w}_2)}_{\text{in } W}.$$

So, $\alpha_1 \mathbf{v}_1 + \alpha_2 \mathbf{v}_2 = \mathbf{u}' + \mathbf{w}'$,   where $\mathbf{u}' = \alpha_1 \mathbf{u}_1 + \alpha_2 \mathbf{u}_2 \in U$ and $\mathbf{w}' = \alpha_1 \mathbf{w}_1 + \alpha_2 \mathbf{w}_2 \in W$. Therefore, we must have

$$P_U(\alpha_1 \mathbf{v}_1 + \alpha_2 \mathbf{v}_2) = \mathbf{u}' = \alpha_1 \mathbf{u}_1 + \alpha_2 \mathbf{u}_2 = \alpha_1 P_U(\mathbf{v}_1) + \alpha_2 P_U(\mathbf{v}_2)$$

and

$$P_W(\alpha_1 \mathbf{v}_1 + \alpha_2 \mathbf{v}_2) = \mathbf{w}' = \alpha_1 \mathbf{w}_1 + \alpha_2 \mathbf{w}_2 = \alpha_1 P_W(\mathbf{v}_1) + \alpha_2 P_W(\mathbf{v}_2).$$

12.3.2 An example

Example 12.19 Suppose that $V = \mathbb{R}^3$ with the standard inner product and suppose that $U = \mathrm{Lin}\{(1, 2, -1)^T, (1, 0, 1)^T\}$. We saw in Examples 12.11 and 12.16 that

$$U^\perp = \mathrm{Lin}\{(1, -1, -1)^T\} = \left\{ \begin{pmatrix} r \\ -r \\ -r \end{pmatrix} \;\middle|\; r \in \mathbb{R} \right\}.$$

Theorem 12.13 tells us that if we let $W = U^\perp$, then $\mathbb{R}^3 = U \oplus W$. What is the projection P_U of $\mathbb{R}^3$ onto U, parallel to W? Well, we'll find a general way of answering such questions later, but let's see if we can do it directly in this particular case. The fact that the sum of U and W is direct means that for each $\mathbf{x} = (x, y, z)^T \in \mathbb{R}^3$ there are unique members $\mathbf{u}$ of U and $\mathbf{w}$ of W so that $\mathbf{x} = \mathbf{u} + \mathbf{w}$. In this case, this means there are unique α, β, γ such that

$$\begin{pmatrix} x \\ y \\ z \end{pmatrix} = \underbrace{\alpha \begin{pmatrix} 1 \\ 2 \\ -1 \end{pmatrix} + \beta \begin{pmatrix} 1 \\ 0 \\ 1 \end{pmatrix}}_{\text{in } U} + \underbrace{\gamma \begin{pmatrix} 1 \\ -1 \\ -1 \end{pmatrix}}_{\text{in } W}.$$

This means that α, β, γ satisfy the linear system

$$\alpha + \beta + \gamma = x$$
$$2\alpha - \gamma = y$$
$$-\alpha + \beta - \gamma = z,$$

which has solution

$$\alpha = \frac{x}{6} + \frac{y}{3} - \frac{z}{6}, \quad \beta = \frac{x}{2} + \frac{z}{2}, \quad \gamma = \frac{x}{3} - \frac{y}{3} - \frac{z}{3}.$$

So the projection P_U is given by

$$P_U(\mathbf{x}) = \alpha \begin{pmatrix} 1 \\ 2 \\ -1 \end{pmatrix} + \beta \begin{pmatrix} 1 \\ 0 \\ 1 \end{pmatrix} = \begin{pmatrix} \frac{2}{3}x + \frac{1}{3}y + \frac{1}{3}z \\ \frac{1}{3}x + \frac{2}{3}y - \frac{1}{3}z \\ \frac{1}{3}x - \frac{1}{3}y + \frac{2}{3}z \end{pmatrix}.$$

There must be an easier way! Well, as we'll see soon, there is.

Activity 12.20 Check this solution.

12.3.3   Orthogonal projections

Example 12.19 concerns a special type of projection. It is the projection onto S parallel to $S^\perp$ (for a particular subspace S). Not all projections are of this form (because, generally, there are many ways to write $V = U \oplus W$ where W is *not* $U^\perp$), but this type of projection is called an *orthogonal projection*.

Definition 12.21 For a subspace S of any finite-dimensional inner product space V (such as $\mathbb{R}^n$), the *orthogonal projection* of V onto S is the projection onto S parallel to $S^\perp$.

12.4   Characterising projections and orthogonal projections

12.4.1   Projections are idempotents

Projections have some important properties. We've already seen that they are linear. Another important property is that any projection P (onto some subspace U, parallel to another, W, such that $V = U \oplus W$) satisfies $P^2 = P$. Such a linear transformation is said to be *idempotent* (or we say it is *an idempotent*).

Definition 12.22 The linear transformation T is said to be *idempotent* if $T^2 = T$.

This term also applies to the matrix representing an idempotent linear transformation when $V = \mathbb{R}^n$ or any finite-dimensional vector space.

Definition 12.23 The matrix A is said to be *idempotent* if $A^2 = A$.

Activity 12.24 As an exercise, show that the only eigenvalues of an idempotent matrix A are 0 and 1.

Theorem 12.25 *Any projection is idempotent.*

Proof: This is quite easy to see. Let us take any $\mathbf{v} \in V$ and write $\mathbf{v}$ as $\mathbf{v} = \mathbf{u} + \mathbf{w}$ where $\mathbf{u} \in U$ and $\mathbf{w} \in W$. Then $P_U(\mathbf{v}) = \mathbf{u}$. What we need to show is that $P^2(\mathbf{v}) = P(\mathbf{v})$; in other words, $P(P(\mathbf{v})) = P(\mathbf{v})$, which means $P(\mathbf{u}) = \mathbf{u}$. But, of course, $P(\mathbf{u}) = \mathbf{u}$ because the way $\mathbf{u}$ is written as a vector in U plus a vector in W is $\mathbf{u} = \mathbf{u} + \mathbf{0}$. $\qquad\square$

The fact that $P^2 = P$ means that, for any n, $P^n = P$. For example, $P^3 = P^2 P = PP = P^2 = P$. This is where the name 'idempotent' comes from: 'idem-potent' means all powers are equal ('idem' signifies equal, and 'potent' power).

In fact, if we have any linear transformation P that satisfies $P^2 = P$, then it turns out to be a projection. In other words, a linear transformation is a projection *if and only if* it is an idempotent.

Theorem 12.26 *A linear transformation is a projection if and only if it is an idempotent.*

Proof: We've already seen that any projection is idempotent. Suppose now that we have an idempotent linear transformation P on a vector space V. So $P^2 = P$. Let us define U to be $R(P) = \{P(\mathbf{x}) \mid \mathbf{x} \in V\}$ and let

$$W = N(P) = \{\mathbf{v} \mid P(\mathbf{v}) = \mathbf{0}\}.$$

We'll show two things: (i) that $V = U \oplus W$ and that (ii) P is the projection onto U parallel to W.

For (i), we observe that for any $\mathbf{x} \in V$, $\mathbf{x} = P(\mathbf{x}) + (\mathbf{x} - P(\mathbf{x}))$. Now, $P(\mathbf{x}) \in R(P)$ and, because

$$P(\mathbf{x} - P(\mathbf{x})) = P(\mathbf{x}) - P^2(\mathbf{x}) = P(\mathbf{x}) - P(\mathbf{x}) = \mathbf{0},$$

we see that $\mathbf{x} - P(\mathbf{x}) \in N(P)$. So any $\mathbf{x}$ in V can be expressed as the sum of a vector in $R(P)$ and a vector in $N(P)$ and therefore

$$V = R(P) + N(P).$$

We need to show that the sum is direct. So suppose that $\mathbf{z} \in R(P) \cap N(P)$. Then, for some $\mathbf{y}$, we have $\mathbf{z} = P(\mathbf{y})$ and, furthermore, $P(\mathbf{z}) = \mathbf{0}$. But this implies that $P(P(\mathbf{y})) = \mathbf{0}$. This means $P^2(\mathbf{y}) = \mathbf{0}$. But $P^2 = P$, so $P(\mathbf{y}) = \mathbf{0}$. Thus, $\mathbf{z} = P(\mathbf{y}) = \mathbf{0}$. On the other hand, since, certainly, $\mathbf{0} \in R(P) \cap N(P)$, we have $R(P) \cap N(P) = \{\mathbf{0}\}$ and the sum is direct.

We now need to establish (ii). Suppose that $\mathbf{v} \in V$ and that $\mathbf{v} = \mathbf{u} + \mathbf{w}$, where $\mathbf{u} \in U = R(P)$ and $\mathbf{w} \in W = N(P)$. Because $\mathbf{u} \in R(P)$, there is some $\mathbf{x}$ such that $\mathbf{u} = P(\mathbf{x})$ and, therefore, since $P^2 = P$,

$$P(\mathbf{u}) = P(P(\mathbf{x})) = P^2(\mathbf{x}) = P(\mathbf{x}) = \mathbf{u}.$$

Therefore,

$$P(\mathbf{v}) = P(\mathbf{u} + \mathbf{w}) = P(\mathbf{u}) + P(\mathbf{w}) = \mathbf{u} + \mathbf{0} = \mathbf{u}.$$

This completes the proof. $\square$

For future reference, we summarise what we have just shown in this proof as follows.

- If P is an idempotent linear transformation, $P : V \to V$, then $V = R(P) \oplus N(P)$ and P is a projection from V onto $U = R(P)$ parallel to $W = N(P)$.

Note that, for a projection P onto U parallel to W, $P\mathbf{x} = \mathbf{0}$ if and only if $\mathbf{x}$ takes the form $\mathbf{x} = \mathbf{0} + \mathbf{w}$ for some $\mathbf{w} \in W$. So, $N(P) = W$. We summarise this in the following statement.

- If P is a projection from V to a subspace U parallel to a subspace W, then $U = R(P)$ and $W = N(P)$.

There is a similar characterisation of orthogonal projections.

Theorem 12.27 *If V is a finite-dimensional inner product space and P is a linear transformation, $P : V \to V$, then P is an orthogonal projection if and only if the matrix representing P is idempotent and symmetric.*

Proof: For simplicity, let P denote both the linear transformation and the matrix representing it, and suppose that P is not just idempotent, but also symmetric ($P = P^{\mathrm{T}}$). Then we know, because it's an idempotent, that P is the projection onto $R(P)$ parallel to $N(P)$. Now, because $P = P^{\mathrm{T}}, N(P) = N(P^{\mathrm{T}}) = (R(P))^{\perp}$. So P projects onto $R(P)$ parallel to $(R(P))^{\perp}$ and is therefore an orthogonal projection.

Conversely, it's true that any orthogonal projection will be both idempotent and symmetric. We already know that it must be idempotent, so we now have to show it is symmetric. Well, suppose that P is the orthogonal projection onto U (parallel to $U^{\perp}$). Any $\mathbf{x} \in V$ can be written uniquely as $\mathbf{x} = \mathbf{u} + \mathbf{u}'$, where $\mathbf{u} \in U$ and $\mathbf{u}' \in U^{\perp}$, and the projection P is such that $P\mathbf{x} = \mathbf{u} \in U$. Note, too, that

$$(I - P)\mathbf{x} = \mathbf{x} - P\mathbf{x} = (\mathbf{u} + \mathbf{u}') - \mathbf{u} = \mathbf{u}' \in U^{\perp}.$$

So, it follows that, for any $\mathbf{v}, \mathbf{w} \in V$, $P\mathbf{v} \in U$ and $(I - P)\mathbf{w} \in U^{\perp}$ and hence $\langle P\mathbf{v}, (I - P)\mathbf{w} \rangle = 0$. That is, $(P\mathbf{v})^{\mathrm{T}}(I - P)\mathbf{w} = 0$, which means $\mathbf{v}^{\mathrm{T}}P^{\mathrm{T}}(I - P)\mathbf{w} = 0$. Now, the fact that this is true for *all* $\mathbf{v}, \mathbf{w}$ means that the matrix $P^{\mathrm{T}}(I - P)$ must be the zero matrix. For, if $\mathbf{e}_i$ denotes, as usual, the ith standard basis vector of $\mathbb{R}^n$, then $\mathbf{e}_i^{\mathrm{T}}P^{\mathrm{T}}(I - P)\mathbf{e}_j$ is simply the (i, j)th entry of the matrix $P^{\mathrm{T}}(I - P)$. So all entries of that matrix are 0. The fact that $P^{\mathrm{T}}(I - P) = 0$ means that $P^{\mathrm{T}} = P^{\mathrm{T}}P$ and we therefore have

$$P = (P^{\mathrm{T}})^{\mathrm{T}} = (P^{\mathrm{T}}P)^{\mathrm{T}} = P^{\mathrm{T}}(P^{\mathrm{T}})^{\mathrm{T}} = P^{\mathrm{T}}P = P^{\mathrm{T}}.$$

In other words, P is symmetric. Admittedly, this is a rather sneaky proof, but it works! Notice, by the way, that it also follows immediately (though we already know this) that P is idempotent, for

$$P^2 = PP = P^{\mathrm{T}}P = P.$$

$\square$

12.5   Orthogonal projection onto the range of a matrix

Let's start with a simple observation. Suppose that A is an $m \times n$ real matrix of rank n. Then the matrix $A^{\mathrm{T}}A$ is an $n \times n$ matrix, and (as we have seen in Section 12.2.2) $A^{\mathrm{T}}A$ is invertible.

Therefore, we can compute the matrix

$$P = A(A^\mathsf{T} A)^{-1} A^\mathsf{T}.$$

It turns out that this matrix P has a very useful property.

Theorem 12.28 *Suppose A is an $m \times n$ real matrix of rank n. Then the matrix $P = A(A^\mathsf{T} A)^{-1} A^\mathsf{T}$ represents the orthogonal projection of $\mathbb{R}^m$ onto the range $R(A)$ of A.*

Proof: We show three things: (i) P is idempotent, (ii) P is symmetric, (iii) $R(P) = R(A)$. Then, (i) and (ii) establish that P is the orthogonal projection onto $R(P)$ and, with (iii), it is therefore the orthogonal projection onto $R(A)$.

First,

$$\begin{aligned}
P^2 &= (A(A^\mathsf{T} A)^{-1} A^\mathsf{T})(A(A^\mathsf{T} A)^{-1} A^\mathsf{T}) \\
&= A(A^\mathsf{T} A)^{-1}(A^\mathsf{T} A)(A^\mathsf{T} A)^{-1} A^\mathsf{T} \\
&= A(A^\mathsf{T} A)^{-1} A^\mathsf{T} \\
&= P.
\end{aligned}$$

Next,

$$\begin{aligned}
P^\mathsf{T} &= (A(A^\mathsf{T} A)^{-1} A^\mathsf{T})^\mathsf{T} \\
&= (A^\mathsf{T})^\mathsf{T} \left((A^\mathsf{T} A)^{-1}\right)^\mathsf{T} A^\mathsf{T} \\
&= A \left((A^\mathsf{T} A)^\mathsf{T}\right)^{-1} A^\mathsf{T} \\
&= A(A^\mathsf{T} A)^{-1} A^\mathsf{T} \\
&= P.
\end{aligned}$$

Now, clearly, since

$$P\mathbf{x} = A(A^\mathsf{T} A)^{-1} A^\mathsf{T}\mathbf{x} = A\left((A^\mathsf{T} A)^{-1} A^\mathsf{T}\mathbf{x}\right),$$

any vector of the form $P\mathbf{x}$ is also of the form $A\mathbf{y}$ for some $\mathbf{y}$. That is, $R(P) \subseteq R(A)$. What we need to do, therefore, to show that $R(P) = R(A)$ is to prove that $R(A) \subseteq R(P)$. So, suppose $\mathbf{z} \in R(A)$, so $\mathbf{z} = A\mathbf{x}$ for some $\mathbf{x}$. Now,

$$P\mathbf{z} = P A\mathbf{x} = \left[A(A^\mathsf{T} A)^{-1} A^\mathsf{T}\right] A\mathbf{x} = A(A^\mathsf{T} A)^{-1}(A^\mathsf{T} A)\mathbf{x} = A\mathbf{x} = \mathbf{z},$$

so $\mathbf{z} = P\mathbf{z} \in R(P)$. This shows that $R(A) \subseteq R(P)$, and we are done.   □

Example 12.29 In Example 12.19, we determined the orthogonal projection of $\mathbb{R}^3$ onto $U = \text{Lin}\{(1, 2, -1)^\mathsf{T}, (1, 0, 1)^\mathsf{T}\}$. We found that this is given by

$$P(\mathbf{x}) = \begin{pmatrix} \frac{2}{3}x + \frac{1}{3}y + \frac{1}{3}z \\ \frac{1}{3}x + \frac{2}{3}y - \frac{1}{3}z \\ \frac{1}{3}x - \frac{1}{3}y + \frac{2}{3}z \end{pmatrix}.$$

The calculation we performed there was quite laborious, but Theorem 12.28 makes life easier. What we want is the orthogonal projection of $\mathbb{R}^3$ onto $R(A)$, where A is the matrix (of rank 2)

$$A = \begin{pmatrix} 1 & 1 \\ 2 & 0 \\ -1 & 1 \end{pmatrix}.$$

By Theorem 12.28, this projection is represented by the matrix $P = A(A^{\mathrm{T}}A)^{-1}A^{\mathrm{T}}$. Now,

$$A^{\mathrm{T}}A = \begin{pmatrix} 6 & 0 \\ 0 & 2 \end{pmatrix},$$

so

$$P = A(A^{\mathrm{T}}A)^{-1}A^{\mathrm{T}}$$

$$= \begin{pmatrix} 1 & 1 \\ 2 & 0 \\ -1 & 1 \end{pmatrix} \frac{1}{6} \begin{pmatrix} 1 & 0 \\ 0 & 3 \end{pmatrix} \begin{pmatrix} 1 & 2 & -1 \\ 1 & 0 & 1 \end{pmatrix}$$

$$= \frac{1}{6} \begin{pmatrix} 1 & 1 \\ 2 & 0 \\ -1 & 1 \end{pmatrix} \begin{pmatrix} 1 & 2 & -1 \\ 3 & 0 & 3 \end{pmatrix}$$

$$= \frac{1}{6} \begin{pmatrix} 4 & 2 & 2 \\ 2 & 4 & -2 \\ 2 & -2 & 4 \end{pmatrix}$$

$$= \frac{1}{3} \begin{pmatrix} 2 & 1 & 1 \\ 1 & 2 & -1 \\ 1 & -1 & 2 \end{pmatrix}.$$

So the projection is given by

$$P\mathbf{x} = \frac{1}{3} \begin{pmatrix} 2 & 1 & 1 \\ 1 & 2 & -1 \\ 1 & -1 & 2 \end{pmatrix} \begin{pmatrix} x \\ y \\ z \end{pmatrix},$$

which is exactly the same as we determined earlier.

12.6   Minimising the distance to a subspace

Suppose that U is a subspace of $\mathbb{R}^m$ and suppose that $\mathbf{v} \in \mathbb{R}^m$. What is the smallest distance between $\mathbf{v}$ and any member of U? Well, obviously, if $\mathbf{v} \in U$, then this smallest distance is 0, since $\mathbf{v}$ has distance 0 from itself. But, generally, for $\mathbf{v} \notin U$, the problem is to find $\mathbf{u} \in U$ such that the distance $\|\mathbf{v} - \mathbf{u}\|$ is as small as possible, over all choices of $\mathbf{u}$ from U. You are probably familiar with the fact that if you have a line in two dimensions and a point p not on the line, then the point on the line closest to p is obtained by 'taking a perpendicular' to the line through p. Where that perpendicular hits the line is the point on the line closest to p. Well, essentially, this is true in general. If we want to find the point of U closest to $\mathbf{v}$, it will be $P\mathbf{v}$ where P is the orthogonal projection of $\mathbb{R}^m$ onto U. Let's prove this.

Theorem 12.30 *Suppose U is a subspace of $\mathbb{R}^m$, that $\mathbf{v} \in \mathbb{R}^m$, and that P is the orthogonal projection of $\mathbb{R}^m$ onto U. Then for all $\mathbf{u} \in U$,*

$$\|\mathbf{v} - \mathbf{u}\| \geq \|\mathbf{v} - P\mathbf{v}\|.$$

That is, $P\mathbf{v}$ is the closest point in U to $\mathbf{v}$.

Proof: For any $\mathbf{u} \in U$, we have

$$\|\mathbf{u} - \mathbf{v}\| = \|(\mathbf{u} - P\mathbf{v}) + (P\mathbf{v} - \mathbf{v})\|.$$

Now,

$$P(P\mathbf{v} - \mathbf{v}) = P^2\mathbf{v} - P\mathbf{v} = P\mathbf{v} - P\mathbf{v} = \mathbf{0},$$

so $P\mathbf{v} - \mathbf{v} \in N(P) = U^\perp$. Also, $\mathbf{u} - P\mathbf{v} \in U$ because $P\mathbf{v} \in U$ and U is a subspace. So the vectors $\mathbf{u} - P\mathbf{v}$ and $P\mathbf{v} - \mathbf{v}$ are orthogonal. By the generalised Pythagoras theorem,

$$\|\mathbf{u} - \mathbf{v}\|^2 = \|\mathbf{u} - P\mathbf{v}\|^2 + \|P\mathbf{v} - \mathbf{v}\|^2.$$

Since $\|\mathbf{u} - P\mathbf{v}\|^2 \geq 0$, this implies

$$\|\mathbf{u} - \mathbf{v}\|^2 \geq \|P\mathbf{v} - \mathbf{v}\|^2,$$

as required.   □

12.7   Fitting functions to data: least squares approximation

12.7.1   The idea

Suppose we want to find an equation that models the relationship between two quantities X and Y, of the form $Y = f(X)$. In the simplest case, we might try to model the relationship by assuming that Y is related to X linearly, so that, for some constants a and b, $Y = a + bX$. Now, suppose we have some data which provides pairs of values of X and Y. So we have, say, m pairs

$$(X_1, Y_1), (X_2, Y_2), \ldots, (X_m, Y_m).$$

For, what we want is to find a, b so that, for each i, $Y_i = a + bX_i$. But this might not be possible. It could be that there is some 'noise' or measurement errors in some of the X_i and Y_i values. Or it could be that the true relationship between them is more complex. In any case, suppose we still want to find a linear relationship that is *approximately* correct. That is, we want to find a, b so that $Y = a + bX$ fits the data as well as possible.

Usually, the appropriate way to measure how good a fit a given model $Y = a + bX$ will give can be obtained by measuring the error, $\sum_{i=1}^{m}(Y_i - (a + bX_i))^2$. If this is small, then the fit is good. And what we want to do is find a and b for which this measure of error is as small as it can be. Such values of a and b are called a *least squares solution*. (They give the *least* value of the error, which depends on the *squares* of how far Y_i is from $a + bX_i$.)

There are a number of approaches to finding the least squares solution. In statistics, you might come across formulas that you learn. We can often also find the least squares solution using calculus. But what we want to do here is use the linear algebra we've developed to show how to find a least squares solution. (The method we present can also be adapted to handle more complicated 'fitting' problems.)

12.7.2   A linear algebra view

The equations $Y_i = a + bX_i$ for $i = 1$ to m can be written in matrix form as

$$\begin{pmatrix} 1 & X_1 \\ 1 & X_2 \\ \vdots & \vdots \\ 1 & X_m \end{pmatrix} \begin{pmatrix} a \\ b \end{pmatrix} = \begin{pmatrix} Y_1 \\ Y_2 \\ \vdots \\ Y_m \end{pmatrix}.$$

This can be written $A\mathbf{z} = \mathbf{b}$. As noted, this might not have a solution. What we want to do instead is find $\mathbf{z} = \begin{pmatrix} a \\ b \end{pmatrix}$ so that the least squares measure of error is as small as possible. Now, the least squares error is

$$\sum_{i=1}^{m} (Y_i - (a + bX_i))^2,$$

which is the same as

$$\|\mathbf{b} - A\mathbf{z}\|^2.$$

So what we need is for $A\mathbf{z}$ to be the closest point of the form $A\mathbf{y}$ to $\mathbf{b}$. That is, $A\mathbf{z}$ has to be the closest point in $R(A)$ to $\mathbf{b}$. But we know from Theorem 12.30 that this closest point in $R(A)$ is $P\mathbf{b}$, where P is the orthogonal projection onto $R(A)$. Assuming that A has rank 2, we also know, from Theorem 12.28, that $P = A(A^T A)^{-1} A^T$. So what we want is

$$A\mathbf{z} = P\mathbf{b} = A(A^T A)^{-1} A^T \mathbf{b}.$$

One solution to this (and there may be others) is

$$\mathbf{z} = (A^T A)^{-1} A^T \mathbf{b}.$$

This is therefore a least squares solution.

12.7.3 Examples

Example 12.31 Suppose we want to find the best fit (in the least squares sense) relationship of the form $Y = a + bX$ to the following data:

X	0	3	6
Y	1	4	5

In matrix form, what we want is a least squares solution to the system

$$a = 1$$
$$a + 3b = 4$$
$$a + 6b = 5.$$

(You can easily see no exact solution exists.) This system is $A\mathbf{z} = \mathbf{b}$, where

$$A = \begin{pmatrix} 1 & 0 \\ 1 & 3 \\ 1 & 6 \end{pmatrix}, \quad \mathbf{z} = \begin{pmatrix} a \\ b \end{pmatrix}, \quad \mathbf{b} = \begin{pmatrix} 1 \\ 4 \\ 5 \end{pmatrix}.$$

So a least squares solution is

$$\mathbf{z} = (A^T A)^{-1} A^T \mathbf{b} = \begin{pmatrix} 4/3 \\ 2/3 \end{pmatrix}.$$

(We've omitted the calculations, but you can check this.) So a best-fit linear relationship is

$$Y = \frac{4}{3} + \frac{2}{3} X.$$

Activity 12.32 Check the calculation in this example.

In principle, the least squares method can be used to fit many types of model to data. Here's another example.

Example 12.33 Quantities X, Y are related by a rule of the form

$$Y = \frac{m}{X} + c$$

for some constants m and c. Use the following data to estimate m and c by the least squares method:

X	1/5	1/4	1/3	1/2	1
Y	4	3	2	2	1

This is not a linear relationship between X and Y, but when we use the given values of X and Y we do still get a linear system for m and c. For, what we need is the least squares solution to the system

$$\frac{m}{1/5} + c = 4$$

$$\frac{m}{1/4} + c = 3$$

$$\frac{m}{1/3} + c = 2$$

$$\frac{m}{1/2} + c = 2$$

$$\frac{m}{1} + c = 1.$$

In matrix form, this is $A\mathbf{z} = \mathbf{b}$, where

$$A = \begin{pmatrix} 5 & 1 \\ 4 & 1 \\ 3 & 1 \\ 2 & 1 \\ 1 & 1 \end{pmatrix}, \quad \mathbf{z} = \begin{pmatrix} m \\ c \end{pmatrix}, \quad \mathbf{b} = \begin{pmatrix} 4 \\ 3 \\ 2 \\ 2 \\ 1 \end{pmatrix}.$$

So a least squares solution is

$$\mathbf{z} = (A^{\mathrm{T}} A)^{-1} A^{\mathrm{T}} \mathbf{b} = \begin{pmatrix} 7/10 \\ 3/10 \end{pmatrix}.$$

Therefore, the best fit is

$$Y = \frac{0.7}{X} + 0.3.$$

More complex relationships can also be examined. Suppose, for instance, that we believe that X and Y are related by

$$Y = a + bX + cX^2,$$

for some constants a, b, c. Suppose we have m data pairs (X_i, Y_i). Then what we want are values of a, b and c which are the best fit to the system

$$a + bX_1 + cX_1^2 = Y_1$$

$$\vdots \quad \vdots$$

$$a + bX_m + cX_m^2 = Y_m.$$

In matrix form, this is $A\mathbf{z} = \mathbf{b}$, where

$$A = \begin{pmatrix} 1 & X_1 & X_1^2 \\ 1 & X_2 & X_2^2 \\ \vdots & \vdots & \vdots \\ 1 & X_m & X_m^2 \end{pmatrix}, \quad \mathbf{z} = \begin{pmatrix} a \\ b \\ c \end{pmatrix}, \quad \mathbf{b} = \begin{pmatrix} Y_1 \\ \vdots \\ Y_m \end{pmatrix}.$$

Then, assuming A has rank 3, the theory above tells us that a least squares solution will be $\mathbf{z} = (A^{\mathrm{T}} A)^{-1} A^{\mathrm{T}} \mathbf{b}$.

12.8 Learning outcomes

You should now be able to:

- explain what is meant by the sum of two subspaces of a vector space
- explain what it means to say that a sum of two subspaces is a direct sum
- demonstrate that you know how to prove that a sum is direct
- show that a sum of subspaces is direct
- state the definition of the orthogonal complement of a subspace and be able to prove properties of the orthogonal complement
- determine the orthogonal complement of a subspace
- demonstrate that you know that, for a matrix A, $R(A)^{\perp} = N(A^{\mathrm{T}})$ and $N(A)^{\perp} = R(A^{\mathrm{T}})$; be able to use these results

- state precisely what is meant by a projection and an orthogonal projection
- show that projections are linear transformations; show that a matrix represents a projection if and only if it is idempotent; and that a matrix represents an orthogonal projection if and only if it is symmetric and idempotent
- show that the matrix of an orthogonal projection onto $R(A)$, for a given $m \times n$ matrix A of rank n, is $P = A(A^{\mathrm{T}}A)^{-1}A^{\mathrm{T}}$, and be able to use this to determine such projections
- demonstrate an understanding of the rationale behind least squares approximation
- explain why a least squares solution to $A\mathbf{z} = \mathbf{b}$ when A is an $m \times n$ matrix of rank n is $\mathbf{z} = (A^{\mathrm{T}}A)^{-1}A^{\mathrm{T}}\mathbf{b}$; and use this in numerical examples to determine a least squares solution.

12.9   Comments on activities

Activity 12.2 Since $\mathbf{0} \in U$ and $\mathbf{0} \in W$, we have $\mathbf{0} = \mathbf{0} + \mathbf{0} \in U + W$ and hence $U + W \neq \emptyset$. Suppose that $\mathbf{v}, \mathbf{v}' \in U + W$, so for some $\mathbf{u}, \mathbf{u}' \in U$ and $\mathbf{w}, \mathbf{w}' \in W$, $\mathbf{v} = \mathbf{u} + \mathbf{w}$ and $\mathbf{v}' = \mathbf{u}' + \mathbf{w}'$. For scalars α, β, we have

$$\alpha \mathbf{v} + \beta \mathbf{v}' = \alpha(\mathbf{u} + \mathbf{w}) + \beta(\mathbf{u}' + \mathbf{w}') = \underbrace{(\alpha \mathbf{u} + \beta \mathbf{u}')}_{\in U} + \underbrace{(\alpha \mathbf{w} + \beta \mathbf{w}')}_{\in W},$$

which is in $U + W$.

Activity 12.3 Since $\mathrm{Lin}\{\mathbf{u}\}$ is the set of all vectors of the form $\alpha \mathbf{u}$ and since $\mathrm{Lin}\{\mathbf{w}\}$ is the set of all vectors of the form $\beta \mathbf{w}$, it follows that

$$\mathrm{Lin}\{\mathbf{u}\} + \mathrm{Lin}\{\mathbf{w}\} = \{\mathbf{x} + \mathbf{y} \mid \mathbf{x} \in \mathrm{Lin}\{\mathbf{u}\}, \mathbf{y} \in \mathrm{Lin}\{\mathbf{w}\}\}$$
$$= \{\alpha \mathbf{u} + \beta \mathbf{w} \mid \alpha, \beta \in \mathbb{R}\}$$
$$= \mathrm{Lin}\{\mathbf{u}, \mathbf{w}\}.$$

Activity 12.9 We have $S^{\perp} \neq \emptyset$ because $\langle \mathbf{0}, \mathbf{s} \rangle = 0$ for all $\mathbf{s} \in S$ and hence $\mathbf{0} \in S^{\perp}$. Suppose $\mathbf{u}, \mathbf{v} \in S^{\perp}$. So, for all $\mathbf{s} \in S$, $\langle \mathbf{u}, \mathbf{s} \rangle = \langle \mathbf{v}, \mathbf{s} \rangle = 0$. Then, for scalars α, β, for all $\mathbf{s} \in S$,

$$\langle \alpha \mathbf{u} + \beta \mathbf{v}, \mathbf{s} \rangle = \alpha \langle \mathbf{u}, \mathbf{s} \rangle + \beta \langle \mathbf{v}, \mathbf{s} \rangle = \alpha 0 + \beta 0 = 0,$$

so $\alpha \mathbf{u} + \beta \mathbf{v} \in S^{\perp}$.

Activity 12.12 You could find the equation of the plane using

$$\begin{vmatrix} 1 & 1 & x \\ 2 & 0 & y \\ -1 & 1 & z \end{vmatrix} = 2x - 2y - 2z = 0.$$

So the plane has Cartesian equation, $x - y - z = 0$ with a normal vector $\mathbf{n} = (1, -1, -1)^\mathrm{T}$ as a basis of $S^\perp$.

Activity 12.18 If the sum is not direct, then it won't be the case that every vector can be written uniquely in the form $\mathbf{u} + \mathbf{w}$. If

$$\mathbf{v} = \mathbf{u} + \mathbf{w} = \mathbf{u}' + \mathbf{w}'$$

are two different such expressions for $\mathbf{v}$, then it is not possible to define $P_U(\mathbf{v})$ without ambiguity: is it $\mathbf{u}$ or $\mathbf{u}'$? The definition does not make sense in this case.

Activity 12.24 If λ is an eigenvalue of A, then $A\mathbf{v} = \lambda\mathbf{v}$, where $\mathbf{v} \neq \mathbf{0}$ is a corresponding eigenvector of A. If A is idempotent, then also $A\mathbf{v} = A^2\mathbf{v} = A(\lambda\mathbf{v}) = \lambda(A\mathbf{v}) = \lambda(\lambda\mathbf{v}) = \lambda^2\mathbf{v}$, so we have $\lambda^2\mathbf{v} = \lambda\mathbf{v}$ or $(\lambda^2 - \lambda)\mathbf{v} = \mathbf{0}$. Since $\mathbf{v} \neq \mathbf{0}$, we conclude that $\lambda^2 - \lambda = \lambda(\lambda - 1) = 0$ with $\lambda = 0$ or $\lambda = 1$ as the only solutions.

12.10 Exercises

Exercise 12.1 Suppose S is a subspace of $\mathbb{R}^n$. Prove that

$$\dim(S) + \dim(S^\perp) = n.$$

Exercise 12.2 Suppose S is a subspace of $\mathbb{R}^n$. Prove that $S \subseteq (S^\perp)^\perp$. Prove also that $\dim(S) = \dim((S^\perp)^\perp)$. (You may assume the result of the previous exercise.) Hence, deduce that $(S^\perp)^\perp = S$.

Exercise 12.3 What is the orthogonal projection of $\mathbb{R}^4$ onto the subspace spanned by the vectors $(1, 0, 1, 0)^\mathrm{T}$ and $(1, 2, 1, 2)^\mathrm{T}$?

Exercise 12.4 Let A be an $n \times n$ idempotent matrix which is diagonalisable. Show that $R(A)$, the range of A, is equal to the eigenspace corresponding to the eigenvalue $\lambda = 1$.

Exercise 12.5 Consider the matrix

$$A = \begin{pmatrix} 5 & -8 & -4 \\ 3 & -5 & -3 \\ -1 & 2 & 2 \end{pmatrix},$$

which you diagonalised in Exercise 9.10. Show that A is idempotent.

Let T denote the linear transformation $T : \mathbb{R}^3 \to \mathbb{R}^3$ given by $T(\mathbf{x}) = A\mathbf{x}$. Deduce that T is a projection. Show that T is the projection from $\mathbb{R}^3$ onto the eigenspace corresponding to the eigenvalue $\lambda = 1$, parallel to the eigenspace corresponding to $\lambda = 0$. Is this an orthogonal projection?

Exercise 12.6 Find a least squares fit by a function of the form $Y = a + bX$ to the following data:

X	-1	0	1	2
Y	0	1	3	9

Exercise 12.7 Suppose we want to model the relationship between X and Y by

$$Y = a + bX + cX^2$$

for some constants a, b, c. Find a least squares solution for a, b, c given the following data:

X	0	1	2	3
Y	3	2	4	4

12.11   Problems

Problem 12.1 Suppose that

$$X = \text{Lin} \left\{ \begin{pmatrix} 1 \\ 0 \\ 1 \\ 0 \end{pmatrix}, \begin{pmatrix} 0 \\ 0 \\ 0 \\ 1 \end{pmatrix} \right\} \quad \text{and} \quad Y = \text{Lin} \left\{ \begin{pmatrix} 0 \\ 1 \\ 0 \\ 0 \end{pmatrix}, \begin{pmatrix} 1 \\ 0 \\ 1 \\ -1 \end{pmatrix} \right\}.$$

Is the sum $X + Y$ direct? If so, why, and if not, why not? Find a basis for $X + Y$.

Problem 12.2 Let

$$Y = \text{Lin} \left\{ \begin{pmatrix} 1 \\ 3 \\ -1 \\ 1 \end{pmatrix}, \begin{pmatrix} 1 \\ 4 \\ 0 \\ 2 \end{pmatrix} \right\} \subset \mathbb{R}^4.$$

Find a basis of $Y^\perp$.

Problem 12.3 Let U and V be subspaces of an inner product space X, and let $U^\perp$ and $V^\perp$ be their orthogonal complements. Prove that

$$(U + V)^\perp = U^\perp \cap V^\perp.$$

Problem 12.4 Suppose that $\mathbf{u}, \mathbf{w} \in \mathbb{R}^2$ are the vectors

$$\mathbf{u} = \begin{pmatrix} -1 \\ 2 \end{pmatrix}, \quad \mathbf{w} = \begin{pmatrix} -3 \\ 5 \end{pmatrix}.$$

Using the definition of a direct sum, show that $\mathbb{R}^2 = \text{Lin}\{\mathbf{u}\} \oplus \text{Lin}\{\mathbf{w}\}$.

(a) Find the projection P of $\mathbb{R}^2$ onto $U = \text{Lin}\{\mathbf{u}\}$ parallel to $W = \text{Lin}\{\mathbf{w}\}$. Find the image $P(\mathbf{e}_1)$ of the vector $\mathbf{e}_1 = (1, 0)^\mathrm{T}$.

(b) Find the orthogonal projection T from $\mathbb{R}^2$ onto $\text{Lin}\{\mathbf{u}\}$. Then find the image of $\mathbf{e}_1 = (1, 0)^\mathrm{T}$ under this linear transformation.

Problem 12.5 Suppose $p \geq 1$ and that the $n \times n$ real matrix A satisfies $A^{p+1} = A^p$. Prove that $A^j = A^p$ for all $j \geq p$ and that

$$\mathbb{R}^n = R(A^p) \oplus N(A^p).$$

Problem 12.6 Let $\mathbf{x} \in \mathbb{R}^n$ and let S be the subspace $\text{Lin}\{\mathbf{x}\}$ spanned by $\mathbf{x}$. Show that the orthogonal projection matrix P of $\mathbb{R}^n$ onto S is

$$P = \frac{1}{\|\mathbf{x}\|^2} \mathbf{x}\mathbf{x}^\mathrm{T}.$$

Problem 12.7 If $\mathbf{z} = (2, -3, 2, -1)^\mathrm{T}$, find the matrix of the orthogonal projection of $\mathbb{R}^4$ onto $\text{Lin}\{\mathbf{z}\}$.

Problem 12.8 Let X be the subspace of $\mathbb{R}^3$ spanned by the vectors $(0, 1, 1)^\mathrm{T}$ and $(2, 1, -1)^\mathrm{T}$. Find a basis of $X^\perp$.

Find the matrix P representing the orthogonal projection of $\mathbb{R}^3$ onto X. Is P diagonalisable?

Find an eigenvector of P corresponding to the eigenvalue 0 and an eigenvector corresponding to the eigenvalue 1.

Problem 12.9 Show that the matrix

$$P = \frac{1}{9} \begin{pmatrix} 1 & 2 & -2 & 0 \\ 2 & 7 & -1 & 3 \\ -2 & -1 & 7 & 3 \\ 0 & 3 & 3 & 3 \end{pmatrix}$$

is idempotent. Why can you conclude that P represents an orthogonal projection from $\mathbb{R}^4$ to a subspace Y of $\mathbb{R}^4$?

State what is meant by an orthogonal projection. Find subspaces Y and $Y^\perp$ such that P is the orthogonal projection of $\mathbb{R}^4$ onto Y. (Write down a basis for each subspace.)

Problem 12.10 Suppose A is an $n \times n$ diagonalisable matrix and that the only eigenvalues of A are 0 and 1. Show that A is idempotent.

Deduce that the linear transformation defined by $T(\mathbf{x}) = A\mathbf{x}$ is a projection from $\mathbb{R}^n$ onto the eigenspace corresponding to the eigenvalue $\lambda = 1$ parallel to the eigenspace corresponding to $\lambda = 0$.

Problem 12.11 Let U be the plane in $\mathbb{R}^3$ given by

$$U = \left\{ \begin{pmatrix} x \\ y \\ z \end{pmatrix} \ \middle| \ 2x - y - 2z = 0 \right\}.$$

Find the nearest point (position vector) in U to the point whose position vector is $\mathbf{p} = (1, 1, 2)^{\mathrm{T}}$.

Problem 12.12 Quantities x, y are known to be related by a rule of the form $y = ax + b$ for some constants a and b. Readings are taken of y at various values of x, resulting in the following measurements:

x	2	4	5	6
y	13	17	22	25

Find the least squares estimate of a and b.

Problem 12.13 Suppose we want to find the least-squares line $y = m^*x + c^*$ through the data points $(x_1, y_1), (x_2, y_2), \ldots, (x_n, y_n)$. Show that the parameters m^* and c^* of the least-squares line are as follows:

$$m^* = \frac{n \sum_{i=1}^n x_i y_i - \sum_{i=1}^n x_i \sum_{i=1}^n y_i}{n \sum_{i=1}^n x_i^2 - \left(\sum_{i=1}^n x_i \right)^2},$$

$$c^* = \frac{\sum_{i=1}^n y_i \sum_{i=1}^n x_i^2 - \sum_{i=1}^n x_i \sum_{i=1}^n x_i y_i}{n \sum_{i=1}^n x_i^2 - \left(\sum_{i=1}^n x_i \right)^2}.$$

(These formulae might be familiar from statistics courses.)

13

Complex matrices and vector spaces

A complex matrix is a matrix whose entries are complex numbers. A complex vector space is one for which the scalars are complex numbers. We shall see that many of the results we have established for real matrices and real vector spaces carry over immediately to complex ones, but there are also some significant differences.

In this chapter, we explore these similarities and differences. We look at eigenvalues and eigenvectors of a complex matrix and investigate unitary diagonalisation, the complex analogue of orthogonal diagonalisation. Certain results for real matrices and vector spaces (such as the result that the eigenvalues of a symmetric matrix are real) are easily seen as special cases of their complex counterparts.

We begin with a careful review of complex numbers.

13.1 Complex numbers

Consider the two quadratic polynomials, $p(x) = x^2 - 3x + 2$ and $q(x) = x^2 + x + 1$. If you sketch the graph of $p(x)$, you will find that the graph intersects the x axis at the two real solutions (or roots) of the equation $p(x) = 0$, and that the polynomial factorises into two linear factors: $p(x) = x^2 - 3x + 2 = (x - 1)(x - 2)$. Sketching the graph of $q(x)$, you will find that it does not intersect the x axis. The equation $q(x) = 0$ has no solution in the real numbers, and it cannot be factorised over the reals. Such a polynomial is said to be *irreducible*. In order to solve this equation, we need to use *complex numbers*.

13.1.1   Complex numbers

We begin by defining an *imaginary* number, which we denote by the letter i and which has the property that $i^2 = -1$. The term 'imaginary' is historical, and not an indication that this is a figment of someone's imagination.

Definition 13.1 A complex number is a number of the form $z = a + ib$, where a and b are real numbers, and $i^2 = -1$. The set of all such numbers is

$$\mathbb{C} = \{a + ib \mid a, b \in \mathbb{R}\}.$$

If $z = a + ib$ is a complex number, then the real number a is known as the real part of z, denoted $\mathrm{Re}(z)$, and the real number b is the imaginary part of z, denoted $\mathrm{Im}(z)$. Note that $\mathrm{Im}(z)$ is a *real* number.

    If $b = 0$, then $z = a + ib$ is just the real number a, so $\mathbb{R} \subseteq \mathbb{C}$. If $a = 0$, then $z = ib$ is said to be *purely imaginary*.

    The quadratic polynomial $q(z) = x^2 + x + 1$ can be factorised over the complex numbers, because the equation $q(z) = 0$ has two complex solutions. Solving in the usual way, we have

$$x = \frac{-1 \pm \sqrt{-3}}{2}.$$

We write $\sqrt{-3} = \sqrt{(-1)3} = \sqrt{-1}\,\sqrt{3} = i\sqrt{3}$, so that the solutions are

$$w = -\frac{1}{2} + i\frac{\sqrt{3}}{2} \qquad \text{and} \qquad \overline{w} = -\frac{1}{2} - i\frac{\sqrt{3}}{2}.$$

Notice the form of these two solutions. They are what is called a *conjugate pair*. We have the following definition:

Definition 13.2 (Complex conjugate) If $z = a + ib$ is a complex number, then the *complex conjugate* of z is the complex number $\overline{z} = a - ib$.

We can see by the application of the quadratic formula that the roots of an irreducible quadratic polynomial with real coefficients will always be a conjugate pair of complex numbers.

13.1.2   Algebra of complex numbers

Addition and *multiplication* of complex numbers are defined by treating the numbers as polynomials in i, and using $i^2 = -1$.

Example 13.3 If $z = (1 + i)$ and $w = (4 - 2i)$, then

$$z + w = (1 + i) + (4 - 2i) = (1 + 4) + i(1 - 2) = 5 - i$$

and

$$zw = (1 + i)(4 - 2i) = 4 + 4i - 2i - 2i^2 = 6 + 2i.$$

If $z \in \mathbb{C}$, then $z\bar{z}$ is a real number:

$$z\bar{z} = (a + ib)(a - ib) = a^2 + b^2.$$

Activity 13.4 Carry out the multiplication to verify that $z\bar{z} = a^2 + b^2$.

Division of complex numbers is then defined by

$$\frac{z}{w} = \frac{z\bar{w}}{w\bar{w}},$$

noting that $w\bar{w}$ is real.

Example 13.5

$$\frac{1 + i}{4 - 2i} = \frac{(1 + i)(4 + 2i)}{(4 - 2i)(4 + 2i)} = \frac{2 + 6i}{16 + 4} = \frac{1}{10} + \frac{3}{10}i.$$

We now look at some properties of the complex conjugate. A complex number is real if and only if $z = \bar{z}$. Indeed, if $z = a + ib$, then $z = \bar{z}$ if and only if $b = 0$.

The complex conjugate of a complex number satisfies the following properties:

- $z + \bar{z} = 2\,\text{Re}(z)$ is real,
- $z - \bar{z} = 2i\,\text{Im}(z)$ is purely imaginary,
- $\bar{\bar{z}} = z$,
- $\overline{z + w} = \bar{z} + \bar{w}$,
- $\overline{zw} = \bar{z}\,\bar{w}$,
- $\overline{\left(\dfrac{z}{w}\right)} = \dfrac{\bar{z}}{\bar{w}}.$

Activity 13.6 Let $z = a + ib$, $w = c + id$ and verify all of the above properties.

13.1.3 Roots of polynomials

The *Fundamental Theorem of Algebra* asserts that a polynomial of degree n with complex coefficients has n complex roots (not necessarily distinct), and can therefore be factorised into n linear factors. If the

coefficients are restricted to real numbers, the polynomial can be factorised into a product of linear and irreducible quadratic factors over $\mathbb{R}$
and into a product of *linear* factors over $\mathbb{C}$. The proof of the *Fundamental Theorem of Algebra* is beyond the scope of this text. However,
we note the following useful result:

Theorem 13.7 *Complex roots of polynomials with real coefficients
appear in conjugate pairs.*

Proof: Let $P(x) = a_0 + a_1 x + \cdots + a_n x^n$, $a_i \in \mathbb{R}$, be a polynomial
of degree n. We shall show that if z is a root of $P(x)$, then so is $\bar{z}$.

Let z be a complex number such that $P(z) = 0$, then

$$a_0 + a_1 z + a_2 z^2, + \cdots + a_n z^n = 0.$$

Conjugating both sides of this equation,

$$\overline{a_0 + a_1 z + a_2 z^2 + \cdots + a_n z^n} = \bar{0} = 0.$$

Since 0 is a real number, it is equal to its complex conjugate. We
now use the following properties of the complex conjugate: that
the complex conjugate of the sum is the sum of the conjugates, and
the complex conjugate of a product is the product of the conjugates. We
have

$$\overline{a_0} + \overline{a_1 z} + \overline{a_2 z^2} + \cdots + \overline{a_n z^n} = 0,$$

and

$$\overline{a_0} + \overline{a_1}\bar{z} + \overline{a_2}\bar{z}^2 + \cdots + \overline{a_n}\bar{z}^n = 0.$$

Since the coefficients a_i are real numbers, this becomes

$$a_0 + a_1 \bar{z} + a_2 \bar{z}^2 + \cdots + a_n \bar{z}^n = 0.$$

That is, $P(\bar{z}) = 0$, so the number $\bar{z}$ is also a root of $P(x)$. $\qquad\square$

Example 13.8 Let us consider the polynomial

$$x^3 - 2x^2 - 2x - 3 = (x - 3)(x^2 + x + 1).$$

If

$$w = -\frac{1}{2} + i\frac{\sqrt{3}}{2},$$

then

$$x^3 - 2x^2 - 2x - 3 = (x - 3)(x - w)(x - \bar{w}).$$

Activity 13.9 Multiply out the last two factors above to check that
their product is the irreducible quadratic $x^2 + x + 1$.

Figure 13.1
Complex plane or
Argand diagram

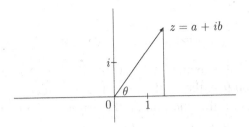

13.1.4 The complex plane

The following theorem shows that a complex number is uniquely determined by its real and imaginary parts.

Theorem 13.10 *Two complex numbers are equal if and only if their real and imaginary parts are equal.*

Proof: Two complex numbers with the same real parts and the same imaginary parts are clearly the same complex number, so we only need to prove this statement in one direction. Let $z = a + ib$ and $w = c + id$. If $z = w$, we will show that their real and imaginary parts are equal. We have $a + ib = c + id$, therefore $a - c = i(d - b)$. Squaring both sides, we obtain $(a - c)^2 = i^2(d - b)^2 = -(d - b)^2$. But $a - c$ and $(d - b)$ are real numbers, so their squares are non-negative. The only way in which this equality can hold is if we have $a - c = d - b = 0$; that is, $a = c$ and $b = d$. $\qquad\square$

As a result of this theorem, we can think of the complex numbers geometrically, as points in a plane. For, we can associate the vector $(a, b)^{\mathrm{T}}$ uniquely to each complex number $z = a + ib$, and all the properties of a two-dimensional real vector space apply. A complex number $z = a + ib$ is represented as a point (a, b) in the complex plane: we draw two axes, a horizontal axis to represent the real parts of complex numbers and a vertical axis to represent the imaginary parts of complex numbers, as in Figure 13.1. Points on the horizontal axis represent real numbers, and points on the vertical axis represent purely imaginary numbers.

Activity 13.11 Plot $z = 2 + 2i$ and $w = 1 - i\sqrt{3}$ in the complex plane.

13.1.5 Polar form

If the complex number $z = a + ib$ is plotted as a point (a, b) in the complex plane, then we can determine the polar coordinates of this point. We have

$$a = r \cos \theta, \quad b = r \sin \theta,$$

where $r = \sqrt{a^2 + b^2}$ is the length of the line joining the origin to the point (a, b), and θ is the angle measured anticlockwise from the real (horizontal) axis to the line joining the origin to the point (a, b). Then we can write $z = a + ib = r \cos \theta + i r \sin \theta$.

Definition 13.12 The *polar form* of the complex number z is

$$z = r(\cos \theta + i \sin \theta).$$

The length $r = \sqrt{a^2 + b^2}$ is called the *modulus* of z, denoted $|z|$, and the angle θ is called the *argument* of z.

Note the following properties:

- z and $\bar{z}$ are reflections in the real axis. If θ is the argument of z, then $-\theta$ is the argument of $\bar{z}$.
- $|z|^2 = z\bar{z}$.
- θ and $\theta + 2n\pi$ give the same complex number.

We define the *principal argument* of z to be the argument in the range $-\pi < \theta \le \pi$, and it is often denoted $\text{Arg}(z)$.

Activity 13.13 Express $z = 2 + 2i$, $w = 1 - i\sqrt{3}$ in polar form.

Activity 13.14 Describe the following sets of complex numbers:

(a) $\{z \mid |z| = 3\}$,
(b) $\{z \mid \text{Arg}(z) = \pi/4\}$.

Multiplication and division using polar coordinates gives

$$zw = r(\cos \theta + i \sin \theta) \cdot \rho(\cos \phi + i \sin \phi)$$
$$= r\rho(\cos(\theta + \phi) + i \sin(\theta + \phi))$$
$$\frac{z}{w} = \frac{r}{\rho}(\cos(\theta - \phi) + i \sin(\theta - \phi)).$$

Activity 13.15 Show these by performing the multiplication and the division as defined earlier, and by using the facts (trigonometric identities) that $\cos(\theta + \phi) = \cos \theta \cos \phi - \sin \theta \sin \phi$ and $\sin(\theta + \phi) = \sin \theta \cos \phi + \cos \theta \sin \phi$.

We consider explicitly a special case of the multiplication result above, in which $w = z$. Suppose that $z = r(\cos \theta + i \sin \theta)$. If we apply the above multiplication rule to determine $z^2 = zz$, we have

$$z^2 = zz$$
$$= (r(\cos\theta + i\sin\theta))(r(\cos\theta + i\sin\theta))$$
$$= r^2(\cos^2\theta + i^2\sin^2\theta + 2i\sin\theta\cos\theta)$$
$$= r^2(\cos^2\theta - \sin^2\theta + 2i\sin\theta\cos\theta)$$
$$= r^2(\cos 2\theta + i\sin 2\theta).$$

Here we have used the double angle formulae for $\cos 2\theta$ and $\sin 2\theta$.

Applying the product rule n times, where n is a positive integer, we have

$$z^n = \underbrace{z \cdots z}_{n \text{ times}}$$
$$= (r(\cos\theta + i\sin\theta))^n$$
$$= r^n\left(\cos(\underbrace{\theta + \cdots + \theta}_{n \text{ times}}) + i\sin(\underbrace{\theta + \cdots + \theta}_{n \text{ times}})\right).$$

From the two expressions on the right, we conclude *DeMoivre's formula (or theorem)*.

Theorem 13.16 (DeMoivre's theorem)

$$(\cos\theta + i\sin\theta)^n = \cos n\theta + i\sin n\theta.$$

13.1.6 Exponential form and Euler's formula

You may be aware that standard functions of a real variable can often be defined by power series (or Taylor or Maclaurin expansions). These power series definitions can also be used when the variable is complex. In particular, we have

$$e^z = 1 + z + \frac{z^2}{2!} + \frac{z^3}{3!} + \cdots$$

$$\sin z = z - \frac{z^3}{3!} + \frac{z^5}{5!} - \cdots$$

$$\cos z = 1 - \frac{z^2}{2!} + \frac{z^4}{4!} - \cdots$$

If we use the expansion for e^z to expand $e^{i\theta}$, and then factor out the real and imaginary parts, we find:

$$e^{i\theta} = 1 + (i\theta) + \frac{(i\theta)^2}{2!} + \frac{(i\theta)^3}{3!} + \frac{(i\theta)^4}{4!} + \frac{(i\theta)^5}{5!} + \cdots$$

$$= 1 + i\theta - \frac{\theta^2}{2!} - i\frac{\theta^3}{3!} + \frac{\theta^4}{4!} + i\frac{\theta^5}{5!} - \cdots$$

$$= \left(1 - \frac{\theta^2}{2!} + \frac{\theta^4}{4!} - \cdots\right) + i\left(\theta - \frac{\theta^3}{3!} + \frac{\theta^5}{5!} - \cdots\right).$$

From this, we may conclude *Euler's formula*, which is as follows:

$$e^{i\theta} = \cos\theta + i\sin\theta.$$

Using this result, we obtain the *exponential form* of a complex number.

Definition 13.17 The *exponential form* of a complex number $z = a + ib$ is

$$z = re^{i\theta},$$

where $r = |z|$ is the modulus of z and θ is the argument of z.

Example 13.18 Using the exponential form, we can write

$$e^{i\pi} + 1 = 0,$$

which combines the numbers e, π and i in a single expression.

If $z = re^{i\theta}$, then its complex conjugate is given by $\bar{z} = re^{-i\theta}$. This is because, if $z = re^{i\theta} = r(\cos\theta + i\sin\theta)$, then

$$\bar{z} = r(\cos\theta - i\sin\theta) = r(\cos(-\theta) + i\sin(-\theta)) = re^{-i\theta}.$$

We can use either the exponential form, $z = re^{i\theta}$, or the standard form, $z = a + ib$, according to the application or computation we are doing. For example, addition is simplest in the form $z = a + ib$, but multiplication and division are simpler when working with the exponential form. To change a complex number between $re^{i\theta}$ and $a + ib$, we use Euler's formula and the complex plane (polar form).

Example 13.19 Here are two examples of converting from exponential form to standard form:

$$e^{i2\pi/3} = \cos\left(\frac{2\pi}{3}\right) + i\sin\left(\frac{2\pi}{3}\right) = -\frac{1}{2} + i\frac{\sqrt{3}}{2}.$$

$$e^{2+i\sqrt{3}} = e^2 e^{i\sqrt{3}} = e^2\cos\sqrt{3} + ie^2\sin\sqrt{3}.$$

Activity 13.20 Write each of the following complex numbers in the form $a + ib$:

$$e^{i\pi/2}, \qquad e^{i3\pi/2}, \qquad e^{i5\pi/4}, \qquad e^{i11\pi/6}, \qquad e^{2-i}, \qquad e^{-3}.$$

Example 13.21 Let $z = 2 + 2i = 2\sqrt{2}\,e^{i\frac{\pi}{4}}$ and $w = 1 - i\sqrt{3} = 2e^{-i\frac{\pi}{3}}$. Then

$$w^6 = (1 - i\sqrt{3})^6 = (2e^{-i\frac{\pi}{3}})^6 = 2^6 e^{-i2\pi} = 64,$$

$$zw = (2\sqrt{2}e^{i\frac{\pi}{4}})(2e^{-i\frac{\pi}{3}}) = 4\sqrt{2}e^{-i\frac{\pi}{12}}, \qquad \frac{z}{w} = \sqrt{2}e^{i\frac{7\pi}{12}}.$$

You can see that these calculations are relatively straightforward using the exponential form of the complex numbers, but they would be quite involved if we used the standard form.

Notice that in Example 13.21, we are using certain properties of the complex exponential function, specifically that, if $z, w \in \mathbb{C}$, then

$$e^{z+w} = e^z e^w \qquad \text{and} \qquad (e^z)^n = e^{nz} \qquad \text{for } n = 1, 2, 3, \dots.$$

This last property is, in fact, DeMoivre's theorem, and it is easily generalised to include all integers.

Use of the exponential form sometimes makes solving equations easier, as the following example shows:

Example 13.22 We solve the equation $z^6 = -1$ to find the 6th roots of -1.

Writing $z = re^{i\theta}$, we have $z^6 = (re^{i\theta})^6 = r^6 e^{i6\theta}$, and

$$-1 = e^{i\pi} = e^{i(\pi + 2n\pi)} \qquad \text{for } n \in \mathbb{Z}.$$

So we need to solve

$$r^6 e^{i6\theta} = e^{i(\pi + 2n\pi)}.$$

Using the fact that r is a real positive number, we have $r = 1$ and $6\theta = \pi + 2n\pi$, so

$$\theta = \frac{\pi}{6} + \frac{2n\pi}{6}.$$

This will give the six complex roots by taking $n = 0, 1, 2, 3, 4, 5$.

Activity 13.23 Show this. Write down the sixth roots of -1 and show that any one raised to the power 6 is equal to -1. Show that $n = 6$ gives the same root as $n = 0$. Use this to factor the polynomial $x^6 + 1$ into linear factors over the complex numbers, and into irreducible quadratics over the real numbers.

13.2   Complex vector spaces

A vector space where the scalars are complex numbers is called a complex vector space. The following definition is the same as Definition 5.1 except that the scalars are complex numbers.

Definition 13.24 (Complex vector space) A complex *vector space V* is a non-empty set of objects, called *vectors*, equipped with an addition operation and a scalar multiplication operation such that for all α, $\beta \in \mathbb{C}$ and all $\mathbf{u}$, $\mathbf{v}$, $\mathbf{w} \in V$:

1. $\mathbf{u} + \mathbf{v} \in V$ (*closure* under addition).
2. $\mathbf{u} + \mathbf{v} = \mathbf{v} + \mathbf{u}$ (the *commutative* law for addition).
3. $\mathbf{u} + (\mathbf{v} + \mathbf{w}) = (\mathbf{u} + \mathbf{v}) + \mathbf{w}$ (the *associative* law for addition).
4.   There is a single member $\mathbf{0}$ of V, called the *zero vector*, such that for all $\mathbf{v} \in V$, $\mathbf{v} + \mathbf{0} = \mathbf{v}$.
5.   For every $\mathbf{v} \in V$, there is an element $\mathbf{w} \in V$ (usually written as $-\mathbf{v}$), called the *negative* of $\mathbf{v}$, such that $\mathbf{v} + \mathbf{w} = \mathbf{0}$.
6. $\alpha\mathbf{v} \in V$ (*closure* under scalar multiplication).
7. $\alpha(\mathbf{u} + \mathbf{v}) = \alpha\mathbf{u} + \alpha\mathbf{v}$ (*distributive* law).
8. $(\alpha + \beta)\mathbf{v} = \alpha\mathbf{v} + \beta\mathbf{v}$ (*distributive* law).
9. $\alpha(\beta\mathbf{v}) = (\alpha\beta)\mathbf{v}$ (*associative* law).
10. $1\mathbf{v} = \mathbf{v}$.

Example 13.25 The set $\mathbb{C}^n$ of n-tuples of complex numbers is a complex vector space. Just as in $\mathbb{R}^n$, we will write a vector as a column,

$$\mathbf{v} = \begin{pmatrix} v_1 \\ v_2 \\ \vdots \\ v_n \end{pmatrix} \qquad v_i \in \mathbb{C}.$$

Addition and scalar multiplication are defined component-wise, exactly as in $\mathbb{R}^n$.

Example 13.26 The set $M_2(\mathbb{C})$ of 2×2 matrices with complex entries is a complex vector space under matrix addition and scalar multiplication.

Most of the results established in Chapter 3 for real vector spaces carry over immediately to a complex vector space V. All that is necessary is to change any reference from real numbers to complex numbers. A linear combination of vectors has the same meaning, except that the coefficients are complex numbers. That is, $\mathbf{w} \in V$ is a linear combination of $\mathbf{v}_1, \mathbf{v}_2, \ldots, \mathbf{v}_k \in V$ if

$$\mathbf{w} = a_1\mathbf{v}_1 + a_2\mathbf{v}_2 + \cdots + a_k\mathbf{v}_k \qquad a_i \in \mathbb{C}.$$

The concepts of subspace, linear span, linear independence, basis and dimension carry over in the same way. Theorems about $\mathbb{R}^n$ continue to hold with $\mathbb{R}^n$ changed to $\mathbb{C}^n$.

Example 13.27 Suppose that, for $i = 1, 2, \ldots, n$, the vector $\mathbf{e}_i$ has every entry equal to 0 except for the ith, which is 1. Then the vectors $\mathbf{e}_1, \mathbf{e}_2, \ldots, \mathbf{e}_n$ form a basis of $\mathbb{C}^n$. For any $\mathbf{z} = (z_1, z_2, \ldots, z_n)^T \in \mathbb{C}^n$,

$$\mathbf{z} = z_1\mathbf{e}_1 + z_2\mathbf{e}_2 + \cdots + z_n\mathbf{e}_n.$$

The basis $\{\mathbf{e}_1, \mathbf{e}_2, \ldots, \mathbf{e}_n\}$ is called the *standard basis* of $\mathbb{C}^n$, and $\mathbb{C}^n$ is an n-dimensional complex vector space.

Activity 13.28 $\mathbb{C}^n$ can also be considered as a $2n$-dimensional real vector space. Why? What is a basis for this space?

13.3 Complex matrices

We will refer to a matrix whose entries are complex numbers as a *complex matrix* for short, as opposed to a real matrix (one whose entries are real numbers). Sometimes, we will just use the term *matrix* for either, when this will not cause any confusion. If A is an $m \times n$ complex matrix, then we denote by $\overline{A}$ the $m \times n$ matrix whose (i, j) entry is the complex conjugate of the (i, j) entry of A. That is, if $A = (a_{ij})$, then $\overline{A} = (\overline{a_{ij}})$.

We can use row reduction to solve a system of equations $A\mathbf{x} = \mathbf{b}$, where A is an $m \times n$ complex matrix, $\mathbf{x} \in \mathbb{C}^n$ and $\mathbf{b} \in \mathbb{C}^m$. Results concerning the range and null space of a matrix which we established in previous chapters carry over immediately to complex matrices with the appropriate modifications. The null space is a subspace of $\mathbb{C}^n$ and the range, or column space, of the matrix is a subspace of $\mathbb{C}^m$.

The concepts of eigenvector and eigenvalue are the same for complex matrices as for real ones, and the same method is used to find them. In particular, by working in $\mathbb{C}^n$ rather than $\mathbb{R}^n$ we can now sometimes diagonalise real matrices with complex eigenvalues, as the following example shows:

Example 13.29 We find the eigenvalues and corresponding eigenvectors for the matrix

$$A = \begin{pmatrix} 0 & 1 \\ -1 & 0 \end{pmatrix}.$$

The characteristic equation is

$$|A - \lambda I| = \begin{vmatrix} -\lambda & 1 \\ -1 & -\lambda \end{vmatrix} = \lambda^2 + 1 = 0,$$

with complex roots $\lambda = \pm i$. We now find the corresponding eigenvectors.

For $\lambda_1 = i$, we solve $(A - iI)\mathbf{x} = \mathbf{0}$ by row reducing the coefficient matrix,

$$(A - iI) = \begin{pmatrix} -i & 1 \\ -1 & -i \end{pmatrix} \longrightarrow \begin{pmatrix} 1 & i \\ 0 & 0 \end{pmatrix}, \quad \text{so that } \mathbf{v}_1 = \begin{pmatrix} -i \\ 1 \end{pmatrix}.$$

In the same way, for $\lambda_2 = -i$,

$$(A + iI) = \begin{pmatrix} i & 1 \\ -1 & i \end{pmatrix} \longrightarrow \begin{pmatrix} 1 & -i \\ 0 & 0 \end{pmatrix}, \quad \text{so that } \mathbf{v}_2 = \begin{pmatrix} i \\ 1 \end{pmatrix}.$$

We can check that these eigenvectors are correct by showing that $A\mathbf{v}_1 = i\mathbf{v}_1$ and $A\mathbf{v}_2 = -i\mathbf{v}_2$.

Activity 13.30 Check that $A\mathbf{v}_1 = i\mathbf{v}_1$ and $A\mathbf{v}_2 = -i\mathbf{v}_2$.

Can we now diagonalise the matrix A in the same way as for real matrices? The answer is yes. If P is the matrix whose columns are the eigenvectors, and D is the diagonal matrix of corresponding eigenvalues, we will show that $P^{-1}AP = D$. We set

$$P = \begin{pmatrix} -i & i \\ 1 & 1 \end{pmatrix}, \quad D = \begin{pmatrix} i & 0 \\ 0 & -i \end{pmatrix}$$

and find P^{-1} exactly as with real matrices. We have $|P| = -2i$, so that

$$P^{-1} = -\frac{1}{2i} \begin{pmatrix} 1 & -i \\ -1 & -i \end{pmatrix}.$$

Then

$$P^{-1}AP = \frac{1}{2i} \begin{pmatrix} -1 & i \\ 1 & i \end{pmatrix} \begin{pmatrix} 0 & 1 \\ -1 & 0 \end{pmatrix} \begin{pmatrix} -i & i \\ 1 & 1 \end{pmatrix}$$

$$= \frac{1}{2i} \begin{pmatrix} -1 & i \\ 1 & i \end{pmatrix} \begin{pmatrix} 1 & 1 \\ i & -i \end{pmatrix}$$

$$= \frac{1}{2i} \begin{pmatrix} -2 & 0 \\ 0 & 2 \end{pmatrix} = \frac{1}{i} \begin{pmatrix} -1 & 0 \\ 0 & 1 \end{pmatrix}$$

$$= \begin{pmatrix} i & 0 \\ 0 & -i \end{pmatrix} = D.$$

Activity 13.31 Work through all the calculations in this example.

You may have noticed in this example that the eigenvectors of the complex conjugate eigenvalues, $\lambda_1 = i$ and $\lambda_2 = \overline{\lambda_1} = -i$ are complex conjugate vectors, $\mathbf{v}_2 = \overline{\mathbf{v}}_1$. This is true in general for real matrices.

Theorem 13.32 *If A is an $n \times n$ matrix with real entries and if λ is a complex eigenvalue with corresponding eigenvector $\mathbf{v}$, then $\overline{\lambda}$ is also an eigenvalue of A with corresponding eigenvector $\overline{\mathbf{v}}$.*

Proof: Since A is a real matrix, the characteristic equation of A is a polynomial of degree n with real coefficients, and hence any complex roots occur in conjugate pairs. This means that if λ is an eigenvalue of A, then so is $\overline{\lambda}$. If λ is an eigenvalue with corresponding eigenvector $\mathbf{v}$, then $A\mathbf{v} = \lambda\mathbf{v}$. Taking the complex conjugate of both sides, $\overline{A\mathbf{v}} = \overline{\lambda\mathbf{v}}$, which, since A is real, yields $A\overline{\mathbf{v}} = \overline{\lambda}\overline{\mathbf{v}}$. This says that $\overline{\mathbf{v}}$ is an eigenvector corresponding to $\overline{\lambda}$. □

13.4   Complex inner product spaces

13.4.1   The inner product on $\mathbb{C}^n$

The standard inner product of two vectors $\mathbf{x}, \mathbf{y}$ in $\mathbb{R}^n$ is the real number $\langle \mathbf{x}, \mathbf{y} \rangle$ given by

$$\langle \mathbf{x}, \mathbf{y} \rangle = \mathbf{x}^{\mathrm{T}}\mathbf{y} = x_1 y_1 + x_2 y_2 + \cdots + x_n y_n.$$

The norm of a vector $\mathbf{x}$ is given in terms of this inner product by $\|\mathbf{x}\| = \sqrt{\langle \mathbf{x}, \mathbf{x} \rangle}$. This definition of inner product will not work in $\mathbb{C}^n$. For example, if $\mathbf{x} = (1, 0, i)^{\mathrm{T}} \in \mathbb{C}^3$, then clearly $\mathbf{x} \neq \mathbf{0}$, but we would have

$$\|\mathbf{x}\|^2 = x_1^2 + x_2^2 + x_3^2 = 1^2 + 0 + i^2 = 1 - 1 = 0.$$

It seems we need to alter this definition to make it work in a complex vector space. A good guide to what should be done comes from the modulus of a complex number. If $z = a + ib$, then $|z|^2 = z\overline{z} = a^2 + b^2$ is a real non-negative number, and $|z| = 0$ only for $z = 0$.

Definition 13.33 For $\mathbf{x}, \mathbf{y} \in \mathbb{C}^n$, the *standard complex inner product* is defined to be the complex number $\langle \mathbf{x}, \mathbf{y} \rangle$ given by

$$\langle \mathbf{x}, \mathbf{y} \rangle = x_1 \overline{y}_1 + x_2 \overline{y}_2 + \cdots + x_n \overline{y}_n.$$

Example 13.34 If $\mathbf{x} = (1, 4i, 3 + i)^{\mathrm{T}}$ and $\mathbf{y} = (i, -3, 1 - 2i)^{\mathrm{T}}$, then

$$\langle \mathbf{x}, \mathbf{y} \rangle = 1(-i) + 4i(-3) + (3 + i)(1 + 2i) = -i - 12i + (1 + 7i)$$
$$= 1 - 6i.$$

Since

$$\langle \mathbf{x}, \mathbf{x} \rangle = x_1\overline{x}_1 + x_2\overline{x}_2 + \cdots + x_n\overline{x}_n = |x_1|^2 + |x_2|^2 + \cdots + |x_n|^2$$

is the sum of the squares of the moduli of the components of the vector $\mathbf{x}$, the inner product of a complex vector with itself is a non-negative real number. Then, the norm of the vector is $\|\mathbf{x}\| = \sqrt{\langle \mathbf{x}, \mathbf{x} \rangle}$, and $\|\mathbf{x}\| = 0$ if and only if $\mathbf{x}$ is the zero vector. This last statement is part of the following theorem:

Theorem 13.35 *The standard complex inner product*

$$\langle \mathbf{x}, \mathbf{y} \rangle = x_1\overline{y}_1 + x_2\overline{y}_2 + \cdots + x_n\overline{y}_n \qquad (\mathbf{x}, \mathbf{y} \in \mathbb{C}^n)$$

satisfies the following for all $\mathbf{x}, \mathbf{y}, \mathbf{z} \in \mathbb{C}^n$ *and for all* $\alpha, \beta \in \mathbb{C}$:

(i) $\langle \mathbf{x}, \mathbf{y} \rangle = \overline{\langle \mathbf{y}, \mathbf{x} \rangle}$
(ii) $\langle \alpha\mathbf{x} + \beta\mathbf{y}, \mathbf{z} \rangle = \alpha\langle \mathbf{x}, \mathbf{z} \rangle + \beta\langle \mathbf{y}, \mathbf{z} \rangle$
(iii) $\langle \mathbf{x}, \mathbf{x} \rangle \geq 0$, *and* $\langle \mathbf{x}, \mathbf{x} \rangle = 0$ *if and only if* $\mathbf{x} = \mathbf{0}$.

Proof: We have

$$\begin{aligned}
\langle \mathbf{x}, \mathbf{y} \rangle &= x_1\overline{y}_1 + x_2\overline{y}_2 + \cdots + x_n\overline{y}_n \\
&= \overline{y}_1 x_1 + \overline{y}_2 x_2 + \cdots + \overline{y}_n x_n \\
&= \overline{y_1\overline{x}_1} + \overline{y_2\overline{x}_2} + \cdots + \overline{y_n\overline{x}_n} \\
&= \overline{y_1\overline{x}_1 + y_2\overline{x}_2 + \cdots + y_n\overline{x}_n} \\
&= \overline{\langle \mathbf{y}, \mathbf{x} \rangle},
\end{aligned}$$

which proves (i). We leave the proof of (ii) as an exercise. For (iii), note that if $x_j = a_j + ib_j$ is the jth component of $\mathbf{x}$, then

$$\begin{aligned}
\langle \mathbf{x}, \mathbf{x} \rangle &= |x_1|^2 + |x_2|^2 + \cdots + |x_n|^2 \\
&= a_1^2 + b_1^2 + a_2^2 + b_2^2 + \cdots + a_n^2 + b_n^2
\end{aligned}$$

is a sum of squares of real numbers, so $\langle \mathbf{x}, \mathbf{x} \rangle \geq 0$, and $\langle \mathbf{x}, \mathbf{x} \rangle = 0$ if and only if each term a_j^2 and b_j^2 is equal to 0; that is, if and only if $\mathbf{x}$ is the zero vector, $\mathbf{x} = \mathbf{0}$. $\qquad \square$

Activity 13.36 Prove property (ii).

Activity 13.37 Calculate the norm of the vector $\mathbf{x} = (1, 0, i)^{\mathrm{T}} \in \mathbb{C}^3$.

13.4.2   Complex inner product in general

As with real vector spaces, there is a general notion of inner product on complex vector spaces.

Definition 13.38 (Complex inner product) Let V be a vector space over the complex numbers. An *inner product* on V is a mapping from (or operation on) pairs of vectors $\mathbf{x}, \mathbf{y}$ to the complex numbers, the result of which is a complex number denoted $\langle \mathbf{x}, \mathbf{y} \rangle$, which satisfies the following properties:

(i) $\langle \mathbf{x}, \mathbf{y} \rangle = \overline{\langle \mathbf{y}, \mathbf{x} \rangle}$ for all $\mathbf{x}, \mathbf{y} \in V$.

(ii) $\langle \alpha\mathbf{x} + \beta\mathbf{y}, \mathbf{z} \rangle = \alpha \langle \mathbf{x}, \mathbf{z} \rangle + \beta \langle \mathbf{y}, \mathbf{z} \rangle$ for all $\mathbf{x}, \mathbf{y}, \mathbf{z} \in V$ and all $\alpha, \beta \in \mathbb{C}$.

(iii) $\langle \mathbf{x}, \mathbf{x} \rangle \geq 0$ is a real number for all $\mathbf{x} \in V$, and $\langle \mathbf{x}, \mathbf{x} \rangle = 0$ if and only if $\mathbf{x} = \mathbf{0}$, the zero vector of the vector space V.

A vector space with a complex inner product is called a *complex inner product space*. From any complex inner product, we can define a norm by

$$\|\mathbf{x}\| = \sqrt{\langle \mathbf{x}, \mathbf{x} \rangle}.$$

The inner product defined on $\mathbb{C}^n$ in the previous section is clearly an inner product under this general definition.

Two further properties, which follow directly from this definition, are:

- $\langle \mathbf{x}, \alpha\mathbf{y} \rangle = \overline{\alpha} \langle \mathbf{x}, \mathbf{y} \rangle$ for all $\mathbf{x}, \mathbf{y} \in V$ and all $\alpha \in \mathbb{C}$.
- $\langle \mathbf{x}, \mathbf{y} + \mathbf{z} \rangle = \langle \mathbf{x}, \mathbf{y} \rangle + \langle \mathbf{x}, \mathbf{z} \rangle$ for all $\mathbf{x}, \mathbf{y}, \mathbf{z} \in V$.

Activity 13.39 Use the definition to prove these two additional properties.

Example 13.40 (This is a complex version of Example 10.3.) Suppose that V is the vector space consisting of all complex polynomial functions of degree at most n; that is, V consists of all functions $\mathbf{p} : x \mapsto p(x)$ of the form

$$p(x) = a_0 + a_1 x + a_2 x^2 + \cdots + a_n x^n, \qquad a_0, a_1, \ldots, a_n \in \mathbb{C}.$$

The addition and scalar multiplication are, as before, defined pointwise. Let $x_1, x_2, \ldots, x_{n+1}$ be $n + 1$ fixed, different, complex numbers, and define, for $\mathbf{p}, \mathbf{q} \in V$,

$$\langle \mathbf{p}, \mathbf{q} \rangle = \sum_{i=1}^{n+1} p(x_i)\overline{q(x_i)}.$$

Then this is an inner product. To see this, we check the properties in the definition of an inner product. Property (i) follows from properties of complex numbers and the complex conjugate:

$$\langle \mathbf{p}, \mathbf{q} \rangle = \sum_{i=1}^{n+1} p(x_i)\overline{q(x_i)} = \sum_{i=1}^{n+1} \overline{q(x_i)}p(x_i) = \sum_{i=1}^{n+1} \overline{q(x_i)\overline{p(x_i)}} = \overline{\langle \mathbf{q}, \mathbf{p} \rangle}.$$

For (iii), we have

$$\langle \mathbf{p}, \mathbf{p} \rangle = \sum_{i=1}^{n+1} p(x_i)\overline{p(x_i)} = \sum_{i=1}^{n+1} |p(x_i)|^2 \geq 0,$$

since it is the sum of squares of real numbers. The rest of the argument proceeds in exactly the same way as before. If $\mathbf{p}$ is the zero vector of the vector space (which is the identically-zero function), then $\langle \mathbf{p}, \mathbf{p} \rangle = 0$. To complete the verification of (iii), we need to check that, if $\langle \mathbf{p}, \mathbf{p} \rangle = 0$, then $\mathbf{p}$ must be the zero function. Now, $\langle \mathbf{p}, \mathbf{p} \rangle = 0$ must mean that $p(x_i) = 0$ for $i = 1, 2, \ldots, n+1$, so $p(x)$ has $n+1$ different roots. But $p(x)$ has degree no more than n, so $\mathbf{p}$ must be the identically-zero function. (The fact that a non-zero polynomial of degree n has no more than n distinct roots is just as true for complex numbers as it is for real numbers.) As before, part (ii) is left to you.

13.4.3   Orthogonal vectors

The definition of orthogonal vectors for real inner product spaces carries over exactly to complex ones:

Definition 13.41 Two vectors $\mathbf{x}$, $\mathbf{y}$ in a complex inner product space are said to be *orthogonal* if

$$\langle \mathbf{x}, \mathbf{y} \rangle = 0.$$

We write $\mathbf{x} \perp \mathbf{y}$.

A set of vectors $\{\mathbf{v}_1, \mathbf{v}_2, \ldots, \mathbf{v}_n\}$ in a complex inner product space V is *orthogonal* if $\langle \mathbf{v}_i, \mathbf{v}_j \rangle = 0$ for $i \neq j$. It is *orthonormal* if each vector is also a unit vector; that is, $\langle \mathbf{v}_i, \mathbf{v}_i \rangle = 1$.

Just as in $\mathbb{R}^n$, an orthogonal set of non-zero vectors in $\mathbb{C}^n$ is linearly independent. The proof is essentially the same as the one given for Theorem 10.14, but we state and prove it for a complex inner product space to illustrate the modifications. Notice that it is useful to think ahead about the order in which we choose to place the vectors in the inner product so that the proof is as straightforward as possible.

Theorem 13.42 *Suppose that V is a complex inner product space and that vectors $\mathbf{v}_1, \mathbf{v}_2, \ldots, \mathbf{v}_k \in V$ are pairwise orthogonal ($\langle \mathbf{v}_i, \mathbf{v}_j \rangle = 0$ for $i \neq j$), and none is the zero-vector. Then $\{\mathbf{v}_1, \mathbf{v}_2, \ldots, \mathbf{v}_k\}$ is a linearly independent set of vectors.*

Proof: We need to show that if

$$\alpha_1 \mathbf{v}_1 + \alpha_2 \mathbf{v}_2 + \cdots + \alpha_k \mathbf{v}_k = \mathbf{0}, \qquad \alpha_i \in \mathbb{C},$$

then $\alpha_1 = \alpha_2 = \cdots = \alpha_k = 0$. Let i be any integer between 1 and k. Then

$$\langle \alpha_1 \mathbf{v}_1 + \alpha_2 \mathbf{v}_2 + \cdots + \alpha_k \mathbf{v}_k, \mathbf{v}_i \rangle = \langle \mathbf{0}, \mathbf{v}_i \rangle = 0.$$

But

$$\langle \alpha_1 \mathbf{v}_1 + \cdots + \alpha_k \mathbf{v}_k, \mathbf{v}_i \rangle$$
$$= \alpha_1 \langle \mathbf{v}_1, \mathbf{v}_i \rangle + \cdots + \alpha_{i-1} \langle \mathbf{v}_{i-1}, \mathbf{v}_i \rangle + \alpha_i \langle \mathbf{v}_i, \mathbf{v}_i \rangle$$
$$+ \alpha_{i+1} \langle \mathbf{v}_{i+1}, \mathbf{v}_i \rangle + \cdots + \alpha_k \langle \mathbf{v}_k, \mathbf{v}_i \rangle.$$

Since $\langle \mathbf{v}_i, \mathbf{v}_j \rangle = 0$ for $j \neq i$, this equals $\alpha_i \langle \mathbf{v}_i, \mathbf{v}_i \rangle$, which is $\alpha_i \|\mathbf{v}_i\|^2$. So we have $\alpha_i \|\mathbf{v}_i\|^2 = 0$. Since $\mathbf{v}_i \neq \mathbf{0}$, $\|\mathbf{v}_i\|^2 \neq 0$ and hence $\alpha_i = 0$. But i was any integer in the range 1 to k, so we deduce that

$$\alpha_1 = \alpha_2 = \cdots = \alpha_k = 0,$$

as required. $\qquad\square$

Example 13.43 The vectors

$$\mathbf{v}_1 = \frac{1}{\sqrt{2}} \begin{pmatrix} 1 \\ i \end{pmatrix}, \qquad \mathbf{v}_2 = \frac{1}{\sqrt{2}} \begin{pmatrix} i \\ 1 \end{pmatrix},$$

form an orthonormal basis of $\mathbb{C}^2$. To show this, we calculate the inner products. They are orthogonal since

$$\langle \mathbf{v}_1, \mathbf{v}_2 \rangle = \frac{1}{2} \left\langle \begin{pmatrix} 1 \\ i \end{pmatrix}, \begin{pmatrix} i \\ 1 \end{pmatrix} \right\rangle = \frac{1}{2}(1(-i) + i(1)) = 0$$

and the norm of each vector is 1 since we have

$$\langle \mathbf{v}_1, \mathbf{v}_1 \rangle = \frac{1}{2} \left\langle \begin{pmatrix} 1 \\ i \end{pmatrix}, \begin{pmatrix} 1 \\ i \end{pmatrix} \right\rangle = \frac{1}{2}(1(1) + (i)(-i)) = 1$$

and $\qquad \langle \mathbf{v}_2, \mathbf{v}_2 \rangle = \frac{1}{2} \left\langle \begin{pmatrix} i \\ 1 \end{pmatrix}, \begin{pmatrix} i \\ 1 \end{pmatrix} \right\rangle = \frac{1}{2}((i)(-i) + 1(1)) = 1.$

It is a basis of $\mathbb{C}^2$ since they are linearly independent and $\mathbb{C}^2$ has dimension 2.

If $\{\mathbf{v}_1, \mathbf{v}_2, \ldots, \mathbf{v}_k\}$ is a basis of a complex inner product space V, then, just as for a real vector space, we can apply the Gram–Schmidt orthonormalisation process to obtain an orthonormal basis of V.

Example 13.44 Let V be the linear span, $V = \text{Lin}\{\mathbf{v}_1, \mathbf{v}_2\}$, of the vectors $\mathbf{v}_1, \mathbf{v}_2$ in $\mathbb{C}^n$, where

$$\mathbf{v}_1 = \begin{pmatrix} 1 \\ i \\ 0 \end{pmatrix}, \quad \mathbf{v}_2 = \begin{pmatrix} 1 \\ 2 \\ 1+i \end{pmatrix}.$$

We will find an orthonormal basis for V. First, we find a unit vector parallel to $\mathbf{v}_1$. Since $\langle \mathbf{v}_1, \mathbf{v}_1 \rangle = 1(1) + i(-i) = 2$, we set $\mathbf{u}_1 = (1/\sqrt{2})\mathbf{v}_1$. Then we set

$$\mathbf{w}_2 = \begin{pmatrix} 1 \\ 2 \\ 1+i \end{pmatrix} - \left\langle \begin{pmatrix} 1 \\ 2 \\ 1+i \end{pmatrix}, \frac{1}{\sqrt{2}} \begin{pmatrix} 1 \\ i \\ 0 \end{pmatrix} \right\rangle \frac{1}{\sqrt{2}} \begin{pmatrix} 1 \\ i \\ 0 \end{pmatrix}.$$

Calculating the inner product, we have

$$\mathbf{w}_2 = \begin{pmatrix} 1 \\ 2 \\ 1+i \end{pmatrix} - \frac{1-2i}{2} \begin{pmatrix} 1 \\ i \\ 0 \end{pmatrix}$$

$$= \begin{pmatrix} 1 \\ 2 \\ 1+i \end{pmatrix} - \begin{pmatrix} \frac{1}{2}-i \\ 1+\frac{1}{2}i \\ 0 \end{pmatrix} = \begin{pmatrix} \frac{1}{2}+i \\ 1-\frac{1}{2}i \\ 1+i \end{pmatrix}.$$

We need a unit vector in this direction. To make calculations easier, we use the parallel vector

$$\widehat{\mathbf{w}}_2 = \begin{pmatrix} 1+2i \\ 2-i \\ 2+2i \end{pmatrix},$$

and check that $\widehat{\mathbf{w}}_2 \perp \mathbf{v}_1$:

$$\langle \widehat{\mathbf{w}}_2, \mathbf{v}_1 \rangle = (1+2i)(1) + (2-i)(-i) + (2+2i)(0) = 0.$$

We find $\|\widehat{\mathbf{w}}_2\| = \sqrt{18}$, so that

$$\mathbf{u}_1 = \frac{1}{\sqrt{2}} \begin{pmatrix} 1 \\ i \\ 0 \end{pmatrix}, \quad \mathbf{u}_2 = \frac{1}{3\sqrt{2}} \begin{pmatrix} 1+2i \\ 2-i \\ 2+2i \end{pmatrix}$$

form an orthonormal basis of V.

Suppose we now wish to find the coordinates of $\mathbf{v}_2$ in this new basis. As in $\mathbb{R}^n$ (see Theorem 10.20), the coordinates a_i of any vector $\mathbf{v} \in V$ with respect to the orthonormal basis $\{\mathbf{u}_1, \mathbf{u}_2\}$ are given by $a_i = \langle \mathbf{v}, \mathbf{u}_i \rangle$. Here the order is important, because $\langle \mathbf{u}_i, \mathbf{v} \rangle = \overline{a_i}$. We have

$$a_1 = \left\langle \begin{pmatrix} 1 \\ 2 \\ 1+i \end{pmatrix}, \frac{1}{\sqrt{2}} \begin{pmatrix} 1 \\ i \\ 0 \end{pmatrix} \right\rangle = \frac{1-2i}{\sqrt{2}}$$

$$a_2 = \left\langle \begin{pmatrix} 1 \\ 2 \\ 1+i \end{pmatrix}, \frac{1}{3\sqrt{2}} \begin{pmatrix} 1+2i \\ 2-i \\ 2+2i \end{pmatrix} \right\rangle = \frac{9}{3\sqrt{2}} = \frac{3}{\sqrt{2}}$$

so that

$$\mathbf{v}_2 = \frac{1 - 2i}{\sqrt{2}}\,\mathbf{u}_1 + \frac{3}{\sqrt{2}}\,\mathbf{u}_2.$$

Activity 13.45 Check all the calculations in this example.

13.5 Hermitian conjugates

13.5.1 The Hermitian conjugate

If A is a complex matrix, the Hermitian conjugate, which we denote by A^*, is the matrix $\overline{A}^{\mathrm{T}}$, the result of taking the complex conjugate of every entry of A and then transposing the matrix. Whereas the transpose of a real matrix played an important role in the orthogonal diagonalisation of real matrices, we shall see that it is the Hermitian conjugate which we need for complex matrices.

Definition 13.46 (Hermitian conjugate) If A is an $m \times n$ matrix with complex entries, then the *Hermitian conjugate* of A, denoted by A^*, is defined by

$$A^* = \overline{A}^{\mathrm{T}}.$$

That is, if $A = (a_{ij})$, then $\overline{A} = (\overline{a_{ij}})$ and $A^* = \overline{A}^{\mathrm{T}} = (\overline{a_{ji}})$.

Example 13.47

If $A = \begin{pmatrix} i & 5 - 3i & 2 + i \\ 3 & 1 + 2i & 4 - 9i \end{pmatrix}$, then $A^* = \begin{pmatrix} -i & 3 \\ 5 + 3i & 1 - 2i \\ 2 - i & 4 + 9i \end{pmatrix}$.

If $\mathbf{x}$, $\mathbf{y}$ are vectors in $\mathbb{C}^n$, then we can express the standard complex inner product in terms of matrix multiplication as

$$\langle \mathbf{x}, \mathbf{y} \rangle = x_1 \overline{y}_1 + x_2 \overline{y}_2 + \cdots + x_n \overline{y}_n = \overline{y}_1 x_1 + \overline{y}_2 x_2 + \cdots + \overline{y}_n x_n = \mathbf{y}^* \mathbf{x}.$$

Unfortunately, this is not quite as neat as the corresponding expression for the inner product on $\mathbb{R}^n$ as a matrix product. (How are they different?) However, we do have.

$$\langle \mathbf{x}, \mathbf{x} \rangle = \mathbf{x}^* \mathbf{x} = \|\mathbf{x}\|^2.$$

Compare these properties of the Hermitian conjugate of a matrix with those of the transpose of a real matrix:

$$(A^*)^* = A, \qquad (A + B)^* = A^* + B^*, \qquad (AB)^* = B^* A^*.$$

Because the two operations involved in forming a Hermitian conjugate – taking the conjugate and taking the transpose – commute with each other (meaning it doesn't matter in which order you perform the operations), the first two properties follow immediately from the definition of the Hermitian conjugate. Let us look more closely at the last property.

We will prove that $(AB)^* = B^*A^*$ by showing that the entries are the same. If $A = (a_{ij})$ and $B = (b_{ij})$, the (i, j) entry of AB is

$$a_{i1}b_{1j} + a_{i2}b_{2j} + \cdots + a_{in}b_{nj},$$

so the (j, i) entry of $(AB)^*$ is

$$\overline{a_{i1}b_{1j} + a_{i2}b_{2j} + \cdots + a_{in}b_{nj}}.$$

Now look at the (j, i) entry of B^*A^*, which is the matrix product of the jth row of B^* with the ith column of A^*. The jth row of B^* is given by the complex conjugate of the jth column of B, $(\overline{b}_{1j}, \overline{b}_{2j}, \ldots, \overline{b}_{nj})$, and the ith column of A^* is the complex conjugate of the ith row of A, which is $(\overline{a}_{i1}, \overline{a}_{i2}, \ldots, \overline{a}_{in})^{\mathrm{T}}$. Thus, the (j, i) entry of B^*A^* is

$$\overline{b}_{1j}\overline{a}_{i1} + \overline{b}_{2j}\overline{a}_{i2} + \cdots + \overline{b}_{nj}\overline{a}_{in},$$

which is equal to the expression we obtained for the (j, i) entry of $(AB)^*$.

If A is a matrix with real entries, then $A^* = \overline{A}^{\mathrm{T}} = A^{\mathrm{T}}$. Therefore, the proof we have just given includes the familiar result for real matrices A and B, that $(AB)^{\mathrm{T}} = B^{\mathrm{T}}A^{\mathrm{T}}$.

Activity 13.48 Show that for any complex matrix A and any complex number k,

$$(kA)^* = \overline{k}A^*.$$

What is the analogous result for real matrices and real numbers?

Often, the term *adjoint* is used instead of Hermitian conjugate. But we have already used that terminology for something completely different in Chapter 3 (in the context of finding the inverse of a matrix), so we will avoid it: but we wanted to let you know, to avoid confusion.

13.5.2   Hermitian matrices

Recall that a real matrix A is *symmetric* if $A = A^{\mathrm{T}}$. The complex analogue is a *Hermitian* matrix.

Definition 13.49 (Hermitian matrix) An $n \times n$ complex matrix A is *Hermitian* if and only if

$$A = A^*.$$

A Hermitian matrix with real entries is a symmetric matrix, since $A = A^* = A^T$. So what does a Hermitian matrix look like? If $A = (a_{ij})$ is equal to $A^* = (\overline{a_{ji}})$, then the diagonal entries must be real numbers, since they satisfy $a_{ii} = \overline{a_{ii}}$. The corresponding entries across the main diagonal must be complex conjugates of one another.

Example 13.50 The matrix

$$A = \begin{pmatrix} 1 & 1+2i & 4-i \\ 1-2i & -3 & i \\ 4+i & -i & 2 \end{pmatrix}$$

is a Hermitian matrix.

Activity 13.51 Check that $A^* = \overline{A}^T = A$.

When we looked at orthogonal diagonalisation of symmetric matrices, we stated (in the proof of Theorem 11.5) that the eigenvalues of a symmetric matrix are real. We can now prove this. The result is a corollary of the following theorem:

Theorem 13.52 *If A is a Hermitian matrix, then the eigenvalues of A are real.*

Proof: Suppose λ is an eigenvalue of A with corresponding eigenvector **v**. Then $A\mathbf{v} = \lambda\mathbf{v}$ and $\mathbf{v} \neq \mathbf{0}$. We multiply this equality on the left by the Hermitian conjugate of **v**, obtaining

$$\mathbf{v}^* A\mathbf{v} = \mathbf{v}^* \lambda\mathbf{v} = \lambda\mathbf{v}^*\mathbf{v} = \lambda\|\mathbf{v}\|^2,$$

where the norm of **v** is a positive real number. On the other hand, taking the complex conjugate transpose of both sides of $A\mathbf{v} = \lambda\mathbf{v}$, we have

$$(A\mathbf{v})^* = (\lambda\mathbf{v})^*,$$

which gives

$$\mathbf{v}^* A^* = \overline{\lambda}\mathbf{v}^*.$$

We then multiply this last equality on the right by **v** to get

$$\mathbf{v}^* A^*\mathbf{v} = \overline{\lambda}\mathbf{v}^*\mathbf{v} = \overline{\lambda}\|\mathbf{v}\|^2.$$

Since A is Hermitian, $\mathbf{v}^* A\mathbf{v} = \mathbf{v}^* A^*\mathbf{v}$, and therefore it follows that

$$\lambda\|\mathbf{v}\|^2 = \overline{\lambda}\|\mathbf{v}\|^2.$$

Since $\|\mathbf{v}\|^2 \neq 0$, we conclude that $\lambda = \overline{\lambda}$; that is, λ is real. $\square$

This has as an immediate consequence the following important fact that we used in Chapter 11 to prove the Spectral theorem

(Theorem 11.5); that is, Theorem 11.7 is the following corollary of Theorem 13.52:

Corollary 13.53 *If A is a real symmetric matrix, then the eigenvalues of A are real.*

As with real symmetric matrices, it is also true for Hermitian matrices that eigenvectors corresponding to different eigenvalues are orthogonal.

Theorem 13.54 *If the matrix A is Hermitian, then eigenvectors corresponding to distinct eigenvalues are orthogonal.*

Activity 13.55 Prove this theorem. Look at the proof of Theorem 11.8 and rework it for the complex case.

13.5.3 Unitary matrices

The counterpart for complex matrices to an orthogonal matrix is a *unitary* matrix.

Definition 13.56 An $n \times n$ complex matrix P is said to be *unitary* if and only if $PP^* = P^*P = I$; that is, if P has inverse P^*.

Example 13.57 The matrix

$$P = \begin{pmatrix} 1/\sqrt{2} & i/\sqrt{2} \\ i/\sqrt{2} & 1/\sqrt{2} \end{pmatrix}$$

is a unitary matrix,

Activity 13.58 Check this.

An immediate consequence of this definition is that if P is a unitary matrix, then so is P^*.

Activity 13.59 Show this. Show, also, that if A and B are unitary matrices, then so is their product AB.

A unitary matrix P with real entries is an *orthogonal* matrix, since then $P^* = P^T$. Recall that a matrix is orthogonal if and only if its columns are an orthonormal basis of $\mathbb{R}^n$. We prove the analogous result for unitary matrices.

Theorem 13.60 *The $n \times n$ matrix P is unitary if and only if the columns of P are an orthonormal basis of $\mathbb{C}^n$.*

Proof: The proof of this theorem follows the same argument as the proof of Theorem 10.21. It is an 'if and only if' statement, so we must prove it in both directions.

Let $\mathbf{x}_1, \mathbf{x}_2, \ldots, \mathbf{x}_n$ be the columns of the matrix P. Then the rows of P^* are the complex conjugate transposes of these vectors.

If $I = P^*P$, we have

$$\begin{pmatrix} 1 & 0 & \cdots & 0 \\ 0 & 1 & \cdots & 0 \\ \vdots & \vdots & \ddots & \vdots \\ 0 & 0 & \cdots & 1 \end{pmatrix} = \begin{pmatrix} \mathbf{x}_1^* \\ \mathbf{x}_2^* \\ \vdots \\ \mathbf{x}_n^* \end{pmatrix} (\mathbf{x}_1 \; \mathbf{x}_2 \; \cdots \; \mathbf{x}_n)$$

$$= \begin{pmatrix} \mathbf{x}_1^*\mathbf{x}_1 & \mathbf{x}_1^*\mathbf{x}_2 & \cdots & \mathbf{x}_1^*\mathbf{x}_n \\ \mathbf{x}_2^*\mathbf{x}_1 & \mathbf{x}_2^*\mathbf{x}_2 & \cdots & \mathbf{x}_2^*\mathbf{x}_n \\ \vdots & \vdots & \ddots & \vdots \\ \mathbf{x}_n^*\mathbf{x}_1 & \mathbf{x}_n^*\mathbf{x}_2 & \cdots & \mathbf{x}_n^*\mathbf{x}_n \end{pmatrix}.$$

Equating the entries of these matrices, we have $\mathbf{x}_i^*\mathbf{x}_j = \langle \mathbf{x}_j, \mathbf{x}_i \rangle = 0$ if $i \neq j$ and $\mathbf{x}_i^*\mathbf{x}_i = \langle \mathbf{x}_i, \mathbf{x}_i \rangle = 1$ if $i = j$, which means precisely that the columns $\{\mathbf{x}_1, \mathbf{x}_2, \ldots, \mathbf{x}_n\}$ are an orthonormal set of vectors. They are therefore linearly independent, and since there are n of them, they are a basis of $\mathbb{C}^n$.

Conversely, if the columns of P are an orthonormal basis of $\mathbb{C}^n$, then the matrix product P^*P as shown above must be the identity matrix, so that $P^*P = I$. This says that $P^* = P^{-1}$, so also $PP^* = I$. $\square$

Since P^* is also unitary, this result applies to the rows of the matrix P.

Just as for an orthogonal matrix, the linear transformation defined by a unitary matrix P is an *isometry*, meaning that it preserves the inner product, and therefore the length of any vector. In fact, this characterises a unitary matrix, as the following theorem shows:

Theorem 13.61 *The matrix P is unitary if and only if the linear transformation defined by P preserves the standard complex inner product; that is, $\langle P\mathbf{x}, P\mathbf{y} \rangle = \langle \mathbf{x}, \mathbf{y} \rangle$ for all $\mathbf{x}, \mathbf{y} \in \mathbb{C}^n$.*

Proof: If P is a unitary matrix, then

$$\langle P\mathbf{x}, P\mathbf{y} \rangle = (P\mathbf{y})^*(P\mathbf{x}) = \mathbf{y}^*P^*P\mathbf{x} = \mathbf{y}^*I\mathbf{x} = \mathbf{y}^*\mathbf{x} = \langle \mathbf{x}, \mathbf{y} \rangle,$$

so P preserves the inner product.

Conversely, assume we have a matrix P for which

$$\langle P\mathbf{x}, P\mathbf{y} \rangle = \langle \mathbf{x}, \mathbf{y} \rangle$$

for all $\mathbf{x}, \mathbf{y} \in \mathbb{C}^n$. Let $\{\mathbf{e}_1, \mathbf{e}_2, \ldots, \mathbf{e}_n\}$ denote (as usual) the standard basis on $\mathbb{C}^n$. Then $P\mathbf{e}_i = \mathbf{v}_i$, where $\mathbf{v}_1, \mathbf{v}_2, \ldots, \mathbf{v}_n$ are the columns of P. We have

$$\langle \mathbf{v}_i, \mathbf{v}_j \rangle = \langle P\mathbf{e}_i, P\mathbf{e}_j \rangle = \langle \mathbf{e}_i, \mathbf{e}_j \rangle,$$

from which we deduce that the columns of P are an orthonormal basis of $\mathbb{C}^n$, and therefore P is a unitary matrix. □

13.6   Unitary diagonalisation and normal matrices

Recalling the definition of orthogonal diagonalisation (Section 11.1), the following definition will come as no surprise:

Definition 13.62 A matrix A is said to be *unitarily diagonalisable* if there is a unitary matrix P such that $P^*AP \doteq D$, where D is a diagonal matrix.

Suppose the matrix A can be unitarily diagonalised, with $P^*AP = D$. Since the matrix P diagonalises A, the columns of P are a basis of $\mathbb{C}^n$ consisting of eigenvectors of A. Since P is unitary, the columns of P are an orthonormal basis of $\mathbb{C}^n$. That is, if P unitarily diagonalises A, then the columns of P are an orthonormal basis of $\mathbb{C}^n$ consisting of eigenvectors of A.

Conversely, if the eigenvectors of A are an orthonormal basis of $\mathbb{C}^n$, then the matrix P whose columns are these basis vectors is unitary. Since the vectors are eigenvectors of A, we have $AP = PD$, where D is the diagonal matrix of corresponding eigenvalues. Since $P^{-1} = P^*$, we have $P^*AP = D$, so that A is unitarily diagonalised. This is summarised in the following theorem:

Theorem 13.63 *The matrix A can be unitarily diagonalised if and only if there is an orthonormal basis of $\mathbb{C}^n$ consisting of eigenvectors of A.*

For real matrices, only a symmetric matrix can be orthogonally diagonalised. Considering what we have done so far, it is natural to ask if there is an analogous result for complex matrices, but the result for complex matrices is quite different. Whereas it is true that a Hermitian matrix can be unitarily diagonalised, these are not the only matrices for which this is true. There is a much larger class of complex matrices which can be unitarily diagonalised. These are the *normal* matrices.

Definition 13.64 (Normal matrix) An $n \times n$ complex matrix A is called *normal* if

$$AA^* = A^*A.$$

Every Hermitian matrix is normal since $AA^* = AA = A^*A$. Also, every unitary matrix is normal since $AA^* = I = A^*A$.

Furthermore, every diagonal matrix is normal. To see this, let $D = \text{diag}(d_1, \ldots, d_n)$, meaning d_i is the entry in the (i, i) position and all other entries are zero. Then D^* is the diagonal matrix with $\overline{d_i}$ in the (i, i) position and zeros elsewhere. Therefore,

$$DD^* = \text{diag}(|d_1|^2, |d_2|^2, \ldots, |d_n|^2) = D^*D.$$

This shows that D is normal, and also that the entries of the diagonal matrix DD^* are real. Diagonal matrices provide some simple examples of matrices that are normal, but neither Hermitian nor unitary.

Activity 13.65 Write down a diagonal matrix which is not Hermitian and not unitary.

We state the following important result:

Theorem 13.66 *The matrix A is unitarily diagonalisable if and only if A is normal.*

We will prove this theorem in one direction only: if A is unitarily diagonalisable, then A is normal. This means that only normal matrices can be unitarily diagonalised. The proof that if A is normal then A can be unitarily diagonalised requires additional theory and will not be given in this book.

Proof: [that only normal matrices can be unitarily diagonalised.] Suppose A can be unitarily diagonalised. Then there is a unitary matrix P and a diagonal matrix D such that $P^*AP = D$. Solving for A, we have $A = PDP^*$. Then

$$AA^* = (PDP^*)(PDP^*)^* = (PDP^*)(PD^*P^*) = PD(P^*P)D^*P^*$$
$$= P(DD^*)P^*.$$

In the same way,

$$A^*A = (PDP^*)^*(PDP^*) = (PD^*P^*)(PDP^*) = PD^*(P^*P)DP^*$$
$$= P(D^*D)P.$$

Since D is diagonal, it is normal, so that $P(DD^*)P^* = P(D^*D)P$, from which we conclude that A is normal. $\square$

How do we unitarily diagonalise a normal matrix A? We carry out the same steps as for orthogonal diagonalisation. First, we solve the characteristic equation of A to find the eigenvalues. For each eigenvalue λ, we find an orthonormal basis for the corresponding eigenspace, using

Gram–Schmidt if necessary. Then the set of all such eigenvectors is an orthonormal basis of $\mathbb{C}^n$. That this is always possible is the content of the above theorem. We form the matrix P with these eigenvectors as the columns. Then P is unitary, and $P^*AP = D$, where D is the diagonal matrix of corresponding eigenvalues.

All the examples you have seen of orthogonal diagonalisation are examples of unitary diagonalisation in the case where A is a real symmetric matrix (Section 11.1). We now give an example for a complex matrix.

Example 13.67 The matrix

$$A = \begin{pmatrix} 1 & 2+i \\ 2-i & 5 \end{pmatrix}$$

is Hermitian and can therefore be unitarily diagonalised. The eigenvalues are given by

$$|A - \lambda I| = \begin{vmatrix} 1 - \lambda & 2+i \\ 2-i & 5-\lambda \end{vmatrix} = \lambda^2 - 6\lambda + 5 - 5 = 0.$$

So the eigenvalues are 0 and 6. (As expected, these are real numbers.) We now find the corresponding eigenvectors by row reducing the matrices $(A - \lambda I)$.

For $\lambda_1 = 0$,

$$A = \begin{pmatrix} 1 & 2+i \\ 2-i & 5 \end{pmatrix} \longrightarrow \begin{pmatrix} 1 & 2+i \\ 0 & 0 \end{pmatrix},$$

so we let

$$\mathbf{v}_1 = \begin{pmatrix} 2+i \\ -1 \end{pmatrix}.$$

For $\lambda_2 = 6$,

$$A - 6I = \begin{pmatrix} -5 & 2+i \\ 2-i & -1 \end{pmatrix} \longrightarrow \begin{pmatrix} 1 & -(2+i)/5 \\ 0 & 0 \end{pmatrix},$$

so we may take

$$\mathbf{v}_2 = \begin{pmatrix} 2+i \\ 5 \end{pmatrix}$$

as an eigenvector. These two eigenvectors are orthogonal. The vector $\mathbf{v}_1$ has $\|\mathbf{v}_1\|^2 = 6$. The vector $\mathbf{v}_2$ has norm equal to $\sqrt{30}$. If we set

$$P = \begin{pmatrix} (2+i)/\sqrt{6} & (2+i)/\sqrt{30} \\ -1/\sqrt{6} & 5/\sqrt{30} \end{pmatrix} \quad \text{and} \quad D = \begin{pmatrix} 0 & 0 \\ 0 & 6 \end{pmatrix},$$

then $P^*AP = P^{-1}AP = D$.

Activity 13.68 Check all the calculations in this example.

We have already seen that all Hermitian matrices are normal. We also know that the eigenvalues of a Hermitian matrix are real. So all Hermitian matrices are normal, with real eigenvalues. We can now prove the converse; namely, that if a normal matrix has real eigenvalues, then it is Hermitian.

Theorem 13.69 *Let A be a normal matrix. If all of the eigenvalues of A are real, then A is Hermitian.*

Proof: Since A is normal, it can be unitarily diagonalised. Let P be a unitary matrix such that $P^*AP = D$, where D is the diagonal matrix of eigenvalues of A. Since A has real eigenvalues, $D^* = D$. Then $A = PDP^*$ and

$$A^* = (PDP^*)^* = PD^*P^* = PDP^* = A,$$

which shows that A is Hermitian. $\qquad\qquad\qquad\qquad\qquad\qquad\square$

13.7 Spectral decomposition

13.7.1 The spectral decomposition of a matrix

We will now look at the unitary diagonalisation of a matrix A in a different way. Let P be a unitary matrix such that $P^*AP = D$, where D is the diagonal matrix of eigenvalues of A, and the columns of P are the corresponding eigenvectors, $\mathbf{x}_1, \mathbf{x}_2, \ldots, \mathbf{x}_n$. Since P is unitary, we have $P^*P = PP^* = I$. We used the equality $P^*P = I$ in the proof of Theorem 13.60 to show that the column vectors of P are an orthonormal basis of $\mathbb{C}^n$. That is,

$$I = P^*P = \begin{pmatrix} \mathbf{x}_1^* \\ \mathbf{x}_2^* \\ \vdots \\ \mathbf{x}_n^* \end{pmatrix} (\mathbf{x}_1 \ \mathbf{x}_2 \ \cdots \ \mathbf{x}_n) = \begin{pmatrix} \mathbf{x}_1^*\mathbf{x}_1 & \mathbf{x}_1^*\mathbf{x}_2 & \cdots & \mathbf{x}_1^*\mathbf{x}_n \\ \mathbf{x}_2^*\mathbf{x}_1 & \mathbf{x}_2^*\mathbf{x}_2 & \cdots & \mathbf{x}_2^*\mathbf{x}_n \\ \vdots & \vdots & \ddots & \vdots \\ \mathbf{x}_n^*\mathbf{x}_1 & \mathbf{x}_n^*\mathbf{x}_2 & \cdots & \mathbf{x}_n^*\mathbf{x}_n \end{pmatrix},$$

where the entry $\mathbf{x}_i^*\mathbf{x}_j = \langle \mathbf{x}_j, \mathbf{x}_i \rangle$ is a complex number (either 1 or 0 in this case). But what information can we derive from the other product, $PP^* = I$ in terms of the *column* vectors? Carrying out the matrix multiplication, we have

$$I = PP^* = (\mathbf{x}_1 \ \mathbf{x}_2 \ \cdots \ \mathbf{x}_n) \begin{pmatrix} \mathbf{x}_1^* \\ \mathbf{x}_2^* \\ \vdots \\ \mathbf{x}_n^* \end{pmatrix} = \mathbf{x}_1\mathbf{x}_1^* + \mathbf{x}_2\mathbf{x}_2^* + \cdots + \mathbf{x}_n\mathbf{x}_n^*,$$

where this time $E_i = \mathbf{x}_i\mathbf{x}_i^*$ is an $n \times n$ matrix. It is the matrix product of the $n \times 1$ column vector $\mathbf{x}_i$ with the $1 \times n$ row vector $\mathbf{x}_i^*$. Using the matrices E_i, the above equality can be written as

$$I = E_1 + E_2 + \cdots + E_n.$$

This result is true for any unitary matrix P, but it is most interesting when the columns of P are the eigenvectors of a matrix A. The connection with the matrix A is the following theorem:

Theorem 13.70 (Spectral decomposition) *Let A be a normal matrix and let $\{\mathbf{x}_1, \mathbf{x}_2, \ldots, \mathbf{x}_n\}$ be an orthonormal set of eigenvectors of A with corresponding eigenvalues $\lambda_1, \lambda_2, \ldots, \lambda_n$. Then*

$$A = \lambda_1 \mathbf{x}_1 \mathbf{x}_1^* + \lambda_2 \mathbf{x}_2 \mathbf{x}_2^* + \cdots + \lambda_n \mathbf{x}_n \mathbf{x}_n^*.$$

Proof. If P is the unitary matrix whose columns are the eigenvectors of A, then, as we have seen, we can write

$$I = \mathbf{x}_1 \mathbf{x}_1^* + \mathbf{x}_2 \mathbf{x}_2^* + \cdots + \mathbf{x}_n \mathbf{x}_n^*.$$

Multiplying both sides of this equality by the matrix A, we have

$$
\begin{aligned}
A = AI &= A(\mathbf{x}_1 \mathbf{x}_1^* + \mathbf{x}_2 \mathbf{x}_2^* + \cdots + \mathbf{x}_n \mathbf{x}_n^*) \\
&= A\mathbf{x}_1 \mathbf{x}_1^* + A\mathbf{x}_2 \mathbf{x}_2^* + \cdots + A\mathbf{x}_n \mathbf{x}_n^* \\
&= \lambda_1 \mathbf{x}_1 \mathbf{x}_1^* + \lambda_2 \mathbf{x}_2 \mathbf{x}_2^* + \cdots + \lambda_n \mathbf{x}_n \mathbf{x}_n^*.
\end{aligned}
$$

$\square$

The *spectral decomposition* of A is the formula

$$A = \lambda_1 E_1 + \lambda_2 E_2 + \cdots + \lambda_n E_n,$$

where $E_i = \mathbf{x}_i \mathbf{x}_i^*$.

Example 13.71 In Example 13.67, we unitarily diagonalised the matrix

$$A = \begin{pmatrix} 1 & 2+i \\ 2-i & 5 \end{pmatrix}.$$

We found $P^*AP = P^{-1}AP = D$ for the matrices

$$P = \begin{pmatrix} (2+i)/\sqrt{6} & (2+i)/\sqrt{30} \\ -1/\sqrt{6} & 5/\sqrt{30} \end{pmatrix} \quad \text{and} \quad D = \begin{pmatrix} 0 & 0 \\ 0 & 6 \end{pmatrix}.$$

Therefore, the spectral decomposition of A is $A = 0E_1 + 6E_2$. Clearly, we only need to calculate the matrix $E_2 = \mathbf{x}_2 \mathbf{x}_2^*$. We have

$$
\begin{aligned}
A &= 6\frac{1}{30} \begin{pmatrix} 2+i \\ 5 \end{pmatrix} (2-i \quad 5) = \frac{1}{5} \begin{pmatrix} 5 & 5(2+i) \\ 5(2-i) & 25 \end{pmatrix} \\
&= \begin{pmatrix} 1 & 2+i \\ 2-i & 5 \end{pmatrix},
\end{aligned}
$$

as expected. So the spectral decomposition of this matrix is just the matrix A itself.

Example 13.72 In Example 11.14, we orthogonally diagonalised the matrix

$$B = \begin{pmatrix} 2 & 1 & 1 \\ 1 & 2 & 1 \\ 1 & 1 & 2 \end{pmatrix}.$$

If we let

$$P = \begin{pmatrix} -\frac{1}{\sqrt{6}} & -\frac{1}{\sqrt{2}} & \frac{1}{\sqrt{3}} \\ -\frac{1}{\sqrt{6}} & \frac{1}{\sqrt{2}} & \frac{1}{\sqrt{3}} \\ \frac{2}{\sqrt{6}} & 0 & \frac{1}{\sqrt{3}} \end{pmatrix} \quad \text{and} \quad D = \begin{pmatrix} 1 & 0 & 0 \\ 0 & 1 & 0 \\ 0 & 0 & 4 \end{pmatrix},$$

then $P^T = P^{-1}$ and $P^T B P = D$.

The spectral decomposition of B is $B = 1E_1 + 1E_2 + 4E_3$, where the matrices $E_i = \mathbf{x}_i \mathbf{x}_i^*$ are obtained from the orthonormal eigenvectors of B (which are the columns of P). We have,

$$E_1 = \left(\frac{1}{\sqrt{6}}\right)^2 \begin{pmatrix} -1 \\ -1 \\ 2 \end{pmatrix} (-1 \quad -1 \quad 2) = \frac{1}{6} \begin{pmatrix} 1 & 1 & -2 \\ 1 & 1 & -2 \\ -2 & -2 & 4 \end{pmatrix}$$

and, similarly,

$$E_2 = \frac{1}{2} \begin{pmatrix} 1 & -1 & 0 \\ -1 & 1 & 0 \\ 0 & 0 & 0 \end{pmatrix}, \qquad E_3 = \frac{1}{3} \begin{pmatrix} 1 & 1 & 1 \\ 1 & 1 & 1 \\ 1 & 1 & 1 \end{pmatrix}.$$

The spectral decomposition of B is

$$B = \frac{1}{6} \begin{pmatrix} 1 & 1 & -2 \\ 1 & 1 & -2 \\ -2 & -2 & 4 \end{pmatrix} + \frac{1}{2} \begin{pmatrix} 1 & -1 & 0 \\ -1 & 1 & 0 \\ 0 & 0 & 0 \end{pmatrix} + \frac{4}{3} \begin{pmatrix} 1 & 1 & 1 \\ 1 & 1 & 1 \\ 1 & 1 & 1 \end{pmatrix}.$$

Activity 13.73 Check all the calculations for this example.

13.7.2 The matrices E_i

Let's take a closer look at the matrices E_i. As the following theorem shows, we will find that they are Hermitian, they are idempotent and they satisfy $E_i E_j = 0$ if $i \neq j$ (where 0 denotes the zero matrix).

Theorem 13.74 *If* $\{\mathbf{x}_1, \mathbf{x}_2, \ldots, \mathbf{x}_n\}$ *is an orthonormal basis of* $\mathbb{C}^n$, *then the matrices* $E_i = \mathbf{x}_i \mathbf{x}_i^*$ *have the following properties:*

    (i) $E_i^* = E_i$.
    (ii) $E_i^2 = E_i$.
    (iii) $E_i E_j = 0$ *if* $i \neq j$.

Proof: We will prove (ii), and leave properties (i) and (iii) for you to verify. The matrices are *idempotent* since

$$E_i^2 = E_i E_i = (\mathbf{x}_i \mathbf{x}_i^*)(\mathbf{x}_i \mathbf{x}_i^*) = \mathbf{x}_i (\mathbf{x}_i^* \mathbf{x}_i) \mathbf{x}_i^* = \mathbf{x}_i \langle \mathbf{x}_i, \mathbf{x}_i \rangle \mathbf{x}_i^*$$
$$= \mathbf{x}_i \cdot 1 \cdot \mathbf{x}_i^* = E_i.$$

The other two proofs are equally straightforward.   □

Activity 13.75 Show that (i) E_i is Hermitian and that (iii) $E_i E_j = 0$ if $i \neq j$.

The fact that each E_i is an idempotent matrix means that it represents a projection (just as for real matrices). To see what this projection is, look at its action on the orthonormal basis vectors

$$E_i \mathbf{x}_i = (\mathbf{x}_i \mathbf{x}_i^*) \mathbf{x}_i = \mathbf{x}_i (\mathbf{x}_i^* \mathbf{x}_i) = \mathbf{x}_i \langle \mathbf{x}_i, \mathbf{x}_i \rangle = \mathbf{x}_i \cdot 1 = \mathbf{x}_i$$

and

$$E_i \mathbf{x}_j = (\mathbf{x}_i \mathbf{x}_i^*) \mathbf{x}_j = \mathbf{x}_i (\mathbf{x}_i^* \mathbf{x}_j) = \mathbf{x}_i \cdot 0 = \mathbf{0}.$$

If $\mathbf{v}$ is any vector in $\mathbb{C}^n$, $\mathbf{v}$ can be written as a unique linear combination $\mathbf{v} = a_1 \mathbf{x}_1 + a_2 \mathbf{x}_2 + \cdots + a_n \mathbf{x}_n$. Then

$$E_i \mathbf{v} = E_i (a_1 \mathbf{x}_1 + a_2 \mathbf{x}_2 + \cdots + a_n \mathbf{x}_n)$$
$$= a_1 E_i \mathbf{x}_1 + a_2 E_i \mathbf{x}_2 + \cdots + a_{i-1} E_i \mathbf{x}_{i-1} + a_i E_i \mathbf{x}_i$$
$$+ a_{i+1} E_i \mathbf{x}_{i+1} + \cdots + a_n E_i \mathbf{x}_n = a_i \mathbf{x}_i.$$

E_i is the *orthogonal projection* of $\mathbb{C}^n$ onto the subspace spanned by the vector $\mathbf{x}_i$.

Activity 13.76 Look at the previous example and write down the orthogonal projection of $\mathbb{C}^n$ onto the subspace $\text{Lin}\{(1, 1, 1)^T\}$.

Matrices which satisfy properties (ii) and (iii) of Theorem 13.74 have an interesting application. Suppose E_1, E_2, E_3 are three such matrices; that is,

$$E_i E_j = \begin{cases} E_i & \text{for} \quad i = j \\ 0 & \text{for} \quad i \neq j \end{cases}$$

for $i = 1, 2, 3$. Then for *any* real numbers $\alpha_1, \alpha_2, \alpha_3$, and any positive integer n, we will show that

$$(\alpha_1 E_1 + \alpha_2 E_2 + \alpha_3 E_3)^n = \alpha_1^n E_1 + \alpha_2^n E_2 + \alpha_3^n E_3.$$

To establish this result, we will use an inductive argument. For $n = 2$, observe that

$$(\alpha_1 E_1 + \alpha_2 E_2 + \alpha_3 E_3)^2$$
$$= (\alpha_1 E_1 + \alpha_2 E_2 + \alpha_3 E_3)(\alpha_1 E_1 + \alpha_2 E_2 + \alpha_3 E_3)$$
$$= \alpha_1^2 E_1 E_1 + \alpha_2^2 E_2 E_2 + \alpha_3^2 E_3 E_3 \qquad \text{(since } E_i E_j = 0 \text{ for } i \neq j\text{)}$$
$$= \alpha_1^2 E_1 + \alpha_2^2 E_2 + \alpha_3^2 E_3 \qquad \text{(since } E_i E_i = E_i\text{)}.$$

Now assume that the result holds for n,

$$(\alpha_1 E_1 + \alpha_2 E_2 + \alpha_3 E_3)^n = \alpha_1^n E_1 + \alpha_2^n E_2 + \alpha_3^n E_3.$$

We will show that it therefore also holds for $n + 1$. In this way, the result for $n = 2$ above will imply the result is also true for $n = 3$, and so on. We have

$$(\alpha_1 E_1 + \alpha_2 E_2 + \alpha_3 E_3)^{n+1}$$
$$= (\alpha_1 E_1 + \alpha_2 E_2 + \alpha_3 E_3)^n (\alpha_1 E_1 + \alpha_2 E_2 + \alpha_3 E_3)$$
$$= (\alpha_1^n E_1 + \alpha_2^n E_2 + \alpha_3^n E_3)(\alpha_1 E_1 + \alpha_2 E_2 + \alpha_3 E_3)$$
$$= \alpha_1^{n+1} E_1 E_1 + \alpha_2^{n+1} E_2 E_2 + \alpha_3^{n+1} E_3 E_3$$
$$\text{(since } E_i E_j = 0 \text{ for } i \neq j\text{)}$$
$$= \alpha_1^{n+1} E_1 + \alpha_2^{n+1} E_2 + \alpha_3^{n+1} E_3$$
$$\text{(since } E_i E_i = E_i\text{)}.$$

Example 13.77 Continuing with the Example 13.72, for the matrix

$$B = \begin{pmatrix} 2 & 1 & 1 \\ 1 & 2 & 1 \\ 1 & 1 & 2 \end{pmatrix},$$

we have the spectral decomposition $B = E_1 + E_2 + 4E_3$, given by

$$B = \frac{1}{6} \begin{pmatrix} 1 & 1 & -2 \\ 1 & 1 & -2 \\ -2 & -2 & 4 \end{pmatrix} + \frac{1}{2} \begin{pmatrix} 1 & -1 & 0 \\ -1 & 1 & 0 \\ 0 & 0 & 0 \end{pmatrix} + \frac{4}{3} \begin{pmatrix} 1 & 1 & 1 \\ 1 & 1 & 1 \\ 1 & 1 & 1 \end{pmatrix}.$$

Suppose we wish to find a matrix C such that $C^2 = B$. According to this result, if we set $C = \alpha_1 E_1 + \alpha_2 E_2 + \alpha_3 E_3$, for some constants $\alpha_1, \alpha_2, \alpha_3$ to be determined, then

$$C^2 = \alpha_1^2 E_1 + \alpha_2^2 E_2 + \alpha_3^2 E_3 = B = E_1 + E_2 + 4E_3.$$

The constants $\alpha_1 = 1$, $\alpha_2 = 1$ and $\alpha_3 = 2$ will give an appropriate matrix C:

$$C = \frac{1}{6}\begin{pmatrix} 1 & 1 & -2 \\ 1 & 1 & -2 \\ -2 & -2 & 4 \end{pmatrix} + \frac{1}{2}\begin{pmatrix} 1 & -1 & 0 \\ -1 & 1 & 0 \\ 0 & 0 & 0 \end{pmatrix} + \frac{2}{3}\begin{pmatrix} 1 & 1 & 1 \\ 1 & 1 & 1 \\ 1 & 1 & 1 \end{pmatrix}.$$

Activity 13.78 Calculate C and show that $C^2 = B$.

13.8 Learning outcomes

You should now be able to:

- use complex numbers, understand the three forms of a complex number and know how to use them
- explain what is meant by a complex matrix, a complex vector space, a complex inner product, a complex inner product space and translate results between real ones and complex ones
- diagonalise a complex matrix
- find an orthonormal basis of a complex vector space by applying the Gram–Schmidt process if necessary
- state what is meant by the Hermitian conjugate of a complex matrix, a Hermitian matrix and a unitary matrix
- show that a Hermitian matrix has real eigenvalues, and that eigenvectors of a Hermitian matrix corresponding to distinct eigenvalues are orthogonal
- demonstrate that you know how to show that a matrix is unitary if and only if its columns are an orthonormal basis of $\mathbb{C}^n$ and if and only if the linear transformation it defines preserves the inner product
- state what it means to unitarily diagonalise a matrix
- state what is meant by a normal matrix and show that Hermitian, unitary and diagonal matrices are normal
- unitarily diagonalise a normal matrix and show that only normal matrices can be unitarily diagonalised
- explain what is meant by the spectral decomposition of a normal matrix and find the spectral decomposition of a given normal matrix
- demonstrate that you know the properties of the matrices E_i in a spectral decomposition, and that you can use them to obtain an orthogonal projection from $\mathbb{C}^n$ onto a subspace, or to find a matrix B such that $B^n = A$ ($n = 2, 3, \ldots$) for a given normal matrix A.

13.9   Comments on activities

Activity 13.9 We have

$$(x - w)(x - \overline{w}) = x^2 - (w + \overline{w})x + w\overline{w}.$$

Now, $w + \overline{w} = 2\,\mathrm{Re}(w) = 2(-\frac{1}{2})$ and $w\overline{w} = \frac{1}{4} + \frac{3}{4}$ so the product of the last two factors is $x^2 + x + 1$.

Activity 13.11 The points plotted in the complex plane are as follows:

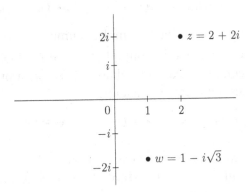

Activity 13.13 Draw the line from the origin to the point z in the diagram above. Do the same for w. For z, $|z| = 2\sqrt{2}$ and $\theta = \frac{\pi}{4}$, so $z = 2\sqrt{2}(\cos(\frac{\pi}{4}) + i\sin(\frac{\pi}{4}))$. The modulus of w is $|w| = 2$ and the argument is $-\frac{\pi}{3}$, so that

$$w = 2\left(\cos\left(-\frac{\pi}{3}\right) + i\sin\left(-\frac{\pi}{3}\right)\right) = 2\left(\cos\left(\frac{\pi}{3}\right) - i\sin\left(\frac{\pi}{3}\right)\right).$$

Activity 13.14 The set (a) consisting of z such that $|z| = 3$ is the circle of radius 3 centered at the origin. The set (b), for which the principal argument of z is $\pi/4$, is the half line from the origin through the point $(1, 1)$.

Activity 13.20 We have

$$e^{i\pi/2} = i, \qquad e^{i3\pi/2} = -i, \qquad e^{i5\pi/4} = -\frac{1}{\sqrt{2}} - i\frac{1}{\sqrt{2}},$$

$$e^{i(11\pi/6)} = e^{-i(\pi/6)} = \frac{\sqrt{3}}{2} - i\frac{1}{2},$$

$$e^{2-i} = e^2 e^{-i} = e^2 \cos(1) - i\, e^2 \sin(1).$$

Finally, e^{-3} is real, so it is already in the form $a + ib$.

Activity 13.23 The roots are:

$$z_1 = e^{i\frac{\pi}{6}}, \qquad z_2 = e^{i\frac{3\pi}{6}}, \qquad z_3 = e^{i\frac{5\pi}{6}},$$

$$z_4 = e^{i\frac{7\pi}{6}}, \qquad z_5 = e^{i\frac{9\pi}{6}}, \qquad z_6 = e^{i\frac{11\pi}{6}}.$$

These roots are in conjugate pairs, and $e^{i\frac{13\pi}{6}} = e^{i\frac{\pi}{6}}$:

$$z_4 = \bar{z}_3 = e^{-i\frac{5\pi}{6}}, \qquad z_5 = \bar{z}_2 = e^{-i\frac{\pi}{2}}, \qquad z_6 = \bar{z}_1 = e^{-i\frac{\pi}{6}}.$$

The polynomial factors as

$$x^6 + 1 = (x - z_1)(x - \bar{z}_1)(x - z_2)(x - \bar{z}_2)(x - z_3)(x - \bar{z}_3).$$

Using the $a + ib$ form of each complex number, for example, $z_1 = \frac{\sqrt{3}}{2} + i\frac{1}{2}$, you can carry out the multiplication of the linear terms pairwise (complex conjugate pairs) to obtain $x^6 + 1$ as a product of irreducible quadratics with real coefficients,

$$x^6 + 1 = (x^2 - \sqrt{3}x + 1)(x^2 + \sqrt{3}x + 1)(x^2 + 1).$$

Activity 13.28 Any vector $\mathbf{v} \in \mathbb{C}^n$ can be separated into real and imaginary parts and written as $\mathbf{v} = \mathbf{a} + i\mathbf{b}$, where $\mathbf{a}, \mathbf{b} \in \mathbb{R}^n$. By only allowing real numbers as scalars, this space has a basis consisting of the n vectors $\{\mathbf{e}_1, \mathbf{e}_2, \ldots, \mathbf{e}_n\}$ together with n vectors $\{\mathbf{u}_1, \mathbf{u}_2, \ldots, \mathbf{u}_n\}$, where $\mathbf{u}_j$ is the vector with every entry equal to 0, except for the jth entry which is equal to the purely imaginary number i. Hence, as a real vector space, $\mathbb{C}^n$ has dimension $2n$.

Activity 13.30 We check $A\mathbf{v}_1 = i\mathbf{v}_1$ and $A\mathbf{v}_2 = -i\mathbf{v}_2$:

$$\begin{pmatrix} 0 & 1 \\ -1 & 0 \end{pmatrix} \begin{pmatrix} -i \\ 1 \end{pmatrix} = \begin{pmatrix} 1 \\ i \end{pmatrix} = i \begin{pmatrix} -i \\ 1 \end{pmatrix}$$

and

$$\begin{pmatrix} 0 & 1 \\ -1 & 0 \end{pmatrix} \begin{pmatrix} i \\ 1 \end{pmatrix} = \begin{pmatrix} 1 \\ -i \end{pmatrix} = -i \begin{pmatrix} i \\ 1 \end{pmatrix}.$$

Activity 13.36 We have

$$\langle \alpha\mathbf{x} + \beta\mathbf{y}, \mathbf{z} \rangle = (\alpha x_1 + \beta y_1)\bar{z}_1 + (\alpha x_2 + \beta y_2)\bar{z}_2 + \cdots + (\alpha x_n + \beta y_n)\bar{z}_n$$
$$= \alpha(x_1\bar{z}_1 + x_2\bar{z}_2 + \cdots + x_n\bar{z}_n) + \beta(y_1\bar{z}_1 + y_2\bar{z}_2 + \cdots + y_n\bar{z}_n)$$
$$= \alpha\langle \mathbf{x}, \mathbf{z} \rangle + \beta\langle \mathbf{y}, \mathbf{z} \rangle.$$

Activity 13.37 The norm of $\mathbf{x} = (1, 0, i)^T$ is $\sqrt{2}$ because

$$\|\mathbf{x}\|^2 = \langle \mathbf{x}, \mathbf{x} \rangle = 1(1) + 0 + (i)(-i) = 2.$$

Activity 13.39 To show that $\langle \mathbf{x}, \alpha\mathbf{y} \rangle = \overline{\alpha}\langle \mathbf{x}, \mathbf{y} \rangle$ for all $\mathbf{x}, \mathbf{y} \in V$ and all $\alpha \in \mathbb{C}$, we use properties (i) and (ii) of the definition,

$$\langle \mathbf{x}, \alpha\mathbf{y} \rangle = \overline{\langle \alpha\mathbf{y}, \mathbf{x} \rangle} = \overline{\alpha\langle \mathbf{y}, \mathbf{x} \rangle} = \overline{\alpha}\,\overline{\langle \mathbf{y}, \mathbf{x} \rangle} = \overline{\alpha}\langle \mathbf{x}, \mathbf{y} \rangle.$$

That $\langle \mathbf{x}, \mathbf{y} + \mathbf{z} \rangle = \langle \mathbf{x}, \mathbf{y} \rangle + \langle \mathbf{x}, \mathbf{z} \rangle$ for all $\mathbf{x}, \mathbf{y}, \mathbf{z} \in V$ is proved in a similar way using properties (i) and (ii) of the definition. We have

$$\langle \mathbf{x}, \mathbf{y} + \mathbf{z} \rangle = \overline{\langle \mathbf{y} + \mathbf{z}, \mathbf{x} \rangle} = \overline{\langle \mathbf{y}, \mathbf{x} \rangle + \langle \mathbf{z}, \mathbf{x} \rangle} = \overline{\langle \mathbf{y}, \mathbf{x} \rangle} + \overline{\langle \mathbf{z}, \mathbf{x} \rangle}$$
$$= \langle \mathbf{x}, \mathbf{y} \rangle + \langle \mathbf{x}, \mathbf{z} \rangle.$$

Activity 13.48 If $A = (a_{ij})$, then

$$(kA)^* = (\overline{ka_{ji}}) = (\overline{k}\,\overline{a_{ji}}) = \overline{k}(\overline{a_{ji}}) = \overline{k}A^*.$$

The analogous result for a real matrix A and a real number k, is simply $(kA)^{\mathrm{T}} = kA^{\mathrm{T}}$.

Activity 13.55 Suppose that λ and μ are any two different eigenvalues of A and that $\mathbf{x}, \mathbf{y}$ are corresponding eigenvectors. Then $A\mathbf{x} = \lambda\mathbf{x}$ and $A\mathbf{y} = \mu\mathbf{y}$. Then

$$\mathbf{y}^*A\mathbf{x} = \mathbf{y}^*\lambda\mathbf{x} = \langle \lambda\mathbf{x}, \mathbf{y} \rangle = \lambda\langle \mathbf{x}, \mathbf{y} \rangle.$$

On the other hand, since $A = A^*$

$$\mathbf{y}^*A\mathbf{x} = \mathbf{y}^*A^*\mathbf{x} = (A\mathbf{y})^*\mathbf{x} = (\mu\mathbf{y})^*\mathbf{x} = \overline{\mu}\mathbf{y}^*\mathbf{x} = \overline{\mu}\langle \mathbf{x}, \mathbf{y} \rangle.$$

But the eigenvalues of a Hermitian matrix are real (Theorem 13.52), so $\overline{\mu} = \mu$, and we can conclude from these two expressions for $\mathbf{y}^*A\mathbf{x}$ that $\lambda\langle \mathbf{x}, \mathbf{y} \rangle = \mu\langle \mathbf{x}, \mathbf{y} \rangle$, or

$$(\lambda - \mu)\langle \mathbf{x}, \mathbf{y} \rangle = 0.$$

Since $\lambda \neq \mu$, we deduce that $\langle \mathbf{x}, \mathbf{y} \rangle = 0$. That is, $\mathbf{x}$ and $\mathbf{y}$ are orthogonal.

Activity 13.59 From the definition, $PP^* = P^*P = I$. Since $(P^*)^* = P$, we have $(P^*)^*P^* = P^*(P^*)^* = I$ so P^* is unitary.

If A and B are unitary matrices, then

$$(AB)(AB)^* = ABB^*A^* = AA^* = I,$$

and, similarly, $(AB)^*(AB) = I$, which shows that AB is unitary.

Activity 13.65 The matrix

$$\begin{pmatrix} 2 & 0 \\ 0 & i \end{pmatrix}$$

(which is normal since it is diagonal) is not Hermitian since the diagonal entries are not real, and it is not unitary since $AA^* \neq I$.

Activity 13.75 To prove (i) $E_i^* = E_i$, we have

$$E_i^* = (\mathbf{x}_i \mathbf{x}_i^*)^* = (\mathbf{x}_i^*)^* \mathbf{x}_i^* = \mathbf{x}_i \mathbf{x}_i^* = E_i.$$

For (iii), if $i \neq j$, then

$$E_i E_j = (\mathbf{x}_i \mathbf{x}_i^*)(\mathbf{x}_j \mathbf{x}_j^*) = \mathbf{x}_i (\mathbf{x}_i^* \mathbf{x}_j) \mathbf{x}_j^* = \mathbf{x}_i \langle \mathbf{x}_j, \mathbf{x}_i \rangle \mathbf{x}_j^* = \mathbf{x}_i \cdot 0 \cdot \mathbf{x}_j^* = 0.$$

Activity 13.76 The orthogonal projection is given by the matrix $E_3 = \mathbf{x}_3 \mathbf{x}_3^*$, where $\mathbf{x}_3$ is the unit vector parallel to $(1, 1, 1)^{\mathrm{T}}$. That is, $P : \mathbb{C}^3 \to \mathrm{Lin}\{(1, 1, 1)^{\mathrm{T}}\}$ is given by $P(\mathbf{v}) = E_3 \mathbf{v}$, where

$$E_3 = \frac{1}{3} \begin{pmatrix} 1 & 1 & 1 \\ 1 & 1 & 1 \\ 1 & 1 & 1 \end{pmatrix}.$$

Compare this with the method you learned in the previous chapter, which will, of course, give you the same solution.

Activity 13.78 The matrix C is

$$C = \frac{1}{3} \begin{pmatrix} 4 & 1 & 1 \\ 1 & 4 & 1 \\ 1 & 1 & 4 \end{pmatrix}.$$

The calculation of C^2 is straightforward, and $C^2 = B$.

13.10   Exercises

Exercise 13.1 Find the four complex roots of the equation $z^4 = -4$ and express them in the form $a + ib$. Use these to write $z^4 + 4$ as a product of quadratic factors with real coefficients.

Exercise 13.2 Suppose the complex matrix A and vector $\mathbf{b}$ are as follows:

$$A = \begin{pmatrix} 1 & i \\ 1+i & -1 \end{pmatrix}, \qquad \mathbf{b} = \begin{pmatrix} 1 \\ i \end{pmatrix}.$$

Calculate the determinant of A, $|A|$. Find the solution of the system of equations $A\mathbf{x} = \mathbf{b}$.

Exercise 13.3 Consider the real vector space $M_2(\mathbb{R})$ of 2×2 matrices with real entries and the complex vector space $M_2(\mathbb{C})$ of 2×2 matrices with complex entries, and the subsets $W_1 \subset M_2(\mathbb{R})$ and $W_2 \subset M_2(\mathbb{C})$:

$$W_1 = \left\{ \begin{pmatrix} a^2 & 0 \\ 0 & b^2 \end{pmatrix} \mid a, b \in \mathbb{R} \right\}, \quad W_2 = \left\{ \begin{pmatrix} a^2 & 0 \\ 0 & b^2 \end{pmatrix} \mid a, b \in \mathbb{C} \right\}.$$

Show that W_1 is not a subspace of $M_2(\mathbb{R})$, but that W_2 is a subspace of $M_2(\mathbb{C})$.

Find a basis and state the dimension of the subspace $\text{Lin}(S)$ in $M_2(\mathbb{R})$ and in $M_2(\mathbb{C})$, where

$$S = \left\{ \begin{pmatrix} 2 & 1 \\ 0 & 2 \end{pmatrix}, \begin{pmatrix} 1 & 2 \\ 0 & 1 \end{pmatrix}, \begin{pmatrix} 3 & 3 \\ 0 & 3 \end{pmatrix}, \begin{pmatrix} -1 & -1 \\ 0 & -1 \end{pmatrix} \right\}.$$

Exercise 13.4 Find the eigenvalues of the matrix

$$A = \begin{pmatrix} 1 & i \\ 1+i & -1 \end{pmatrix}.$$

Express each of the eigenvalues in the form $a + ib$.

Exercise 13.5 Diagonalise the matrix

$$A = \begin{pmatrix} 3 & -1 \\ 2 & 1 \end{pmatrix}.$$

Exercise 13.6 Let

$$A = \begin{pmatrix} 5 & 5 & -5 \\ 3 & 3 & -5 \\ 4 & 0 & -2 \end{pmatrix} \quad \text{and} \quad v = \begin{pmatrix} 0 \\ 1 \\ 1 \end{pmatrix}.$$

(a) Show that v is an eigenvector of A and find the corresponding eigenvalue.

(b) Show that $\lambda = 4 + 2i$ is an eigenvalue of A and find a corresponding eigenvector.

(c) Deduce a third eigenvalue and corresponding eigenvector of A. Hence write down an invertible matrix P and a diagonal matrix D such that $P^{-1}AP = D$.

Is it possible to unitarily diagonalise the matrix A? Do this if it is possible, or explain why it is not possible.

Exercise 13.7 Show that the vectors

$$v_1 = \frac{1}{\sqrt{2}} \begin{pmatrix} 1 \\ i \\ 0 \end{pmatrix} \quad \text{and} \quad v_2 = \frac{1}{3\sqrt{2}} \begin{pmatrix} 1+2i \\ 2-i \\ 2+2i \end{pmatrix}$$

form an orthonormal set, $S = \{v_1, v_2\}$.

Extend S to an orthonormal basis of $\mathbb{C}^3$.

Exercise 13.8 What is a *unitary* matrix? What does it mean to say that a matrix is *unitarily diagonalisable*? Prove that if A is unitarily diagonalisable, then $AA^* = A^*A$.

Exercise 13.9 Prove that if A is a unitary matrix, then all eigenvalues of A have a modulus equal to 1.

Exercise 13.10 Let

$$A = \begin{pmatrix} 7 & 0 & 9 \\ 0 & 2 & 0 \\ 9 & 0 & 7 \end{pmatrix}.$$

Express A in the form

$$A = \lambda_1 E_1 + \lambda_2 E_2 + \lambda_3 E_3,$$

where $\lambda_1, \lambda_2, \lambda_3$ are the eigenvalues of A, and E_1, E_2, E_3 are symmetric idempotent matrices such that if $i \neq j$, then $E_i E_j$ is the zero matrix.
  Determine a matrix B such that $B^3 = A$.

Exercise 13.11 (a) If A is any $m \times k$ complex matrix, prove that the matrix A^*A is Hermitian and normal.
(b) Prove that, for any $m \times k$ matrix A of rank k, the matrix A^*A is positive definite, meaning that $\mathbf{v}^*(A^*A)\mathbf{v} > 0$ for all $\mathbf{v} \in \mathbb{C}^n$, $\mathbf{v} \neq \mathbf{0}$. Prove also that A^*A is invertible.

Exercise 13.12 Explain why you know that the matrix

$$B = \begin{pmatrix} 1 & i & 0 \\ i & 1 & 0 \\ 0 & 0 & 1 \end{pmatrix}$$

can be unitarily diagonalised before finding any eigenvalues and eigenvectors.
  Then find a unitary matrix P and a diagonal matrix D such that $P^*BP = D$.
  Write down the spectral decomposition of B.

13.11   Problems

Problem 13.1 Consider the complex numbers

$$z = \sqrt{3} - i, \quad w = 1 + i, \quad q = \frac{(\sqrt{3} - i)^6}{(1 + i)^{10}}.$$

Plot z and w as points in the complex plane. Express them in exponential form and hence evaluate q. Express q in the form $a + ib$.

Problem 13.2 Write each of the following complex numbers in the form $a + ib$:

$$e^{-i\frac{3\pi}{2}}, \quad e^{i\frac{3\pi}{4}}, \quad e^{i\frac{11\pi}{3}}, \quad e^{1+i}, \quad e^{-1}, \quad e^{-3+i2\sqrt{5}}, \quad 4e^{i\frac{7\pi}{6}}.$$

Problem 13.3 Find the roots w and $\bar{w}$ of the equation $x^2 - 4x + 7 = 0$.

For these values of w and $\bar{w}$, find the real and imaginary parts of the following functions,

$$f(t) = e^{wt}, \quad t \in \mathbb{R}; \qquad g(t) = w^t, \quad t \in \mathbb{Z}^+.$$

Problem 13.4 Let $y(t)$ be the function

$$y(t) = Ae^{\lambda t} + Be^{\bar{\lambda} t},$$

where $A, B \in \mathbb{C}$ are constants and $\lambda = a + ib$. Show that $y(t)$ can be written in the alternative forms

$$y(t) = e^{at}(Ae^{ibt} + Be^{-ibt}) = e^{at}(\widehat{A}\cos bt + \widehat{B}\sin bt),$$

where $\widehat{A} = A + B$ and $\widehat{B} = i(A - B)$. How can A and B be chosen so that $\widehat{A}$ and $\widehat{B}$ are real? For $\widehat{A}$ and $\widehat{B}$ real, show that $y(t)$ can be written as

$$y(t) = e^{at}(\widehat{A}\cos bt + \widehat{B}\sin bt) = Ce^{at}\cos(bt - \phi),$$

where $C = \sqrt{\widehat{A}^2 + \widehat{B}^2} = 2\sqrt{AB}$ and ϕ satisfies $\tan\phi = (\widehat{B}/\widehat{A})$.

Problem 13.5 Show that for any $z \in \mathbb{C}$, the expressions $e^{zt} + e^{\bar{z}t}$, $t \in \mathbb{R}$, and $z^t + \bar{z}^t, t \in \mathbb{Z}^+$, are both real.

Problem 13.6 Find the three complex roots of the equation $z^3 = -1$. Illustrate the roots as points in the complex plane.

Find the roots of $z^4 = -1$ and illustrate them on another graph of the complex plane. Without actually solving the equations, illustrate the roots of $x^5 = -1$ and $x^6 = 64$ as points in the complex plane.

Problem 13.7 Let

$$C = \begin{pmatrix} 1 & -1 \\ 1 & 1 \end{pmatrix}.$$

(a) Find the eigenvalues of the matrix C.
   Diagonalise C by finding complex matrices P (invertible) and D (diagonal) such that $P^{-1}CP = D$.
(b) Let $y_1(t)$, $y_2(t)$ be functions of the real variable t which satisfy the system of differential equations, $\mathbf{y}' = C\mathbf{y}$ with $\mathbf{y} = (y_1, y_2)^{\mathrm{T}}$:

$$y_1'(t) = y_1(t) - y_2(t)$$
$$y_2'(t) = y_1(t) + y_2(t)$$

and the initial conditions $y_1(0) = 0$ and $y_2(0) = 1$.

Set $\mathbf{y} = P\mathbf{z}$ and use the diagonalisation from part (a) to find the solutions $y_1(t)$, $y_2(t)$ in complex form.

Simplify your solutions using Euler's formula to express $y_1(t)$ and $y_2(t)$ as real functions of t.

Problem 13.8 Consider the following subspaces of $\mathbb{C}^3$:

$$U = \mathrm{Lin}\left\{ \begin{pmatrix} 1 \\ 0 \\ i \end{pmatrix}, \begin{pmatrix} 0 \\ -i \\ -1 \end{pmatrix} \right\}, \qquad W = \mathrm{Lin}\left\{ \begin{pmatrix} 0 \\ i \\ -1 \end{pmatrix}, \begin{pmatrix} 1 \\ 0 \\ -i \end{pmatrix} \right\}.$$

Find vectors $\mathbf{x}, \mathbf{y}, \mathbf{z}$ in $\mathbb{C}^3$ which satisfy all of the following:

(i)   the vector $\mathbf{x}$ is in both subspaces U and W; that is, $\mathbf{x} \in U \cap W$;
(ii) the set $\{\mathbf{x}, \mathbf{y}\}$ is an orthonormal basis of U;
(iii) the set $\{\mathbf{x}, \mathbf{z}\}$ is an orthonormal basis of W.

Is your set $\{\mathbf{x}, \mathbf{y}, \mathbf{z}\}$ a basis of $\mathbb{C}^3$? Is it an orthonormal basis of $\mathbb{C}^3$? Justify your answers.

Problem 13.9 Show that the vectors

$$\mathbf{v}_1 = \begin{pmatrix} 1 \\ i \\ 1 \end{pmatrix} \qquad \text{and} \qquad \mathbf{v}_2 = \begin{pmatrix} 0 \\ 2 - i \\ 1 + 2i \end{pmatrix}.$$

are orthogonal in $\mathbb{C}^3$ with the standard inner product. Write down an orthonormal basis $\{\mathbf{u}_1, \mathbf{u}_2\}$ for the linear span of these vectors, $\mathrm{Lin}\{\mathbf{v}_1, \mathbf{v}_2\}$.

Extend this to an orthonormal basis B of $\mathbb{C}^3$, by finding an appropriate vector $\mathbf{u}_3$. (Try this two ways: a Gram–Schmidt process, solving a system of linear equations.)

Express the vector $\mathbf{a} = (1, 1, i)^{\mathrm{T}}$ as a linear combination of $\mathbf{u}_1, \mathbf{u}_2, \mathbf{u}_3$.

Find the matrix of the orthogonal projection of $\mathbb{C}^3$ onto $\mathrm{Lin}\{\mathbf{u}_1\}$. Call this matrix E_1. Show that E_1 is both Hermitian and idempotent. Find $E_1\mathbf{a}$.

Find the matrices E_2 and E_3 of the orthogonal projections of $\mathbb{C}^3$ onto $\mathrm{Lin}\{\mathbf{u}_2\}$ and $\mathrm{Lin}\{\mathbf{u}_3\}$ respectively. Then express the identity matrix I as a linear combination of the matrices E_1, E_2 and E_3.

Problem 13.10 Find an orthonormal basis of $\mathbb{C}^3$ consisting of eigenvectors of the matrix

$$A = \begin{pmatrix} 2 & 1+i & 0 \\ 1-i & 3 & 0 \\ 0 & 0 & 5 \end{pmatrix}.$$

Is the matrix A Hermitian? Is A normal? (Justify your answers.)

Problem 13.11 What does it mean to say that a complex matrix A is normal? Write down the equations that a, b, c, d must satisfy for the 2×2 real matrix

$$\begin{pmatrix} a & b \\ c & d \end{pmatrix}, \qquad a, b, c, d \in \mathbb{R},$$

to be normal. Show that every 2×2 real normal matrix is either symmetric or a multiple of an orthogonal matrix.

Problem 13.12 Show that the matrix

$$A = \begin{pmatrix} 2 & 1 \\ -1 & 2 \end{pmatrix}.$$

is normal, but that it is not Hermitian.

Find the eigenvalues and corresponding eigenvectors of A. Find a unitary matrix P and a diagonal matrix D such that $P^*AP = D$.

Write down the spectral decomposition of A.

Problem 13.13 Consider the following matrix A, where z is a complex number:

$$A = \begin{pmatrix} 0 & i \\ 1 & z \end{pmatrix}.$$

For which values of z is the matrix A unitary? For which values of z is A normal?

Problem 13.14 Unitarily diagonalise the matrix

$$A = \begin{pmatrix} 2 & i & 0 \\ -i & 2 & 0 \\ 0 & 0 & 5i \end{pmatrix}.$$

Is the matrix A Hermitian? Is A normal? (Justify your answers.)

Problem 13.15 A complex matrix A is called *skew-Hermitian* if $A^* = -A$. Prove the following three statements about skew-Hermitian matrices:

(1) The non-zero eigenvalues of a skew-Hermitian matrix are all purely imaginary.
(2) If A is skew-Hermitian, then eigenvectors corresponding to distinct eigenvalues are mutually orthogonal.
(3) Skew-Hermitian matrices are normal.

Problem 13.16 Show that the matrix

$$A = \begin{pmatrix} i & 1 & 0 \\ -1 & 0 & -1 \\ 0 & 1 & i \end{pmatrix}$$

is skew-Hermitian. (See Problem 13.15.)

Find the spectral decomposition of A and deduce the spectral decomposition of A^*.

Problem 13.17 Let A be a normal $n \times n$ matrix and let P be a unitary matrix such that $P^*AP = D$, where $D = \text{diag}(\lambda_1, \lambda_2, \ldots, \lambda_n)$ and the columns of P are the corresponding eigenvectors, $\mathbf{x}_1, \mathbf{x}_2, \ldots, \mathbf{x}_n$. Show that for a positive integer k

$$A^k = \lambda_1^k E_1 + \lambda_2^k E_2 + \cdots + \lambda_n^k E_n = PD^k P^*,$$

where $E_i = \mathbf{x}_i \mathbf{x}_i^*$.

Find the spectral decomposition of the following matrix A. (See Problem 13.10.)

$$A = \begin{pmatrix} 2 & 1+i & 0 \\ 1-i & 3 & 0 \\ 0 & 0 & 5 \end{pmatrix}.$$

Deduce the spectral decomposition of A^3, and use it to find the matrix A^3.

Comments on exercises

Chapter 1 exercises

Exercise 1.1 (a) $A\mathbf{d} = \begin{pmatrix} 4 \\ -3 \end{pmatrix}$.

(b) AB is $(2 \times 3)(3 \times 3)$ giving a 2×3 matrix. C is a 3×2 matrix, so $AB + C$ is not defined.

(c) $A + C^{\mathrm{T}} = \begin{pmatrix} 1 & 3 & 5 \\ -1 & 1 & 0 \end{pmatrix} + \begin{pmatrix} 1 & 3 & -1 \\ 1 & 2 & 4 \end{pmatrix} = \begin{pmatrix} 2 & 6 & 4 \\ 0 & 3 & 4 \end{pmatrix}$.

(d) $C^{\mathrm{T}}C = \begin{pmatrix} 1 & 3 & -1 \\ 1 & 2 & 4 \end{pmatrix} \begin{pmatrix} 1 & 1 \\ 3 & 2 \\ -1 & 4 \end{pmatrix} = \begin{pmatrix} 11 & 3 \\ 3 & 21 \end{pmatrix}$.

(e) $BC = \begin{pmatrix} 1 & 0 & 1 \\ 2 & 1 & 1 \\ 1 & 1 & -1 \end{pmatrix} \begin{pmatrix} 1 & 1 \\ 3 & 2 \\ -1 & 4 \end{pmatrix} = \begin{pmatrix} 0 & 5 \\ 4 & 8 \\ 5 & -1 \end{pmatrix}$.

(f) $\mathbf{d}^{\mathrm{T}}B = (2 \quad -1 \quad 1) \begin{pmatrix} 1 & 0 & 1 \\ 2 & 1 & 1 \\ 1 & 1 & -1 \end{pmatrix} = (1 \quad 0 \quad 0)$.

(g) $C\mathbf{d}$ is not defined.

(h) $\mathbf{d}^{\mathrm{T}}\mathbf{d} = (2 \quad -1 \quad 1) \begin{pmatrix} 2 \\ -1 \\ 1 \end{pmatrix} = (6)$.

(i) $\mathbf{d}\mathbf{d}^{\mathrm{T}} = \begin{pmatrix} 2 \\ -1 \\ 1 \end{pmatrix} (2 \quad -1 \quad 1) = \begin{pmatrix} 4 & -2 & 2 \\ -2 & 1 & -1 \\ 2 & -1 & 1 \end{pmatrix}$.

Exercise 1.2 The matrix A must be 2×2. Let $A = \begin{pmatrix} a & b \\ c & d \end{pmatrix}$. Then the equation

$$\begin{pmatrix} 1 & 7 \\ 5 & 0 \\ 9 & 3 \end{pmatrix} \begin{pmatrix} a & b \\ c & d \end{pmatrix} = \begin{pmatrix} -4 & 14 \\ 15 & 0 \\ 24 & x \end{pmatrix}$$

holds if and only if

$$a + 7c = -4, \quad 5a = 15, \quad \text{and} \quad 9a + 3c = 24$$

and

$$b + 7d = 14, \quad 5b = 0, \quad \text{and} \quad 9b + 3d = x.$$

Solving the first two equations for a and c, we have $a = 3$ and $c = -1$. This solution also satisfies the third equation, $9a + 3c = 9(3) = 3(-1) = 24$, so the matrix A does exist. Solving the second set of equations for b and d, we find $b = 0$ and $d = 2$, therefore $x = 9b + 3d = 6$. The matrix A is

$$A = \begin{pmatrix} 3 & 0 \\ -1 & 2 \end{pmatrix}.$$

Exercise 1.3 By definition, to prove that the matrix AB is invertible you have to show that there exists a matrix, C, such that

$$(AB)C = C(AB) = I.$$

You are given that $C = B^{-1}A^{-1}$. Since both A and B are invertible matrices, you know that both A^{-1} and B^{-1} exist and both are $n \times n$, so the matrix product $B^{-1}A^{-1}$ is defined. So all you need to do is to show that if you multiply AB on the left or on the right by the matrix $B^{-1}A^{-1}$, then you will obtain the identity matrix, I.

$$\begin{aligned}
(AB)(B^{-1}A^{-1}) &= A(BB^{-1})A^{-1} \quad \text{(matrix multiplication is associative)} \\
&= AIA^{-1} \quad \text{(by the definition of } B^{-1}) \\
&= AA^{-1} \quad \text{(since } AI = A \text{ for any matrix } A) \\
&= I \quad \text{(by the definition of } A^{-1}).
\end{aligned}$$

In the same way,

$$(B^{-1}A^{-1})(AB) = (B^{-1})(A^{-1}A)(B) = B^{-1}IB = B^{-1}B = I.$$

Hence $B^{-1}A^{-1}$ is the inverse of the matrix AB.

Exercise 1.4 Use the method shown on page 19 to find the inverse matrix. Then solve for A and you should obtain

$$A = \begin{pmatrix} 1 & 0 \\ \frac{1}{2} & -2 \end{pmatrix}.$$

Exercise 1.5 Begin by taking the inverse of both sides of this equation and then use A^{-1}. A is square, since it is invertible. If A is $n \times n$, then B must also be $n \times n$ for AB to be defined and invertible. Simplifying the equation, you can deduce that $B = \frac{1}{2}I$, where I is the $n \times n$ identity matrix.

Exercise 1.6 Since B^{T} is a $k \times m$ matrix, $B^{\mathrm{T}}B$ is $k \times k$. Furthermore, $(B^{\mathrm{T}}B)^{\mathrm{T}} = B^{\mathrm{T}}(B^{\mathrm{T}})^{\mathrm{T}} = B^{\mathrm{T}}B$, which shows that it is symmetric.

Exercise 1.7 The expression $(A^{\mathrm{T}}A)^{-1}A^{\mathrm{T}}(B^{-1}A^{\mathrm{T}})^{\mathrm{T}}B^{\mathrm{T}}B^2B^{-1}$ simplifies to B. Be careful how you do this: you cannot assume the existence of a matrix A^{-1}; indeed, if $m \neq n$, such a matrix cannot exist.

Exercise 1.8 (a) A vector equation of the line in $\mathbb{R}^2$ is

$$\mathbf{x} = \begin{pmatrix} x \\ y \end{pmatrix} = \begin{pmatrix} 3 \\ 1 \end{pmatrix} + t\begin{pmatrix} 5 \\ -3 \end{pmatrix}, \quad t \in \mathbb{R}.$$

(b) A vector equation of the line in $\mathbb{R}^5$ is

$$\mathbf{x} = \begin{pmatrix} x_1 \\ x_2 \\ x_3 \\ x_4 \\ x_5 \end{pmatrix} = \begin{pmatrix} 3 \\ 1 \\ -1 \\ 2 \\ 5 \end{pmatrix} + t\begin{pmatrix} 5 \\ -3 \\ -1 \\ 1 \\ 4 \end{pmatrix}, \quad t \in \mathbb{R}.$$

The point $(4, 3, 2, 1, 4)$ is not on this line as there is no value of t for which

$$\begin{pmatrix} 4 \\ 3 \\ 2 \\ 1 \\ 4 \end{pmatrix} = \begin{pmatrix} 3 \\ 1 \\ -1 \\ 2 \\ 5 \end{pmatrix} + t\begin{pmatrix} 5 \\ -3 \\ -1 \\ 1 \\ 4 \end{pmatrix}.$$

For example, in order to have $x_4 = 1$, then $t = -1$, and none of the other components corresponds to this value of t.

Exercise 1.9 To obtain the vector equation of the line, set

$$t = \frac{x-1}{3} = y + 2 = \frac{5-z}{4}$$

and solve for x, y, z. You should easily obtain $x = 1 + 3t$, $y = t - 2$ and $z = 5 - 4t$, so the equation is

$$\begin{pmatrix} x \\ y \\ z \end{pmatrix} = \begin{pmatrix} 1 \\ -2 \\ 5 \end{pmatrix} + t\begin{pmatrix} 3 \\ 1 \\ -4 \end{pmatrix}, \quad t \in \mathbb{R}.$$

Exercise 1.10 The vector equations of the three lines are

$$L_1: \quad \mathbf{x} = \begin{pmatrix} x \\ y \\ z \end{pmatrix} = \begin{pmatrix} 1 \\ 3 \\ 2 \end{pmatrix} + t \begin{pmatrix} -1 \\ 5 \\ 4 \end{pmatrix}, \quad t \in \mathbb{R},$$

$$L_2: \quad \mathbf{x} = \begin{pmatrix} 8 \\ 0 \\ -3 \end{pmatrix} + s \begin{pmatrix} 6 \\ 2 \\ -1 \end{pmatrix}, \quad s \in \mathbb{R},$$

$$L_3: \quad \mathbf{x} = \begin{pmatrix} 9 \\ 3 \\ 1 \end{pmatrix} + q \begin{pmatrix} 2 \\ -10 \\ -8 \end{pmatrix}, \quad q \in \mathbb{R},$$

taking $(9, 3, 1)^T - (7, 13, 9)^T = (2, -10, -8)^T$ as the direction vector.

The lines L_1 and L_3 are parallel, since their directions vectors are scalar multiples, $-2(-1, 5, 4)^T = (2, -10, -8)^T$. The lines are not coincident since the point $(9, 3, 1)$ does not lie on L_1 (check this).

The lines L_1 and L_2 either intersect or are skew. They will intersect if there are constants $s, t \in \mathbb{R}$ satisfying the following equations:

$$\left. \begin{array}{l} 1 - t = 8 + 6s \\ 3 + 5t = 2s \\ 2 + 4t = -3 - s \end{array} \right\} \iff \left. \begin{array}{l} 6s + t = -7 \\ 2s - 5t = 3 \\ s + 4t = -5 \end{array} \right\}.$$

Use two of these equations to solve for s and t, then check the solution in the third equation. The three equations have the solution $s = -1$ and $t = -1$, so the two lines intersect in the point $(2, -2, -2)$. The angle of intersection is the acute angle between their direction vectors. Since

$$\left\langle \begin{pmatrix} -1 \\ 5 \\ 4 \end{pmatrix}, \begin{pmatrix} 6 \\ 2 \\ -1 \end{pmatrix} \right\rangle = 0,$$

the angle between them is $\dfrac{\pi}{2}$ and the two lines are perpendicular.

The lines L_2 and L_3 are skew. To show this, you need to show that there is no solution, $q, s \in \mathbb{R}$, to the system of equations

$$\left. \begin{array}{l} 9 + 2q = 8 + 6s \\ 3 - 10q = 2s \\ 1 - 8q = -3 - s \end{array} \right\} \iff \left. \begin{array}{l} 6s - 2q = 1 \\ 2s + 10q = 3 \\ s - 8q = -4 \end{array} \right\}.$$

Solving the first two equations, you obtain $q = \frac{1}{4}$ and $s = \frac{1}{4}$, but this solution does not satisfy the third equation, so the lines do not intersect and are skew.

Exercise 1.11 The plane which contains the two parallel lines contains the points $(1, 3, 2)$ and $(9, 3, 1)$ and the direction vector of the two lines, $\mathbf{v} = (-1, 5, 4)^{\mathrm{T}}$. So it also contains the line through the two points, which has direction $\mathbf{w} = (8, 0, -1)^{\mathrm{T}}$. Therefore, a vector equation of the plane is

$$\mathbf{x} = \begin{pmatrix} x \\ y \\ z \end{pmatrix} = \begin{pmatrix} 1 \\ 3 \\ 2 \end{pmatrix} + s \begin{pmatrix} -1 \\ 5 \\ 4 \end{pmatrix} + t \begin{pmatrix} 8 \\ 0 \\ -1 \end{pmatrix} = \mathbf{p} + s\mathbf{v} + t\mathbf{w}, \quad s, t \in \mathbb{R}.$$

There are two ways to obtain the Cartesian equation from this. You can find a normal vector $\mathbf{n} = (a, b, c)^{\mathrm{T}}$ by solving the two linear equations obtained from $\langle \mathbf{n}, \mathbf{v} \rangle = 0$ and $\langle \mathbf{n}, \mathbf{w} \rangle = 0$, namely

$$-a + 5b + 4c = 0 \quad \text{and} \quad 8a - c = 0.$$

There will be infinitely many solutions, and you can just choose one.

Or, you can write down the three component equations resulting from the vector equation and eliminate s and t. This last method is more easily accomplished by first finding the Cartesian equation of a parallel plane through the origin:

$$\begin{pmatrix} x \\ y \\ z \end{pmatrix} = s \begin{pmatrix} -1 \\ 5 \\ 4 \end{pmatrix} + t \begin{pmatrix} 8 \\ 0 \\ -1 \end{pmatrix} \quad \Longleftrightarrow \quad \left. \begin{aligned} x &= -s + 8t \\ y &= 5s \\ z &= 4s - t \end{aligned} \right\}.$$

Eliminating s and t from these equations yields

$$5x - 31y + 40z = 0,$$

and you should check that this is correct by showing that the vector $\mathbf{n} = (5, -31, 40)^{\mathrm{T}}$ is orthogonal to both $\mathbf{v}$ and $\mathbf{w}$. Then, since $5(1) - 31(3) + 40(2) = -8$, a Cartesian equation of the plane is

$$5x - 31y + 40z = -8.$$

Again, you can check that the point $(9, 3, 1)$ also satisfies this equation. You should try both methods.

Exercise 1.12 The normal to the plane is orthogonal to the direction of the line, as

$$\left\langle \begin{pmatrix} 2 \\ 0 \\ 1 \end{pmatrix}, \begin{pmatrix} -1 \\ 4 \\ 2 \end{pmatrix} \right\rangle = 0.$$

Since the point $(2, 3, 1)$ on the line does not satisfy the equation of the plane, the line is parallel to the plane. Therefore, it makes sense to ask for the distance of the line from the plane. This can be found by dropping a perpendicular from the line to the plane and measuring its length. A method for doing this is given in the question.

The line through (2, 3, 1) and parallel to the normal to the plane is perpendicular to the plane. A vector equation of the line is

$$\begin{pmatrix} x \\ y \\ z \end{pmatrix} = \begin{pmatrix} 2 \\ 3 \\ 1 \end{pmatrix} + t \begin{pmatrix} 2 \\ 0 \\ 1 \end{pmatrix}, \quad t \in \mathbb{R}.$$

Equating components, we have $x = 2 + 2t$, $y = 3$, and $z = 1 + t$. At the point of intersection of the line with the plane, these components will satisfy the equation of the plane, so that

$$2x + z = 9 \implies 2(2 + 2t) + (1 + t) = 9 \implies 5 + 5t = 9,$$

or $t = \frac{4}{5}$. Then putting this value for t in the line, we find the point of intersection is

$$\mathbf{p} = \begin{pmatrix} 2 \\ 3 \\ 1 \end{pmatrix} + \frac{4}{5} \begin{pmatrix} 2 \\ 0 \\ 1 \end{pmatrix}.$$

The distance between the line and the plane is the distance between this point and the point (2, 3, 1), which is given by the length of the vector

$$\mathbf{v} = \mathbf{p} - \begin{pmatrix} 2 \\ 3 \\ 1 \end{pmatrix} = \frac{4}{5} \begin{pmatrix} 2 \\ 0 \\ 1 \end{pmatrix},$$

so the distance is $\dfrac{4\sqrt{5}}{5}$.

Chapter 2 exercises

Exercise 2.1 For this solution, we'll indicate the row operations.

(a) $(A|\mathbf{b}) = \begin{pmatrix} 1 & -1 & 1 & -3 \\ -3 & 4 & -1 & 2 \\ 1 & -3 & -2 & 7 \end{pmatrix} \xrightarrow[R_3-R_1]{R_2+3R_1} \begin{pmatrix} 1 & -1 & 1 & -3 \\ 0 & 1 & 2 & -7 \\ 0 & -2 & -3 & 10 \end{pmatrix}$

$\xrightarrow{R_3+2R_2} \begin{pmatrix} 1 & -1 & 1 & -3 \\ 0 & 1 & 2 & -7 \\ 0 & 0 & 1 & -4 \end{pmatrix} \xrightarrow[R_2-2R_3]{R_1-R_3} \begin{pmatrix} 1 & -1 & 0 & 1 \\ 0 & 1 & 0 & 1 \\ 0 & 0 & 1 & -4 \end{pmatrix}$

$\xrightarrow{R_1+R_2} \begin{pmatrix} 1 & 0 & 0 & 2 \\ 0 & 1 & 0 & 1 \\ 0 & 0 & 1 & -4 \end{pmatrix},$

from which it follows that

$$\begin{pmatrix} x \\ y \\ z \end{pmatrix} = \begin{pmatrix} 2 \\ 1 \\ -4 \end{pmatrix}.$$

This system has a unique solution. The three planes intersect in one point.

$$\text{(b) } (A|\mathbf{b}) = \begin{pmatrix} 2 & -1 & 3 & 4 \\ 1 & 1 & -1 & 1 \\ 5 & 2 & 0 & 7 \end{pmatrix} \xrightarrow{R_1 \rightleftharpoons R_2} \begin{pmatrix} 1 & 1 & -1 & 1 \\ 2 & -1 & 3 & 4 \\ 5 & 2 & 0 & 7 \end{pmatrix}$$

$$\xrightarrow[R_3-5R_1]{R_2-2R_1} \begin{pmatrix} 1 & 1 & -1 & 1 \\ 0 & -3 & 5 & 2 \\ 0 & -3 & 5 & 2 \end{pmatrix} \xrightarrow{R_3-R_2} \begin{pmatrix} 1 & 1 & -1 & 1 \\ 0 & -3 & 5 & 2 \\ 0 & 0 & 0 & 0 \end{pmatrix}$$

$$\xrightarrow{-\frac{1}{3}R_2} \begin{pmatrix} 1 & 1 & -1 & 1 \\ 0 & 1 & -\frac{5}{3} & -\frac{2}{3} \\ 0 & 0 & 0 & 0 \end{pmatrix} \xrightarrow{R_1-R_2} \begin{pmatrix} 1 & 0 & \frac{2}{3} & \frac{5}{3} \\ 0 & 1 & -\frac{5}{3} & -\frac{2}{3} \\ 0 & 0 & 0 & 0 \end{pmatrix}.$$

Set $z = t$ and then solve for x and y in terms of t. There are infinitely many solutions:

$$\begin{pmatrix} x \\ y \\ z \end{pmatrix} = \begin{pmatrix} \frac{5}{3} - \frac{2}{3}t \\ -\frac{2}{3} + \frac{5}{3}t \\ t \end{pmatrix} = \begin{pmatrix} \frac{5}{3} \\ -\frac{2}{3} \\ 0 \end{pmatrix} + t \begin{pmatrix} -\frac{2}{3} \\ \frac{5}{3} \\ 1 \end{pmatrix}, \quad t \in \mathbb{R}.$$

The three planes intersect in a line. If you set $z = 3s$, then this line of solutions can be written as

$$\begin{pmatrix} x \\ y \\ z \end{pmatrix} = \begin{pmatrix} \frac{5}{3} \\ -\frac{2}{3} \\ 0 \end{pmatrix} + s \begin{pmatrix} -2 \\ 5 \\ 3 \end{pmatrix}, \quad s \in \mathbb{R}.$$

Exercise 2.2 The system of equations in part (a) is the associated homogeneous system of part (b), so to solve both you can row reduce the augmented matrix and then interpret the solutions of each system. Doing this,

$$(A|\mathbf{b}) = \begin{pmatrix} -1 & 1 & -3 & 6 \\ 3 & -2 & 10 & -10 \\ -2 & 3 & -5 & 9 \end{pmatrix} \longrightarrow \begin{pmatrix} 1 & -1 & 3 & -6 \\ 0 & 1 & 1 & 8 \\ 0 & 1 & 1 & -3 \end{pmatrix}$$

from which you can already see that the system is inconsistent since $y + z = 8$ and $y + z = -3$ is impossible. However, the homogeneous system is always consistent, so continuing the reduction of the matrix A,

$$A \longrightarrow \begin{pmatrix} 1 & -1 & 3 \\ 0 & 1 & 1 \\ 0 & 0 & 0 \end{pmatrix} \longrightarrow \begin{pmatrix} 1 & 0 & 4 \\ 0 & 1 & 1 \\ 0 & 0 & 0 \end{pmatrix}.$$

So the answers are:

(a) $\mathbf{x} = t \begin{pmatrix} -4 \\ -1 \\ 1 \end{pmatrix}$, $t \in \mathbb{R}$, and

(b) no solutions.

Exercise 2.3 Solve the first two equations simultaneously using Gaussian elimination. The general solution takes the form $\mathbf{x} = \mathbf{p} + s\mathbf{w}$, $s \in \mathbb{R}$, where $\mathbf{p} = (1, 0, 0)^T$ and $\mathbf{w} = (0, -1, 1)^T$, which is the equation of the line of intersection of the two planes.

The third plane intersects the first two in the same line. You can determine this by solving the linear system of three equations using Gaussian elimination. Alternatively, you can notice that the line of intersection of the first two planes is in the third plane (since its direction is perpendicular to the normal, and the point $\mathbf{v}$ satisfies the Cartesian equation of the plane), so this must be the intersection of all three planes.

Exercise 2.4 To solve the system of equations $A\mathbf{x} = \mathbf{b}$, where

$$A = \begin{pmatrix} 2 & 3 & 1 & 1 \\ 1 & 2 & 0 & -1 \\ 3 & 4 & 2 & 4 \end{pmatrix}, \qquad \mathbf{b} = \begin{pmatrix} 4 \\ 1 \\ 9 \end{pmatrix},$$

using Gaussian elimination, put the augmented matrix into reduced row echelon form:

$$(A|\mathbf{b}) = \begin{pmatrix} 2 & 3 & 1 & 1 & 4 \\ 1 & 2 & 0 & -1 & 1 \\ 3 & 4 & 2 & 4 & 9 \end{pmatrix} \longrightarrow \begin{pmatrix} 1 & 2 & 0 & -1 & 1 \\ 2 & 3 & 1 & 1 & 4 \\ 3 & 4 & 2 & 4 & 9 \end{pmatrix}$$

$$\longrightarrow \begin{pmatrix} 1 & 2 & 0 & -1 & 1 \\ 0 & -1 & 1 & 3 & 2 \\ 0 & -2 & 2 & 7 & 6 \end{pmatrix} \longrightarrow \begin{pmatrix} 1 & 2 & 0 & -1 & 1 \\ 0 & 1 & -1 & -3 & -2 \\ 0 & 0 & 0 & 1 & 2 \end{pmatrix}$$

$$\longrightarrow \begin{pmatrix} 1 & 2 & 0 & 0 & 3 \\ 0 & 1 & -1 & 0 & 4 \\ 0 & 0 & 0 & 1 & 2 \end{pmatrix} \longrightarrow \begin{pmatrix} 1 & 0 & 2 & 0 & -5 \\ 0 & 1 & -1 & 0 & 4 \\ 0 & 0 & 0 & 1 & 2 \end{pmatrix}.$$

There is one non-leading variable. If $\mathbf{x} = (x, y, z, w)^T$, then set $z = t$. Reading from the matrix (starting from the bottom row), a general solution is

$$\mathbf{x} = \begin{pmatrix} x \\ y \\ z \\ w \end{pmatrix} = \begin{pmatrix} -5 - 2t \\ 4 + t \\ t \\ 2 \end{pmatrix} = \begin{pmatrix} -5 \\ 4 \\ 0 \\ 2 \end{pmatrix} + t \begin{pmatrix} -2 \\ 1 \\ 1 \\ 0 \end{pmatrix} = \mathbf{p} + t\mathbf{v}, \quad t \in \mathbb{R}.$$

Checking the solution by multiplying the matrices, you should find $A\mathbf{p} = \mathbf{b}$ and $A\mathbf{v} = \mathbf{0}$. If your solution is not correct, then you will now know it, in which case you should go back and correct it.

$$A\mathbf{p} = \begin{pmatrix} 2 & 3 & 1 & 1 \\ 1 & 2 & 0 & -1 \\ 3 & 4 & 2 & 4 \end{pmatrix} \begin{pmatrix} -5 \\ 4 \\ 0 \\ 2 \end{pmatrix} = \begin{pmatrix} 4 \\ 1 \\ 9 \end{pmatrix},$$

$$A\mathbf{v} = \begin{pmatrix} 2 & 3 & 1 & 1 \\ 1 & 2 & 0 & -1 \\ 3 & 4 & 2 & 4 \end{pmatrix} \begin{pmatrix} -2 \\ 1 \\ 1 \\ 0 \end{pmatrix} = \begin{pmatrix} 0 \\ 0 \\ 0 \end{pmatrix}.$$

The reduced row echelon form of the matrix A consists of the first four columns of the reduced row echelon form of the augmented matrix above, that is

$$\begin{pmatrix} 1 & 0 & 2 & 0 \\ 0 & 1 & -1 & 0 \\ 0 & 0 & 0 & 1 \end{pmatrix}.$$

(i) No, there is no vector $\mathbf{d}$ for which the system is inconsistent. Since there is a leading one in every row of the reduced row echelon form of A, the system of equations $A\mathbf{x} = \mathbf{d}$ is consistent for all $\mathbf{d} \in \mathbb{R}^3$.
(ii) No, there is no vector $\mathbf{d}$ for which the system has a unique solution. The system of equations $A\mathbf{x} = \mathbf{d}$ has infinitely many solutions for all $\mathbf{d} \in \mathbb{R}^3$ since there will always be a free variable. There is no leading one in the third column.

Exercise 2.5 Reducing the matrix C to reduced row echelon form, beginning with

$$C = \begin{pmatrix} 1 & 2 & -1 & 3 & 8 \\ -3 & -1 & 8 & 6 & 1 \\ -1 & 0 & 3 & 1 & -2 \end{pmatrix} \longrightarrow \begin{pmatrix} 1 & 2 & -1 & 3 & 8 \\ 0 & 5 & 5 & 15 & 25 \\ 0 & 2 & 2 & 4 & 6 \end{pmatrix},$$

you should eventually obtain (after four more steps)

$$\longrightarrow \cdots \longrightarrow \begin{pmatrix} 1 & 0 & -3 & 0 & 4 \\ 0 & 1 & 1 & 0 & -1 \\ 0 & 0 & 0 & 1 & 2 \end{pmatrix}.$$

(a) If C is the augmented matrix of a system of equations $A\mathbf{x} = \mathbf{b}$, so that $C = (A|\mathbf{b})$, then there is one non-leading variable in the third column, so the solutions are

$$\mathbf{x} = \begin{pmatrix} x_1 \\ x_2 \\ x_3 \\ x_4 \end{pmatrix} = \begin{pmatrix} 4 + 3t \\ -1 - t \\ t \\ 2 \end{pmatrix} = \begin{pmatrix} 4 \\ -1 \\ 0 \\ 2 \end{pmatrix} + t \begin{pmatrix} 3 \\ -1 \\ 1 \\ 0 \end{pmatrix}, \quad t \in \mathbb{R}.$$

These solutions are in $\mathbb{R}^4$.

(b) If C is the coefficient matrix of a homogeneous system of equations, $C\mathbf{x} = \mathbf{0}$, then there are two non-leading variables, one in column three and one in column five. Set $x_3 = s$ and $x_5 = t$. Then the solutions are

$$\mathbf{x} = \begin{pmatrix} x_1 \\ x_2 \\ x_3 \\ x_4 \\ x_5 \end{pmatrix} = \begin{pmatrix} 3s - 4t \\ -s + t \\ s \\ -2t \\ t \end{pmatrix} = s \begin{pmatrix} 3 \\ -1 \\ 1 \\ 0 \\ 0 \end{pmatrix} + t \begin{pmatrix} -4 \\ 1 \\ 0 \\ -2 \\ 1 \end{pmatrix}, \quad s, t \in \mathbb{R}.$$

These solutions are in $\mathbb{R}^5$.

(c) To find $\mathbf{d}$,

$$C\mathbf{w} = \begin{pmatrix} 1 & 2 & -1 & 3 & 8 \\ -3 & -1 & 8 & 6 & 1 \\ -1 & 0 & 3 & 1 & -2 \end{pmatrix} \begin{pmatrix} 1 \\ 0 \\ 1 \\ 1 \\ 1 \end{pmatrix} = \begin{pmatrix} 11 \\ 12 \\ 1 \end{pmatrix}.$$

To find all solutions of $C\mathbf{x} = \mathbf{d}$, there is no need to use Gaussian elimination again. You know that $C\mathbf{w} = \mathbf{d}$ and from part (b) you have the solutions of the associated homogeneous system. So using the Principle of Linearity, a general solution of $C\mathbf{x} = \mathbf{d}$ is given by

$$\mathbf{x} = \begin{pmatrix} x_1 \\ x_2 \\ x_3 \\ x_4 \\ x_5 \end{pmatrix} = \begin{pmatrix} 1 \\ 0 \\ 1 \\ 1 \\ 1 \end{pmatrix} + s \begin{pmatrix} 3 \\ -1 \\ 1 \\ 0 \\ 0 \end{pmatrix} + t \begin{pmatrix} -4 \\ 1 \\ 0 \\ -2 \\ 1 \end{pmatrix}, \quad s, t \in \mathbb{R}.$$

Exercise 2.6 Using row operations on the matrix B, you should obtain the reduced row echelon form,

$$B \longrightarrow \cdots \longrightarrow \begin{pmatrix} 1 & 0 & 0 \\ 0 & 1 & 0 \\ 0 & 0 & 1 \\ 0 & 0 & 0 \end{pmatrix}.$$

Therefore, the only solution of $B\mathbf{x} = \mathbf{0}$ is the trivial solution, $\mathbf{x} = \mathbf{0}$. So the null space is the set containing only the zero vector, $\{\mathbf{0}\} \subset \mathbb{R}^3$.

The vector $\mathbf{d} = \mathbf{c}_1 + 2\mathbf{c}_2 - \mathbf{c}_3 = \begin{pmatrix} 0 \\ 4 \\ -5 \\ -8 \end{pmatrix}.$

To find all solutions of $B\mathbf{x} = \mathbf{d}$, you just need to find one particular solution. Using the expression of a matrix product

$$B\mathbf{x} = x_1\mathbf{c}_1 + x_2\mathbf{c}_2 + x_3\mathbf{c}_3$$

(see Theorem 1.38), you can deduce that

$$\mathbf{x} = \begin{pmatrix} 1 \\ 2 \\ -1 \end{pmatrix}$$

is one solution. Since the null space consists of only the zero vector, this solution is the only solution.

Chapter 3 exercises

Exercise 3.1 We begin with A, by row reducing $(A|I)$.

$$(A|I) = \begin{pmatrix} 1 & 2 & -1 & 1 & 0 & 0 \\ 0 & 1 & 2 & 0 & 1 & 0 \\ 3 & 8 & 1 & 0 & 0 & 1 \end{pmatrix}$$

$$\xrightarrow{R_3 - 3R_1} \begin{pmatrix} 1 & 2 & -1 & 1 & 0 & 0 \\ 0 & 1 & 2 & 0 & 1 & 0 \\ 0 & 2 & 4 & -3 & 0 & 1 \end{pmatrix}.$$

The next step, $R_3 - 2R_2$, will yield a row of zeros in the row echelon form of A, and therefore the matrix A is not invertible.

For the matrix B,

$$(B|I) = \begin{pmatrix} -1 & 2 & 1 & 1 & 0 & 0 \\ 0 & 1 & 2 & 0 & 1 & 0 \\ 3 & 1 & 4 & 0 & 0 & 1 \end{pmatrix}$$

$$\xrightarrow{-R_1} \begin{pmatrix} 1 & -2 & -1 & -1 & 0 & 0 \\ 0 & 1 & 2 & 0 & 1 & 0 \\ 3 & 1 & 4 & 0 & 0 & 1 \end{pmatrix}$$

$$\xrightarrow{R_3-3R_1} \begin{pmatrix} 1 & -2 & -1 & -1 & 0 & 0 \\ 0 & 1 & 2 & 0 & 1 & 0 \\ 0 & 7 & 7 & 3 & 0 & 1 \end{pmatrix}$$

$$\xrightarrow{R_3-7R_2} \begin{pmatrix} 1 & -2 & -1 & -1 & 0 & 0 \\ 0 & 1 & 2 & 0 & 1 & 0 \\ 0 & 0 & -7 & 3 & -7 & 1 \end{pmatrix}$$

$$\xrightarrow{-\frac{1}{7}R_3} \begin{pmatrix} 1 & -2 & -1 & -1 & 0 & 0 \\ 0 & 1 & 2 & 0 & 1 & 0 \\ 0 & 0 & 1 & -\frac{3}{7} & 1 & -\frac{1}{7} \end{pmatrix}$$

$$\xrightarrow[R_2-2R_3]{R_1+R_3} \begin{pmatrix} 1 & -2 & 0 & -\frac{10}{7} & 1 & -\frac{1}{7} \\ 0 & 1 & 0 & \frac{6}{7} & -1 & \frac{2}{7} \\ 0 & 0 & 1 & -\frac{3}{7} & 1 & -\frac{1}{7} \end{pmatrix}$$

$$\xrightarrow{R_1+2R_2} \begin{pmatrix} 1 & 0 & 0 & \frac{2}{7} & -1 & \frac{3}{7} \\ 0 & 1 & 0 & \frac{6}{7} & -1 & \frac{2}{7} \\ 0 & 0 & 1 & -\frac{3}{7} & 1 & -\frac{1}{7} \end{pmatrix}.$$

This final matrix shows that

$$B^{-1} = \frac{1}{7} \begin{pmatrix} 2 & -7 & 3 \\ 6 & -7 & 2 \\ -3 & 7 & -1 \end{pmatrix}.$$

Next, check this is correct by multiplying BB^{-1},

$$BB^{-1} = \begin{pmatrix} -1 & 2 & 1 \\ 0 & 1 & 2 \\ 3 & 1 & 4 \end{pmatrix} \frac{1}{7} \begin{pmatrix} 2 & -7 & 3 \\ 6 & -7 & 2 \\ -3 & 7 & -1 \end{pmatrix} = \frac{1}{7} \begin{pmatrix} 7 & 0 & 0 \\ 0 & 7 & 0 \\ 0 & 0 & 7 \end{pmatrix} = I.$$

We'll answer the next part of the question for the matrix B first. To solve $B\mathbf{x} = \mathbf{b}$, we can use this inverse matrix. The unique solution is $\mathbf{x} = B^{-1}\mathbf{b}$,

$$\mathbf{x} = \begin{pmatrix} \frac{2}{7} & -1 & \frac{3}{7} \\ \frac{6}{7} & -1 & \frac{2}{7} \\ -\frac{3}{7} & 1 & -\frac{1}{7} \end{pmatrix} \begin{pmatrix} 1 \\ 1 \\ 5 \end{pmatrix} = \begin{pmatrix} \frac{10}{7} \\ \frac{9}{7} \\ -\frac{1}{7} \end{pmatrix}.$$

Check that this is correct by multiplying $B\mathbf{x}$ to get $\mathbf{b}$.

Since the matrix B is invertible, by Theorem 3.8 we know that $B\mathbf{x} = \mathbf{d}$ is consistent for all $\mathbf{d} \in \mathbb{R}^3$.

For the matrix A, to solve $A\mathbf{x} = \mathbf{b}$ we need to try Gaussian elimination. Row reducing the augmented matrix,

$$(A|\mathbf{b}) = \begin{pmatrix} 1 & 2 & -1 & 1 \\ 0 & 1 & 2 & 1 \\ 3 & 8 & 1 & 5 \end{pmatrix} \longrightarrow \begin{pmatrix} 1 & 2 & -1 & 1 \\ 0 & 1 & 2 & 1 \\ 0 & 2 & 4 & 2 \end{pmatrix}$$

$$\longrightarrow \begin{pmatrix} 1 & 0 & -5 & -1 \\ 0 & 1 & 2 & 1 \\ 0 & 0 & 0 & 0 \end{pmatrix}.$$

This system is consistent with infinitely many solutions. Setting the non-leading variable equal to t, the solutions are

$$\mathbf{x} = \begin{pmatrix} -1 + 5t \\ 1 - 2t \\ t \end{pmatrix} = \begin{pmatrix} -1 \\ 1 \\ 0 \end{pmatrix} + t \begin{pmatrix} 5 \\ -2 \\ 1 \end{pmatrix}, \qquad t \in \mathbb{R}.$$

Since the matrix A is not invertible, there will be vectors $\mathbf{d} \in \mathbb{R}^3$ for which the system is inconsistent. Looking at the row reduction above, an easy choice is $\mathbf{d} = (1, 1, 4)^T$, for then

$$(A|\mathbf{d}) = \begin{pmatrix} 1 & 2 & -1 & 1 \\ 0 & 1 & 2 & 1 \\ 3 & 8 & 1 & 4 \end{pmatrix} \longrightarrow \begin{pmatrix} 1 & 2 & -1 & 1 \\ 0 & 1 & 2 & 1 \\ 0 & 2 & 4 & 1 \end{pmatrix}$$

$$\longrightarrow \begin{pmatrix} 1 & 2 & -1 & 1 \\ 0 & 1 & 2 & 1 \\ 0 & 0 & 0 & 1 \end{pmatrix},$$

which shows that the system $A\mathbf{x} = \mathbf{d}$ is inconsistent.

Exercise 3.2 Note the row operations as you reduce A to reduced row echelon form,

$$A = \begin{pmatrix} 1 & 0 & 2 \\ 0 & 1 & -1 \\ 1 & 4 & -1 \end{pmatrix} \xrightarrow{R_3 - R_1} \begin{pmatrix} 1 & 0 & 2 \\ 0 & 1 & -1 \\ 0 & 4 & -3 \end{pmatrix} \xrightarrow{R_3 - 4R_2} \begin{pmatrix} 1 & 0 & 2 \\ 0 & 1 & -1 \\ 0 & 0 & 1 \end{pmatrix}$$

$$\xrightarrow{R_2 + R_3} \begin{pmatrix} 1 & 0 & 2 \\ 0 & 1 & 0 \\ 0 & 0 & 1 \end{pmatrix} \xrightarrow{R_1 - 2R_3} \begin{pmatrix} 1 & 0 & 0 \\ 0 & 1 & 0 \\ 0 & 0 & 1 \end{pmatrix} = I.$$

There were four row operations required to do this, so express this as $E_4 E_3 E_2 E_1 A = I$. Then $A = (E_1)^{-1}(E_2)^{-1}(E_3)^{-1}(E_4)^{-1}$, where, for example, the last row operation, $R_1 - 2R_3$, is given by the elementary matrix E_4 so the inverse elementary matrix E_4^{-1} performs the operation $R_1 + 2R_3$.

Then the matrix A is given by

$$A = \begin{pmatrix} 1 & 0 & 0 \\ 0 & 1 & 0 \\ 1 & 0 & 1 \end{pmatrix} \begin{pmatrix} 1 & 0 & 0 \\ 0 & 1 & 0 \\ 0 & 4 & 1 \end{pmatrix} \begin{pmatrix} 1 & 0 & 0 \\ 0 & 1 & -1 \\ 0 & 0 & 1 \end{pmatrix} \begin{pmatrix} 1 & 0 & 2 \\ 0 & 1 & 0 \\ 0 & 0 & 1 \end{pmatrix}.$$

You should multiply out this product to check that you do obtain A.

Then using the fact that the determinant of the product is equal to the product of the determinants,

$$|A| = |(E_1)^{-1}| \, |(E_2)^{-1}| \, |(E_3)^{-1}| \, |(E_4)^{-1}|.$$

Each of the above elementary matrices represents a row operation $RO3$, adding a multiple of one row to another, which does not change the value of the determinant. So for each $i = 1, 2, 3, 4$, $|E_i^{-1}| = 1$. Therefore, $|A| = 1$.

To check this, we'll use the cofactor expansion by row 1,

$$|A| = \begin{vmatrix} 1 & 0 & 2 \\ 0 & 1 & -1 \\ 1 & 4 & -1 \end{vmatrix} = 1(-1 + 4) + 2(0 - 1) = 1.$$

Exercise 3.3 (a) Expanding the first determinant using row 1,

$$\begin{vmatrix} 5 & 2 & -4 \\ -3 & 1 & 1 \\ -1 & 7 & 2 \end{vmatrix} = 5(2 - 7) - 2(-6 + 1) - 4(-21 + 1)$$

$$= -25 + 10 + 80 = 65.$$

You should check this by expanding by a different row or column.
(b) The obvious cofactor expansion for this determinant should be using column 3.

$$\begin{vmatrix} 1 & 23 & 6 & -15 \\ 2 & 5 & 0 & 1 \\ 1 & 4 & 0 & 3 \\ 0 & 1 & 0 & 1 \end{vmatrix} = 6 \begin{vmatrix} 2 & 5 & 1 \\ 1 & 4 & 3 \\ 0 & 1 & 1 \end{vmatrix} = 6[2(4 - 3) - 1(5 - 1)] = -12.$$

Exercise 3.4 You will make fewer errors when there is an unknown constant present if you expand by a row or column containing the unknown. So, for example, expanding by column 1:

$$|B| = \begin{vmatrix} 2 & 1 & w \\ 3 & 4 & -1 \\ 1 & -2 & 7 \end{vmatrix} = w(-10) + 1(-5) + 7(5) = -10w + 30.$$

Therefore, $|B| = 0$ if and only if $w = 3$.

Exercise 3.5 Ideally, you want to obtain a leading one as the $(1, 1)$ entry of the determinant. This can be accomplished by initially doing a row operation which will put either a 1 or -1 at the start of one of the rows. There are several choices for this; we will do $R_2 - R_3$ to replace row 2.

$$\begin{vmatrix} 5 & 2 & -4 & -2 \\ -3 & 1 & 5 & 1 \\ -4 & 3 & 1 & 3 \\ 2 & 1 & -1 & 1 \end{vmatrix} = \begin{vmatrix} 5 & 2 & -4 & -2 \\ 1 & -2 & 4 & -2 \\ -4 & 3 & 1 & 3 \\ 2 & 1 & -1 & 1 \end{vmatrix}$$

$$= -\begin{vmatrix} 1 & -2 & 4 & -2 \\ 5 & 2 & -4 & -2 \\ -4 & 3 & 1 & 3 \\ 2 & 1 & -1 & 1 \end{vmatrix} = -\begin{vmatrix} 1 & -2 & 4 & -2 \\ 0 & 12 & -24 & 8 \\ 0 & -5 & 17 & -5 \\ 0 & 5 & -9 & 5 \end{vmatrix}$$

$$= -4\begin{vmatrix} 1 & -2 & 4 & -2 \\ 0 & 3 & -6 & 2 \\ 0 & 0 & 8 & 0 \\ 0 & 5 & -9 & 5 \end{vmatrix} = -4\begin{vmatrix} 3 & -6 & 2 \\ 0 & 8 & 0 \\ 5 & -9 & 5 \end{vmatrix}$$

$$= -32\begin{vmatrix} 3 & 2 \\ 5 & 5 \end{vmatrix} = -160\begin{vmatrix} 3 & 2 \\ 1 & 1 \end{vmatrix} = -160(1) = -160.$$

The question asks you to check this result by evaluating the determinant using column operations. Staring at the determinant, you might notice that the second and fourth columns have all their lower entries equal, so begin by replacing column 2 with $C_2 - C_4$, which will not change the value of the determinant, and then expand by column 2.

$$\begin{vmatrix} 5 & 2 & -4 & -2 \\ -3 & 1 & 5 & 1 \\ -4 & 3 & 1 & 3 \\ 2 & 1 & -1 & 1 \end{vmatrix} = \begin{vmatrix} 5 & 4 & -4 & -2 \\ -3 & 0 & 5 & 1 \\ -4 & 0 & 1 & 3 \\ 2 & 0 & -1 & 1 \end{vmatrix}$$

$$= -4\begin{vmatrix} -3 & 5 & 1 \\ -4 & 1 & 3 \\ 2 & -1 & 1 \end{vmatrix}$$

$$= -4(40) = -160,$$

where the 3×3 determinant was evaluated using a cofactor expansion.

Exercise 3.6 To answer this question, find the determinant of A,

$$|A| = \begin{vmatrix} 7 - \lambda & -15 \\ 2 & -4 - \lambda \end{vmatrix} = (7 - \lambda)(-4 - \lambda) + 30$$

$$= \lambda^2 - 3\lambda + 2 = (\lambda - 1)(\lambda - 2).$$

Therefore, A^{-1} does not exist $\iff |A| = 0 \iff \lambda = 1$ or $\lambda = 2$.

Exercise 3.7 Given that A is 3×3 and $|A| = 7$,

$$|2A| = 2^3|A| = 56,$$
$$|A^2| = |A||A| = 49,$$
$$|2A^{-1}| = 2^3|A^{-1}| = \frac{2^3}{|A|} = \frac{8}{7},$$
$$|(2A)^{-1}| = \frac{1}{|2A|} = \frac{1}{56}.$$

Exercise 3.8 The first thing you should do for each matrix is evaluate its determinant to see if it is invertible (B is the same matrix as in Exercise 3.1, so you can compare this method of finding the inverse matrix to the method using row operations.)

$$|B| = \begin{vmatrix} -1 & 2 & 1 \\ 0 & 1 & 2 \\ 3 & 1 & 4 \end{vmatrix} = 0 + 1(-4 - 3) - 2(-1 - 6) = 7,$$

so B^{-1} exists.

Next find the cofactors,

$$C_{11} = \begin{vmatrix} 1 & 2 \\ 1 & 4 \end{vmatrix} = 2 \qquad\qquad C_{12} = -\begin{vmatrix} 0 & 2 \\ 3 & 4 \end{vmatrix} = -(-6)$$

$$C_{13} = \begin{vmatrix} 0 & 1 \\ 3 & 1 \end{vmatrix} = -3$$

$$C_{21} = -\begin{vmatrix} 2 & 1 \\ 1 & 4 \end{vmatrix} = -(7) \qquad C_{22} = \begin{vmatrix} -1 & 1 \\ 3 & 4 \end{vmatrix} = -7$$

$$C_{23} = -\begin{vmatrix} -1 & 2 \\ 3 & 1 \end{vmatrix} = -(-7)$$

$$C_{31} = \begin{vmatrix} 2 & 1 \\ 1 & 2 \end{vmatrix} = 3 \qquad\qquad C_{32} = -\begin{vmatrix} -1 & 1 \\ 0 & 2 \end{vmatrix} = -(-2)$$

$$C_{33} = \begin{vmatrix} -1 & 2 \\ 0 & 1 \end{vmatrix} = -1.$$

Then,

$$B^{-1} = \frac{1}{7}\begin{pmatrix} 2 & -7 & 3 \\ 6 & -7 & 2 \\ -3 & 7 & -1 \end{pmatrix}.$$

Check that $BB^{-1} = I$. For the matrix C, expanding by column 1,

$$|C| = \begin{vmatrix} 5 & 2 & -1 \\ 1 & 3 & 4 \\ 6 & 5 & 3 \end{vmatrix} = 5(-11) - 2(-21) - 1(-13) = 0,$$

so C is not invertible.

Exercise 3.9 The system of equations is

$$-x + 2y + z = 1$$
$$y + 2z = 1$$
$$3x + y + 4z = 5 .$$

This has as its coefficient matrix the matrix B of Exercise 3.1. We know from the previous exercise that $|B| = 7$. Since the determinant is non-zero, we can use Cramer's rule. Then

$$x = \frac{1}{|B|} \begin{vmatrix} 1 & 2 & 1 \\ 1 & 1 & 2 \\ 5 & 1 & 4 \end{vmatrix} = \frac{1(2) - 2(-6) + 1(-4)}{|B|} = \frac{10}{7},$$

$$y = \frac{1}{|B|} \begin{vmatrix} -1 & 1 & 1 \\ 0 & 1 & 2 \\ 3 & 5 & 4 \end{vmatrix} = \frac{0 + 1(-7) - 2(-8)}{|B|} = \frac{9}{7},$$

$$z = \frac{1}{|B|} \begin{vmatrix} -1 & 2 & 1 \\ 0 & 1 & 1 \\ 3 & 1 & 5 \end{vmatrix} = \frac{0 + 1(-8) - 1(-7)}{|B|} = -\frac{1}{7}.$$

which, of course, agrees with the result in Exercise 3.1.

Exercise 3.10 (a) Electricity is industry i_2, so column 2 gives the amounts of each industry needed to produce \$1 of electicity:

$c_{12} \leftrightarrow$ \$0.30 water, $c_{22} \leftrightarrow$ \$0.10 electricity, $c_{32} \leftrightarrow 0$ gas.

(b) To solve $(I - C)\mathbf{x} = \mathbf{d}$, you can either use Gaussian elimination or the inverse matrix or Cramer's rule. We will find the inverse matrix using the cofactors method.

$$(I - C) = \begin{pmatrix} 0.8 & -0.3 & -0.2 \\ -0.4 & 0.9 & -0.2 \\ 0 & 0 & 0.9 \end{pmatrix}$$

$$\implies (I - C)^{-1} = \frac{1}{0.54} \begin{pmatrix} 0.81 & 0.27 & 0.24 \\ 0.36 & 0.72 & 0.24 \\ 0 & 0 & 0.6 \end{pmatrix} .$$

Then the solution is given by

$$\mathbf{x} = \begin{pmatrix} x_1 \\ x_2 \\ x_3 \end{pmatrix} = \frac{1}{0.54} \begin{pmatrix} 0.81 & 0.27 & 0.24 \\ 0.36 & 0.72 & 0.24 \\ 0 & 0 & 0.6 \end{pmatrix} \begin{pmatrix} 40,000 \\ 100,000 \\ 72,000 \end{pmatrix}$$

$$= \begin{pmatrix} 142,000 \\ 192,000 \\ 80,000 \end{pmatrix}.$$

The weekly production should be \$142,000 water, \$192,000 electricity and \$80,000 gas.

Exercise 3.11 (a) The cross product $\mathbf{u} \times \mathbf{v}$ is

$$\mathbf{w} = \mathbf{u} \times \mathbf{v} = \begin{vmatrix} \mathbf{e}_1 & \mathbf{e}_2 & \mathbf{e}_3 \\ 1 & 2 & 3 \\ 2 & -5 & 4 \end{vmatrix} = 23\mathbf{e}_1 + 2\mathbf{e}_2 - 9\mathbf{e}_3 = \begin{pmatrix} 23 \\ 2 \\ -9 \end{pmatrix}.$$

This vector is perpendicular to both $\mathbf{u}$ and $\mathbf{v}$ since

$$\left\langle \begin{pmatrix} 23 \\ 2 \\ -9 \end{pmatrix}, \begin{pmatrix} 1 \\ 2 \\ 3 \end{pmatrix} \right\rangle = 23 + 4 - 27 = 0,$$

$$\left\langle \begin{pmatrix} 23 \\ 2 \\ -9 \end{pmatrix}, \begin{pmatrix} 2 \\ -5 \\ 4 \end{pmatrix} \right\rangle = 46 - 10 - 36 = 0.$$

(b) You are being asked to show that the inner product of a vector $\mathbf{a} \in \mathbb{R}^3$ with the cross product of two vectors $\mathbf{b}, \mathbf{c} \in \mathbb{R}^3$ is given by the determinant with these three vectors as its rows. To show this, start with an expression for $\mathbf{b} \times \mathbf{c}$

$$\mathbf{b} \times \mathbf{c} = \begin{vmatrix} \mathbf{e}_1 & \mathbf{e}_2 & \mathbf{e}_3 \\ b_1 & b_2 & b_3 \\ c_1 & c_2 & c_3 \end{vmatrix} = \begin{vmatrix} b_2 & b_3 \\ c_2 & c_3 \end{vmatrix} \mathbf{e}_1 - \begin{vmatrix} b_1 & b_3 \\ c_1 & c_3 \end{vmatrix} \mathbf{e}_2 + \begin{vmatrix} b_1 & b_2 \\ c_1 & c_2 \end{vmatrix} \mathbf{e}_3$$

and then take the inner product with $\mathbf{a}$:

$$\langle \mathbf{a}, \mathbf{b} \times \mathbf{c} \rangle = \begin{vmatrix} b_2 & b_3 \\ c_2 & c_3 \end{vmatrix} a_1 - \begin{vmatrix} b_1 & b_3 \\ c_1 & c_3 \end{vmatrix} a_2 + \begin{vmatrix} b_1 & b_2 \\ c_1 & c_2 \end{vmatrix} a_3$$

$$= \begin{vmatrix} a_1 & a_2 & a_3 \\ b_1 & b_2 & b_3 \\ c_1 & c_2 & c_3 \end{vmatrix}.$$

This shows that the inner product is equal to the given 3×3 determinant (as it is equal to its expansion by row 1).

To show that $\mathbf{b} \times \mathbf{c}$ is orthogonal to both $\mathbf{b}$ and $\mathbf{c}$, we just calculate the inner products $\langle \mathbf{b}, \mathbf{b} \times \mathbf{c} \rangle$ and $\langle \mathbf{c}, \mathbf{b} \times \mathbf{c} \rangle$ using the above determinant expression and show that each is equal to 0:

$$\langle \mathbf{b}, \mathbf{b} \times \mathbf{c}\rangle = \begin{vmatrix} b_1 & b_2 & b_3 \\ b_1 & b_2 & b_3 \\ c_1 & c_2 & c_3 \end{vmatrix} = 0 \quad \text{and}$$

$$\langle \mathbf{c}, \mathbf{b} \times \mathbf{c}\rangle = \begin{vmatrix} c_1 & c_2 & c_3 \\ b_1 & b_2 & b_3 \\ c_1 & c_2 & c_3 \end{vmatrix} = 0,$$

since each of these determinants has two equal rows. Hence the vector $\mathbf{b} \times \mathbf{c}$ is perpendicular to both $\mathbf{b}$ and $\mathbf{c}$.

(c) Therefore, the vector $\mathbf{b} \times \mathbf{c} = \mathbf{n}$ is orthogonal to all linear combinations of the vectors $\mathbf{b}$ and $\mathbf{c}$ and so to the plane determined by these vectors; that is, $\mathbf{b} \times \mathbf{c}$ is a normal vector to the plane containing the vectors $\mathbf{b}$ and $\mathbf{c}$.

If $\langle \mathbf{a}, \mathbf{b} \times \mathbf{c}\rangle = \langle \mathbf{a}, \mathbf{n}\rangle = 0$, then $\mathbf{a}$ must be in this plane; the three vectors are coplanar.

All these statements are reversible.

The given vectors are coplanar when

$$\begin{vmatrix} 3 & -1 & 2 \\ t & 5 & 1 \\ -2 & 3 & 1 \end{vmatrix} = 0,$$

so if and only if $t = -4$.

Chapter 4 exercises

Exercise 4.1 Write down the augmented matrix and put it into reduced row echelon form,

$$(A|\mathbf{b}) = \begin{pmatrix} 1 & 5 & 3 & 7 & 1 & 2 \\ 2 & 10 & 3 & 8 & 5 & -5 \\ 1 & 5 & 1 & 3 & 3 & -4 \end{pmatrix} \xrightarrow[R_3-R_1]{R_2-2R_1} \begin{pmatrix} 1 & 5 & 3 & 7 & 1 & 2 \\ 0 & 0 & -3 & -6 & 3 & -9 \\ 0 & 0 & -2 & -4 & 2 & -6 \end{pmatrix}$$

$$\xrightarrow{-\frac{1}{3}R_2} \begin{pmatrix} 1 & 5 & 3 & 7 & 1 & 2 \\ 0 & 0 & 1 & 2 & -1 & 3 \\ 0 & 0 & -2 & -4 & 2 & -6 \end{pmatrix} \xrightarrow{R_3+2R_2} \begin{pmatrix} 1 & 5 & 3 & 7 & 1 & 2 \\ 0 & 0 & 1 & 2 & -1 & 3 \\ 0 & 0 & 0 & 0 & 0 & 0 \end{pmatrix}$$

$$\xrightarrow{R_1-3R_2} \begin{pmatrix} 1 & 5 & 0 & 1 & 4 & -7 \\ 0 & 0 & 1 & 2 & -1 & 3 \\ 0 & 0 & 0 & 0 & 0 & 0 \end{pmatrix}.$$

Set the non-leading variables to arbitrary constants, say, $x_2 = s$, $x_4 = t$, and $x_5 = u$, and *solve* for the leading variables in terms of these parameters. Starting with the bottom row, write down the vector solution:

$$\mathbf{x} = \begin{pmatrix} x_1 \\ x_2 \\ x_3 \\ x_4 \\ x_5 \end{pmatrix} = \begin{pmatrix} -7 - 5s - t - 4u \\ s \\ 3 - 2t + u \\ t \\ u \end{pmatrix}$$

$$\mathbf{x} = \begin{pmatrix} -7 \\ 0 \\ 3 \\ 0 \\ 0 \end{pmatrix} + s \begin{pmatrix} -5 \\ 1 \\ 0 \\ 0 \\ 0 \end{pmatrix} + t \begin{pmatrix} -1 \\ 0 \\ -2 \\ 1 \\ 0 \end{pmatrix} + u \begin{pmatrix} -4 \\ 0 \\ 1 \\ 0 \\ 1 \end{pmatrix}$$

$$= \mathbf{p} + s\mathbf{v}_1 + t\mathbf{v}_2 + u\mathbf{v}_3.$$

The rank of A is 2 since there are two leading ones. The matrix A has five columns, so there are $n - r = 5 - 2 = 3$ vectors $\mathbf{v}_i$. The solution is in the form required by the question.

Verify the solution as asked by performing the matrix multiplication. Do actually carry this out. We will show the first two.

$$A\mathbf{p} = \begin{pmatrix} 1 & 5 & 3 & 7 & 1 \\ 2 & 10 & 3 & 8 & 5 \\ 1 & 5 & 1 & 3 & 3 \end{pmatrix} \begin{pmatrix} -7 \\ 0 \\ 3 \\ 0 \\ 0 \end{pmatrix} = \begin{pmatrix} -7 + 9 \\ -14 + 9 \\ -7 + 3 \end{pmatrix} = \begin{pmatrix} 2 \\ -5 \\ -4 \end{pmatrix}.$$

$$A\mathbf{v}_1 = \begin{pmatrix} 1 & 5 & 3 & 7 & 1 \\ 2 & 10 & 3 & 8 & 5 \\ 1 & 5 & 1 & 3 & 3 \end{pmatrix} \begin{pmatrix} -5 \\ 1 \\ 0 \\ 0 \\ 0 \end{pmatrix} = \begin{pmatrix} -5 + 3 \\ -10 + 10 \\ -5 + 5 \end{pmatrix} = \begin{pmatrix} 0 \\ 0 \\ 0 \end{pmatrix}.$$

You can use any solution $\mathbf{x}$ (so any values of $s, t, u \in \mathbb{R}$) to write $\mathbf{b}$ as a linear combination of the columns of A, so this can be done in infinitely many ways. In particular, taking $\mathbf{x} = \mathbf{p}$, and letting $\mathbf{c}_i$ indicate column i of the coefficient matrix A,

$$A\mathbf{p} = -7\mathbf{c}_1 + 3\mathbf{c}_3.$$

You should write this out in detail and check that the sum of the vectors does add to the vector $\mathbf{b}$. Notice that this combination uses only the columns corresponding to the leading variables:

$$-7 \begin{pmatrix} 1 \\ 2 \\ 1 \end{pmatrix} + 3 \begin{pmatrix} 3 \\ 3 \\ 1 \end{pmatrix} = \begin{pmatrix} 2 \\ -5 \\ -4 \end{pmatrix}.$$

Similarly, since $A\mathbf{v}_1 = \mathbf{0}$, $A\mathbf{v}_2 = \mathbf{0}$ and $A\mathbf{v}_3 = \mathbf{0}$, any linear combination of these three vectors will give a vector $\mathbf{w} = s\mathbf{v}_1 + t\mathbf{v}_2 + u\mathbf{v}_3$ for which $A\mathbf{w} = \mathbf{0}$, and you can rewrite $A\mathbf{w}$ as a linear combination of the columns of A. For example, taking $\mathbf{v}_1$ (so $s = 1$, $t = u = 0$),

$$-5\mathbf{c}_1 + \mathbf{c}_2 = -5 \begin{pmatrix} 1 \\ 2 \\ 1 \end{pmatrix} + \begin{pmatrix} 5 \\ 10 \\ 5 \end{pmatrix} = \begin{pmatrix} 0 \\ 0 \\ 0 \end{pmatrix}.$$

Exercise 4.2 Using row operations, the matrix A reduces to echelon form

$$A \longrightarrow \cdots \longrightarrow \begin{pmatrix} 1 & 0 & 1 & 0 & 2 \\ 0 & 1 & -1 & 1 & -1 \\ 0 & 0 & 1 & -1 & 3 \\ 0 & 0 & 0 & 0 & 0 \end{pmatrix}.$$

There are three nonzero rows (three leading ones), so the rank of A is 3.

To find $N(A)$, we need to solve $A\mathbf{x} = \mathbf{0}$, which is a system of four equations in five unknowns. Call them x_1, x_2, x_3, x_4, x_5. Continuing to reduced echelon form,

$$A \longrightarrow \cdots \longrightarrow \begin{pmatrix} 1 & 0 & 0 & 1 & -1 \\ 0 & 1 & 0 & 0 & 2 \\ 0 & 0 & 1 & -1 & 3 \\ 0 & 0 & 0 & 0 & 0 \end{pmatrix}.$$

The leading variables are x_1, x_2, and x_3. Set the non-leading variables $x_4 = s$ and $x_5 = t$. Then the solution is

$$\begin{pmatrix} x_1 \\ x_2 \\ x_3 \\ x_4 \\ x_5 \end{pmatrix} = \begin{pmatrix} -s+t \\ -2t \\ s-3t \\ s \\ t \end{pmatrix} = s \begin{pmatrix} -1 \\ 0 \\ 1 \\ 1 \\ 0 \end{pmatrix} + t \begin{pmatrix} 1 \\ -2 \\ -3 \\ 0 \\ 1 \end{pmatrix}, \qquad s, t \in \mathbb{R}.$$

So the null space consists of all vectors of the form $\mathbf{x} = s\mathbf{v}_1 + t\mathbf{v}_2$, where $\mathbf{v}_1$ and $\mathbf{v}_2$ are the vectors displayed above. It is a subset of $\mathbb{R}^5$.

The range of A can be described as the set of all linear combinations of the columns of A,

$$R(A) = \{\alpha_1 \mathbf{c}_1 + \alpha_2 \mathbf{c}_2 + \alpha_3 \mathbf{c}_3 + \alpha_4 \mathbf{c}_4 + \alpha_5 \mathbf{c}_5 \mid \alpha_i \in \mathbb{R}, \ i = 1, \ldots, 5\},$$

where

$$\mathbf{c}_1 = \begin{pmatrix} 1 \\ 2 \\ 1 \\ 0 \end{pmatrix}, \quad \mathbf{c}_2 = \begin{pmatrix} 0 \\ 1 \\ 3 \\ 3 \end{pmatrix}, \quad \mathbf{c}_3 = \begin{pmatrix} 1 \\ 1 \\ -1 \\ -2 \end{pmatrix}, \quad \mathbf{c}_4 = \begin{pmatrix} 0 \\ 1 \\ 2 \\ 2 \end{pmatrix}, \quad \mathbf{c}_5 = \begin{pmatrix} 2 \\ 3 \\ 2 \\ 0 \end{pmatrix}.$$

This is a subset of $\mathbb{R}^4$. We will find a better way to describe this set when we look at the column space of a matrix in the next two chapters.

Exercise 4.3 $|A| = 3\lambda - 9$.

(a) If $|A| \neq 0$, that is if $\lambda \neq 3$, then the system will have a unique solution. In this case, using Cramer's rule, $z = (3 - 3\mu)/(\lambda - 3)$.

To answer (b) and (c), reduce the augmented matrix to echelon form with $\lambda = 3$

$$(A|\mathbf{b}) = \begin{pmatrix} 1 & 2 & 0 & 2 \\ 5 & 1 & 3 & 7 \\ 1 & -1 & 1 & \mu \end{pmatrix} \longrightarrow \begin{pmatrix} 1 & 2 & 0 & 2 \\ 0 & -9 & 3 & -3 \\ 0 & -3 & 1 & \mu - 2 \end{pmatrix}$$

$$\longrightarrow \begin{pmatrix} 1 & 2 & 0 & 2 \\ 0 & 3 & -1 & 1 \\ 0 & -3 & 1 & \mu - 2 \end{pmatrix} \longrightarrow \begin{pmatrix} 1 & 2 & 0 & 2 \\ 0 & 3 & -1 & 1 \\ 0 & 0 & 0 & \mu - 1 \end{pmatrix}.$$

So if $\lambda = 3$, this system will be inconsistent if $\mu \neq 1$, which answers part (b).

If $\lambda = 3$ and $\mu = 1$, we have (c) infinitely many solutions. Setting $\mu = 1$ and continuing to reduced echelon form,

$$(A|\mathbf{b}) \longrightarrow \cdots \longrightarrow \begin{pmatrix} 1 & 2 & 0 & 2 \\ 0 & 1 & -\frac{1}{3} & \frac{1}{3} \\ 0 & 0 & 0 & 0 \end{pmatrix} \longrightarrow \begin{pmatrix} 1 & 0 & \frac{2}{3} & \frac{4}{3} \\ 0 & 1 & -\frac{1}{3} & \frac{1}{3} \\ 0 & 0 & 0 & 0 \end{pmatrix}.$$

The solution can now be read from the matrix. Setting the non-leading variable z equal to t,

$$\begin{pmatrix} x \\ y \\ z \end{pmatrix} = \begin{pmatrix} \frac{4}{3} - \frac{2}{3}t \\ \frac{1}{3} + \frac{1}{3}t \\ t \end{pmatrix} = \begin{pmatrix} \frac{4}{3} \\ \frac{1}{3} \\ 0 \end{pmatrix} + t \begin{pmatrix} -\frac{2}{3} \\ \frac{1}{3} \\ 1 \end{pmatrix} = \mathbf{p} + t\mathbf{v}, \qquad t \in \mathbb{R}.$$

Exercise 4.4 The matrix B must be 3×4 since the solutions are in $\mathbb{R}^4$ and $\mathbf{c}_1 \in \mathbb{R}^3$. Let

$$B = \begin{pmatrix} 1 \\ 1 & \mathbf{c}_2 & \mathbf{c}_3 & \mathbf{c}_4 \\ 2 \end{pmatrix}.$$

The solution is of the form $x = p + sv_1 + tv_2$, where v_1, v_2 are in $N(B)$; therefore, you know that $Bp = d$, $Bv_1 = 0$ and $Bv_2 = 0$. Regarding the matrix products as linear combinations of the column vectors of B, we obtain

$$Bp = c_1 + 2c_3 = d \qquad Bv_1 = -3c_1 + c_2 = 0$$
$$Bv_2 = c_1 - c_3 + c_4 = 0.$$

Knowing c_1, you just need to solve these for the other three columns:

$$2c_3 = \begin{pmatrix} 3 \\ 5 \\ -2 \end{pmatrix} - \begin{pmatrix} 1 \\ 1 \\ 2 \end{pmatrix} = \begin{pmatrix} 2 \\ 4 \\ -4 \end{pmatrix} \implies c_3 = \begin{pmatrix} 1 \\ 2 \\ -2 \end{pmatrix},$$

$$c_2 = 3\begin{pmatrix} 1 \\ 1 \\ 2 \end{pmatrix} = \begin{pmatrix} 3 \\ 3 \\ 6 \end{pmatrix},$$

$$c_4 = c_3 - c_1 = \begin{pmatrix} 1 \\ 2 \\ -2 \end{pmatrix} - \begin{pmatrix} 1 \\ 1 \\ 2 \end{pmatrix} = \begin{pmatrix} 0 \\ 1 \\ -4 \end{pmatrix}.$$

The matrix B is

$$B = \begin{pmatrix} 1 & 3 & 1 & 0 \\ 1 & 3 & 2 & 1 \\ 2 & 6 & -2 & -4 \end{pmatrix}.$$

You can check your answer by row reducing the augmented matrix $(B|d)$ to obtain the solution of $Bx = d$, and matching it to the solution given.

Exercise 4.5 You might have noticed that this is the same coefficient matrix A as we encountered in Example 4.7 on page 135. You can easily tackle this question by forming the augmented matrix and reducing it using row operations:

$$(A|b) = \begin{pmatrix} 1 & 2 & 1 & a \\ 2 & 3 & 0 & b \\ 3 & 5 & 1 & c \end{pmatrix} \rightarrow \begin{pmatrix} 1 & 2 & 1 & a \\ 0 & -1 & -2 & b - 2a \\ 0 & -1 & -2 & c - 3a \end{pmatrix}.$$

After this first step, it is clear that the system will be consistent if and only if

$$b - 2a = c - 3a, \quad \text{or} \quad a + b - c = 0.$$

Hence, the vector $y = (x, y, z)^\mathsf{T}$ is in $R(A)$ if and only if $x + y - z = 0$. This is the Cartesian equation of a plane in $\mathbb{R}^3$.

The vector $\mathbf{d} = (1, 5, 6)^{\mathrm{T}}$ is in $R(A)$, since its components satisfy the equation. This follows from Example 4.7, because the system there was seen to be consistent. The vector $(1, 5, 4)^{\mathrm{T}}$, for which the system is inconsistent, is not in the plane $R(A)$. (Look at Example 4.9 to see that the augmented matrix in this case has an echelon form which indicates that the system is inconsistent.)

For Activity 4.8 on page 135, you found a general solution of the system of equations $A\mathbf{x} = \mathbf{d}$ to be

$$\mathbf{x} = \begin{pmatrix} 7 \\ -3 \\ 0 \end{pmatrix} + t \begin{pmatrix} 3 \\ -2 \\ 1 \end{pmatrix}, \quad t \in \mathbb{R}.$$

Any solution $\mathbf{x}$ will enable you to write $\mathbf{d}$ as a linear combination of the columns of A. For example, taking first $t = 0$ and then $t = -1$, $\mathbf{d} = 7\mathbf{c}_1 - 3\mathbf{c}_2$ or $\mathbf{d} = 4\mathbf{c}_1 - \mathbf{c}_2 - \mathbf{c}_3$; that is,

$$\begin{pmatrix} 1 \\ 5 \\ 6 \end{pmatrix} = 7 \begin{pmatrix} 1 \\ 2 \\ 3 \end{pmatrix} - 3 \begin{pmatrix} 2 \\ 3 \\ 5 \end{pmatrix} \quad \text{or} \quad \begin{pmatrix} 1 \\ 5 \\ 6 \end{pmatrix} = 4 \begin{pmatrix} 1 \\ 2 \\ 3 \end{pmatrix} - \begin{pmatrix} 2 \\ 3 \\ 5 \end{pmatrix} - \begin{pmatrix} 1 \\ 0 \\ 1 \end{pmatrix}.$$

Exercise 4.6 You need to put the matrix into row echelon form to answer the first question, and into reduced row echelon form for the second,

$$A = \begin{pmatrix} 1 & 1 & 1 \\ 0 & 1 & -2 \\ 2 & -1 & 8 \\ 3 & 1 & 7 \end{pmatrix} \longrightarrow \cdots \longrightarrow \begin{pmatrix} 1 & 0 & 3 \\ 0 & 1 & -2 \\ 0 & 0 & 0 \\ 0 & 0 & 0 \end{pmatrix}.$$

The rank of A is 2. There is one non-leading variable. If you write $\mathbf{x} = (x, y, z)^{\mathrm{T}}$, then setting $z = t$, you will obtain the solution

$$\mathbf{x} = t \begin{pmatrix} -3 \\ 2 \\ 1 \end{pmatrix}, \quad t \in \mathbb{R}.$$

Since there are non-trivial solutions of $A\mathbf{x} = \mathbf{0}$, it is possible to express $\mathbf{0}$ as a linear combination of the columns of A with non-zero coefficients. A non-trivial linear combination of the column vectors which is equal to the zero vector is given by any non-zero vector in the null space. For example, using $t = 1$, the product $A\mathbf{x}$ yields

$$-3\mathbf{c}_1 + 2\mathbf{c}_2 + \mathbf{c}_3 = -3 \begin{pmatrix} 1 \\ 0 \\ 2 \\ 3 \end{pmatrix} + 2 \begin{pmatrix} 1 \\ 1 \\ -1 \\ 1 \end{pmatrix} + \begin{pmatrix} 1 \\ -2 \\ 8 \\ 7 \end{pmatrix} = \begin{pmatrix} 0 \\ 0 \\ 0 \\ 0 \end{pmatrix} = \mathbf{0}.$$

The vector $\mathbf{b}$ is in $R(A)$ if $\mathbf{b}$ is a linear combination of the column vectors of A, which is exactly when $A\mathbf{x} = \mathbf{b}$ is consistent. Notice that the matrix A has rank 2, so the augmented matrix must also have rank 2. Reducing $(A|\mathbf{b})$ using row operations,

$$\begin{pmatrix} 1 & 1 & 1 & 4 \\ 0 & 1 & -2 & 1 \\ 2 & -1 & 8 & a \\ 3 & 1 & 7 & b \end{pmatrix} \rightarrow \begin{pmatrix} 1 & 1 & 1 & 4 \\ 0 & 1 & -2 & 1 \\ 0 & -3 & 6 & a-8 \\ 0 & -2 & 4 & b-12 \end{pmatrix}$$

$$\rightarrow \begin{pmatrix} 1 & 1 & 1 & 4 \\ 0 & 1 & -2 & 1 \\ 0 & 0 & 0 & a-5 \\ 0 & 0 & 0 & b-10 \end{pmatrix}.$$

Therefore, $A\mathbf{x} = \mathbf{b}$ is consistent if and only if $a = 5$ and $b = 10$. In that case, continuing to reduced echelon form,

$$\begin{pmatrix} 1 & 1 & 1 & 4 \\ 0 & 1 & -2 & 1 \\ 0 & 0 & 0 & 0 \\ 0 & 0 & 0 & 0 \end{pmatrix} \rightarrow \begin{pmatrix} 1 & 0 & 3 & 3 \\ 0 & 1 & -2 & 1 \\ 0 & 0 & 0 & 0 \\ 0 & 0 & 0 & 0 \end{pmatrix}.$$

Therefore, a general solution is

$$\mathbf{x} = \begin{pmatrix} 3 \\ 1 \\ 0 \end{pmatrix} + t \begin{pmatrix} -3 \\ 2 \\ 1 \end{pmatrix} \qquad t \in \mathbb{R}.$$

(b) Using row operations,

$$|B| = \begin{vmatrix} -2 & 3 & -2 & 5 \\ 3 & -6 & 9 & -6 \\ -2 & 9 & -1 & 9 \\ 5 & -6 & 9 & -4 \end{vmatrix} = (-3) \begin{vmatrix} 1 & -2 & 3 & -2 \\ -2 & 3 & -2 & 5 \\ -2 & 9 & -1 & 9 \\ 5 & -6 & 9 & -4 \end{vmatrix}$$

$$= (-3) \begin{vmatrix} 1 & -2 & 3 & -2 \\ 0 & -1 & 4 & 1 \\ 0 & 5 & 5 & 5 \\ 0 & 4 & -6 & 6 \end{vmatrix}$$

$$= (30) \begin{vmatrix} 1 & -2 & 3 & -2 \\ 0 & 1 & 1 & 1 \\ 0 & -1 & 4 & 1 \\ 0 & 2 & -3 & 3 \end{vmatrix} = (30) \begin{vmatrix} 1 & -2 & 3 & -2 \\ 0 & 1 & 1 & 1 \\ 0 & 0 & 5 & 2 \\ 0 & 0 & -5 & 1 \end{vmatrix} = 450.$$

Since $\det(B) \neq 0$, the rank of B is 4. Therefore, the main theorem (Theorem 4.1.2) tells us that $B\mathbf{x} = \mathbf{0}$ has only the trivial solution. Therefore, there is no way to write $\mathbf{0}$ as a linear combination of the column vectors of B except the trivial way, with all coefficients equal to 0.

Also, using this theorem, $B\mathbf{x} = \mathbf{b}$ has a unique solution for all $\mathbf{b} \in \mathbb{R}^4$. Therefore, $R(B) = \mathbb{R}^4$. That is, a and b can be any real numbers, the system $B\mathbf{x} = \mathbf{b}$ is always consistent.

Chapter 5 exercises

Exercise 5.1 The set S_1 is a subspace. We have

$$\begin{pmatrix} x \\ y \\ z \end{pmatrix} \in S_1 \iff z = y = 3x \iff \begin{pmatrix} x \\ y \\ z \end{pmatrix} = \begin{pmatrix} x \\ 3x \\ 3x \end{pmatrix} = x \begin{pmatrix} 1 \\ 3 \\ 3 \end{pmatrix}, x \in \mathbb{R}.$$

So, the set S_1 is the linear span of the vector $\mathbf{v} = (1, 3, 3)^T$ and is therefore a subspace of $\mathbb{R}^3$. (This is the line through the origin in the direction of the vector $\mathbf{v} = (1, 3, 3)^T$.)

The set S_2 is a subspace. Since

$$S_2 = \left\{ \begin{pmatrix} x \\ y \\ z \end{pmatrix} \, \middle| \, z + y = 3x \right\} = \left\{ \begin{pmatrix} x \\ y \\ z \end{pmatrix} \, \middle| \, z + y - 3x = 0 \right\},$$

it is a plane through the origin in $\mathbb{R}^3$, and you have shown that a plane through the origin is a subspace (see Activity 5.38). You can also show directly that the set is non-empty and closed under addition and scalar multiplication.

The set S_3 is not a subspace. $\mathbf{0} \in S_3$, but S_3 is not closed under addition. For example,

$$\begin{pmatrix} 1 \\ 1 \\ 3 \end{pmatrix} \in S_3, \quad \begin{pmatrix} 1 \\ 3 \\ 1 \end{pmatrix} \in S_3, \quad \text{but} \quad \begin{pmatrix} 1 \\ 1 \\ 3 \end{pmatrix} + \begin{pmatrix} 1 \\ 3 \\ 1 \end{pmatrix} = \begin{pmatrix} 2 \\ 4 \\ 4 \end{pmatrix} \notin S_3$$

since it does not satisfy the condition $zy = 3x$.

The set S_4 is not a subspace because it is not closed under addition. For example,

$$\begin{pmatrix} 1 \\ 1 \\ 0 \end{pmatrix} \in S_4, \quad \begin{pmatrix} 0 \\ 0 \\ 1 \end{pmatrix} \in S_4, \quad \text{but} \quad \begin{pmatrix} 1 \\ 1 \\ 0 \end{pmatrix} + \begin{pmatrix} 0 \\ 0 \\ 1 \end{pmatrix} = \begin{pmatrix} 1 \\ 1 \\ 1 \end{pmatrix} \notin S_4.$$

What is S_4? For a vector $\mathbf{x}$ to be in S_4, either $x = 0$, $y = 0$ or $z = 0$. So this set consists of the xy-plane (if $z = 0$), the xz-plane, and the

yz-plane. But any vector in $\mathbb{R}^3$ which is not on one of these planes is not in S_4.

Exercise 5.2 If A is an $n \times n$ matrix, then all vectors $\mathbf{x}$ for which $A\mathbf{x}$ is defined must be $n \times 1$ vectors, so the set

$$S = \{\mathbf{x} \mid A\mathbf{x} = \lambda\mathbf{x}\}, \qquad \text{some } \lambda \in \mathbb{R},$$

is a subset of $\mathbb{R}^n$. To show it is a subspace, you have to show it is non-empty and closed under addition and scalar multiplication.

Since $A\mathbf{0} = \lambda\mathbf{0} = \mathbf{0}$, the vector $\mathbf{0} \in S$, so S is non-empty. (In fact, depending on λ, S may well be the vector space which contains only the zero vector; more on this is found in Chapter 8.)

Let $\mathbf{u}, \mathbf{v} \in S$ and $a \in \mathbb{R}$. Then you know that $A\mathbf{u} = \lambda\mathbf{u}$ and $A\mathbf{v} = \lambda\mathbf{v}$. Therefore,

$$A(\mathbf{u} + \mathbf{v}) = A\mathbf{u} + A\mathbf{v} = \lambda\mathbf{u} + \lambda\mathbf{v} = \lambda(\mathbf{u} + \mathbf{v})$$

and

$$A(a\mathbf{u}) = a(A\mathbf{u}) = a(\lambda\mathbf{u}) = \lambda(a\mathbf{u})$$

so $\mathbf{u} + \mathbf{v} \in S$ and $a\mathbf{u} \in S$. Therefore, S is a subspace of $\mathbb{R}^n$.

Exercise 5.3 We are given the vectors

$$\mathbf{v}_1 = \begin{pmatrix} -1 \\ 0 \\ 1 \end{pmatrix}, \quad \mathbf{v}_2 = \begin{pmatrix} 1 \\ 2 \\ 3 \end{pmatrix}, \quad \text{and} \quad \mathbf{u} = \begin{pmatrix} -1 \\ 2 \\ 5 \end{pmatrix}, \quad \mathbf{w} = \begin{pmatrix} 1 \\ 2 \\ 5 \end{pmatrix}.$$

(a) The vector $\mathbf{u}$ can be expressed as a linear combination of $\mathbf{v}_1$ and $\mathbf{v}_2$ if you can find constants s, t such that $\mathbf{u} = s\mathbf{v}_1 + t\mathbf{v}_2$. Now,

$$\begin{pmatrix} -1 \\ 2 \\ 5 \end{pmatrix} = s \begin{pmatrix} -1 \\ 0 \\ 1 \end{pmatrix} + t \begin{pmatrix} 1 \\ 2 \\ 3 \end{pmatrix} \quad \Longleftrightarrow \quad \begin{cases} -1 = -s + t \\ 2 = 2t \\ 5 = s + 2t. \end{cases}$$

From the middle component equation, we find $t = 1$, and substituting this into the top equation yields $s = 2$. Substituting these values for s and t in the bottom component equation gives $5 = 2 + 3(1)$, which is correct, so $\mathbf{u} = 2\mathbf{v}_1 + \mathbf{v}_2$. You can check this using the vectors,

$$2 \begin{pmatrix} -1 \\ 0 \\ 1 \end{pmatrix} + \begin{pmatrix} 1 \\ 2 \\ 3 \end{pmatrix} = \begin{pmatrix} -1 \\ 2 \\ 5 \end{pmatrix}.$$

Attempting this for the vector $\mathbf{w}$,

$$\begin{pmatrix} 1 \\ 2 \\ 5 \end{pmatrix} = s \begin{pmatrix} -1 \\ 0 \\ 1 \end{pmatrix} + t \begin{pmatrix} 1 \\ 2 \\ 3 \end{pmatrix} \iff \begin{cases} 1 = -s + t \\ 2 = 2t \\ 5 = s + 2t \, . \end{cases}$$

This time the top two component equations yield $t = 1$ and $s = 0$, and these values do not satisfy the bottom equation, $5 = s + 3t$, so no solution exists. The vector $\mathbf{w}$ cannot be expressed as a linear combination of $\mathbf{v}_1$ and $\mathbf{v}_2$.

(b) Since $\mathbf{u} = 2\mathbf{v}_1 + \mathbf{v}_2$, $\mathbf{u} \in \text{Lin}\{\mathbf{v}_1, \mathbf{v}_2\}$. Therefore, $\text{Lin}\{\mathbf{v}_1, \mathbf{v}_2, \mathbf{u}\}$ and $\text{Lin}\{\mathbf{v}_1, \mathbf{v}_2\}$ are the same subspace. Any vector $\mathbf{x} = a\mathbf{v}_1 + b\mathbf{v}_2 + c\mathbf{u}$ can be expressed as a linear combination of just $\mathbf{v}_1$ and $\mathbf{v}_2$ by just substituting $2\mathbf{v}_1 + \mathbf{v}_2$ for $\mathbf{u}$. Therefore, this is the linear span of two non-parallel vectors in $\mathbb{R}^3$, so it is a plane in $\mathbb{R}^3$.

Since $\mathbf{w} \notin \text{Lin}\{\mathbf{v}_1, \mathbf{v}_2\}$, the subspace $\text{Lin}\{\mathbf{v}_1, \mathbf{v}_2, \mathbf{w}\}$ must be bigger than just the plane, so it must be all of $\mathbb{R}^3$. To show that $\text{Lin}\{\mathbf{v}_1, \mathbf{v}_2, \mathbf{w}\} = \mathbb{R}^3$, you can establish that any $\mathbf{b} \in \mathbb{R}^3$ can be expressed as a linear combination, $\mathbf{b} = a\mathbf{v}_1 + b\mathbf{v}_2 + c\mathbf{w}$, or equivalently that the system of equations $A\mathbf{x} = \mathbf{b}$ has a solution where A is the matrix whose columns are the vectors $\mathbf{v}_1$, $\mathbf{v}_2$ and $\mathbf{w}$. You can show this by reducing A to row echelon form, or by finding the determinant. Since

$$|A| = \begin{vmatrix} -1 & 1 & 1 \\ 0 & 2 & 2 \\ 1 & 3 & 5 \end{vmatrix} = -4 \neq 0,$$

you know from the main theorem (Theorem 4.5) that $A\mathbf{x} = \mathbf{b}$ has a unique solution for all $\mathbf{b} \in \mathbb{R}^3$.

(c) You know from part (b) that $\{\mathbf{v}_1, \mathbf{v}_2, \mathbf{w}\}$ spans $\mathbb{R}^3$, and therefore so does $\{\mathbf{v}_1, \mathbf{v}_2, \mathbf{u}, \mathbf{w}\}$. But more efficiently, you can take the same approach as in part (b) to show that $\{\mathbf{v}_1, \mathbf{v}_2, \mathbf{u}, \mathbf{w}\}$ spans $\mathbb{R}^3$, and at the same time show that any vector $\mathbf{b} \in \mathbb{R}^3$ can be expressed as a linear combination of $\mathbf{v}_1, \mathbf{v}_2, \mathbf{u}, \mathbf{w}$ in infinitely many ways. If B is the matrix with these four vectors as its columns, then the solutions, $\mathbf{x}$, of $B\mathbf{x} = \mathbf{b}$ will determine the possible linear combinations of the vectors. We put the coefficient matrix B into row echelon form (steps not shown),

$$B = \begin{pmatrix} -1 & 1 & -1 & 1 \\ 0 & 2 & 2 & 2 \\ 1 & 3 & 5 & 5 \end{pmatrix} \longrightarrow \cdots \longrightarrow \begin{pmatrix} 1 & -1 & 1 & -1 \\ 0 & 1 & 1 & 1 \\ 0 & 0 & 0 & 1 \end{pmatrix}.$$

Since there is a leading one in every row, the system $B\mathbf{x} = \mathbf{b}$ is always consistent, so every vector $\mathbf{b} \in \mathbb{R}^3$ can be expressed as a linear combination of $\mathbf{v}_1, \mathbf{v}_2, \mathbf{u}, \mathbf{w}$. Since there is a free variable (in column three), there are infinitely many solutions to $B\mathbf{x} = \mathbf{b}$.

Exercise 5.4 For $\mathbf{v}, \mathbf{w} \in \mathbb{R}^n$, the set $A = \{\mathbf{v}, \mathbf{w}\}$ contains precisely the two vectors, $\mathbf{v}$ and $\mathbf{w}$. The set $B = \text{Lin}\{\mathbf{v}, \mathbf{w}\}$ contains infinitely many vectors, namely all possible linear combinations of $\mathbf{v}$ and $\mathbf{w}$.

Exercise 5.5 Clearly, $\mathcal{P}_n \neq \emptyset$. We need to show that for any $\alpha \in \mathbb{R}$ and $f, g \in \mathcal{P}_n$, $\alpha f \in \mathcal{P}_n$ and $f + g \in \mathcal{P}_n$. Suppose that

$$f(x) = a_0 + a_1 x + a_2 x^2 + \cdots + a_n x^n,$$
$$g(x) = b_0 + b_1 x + b_2 x^2 + \cdots + b_n x^n.$$

Then

$$(f + g)(x) = f(x) + g(x)$$
$$= (a_0 + b_0) + (a_1 + b_1)x + \cdots + (a_n + b_n)x^n,$$

so $f + g$ is also a polynomial of degree at most n and therefore $f + g \in \mathcal{P}_n$. Similarly,

$$(\alpha f)(x) = \alpha f(x) = (\alpha a_0) + (\alpha a_1)x + \cdots + (\alpha a_n)x^n,$$

so $\alpha f \in \mathcal{P}_n$ also. It can be seen that the set of functions $\{1, x, x^2, \ldots, x^n\}$ spans $\mathcal{P}_n$, where 1 denotes the function that is identically equal to 1 (that is, the function f with $f(x) = 1$ for all x).

(Note that the requirement that the polynomials are of degree at most n is important. If you consider the set of polynomials of degree exactly n, then this set is not closed under addition. For example, if $f(x) = 1 + \cdots + 3x^n$ and $g(x) = 2 + \cdots - 3x^n$, then $(f + g)(x) = f(x) + g(x)$ is a polynomial of degree at most $n - 1$.

However, if $n = 1$, then the set of constant functions is a subspace. $\mathbf{0} \in U$ and U is closed under addition and scalar multiplication.)

Exercise 5.6 If U and W are subspaces of a vector space V, then the set

$$U \cap W = \{\mathbf{x} \mid \mathbf{x} \in U \text{ and } \mathbf{x} \in W\}$$

is a subspace. It is non-empty because $\mathbf{0} \in U$ and $\mathbf{0} \in W$, so $\mathbf{0} \in U \cap W$. Let $\mathbf{x}, \mathbf{y} \in U \cap W$, $\alpha \in \mathbb{R}$. Then since U is a subspace and $\mathbf{x}, \mathbf{y} \in U$, both $\mathbf{x} + \mathbf{y}$ and $\alpha \mathbf{x}$ are in U, and the same is true with regard to W. Therefore, both $\mathbf{x} + \mathbf{y}$ and $\alpha \mathbf{x}$ are in $U \cap W$, so this is a subspace.

Now look at the set

$$U \cup W = \{\mathbf{x} \mid \mathbf{x} \in U \text{ or } \mathbf{x} \in W\}.$$

If $U \subseteq W$, then $U \cup W$ is equal to W, which is a subspace. If $W \subseteq U$, then $U \cup W = U$, and this is a subspace.

Now suppose that $U \not\subseteq W$ and $W \not\subseteq U$. Then there is a vector $\mathbf{x} \in U$ which is not in W, and a vector $\mathbf{y} \in W$ with $\mathbf{y} \notin U$. Both of these vectors are in $U \cup W$, but the vector $\mathbf{x} + \mathbf{y} = \mathbf{z}$ is not. Why? You need to show that $\mathbf{z}$ is not in U and not in W. If $\mathbf{z} \in U$, then $\mathbf{z} - \mathbf{x} \in U$ since U is a subspace. But $\mathbf{z} - \mathbf{x} = \mathbf{y}$, which contradicts the assumption that $\mathbf{y} \notin U$. A similar argument shows that $\mathbf{z} \notin W$, so $\mathbf{z} \notin U \cup W$. Therefore, the set $U \cup W$ is not closed under addition and is not a subspace.

An example of this is to consider the xz-plane (vectors $(x, y, z)^\mathrm{T}$ with $y = 0$) and the yz-plane (vectors with $x = 0$) in $\mathbb{R}^3$. Each of these sets is a plane through the origin, so each is a subspace of $\mathbb{R}^3$. Their intersection is the z axis, which is a line through the origin, and therefore a subspace of $\mathbb{R}^3$. Their set theoretic union is just the set of all vectors which are on either plane. For example, the vector $\mathbf{u} = (1, 0, 0)$ is on the xz-plane and the vector $\mathbf{v} = (0, 1, 0)^\mathrm{T}$ is on the yz-plane, but their sum $\mathbf{u} + \mathbf{v} = (1, 1, 0)^\mathrm{T}$ is not on either plane.

Chapter 6 exercises

Exercise 6.1 To show that the vectors $\mathbf{x}_1, \mathbf{x}_2, \mathbf{x}_3$ are linearly independent, you can show that the matrix $A = (\mathbf{x}_1 \ \mathbf{x}_2 \ \mathbf{x}_3)$ has rank 3 using row operations, or by showing $|A| \neq 0$. Either method will show that the only solution of $A\mathbf{x} = 0$ is the trivial solution.

However, since the question also asks you to express the vector $\mathbf{v}$ as a linear combination of the first three, you will need to solve $A\mathbf{x} = \mathbf{v}$. So you can answer the entire question by reducing the augmented matrix, $(A|\mathbf{v})$, to reduced echelon form. Then the first three columns will be the reduced row echelon form of A, which will be the identity matrix, showing that the vectors are linearly independent, and you should obtain the unique solution $\mathbf{x} = (2, -1, 3)^\mathrm{T}$. So

$$\begin{pmatrix} -5 \\ 7 \\ -2 \end{pmatrix} = 2 \begin{pmatrix} 2 \\ 1 \\ -1 \end{pmatrix} - \begin{pmatrix} 3 \\ 4 \\ 6 \end{pmatrix} + 3 \begin{pmatrix} -2 \\ 3 \\ 2 \end{pmatrix}.$$

Exercise 6.2 There are many ways of solving this problem, and there are infinitely many possible such vectors x_3. We can solve it by guessing an appropriate vector x_3 and then showing that the matrix $(x_1 \; x_2 \; x_3)$ has rank 3.

A better approach is to answer the second part of the question first, determine what vectors v form a linearly dependent set with x_1 and x_2 (find the condition on a, b, c as asked) and then write down any vector whose components do not satisfy the condition.

We write the three vectors as the columns of a matrix and row reduce it. We take the matrix to be

$$A = \begin{pmatrix} 1 & 2 & a \\ 1 & 3 & b \\ 2 & 5 & c \end{pmatrix}.$$

Notice that you can choose to order x_1 and x_2 so that the row reduction will be easier since it makes no difference in this question. The vectors $\{x_1, x_2, v\}$ will be linearly dependent if the row echelon form of A has a row of zeros.

$$A = \begin{pmatrix} 1 & 2 & a \\ 1 & 3 & b \\ 2 & 5 & c \end{pmatrix} \xrightarrow[R_3 - 2R_1]{R_2 - R_1} \begin{pmatrix} 1 & 2 & a \\ 0 & 1 & b - a \\ 0 & 1 & c - 2a \end{pmatrix}$$

$$\xrightarrow{R_3 - R_2} \begin{pmatrix} 1 & 2 & a \\ 0 & 1 & b - a \\ 0 & 0 & c - a - b \end{pmatrix}.$$

So the vectors will be linearly dependent if and only if the components of v satisfy

$$a + b - c = 0.$$

So, choose any vector for x_3 which does not satisfy this equation, such as $x_3 = (1, 0, 0)^T$.

Note that this condition is the equation of a plane in $\mathbb{R}^3$ determined by the vectors x_1 and x_2. The set $\{x_1, x_2, v\}$ is linearly dependent if and only if v is the position vector of a point in this plane.

Exercise 6.3 Suppose that S is a linearly independent set of vectors. Then the only linear combination of vectors in S that can equal the zero vector is the trivial linear combination (in which all the coefficients are 0). Now, suppose $R = \{x_1, x_2, \ldots, x_r\}$ is some subset of S and suppose that

$$\alpha_1 x_1 + \alpha_2 x_2 + \cdots + \alpha_r x_r = 0.$$

The x_i for $i = 1, 2, \ldots, r$ are some of the vectors in S. So S will contain these vectors and some others (let's say k others); that is, for some vectors $x_{r+1}, \ldots, x_{r+k}$, S is the set $\{x_1, \ldots, x_r, x_{r+1}, \ldots, x_{k+r}\}$. So we can in fact consider the left-hand side of the equation to be a linear combination of all the vectors in S and we have

$$\alpha_1 x_1 + \alpha_2 x_2 + \cdots + \alpha_r x_r + 0x_{r+1} + \cdots + 0x_{r+k} = 0.$$

By linear independence of S, it follows that all the coefficients are 0 and, in particular,

$$\alpha_1 = \alpha_2 = \cdots = \alpha_r = 0.$$

It follows that R is a linearly independent set of vectors.

Exercise 6.4 This question is testing your understanding of some of the theory you have seen in this chapter. The vector equation

$$a_1 v_1 + a_2 v_2 + \cdots + a_n v_n = 0$$

is equivalent to the matrix equation $A\mathbf{x} = \mathbf{0}$ with $\mathbf{x} = (a_1, a_2, \ldots, a_n)^T$ since $A\mathbf{x} = a_1 v_1 + a_2 v_2 + \cdots + a_n v_n$ (Theorem 1.38). Therefore, $a_1 v_1 + a_2 v_2 + \cdots + a_n v_n = 0$ has only the trivial solution if and only if $A\mathbf{x} = \mathbf{0}$ has only the trivial solution. But we know that $A\mathbf{x} = \mathbf{0}$ has only the trivial solution if and only if $|A| \neq 0$ (by Theorem 4.5).

Exercise 6.5 Let

$$A = \begin{pmatrix} 1 & 0 & 4 & 9 \\ 2 & -1 & -11 & 2 \\ 1 & 3 & 5 & 1 \\ 2 & 4 & -1 & -3 \end{pmatrix},$$

the matrix with columns equal to the given vectors. If we only needed to show that the vectors were linearly dependent, it would suffice to show, using row operations, that $\mathrm{rank}(A) < 4$. But we're asked for more: we have to find an explicit non-trivial linear combination that equals the zero vector. So we need to find a non-trivial solution of $A\mathbf{x} = 0$. To do this, put the matrix A into reduced row echelon form, and then write down the general solution of $A\mathbf{x} = 0$. (You should use row operations to find this. The details are omitted here.) One solution is $\mathbf{x} = (5, -3, 1, -1)^T$. This means that

$$5 \begin{pmatrix} 1 \\ 2 \\ 1 \\ 2 \end{pmatrix} - 3 \begin{pmatrix} 0 \\ -1 \\ 3 \\ 4 \end{pmatrix} + \begin{pmatrix} 4 \\ -11 \\ 5 \\ -1 \end{pmatrix} - \begin{pmatrix} 9 \\ 2 \\ 1 \\ -3 \end{pmatrix} = \begin{pmatrix} 0 \\ 0 \\ 0 \\ 0 \end{pmatrix}.$$

Exercise 6.6 You are being asked to show this directly, using what you have learned in this chapter. Let $\{v_1, v_2, \ldots, v_n\} \subset \mathbb{R}^m$ be any set of vectors in $\mathbb{R}^m$ and let A be the $m \times n$ matrix whose columns are these vectors. Then the reduced row echelon form of A will have at most m leading ones, so it will have at least $n - m \geq 1$ columns without a leading one. Therefore, the system of equations $A\mathbf{x} = \mathbf{0}$ will have non-trivial solutions, and the column vectors of A will be linearly dependent.

Exercise 6.7 To prove that $\{v_1, v_2\}$ is linearly independent, assume that α_1 and α_2 are scalars such that

$$\alpha_1 v_1 + \alpha_2 v_2 = 0. \qquad (*).$$

Then
$$A(\alpha_1 v_1 + \alpha_2 v_2) = 0$$
$$\alpha_1 A v_1 + \alpha_2 A v_2 = 0$$
$$\alpha_1 (2 v_1) + \alpha_2 (5 v_2) = 0$$
$$2\alpha_1 v_1 + 5\alpha_2 v_2 = 0.$$

Add this last equation to -2 times equation $(*)$ to obtain $3\alpha_2 v_2 = 0$. Since $v_2 \neq 0$, we must have $\alpha_2 = 0$. Substituting back into either equation gives $\alpha_1 v_1 = 0$, so that $\alpha_1 = 0$ since $v_1 \neq 0$. This shows that v_1, v_2 are linearly independent.

Generalisation 1: The same proof works for any constants, $A v_1 = \kappa v_1$, $A v_2 = \lambda v_2$ provided $\kappa \neq \lambda$.

Generalisation 2: It also extends to three (or more) non-zero vectors: say, $A v_1 = \kappa v_1$, $A v_2 = \lambda v_2$, $A v_3 = \mu v_3$ with κ, λ, μ distinct constants (that is, no two are equal).

Exercise 6.8 Observe that each set of vectors contains at least two linearly independent vectors since no vector in either set is a scalar multiple of another vector in the set. Write the vectors of each set as the columns of a matrix:

$$B = \begin{pmatrix} -1 & 1 & -1 \\ 0 & 2 & 2 \\ 1 & 3 & 5 \end{pmatrix}, \qquad A = \begin{pmatrix} -1 & 1 & 1 \\ 0 & 2 & 2 \\ 1 & 3 & 5 \end{pmatrix}.$$

$|A| \neq 0$, so W is a basis of $\mathbb{R}^3$ and $\text{Lin}(W) = \mathbb{R}^3$. (Therefore, another basis of $\text{Lin}(W)$ is the standard basis, $\{e_1, e_2, e_3\}$.)

$|B| = 0$, so the set U is linearly dependent and one of the vectors is a linear combination of the other two. Since any two vectors of U are linearly independent, we know that we will need two vectors for a basis and $\text{Lin}(U)$ is a two-dimensional subspace of $\mathbb{R}^3$, which is a plane. So we can take the first two vectors in U to be a basis of $\text{Lin}(U)$.

There are two ways you can find the Cartesian equation of the plane. A vector equation is given by

$$\mathbf{x} = \begin{pmatrix} x \\ y \\ z \end{pmatrix} = s \begin{pmatrix} -1 \\ 0 \\ 1 \end{pmatrix} + t \begin{pmatrix} 1 \\ 2 \\ 3 \end{pmatrix}, \quad s, t \in \mathbb{R},$$

and you can find the Cartesian equation by equating components to obtain three equations in the two unknowns s and t. Eliminating s and t between the three equations, you will obtain a single equation relating x, y, and z. Explicitly, we have $x = -s + t$, $y = 2t$, $z = s + 3t$, so

$$t = \frac{y}{2}, \ s = t - x = \frac{y}{2} - x \quad \text{and so} \quad z = s + 3t = \left(\frac{y}{2} - x\right) + \frac{3}{2}y.$$

Therefore, $x - 2y + z = 0$ is a Cartesian equation of the plane.

Alternatively, you could write the two basis vectors and the vector $\mathbf{x}$ as the columns of a matrix M and, using the fact that $|M| = 0$ if and only if the columns of M are linearly dependent, you have the equation

$$\begin{vmatrix} -1 & 1 & x \\ 0 & 2 & y \\ 1 & 3 & z \end{vmatrix} = -2x + 4y - 2z = 0.$$

Exercise 6.9 The xz-plane is the set of all vectors of the form $(x, 0, z)^{\mathrm{T}}$, so the set of vectors $\{\mathbf{e}_1, \mathbf{e}_3\}$ is a basis.

Exercise 6.10 The first thing you should do is write the vectors in B as the columns of a matrix, call it P, and evaluate the determinant. Since $|P| = -2 \neq 0$, the vectors form a basis of $\mathbb{R}^3$. Since you need to find the coordinates of two different vectors, you need to solve two systems of equations, namely $P\mathbf{x} = \mathbf{w}$ and $P\mathbf{x} = \mathbf{e}_1$, to find the coefficients in the basis B. One efficient method is to find P^{-1} and use this to solve the equations as $\mathbf{x} = P^{-1}\mathbf{w}$ and $\mathbf{x} = P^{-1}\mathbf{e}_1$.

If $P = \begin{pmatrix} 1 & -4 & 3 \\ 1 & 0 & 5 \\ 0 & 3 & 1 \end{pmatrix}$, then $P^{-1} = -\frac{1}{2}\begin{pmatrix} -15 & 13 & -20 \\ -1 & 1 & -2 \\ 3 & -3 & 4 \end{pmatrix}$.

You should find that

$$[\mathbf{w}]_B = \begin{bmatrix} -3 \\ 1 \\ 2 \end{bmatrix} \quad \text{and} \quad [\mathbf{e}_1]_B = -\frac{1}{2}\begin{bmatrix} -15 \\ -1 \\ 3 \end{bmatrix},$$

and check your result.

Exercise 6.11 Let

$$\mathbf{u} = u_1\mathbf{v}_1 + u_2\mathbf{v}_2 + \cdots + u_n\mathbf{v}_n \quad \text{and} \quad \mathbf{w} = w_1\mathbf{v}_1 + w_2\mathbf{v}_2 + \cdots + w_n\mathbf{v}_n,$$

so that

$$[\mathbf{u}]_B = \begin{bmatrix} u_1 \\ u_2 \\ \vdots \\ u_n \end{bmatrix}_B, \quad [\mathbf{w}]_B = \begin{bmatrix} w_1 \\ w_2 \\ \vdots \\ w_n \end{bmatrix}_B.$$

Then

$$\alpha\mathbf{u} + \beta\mathbf{w} = \alpha(u_1\mathbf{v}_1 + u_2\mathbf{v}_2 + \cdots + u_n\mathbf{v}_n)$$
$$+ \beta(w_1\mathbf{v}_1 + w_2\mathbf{v}_2 + \cdots + w_n\mathbf{v}_n)$$
$$= (\alpha u_1 + \beta w_1)\mathbf{v}_1 + (\alpha u_2 + \beta w_2)\mathbf{v}_2 + \cdots + (\alpha u_n + \beta w_n)\mathbf{v}_n.$$

Then,

$$[\alpha\mathbf{u} + \beta\mathbf{w}]_B = \begin{bmatrix} \alpha u_1 + \beta w_1 \\ \alpha u_2 + \beta w_2 \\ \vdots \\ \alpha u_n + \beta w_n \end{bmatrix}_B = \alpha\begin{bmatrix} u_1 \\ u_2 \\ \vdots \\ u_n \end{bmatrix}_B + \beta\begin{bmatrix} w_1 \\ w_2 \\ \vdots \\ w_n \end{bmatrix}_B$$
$$= \alpha[\mathbf{u}]_B + \beta[\mathbf{w}]_B.$$

Exercise 6.12 We will give a detailed answer to this question. To begin, put the matrix A into reduced row echelon form.

$$A = \begin{pmatrix} 1 & 2 & -1 & 3 \\ 2 & 3 & 0 & 1 \\ -4 & -5 & -2 & 3 \end{pmatrix} \to \cdots \to \begin{pmatrix} 1 & 0 & 3 & -7 \\ 0 & 1 & -2 & 5 \\ 0 & 0 & 0 & 0 \end{pmatrix}.$$

A basis of row space consists of the non-zero rows of the reduced row echelon form (written as vectors), so

a basis of $RS(A)$ is
$$\left\{ \begin{pmatrix} 1 \\ 0 \\ 3 \\ -7 \end{pmatrix}, \begin{pmatrix} 0 \\ 1 \\ -2 \\ 5 \end{pmatrix} \right\}.$$

A basis of the column space, $CS(A)$ is given by the vectors of the matrix A which correspond to the columns with the leading ones in the reduced row echelon form of A, so these are the first two columns. Therefore,

a basis of $CS(A)$ is
$$\left\{ \begin{pmatrix} 1 \\ 2 \\ -4 \end{pmatrix}, \begin{pmatrix} 2 \\ 3 \\ -5 \end{pmatrix} \right\}.$$

Since the $CS(A)$ has a basis consisting of two vectors, it is a two-dimensional subspace of $\mathbb{R}^3$, which is a plane. To find a Cartesian equation of this plane, you can use

$$\begin{vmatrix} 1 & 2 & x \\ 2 & 3 & y \\ -4 & -5 & z \end{vmatrix} = 0.$$

(Why does this give you the equation of the plane? Because if the vector $\mathbf{x} = (x, y, z)^{\mathrm{T}}$ is in the plane, then it is a linear combination of the basis vectors, so the three vectors are linearly dependent and the matrix whose columns are these three vectors must have determinant equal to 0.) Expanding the determinant by the last column is easiest, and you should obtain the equation

$$2x - 3y - z = 0.$$

(Note that this must be an equation: don't leave off the '$= 0$' part. This is the equation of a plane through the origin, which is a subspace of $\mathbb{R}^3$.) The next thing you should do is check that your solution is correct. The components of all the column vectors of A should satisfy this equation. For example, $2(1) - 3(2) - (-4) = 0$ and $2(2) - 3(3) - (-5) = 0$. The equation is also satisfied by the last two columns, $(-1, 0, -2)^{\mathrm{T}}$ and $(3, 1, 3)^{\mathrm{T}}$ as you can easily check.

You are asked to state the rank–nullity theorem for matrices, ensuring that you define each term and use it to determine the dimension of the null space, $N(A)$. The theorem can be stated either as

$$\mathrm{rank}(A) + \mathrm{nullity}(A) = n$$

or

$$\dim(R(A)) + \dim(N(A)) = n,$$

where n is the number of columns in the matrix A. If you used the terms rank and nullity, then you must say what these terms mean: $\mathrm{rank}(A) = \dim(R(A))$ and $\mathrm{nullity}(A) = \dim(N(A))$. Since $\dim(CS(A)) = 2$ and $n = 4$, this theorem tells us that $\dim(N(A)) = 2$.

You are now asked for what real values of a the vector

$$\mathbf{b}(a) = \begin{pmatrix} -1 \\ a \\ a^2 \end{pmatrix}$$

is in the range of A, $R(A)$. The range of A is equal to the column space of A, and you already know that this subspace of $\mathbb{R}^3$ is a plane with Cartesian equation

$$2x - 3y - z = 0.$$

The vector $(-1, a, a^2)^T \in R(A)$ if and only if its components satisfy this equation. Substituting, you obtain a quadratic equation in a,

$$2(-1) - 3(a) - (a^2) = 0 \quad \text{or} \quad a^2 + 3a + 2 = 0,$$

which factors: $a^2 + 3a + 2 = (a + 2)(a + 1) = 0$. Therefore, the only solutions are $a = -1$ and $a = -2$. The corresponding vectors are

$$\mathbf{b}(-1) = \begin{pmatrix} -1 \\ -1 \\ 1 \end{pmatrix} \quad \text{and} \quad \mathbf{b}(-2) = \begin{pmatrix} -1 \\ -2 \\ 4 \end{pmatrix}.$$

You might notice that the second vector listed above is equal to -1 times the first column of the matrix A.

There are other ways to obtain this result, but they take longer. For example, you could write

$$\begin{pmatrix} -1 \\ a \\ a^2 \end{pmatrix} = s \begin{pmatrix} 1 \\ 2 \\ -4 \end{pmatrix} + t \begin{pmatrix} 2 \\ 3 \\ -5 \end{pmatrix},$$

giving three equations, one for each component of the vector equation, and eliminate s and t to obtain the same quadratic equation in a.

Exercise 6.13 You have already shown that $A^T A$ is symmetric as an exercise in Chapter 1 (Exercise 1.6). To show it is invertible, we will show that $(A^T A)\mathbf{v} = \mathbf{0}$ has only the trivial solution, $\mathbf{v} = \mathbf{0}$, which implies that $A^T A$ is invertible by Theorem 4.5. To do this, we first need to show that $(A^T A)\mathbf{v} = \mathbf{0}$ implies that $A\mathbf{v} = \mathbf{0}$. This is the difficult part. Then we can deduce from $A\mathbf{v} = \mathbf{0}$ that $\mathbf{v} = \mathbf{0}$, since A has rank k.

We will give two arguments to show that $A\mathbf{v} = \mathbf{0}$. The first is a bit tricky. We multiply $A^T A\mathbf{v} = \mathbf{0}$ on the left by $\mathbf{v}^T$ to get $\mathbf{v}^T A^T A\mathbf{v} = 0$. Now, $\mathbf{v}^T A^T A\mathbf{v} = (A\mathbf{v})^T(A\mathbf{v}) = 0$. But for any vector $\mathbf{w} \in \mathbb{R}^n$, $\mathbf{w}^T \mathbf{w} = \langle \mathbf{w}, \mathbf{w} \rangle = \|\mathbf{w}\|^2$, so we have $\|A\mathbf{v}\|^2 = 0$, which implies that $A\mathbf{v} = \mathbf{0}$.

Alternatively, we can show $A\mathbf{v} = \mathbf{0}$ by asking what $A^T A\mathbf{v} = A^T(A\mathbf{v}) = \mathbf{0}$ implies about the vector $A\mathbf{v}$; that is, in what two subspaces associated with the matrix A^T is it? Since $A^T(A\mathbf{v}) = \mathbf{0}$, $A\mathbf{v} \in N(A^T)$. Also, the vector $A\mathbf{v}$ is a linear combination of the columns of A, hence it is a linear combination of the rows of A^T, so $A\mathbf{v} \in RS(A^T)$. But we have seen that the only vector which is in both the null space and the row space of a matrix is the zero vector, since these subspaces are orthogonal subspaces (of $\mathbb{R}^m$ for A^T). Hence the vector $A\mathbf{v} = \mathbf{0}$.

So $A\mathbf{v} = \mathbf{0}$. But the columns of A are linearly independent (A has full column rank), so $A\mathbf{v} = \mathbf{0}$ has only the trivial solution $\mathbf{v} = \mathbf{0}$.

Therefore, $A^T A \mathbf{v} = \mathbf{0}$ has only the trivial solution and the matrix $A^T A$ is invertible.

The columns of the 3×2 matrix M are linearly independent, since they are not scalar multiples of one another. The 2×2 matrix $M^T M$ is

$$M^T M = \begin{pmatrix} 1 & 3 & 1 \\ -2 & 0 & 1 \end{pmatrix} \begin{pmatrix} 1 & -2 \\ 3 & 0 \\ 1 & 1 \end{pmatrix} = \begin{pmatrix} 11 & -1 \\ -1 & 5 \end{pmatrix},$$

which is symmetric and invertible since $|M^T M| = 54 \neq 0$.

Exercise 6.14 From the given information, you can determine that $k = 3$, since the rows of B (written as vectors) must be in $\mathbb{R}^3$. You cannot determine m, but you can say that $m \geq 2$ because you know that B has rank 2, since its row space is a plane.

Can you determine the null space of B? Yes, because the row space and the null space are orthogonal subspaces of $\mathbb{R}^3$, or simply because you know that the null space consists of all vectors for which $B\mathbf{x} = \mathbf{0}$, so all vectors such that $\langle \mathbf{r}_i, \mathbf{x} \rangle = 0$ for each row, $\mathbf{r}_i$ of B. Therefore, the null space must consist of all vectors on the line through the origin in the direction of the normal vector to the plane. So a basis of this space is given by $\mathbf{n} = (4, -5, 3)^T$ and a general solution of $B\mathbf{x} = \mathbf{0}$ is

$$\mathbf{x} = t \begin{pmatrix} 4 \\ -5 \\ 3 \end{pmatrix}, \quad t \in \mathbb{R}.$$

Exercise 6.15 The subspace W has a basis consisting of the three sequences,

$$\mathbf{y}_1 = \{1, 0, 0, 0, 0 \ldots\}, \quad \mathbf{y}_2 = \{0, 1, 0, 0, 0 \ldots\},$$
$$\mathbf{y}_3 = \{0, 0, 1, 0, 0 \ldots\},$$

so it has dimension 3.

Chapter 7 exercises

Exercise 7.1 The matrix A_T and its reduced row echelon form are

$$A_T = \begin{pmatrix} 1 & 1 & 2 \\ 1 & 0 & 1 \\ 2 & 1 & 3 \end{pmatrix} \longrightarrow \cdots \longrightarrow \begin{pmatrix} 1 & 0 & 1 \\ 0 & 1 & 1 \\ 0 & 0 & 0 \end{pmatrix}.$$

A basis for the null space is $\{(-1, -1, 1)^T\}$, and a basis for the range is

$$\left\{ \begin{pmatrix} 1 \\ 1 \\ 2 \end{pmatrix}, \begin{pmatrix} 1 \\ 0 \\ 1 \end{pmatrix} \right\}.$$

There are other possible answers. To verify the rank–nullity theorem,

$$\text{rank}(T) + \text{nullity}(T) = 2 + 1 = 3 = \dim(\mathbb{R}^3).$$

This linear transformation is not invertible, as A_T^{-1} does not exist.

Exercise 7.2 To sketch the effect of S on the unit square, mark off a unit square on a set of axes. Mark the unit vector in the x direction, e_1, in one colour, and the unit vector in the y direction, e_2, in another colour (or differentiate between them by single and double arrowheads). Now draw the vector images of these, $S(e_1)$ and $S(e_2)$, in the same colours, and complete the image of the unit square with these vectors as its two corresponding sides. Do the same for T.

The linear transformation S is reflection in the line $y = x$. The transformation T is a rotation *clockwise* by an angle $\frac{\pi}{2}$ radians (or a rotation anticlockwise by $\frac{3\pi}{2}$).

ST means first do T and then do S, so this will place the unit square in the second quadrant with $ST(e_1) = (0, -1)^T$ and e_2 back in its original position.

TS means reflect and then rotate, after which the unit square will be in the fourth quadrant, with $TS(e_2) = (0, -1)^T$ and e_1 back in its original place.

Their matrices are

$$A_{ST} = A_S A_T = \begin{pmatrix} 0 & 1 \\ 1 & 0 \end{pmatrix} \begin{pmatrix} 0 & 1 \\ -1 & 0 \end{pmatrix} = \begin{pmatrix} -1 & 0 \\ 0 & 1 \end{pmatrix}$$

and

$$A_{TS} = A_T A_S = \begin{pmatrix} 1 & 0 \\ 0 & -1 \end{pmatrix}.$$

These matrices are not equal, $A_{ST} \neq A_{TS}$. The columns of A_{ST} are $ST(e_1)$ and $ST(e_2)$, and these do match the sketch. Check the columns of TS.

Exercise 7.3 Write the vectors v_i as the columns of a matrix,

$$P_B = \begin{pmatrix} 1 & -1 & 0 \\ 0 & 1 & 1 \\ 1 & 2 & 5 \end{pmatrix}.$$

Since $|P_B| = 2 \neq 0$, the columns are linearly independent and hence form a basis of $\mathbb{R}^3$. P_B is the transition matrix from B coordinates to standard coordinates, $\mathbf{v} = P_B[\mathbf{v}]_B$. Finding P_B^{-1} by the cofactor method, or otherwise, the B coordinates of $\mathbf{u}$ are

$$[\mathbf{u}]_B = P_B^{-1}\mathbf{u} = \frac{1}{2}\begin{pmatrix} 3 & 5 & -1 \\ 1 & 5 & -1 \\ -1 & -3 & 1 \end{pmatrix}\begin{pmatrix} 1 \\ 2 \\ 3 \end{pmatrix} = \begin{bmatrix} 5 \\ 4 \\ -2 \end{bmatrix}_B.$$

Hence $\mathbf{u} = 5\mathbf{v}_1 + 4\mathbf{v}_2 - 2\mathbf{v}_3$.

Using properties of linear transformations,

$$\begin{aligned} S(\mathbf{u}) &= S(5\mathbf{v}_1 + 4\mathbf{v}_2 - 2\mathbf{v}_3) \\ &= 5S(\mathbf{v}_1) + 4S(\mathbf{v}_2) - 2S(\mathbf{v}_3) \\ &= 5\mathbf{e}_1 + 4\mathbf{e}_2 - 2\mathbf{e}_3 \\ &= \begin{pmatrix} 5 \\ 4 \\ -2 \end{pmatrix}. \end{aligned}$$

Since $R(S)$ is spanned by $\{\mathbf{e}_1, \mathbf{e}_2, \mathbf{e}_3\}$, $R(S) = \mathbb{R}^3$ and $N(S) = \{\mathbf{0}\}$. The linear transformation S is the inverse of the linear transformation T with $T(\mathbf{e}_1) = \mathbf{v}_1$, $T(\mathbf{e}_2) = \mathbf{v}_2$, $T(\mathbf{e}_3) = \mathbf{v}_1$, which has matrix P_B, so the matrix A_S is P_B^{-1}.

Exercise 7.4 If we had such a linear transformation, $T : \mathbb{R}^3 \to \mathbb{R}^2$, then

$$N(T) = \left\{ \begin{pmatrix} x \\ y \\ z \end{pmatrix} \,\middle|\, x = y = z \right\} = \left\{ s\begin{pmatrix} 1 \\ 1 \\ 1 \end{pmatrix} \,\middle|\, s \in \mathbb{R} \right\},$$

so that a basis of $N(T)$ is the vector $(1, 1, 1)^T$. The rank–nullity theorem states that the dimension of the range plus the dimension of the null space is equal to the dimension of the domain, $\mathbb{R}^3$. We have nullity$(T) = 1$ and rank$(T) = 2$ since $R(T) = \mathbb{R}^2$. That is, rank$(T) +$ nullity$(T) = 2 + 1 = 3$, so the theorem would be satisfied. Note that this does not *guarantee* the existence of T, but if it did *not* hold, then we would know for sure that such a T could not exist.

Since $T : \mathbb{R}^3 \to \mathbb{R}^2$, we are looking for a 2×3 matrix A such that $T(\mathbf{x}) = A\mathbf{x}$. Given that $T(\mathbf{e}_1)$ and $T(\mathbf{e}_2)$ are the standard basis vectors of $\mathbb{R}^2$, we know that the first two columns of A should be these two vectors. What about the third column? We can obtain this column from the basis of the null space since we already have the first two. If $\mathbf{c}_1, \mathbf{c}_2, \mathbf{c}_3$

denote the columns of A, then $\mathbf{c}_1 + \mathbf{c}_2 + \mathbf{c}_3 = \mathbf{0}$, and therefore

$$T(\mathbf{x}) = A_T\mathbf{x} = \begin{pmatrix} 1 & 0 & -1 \\ 0 & 1 & -1 \end{pmatrix} \begin{pmatrix} x \\ y \\ z \end{pmatrix} = \begin{pmatrix} x - z \\ y - z \end{pmatrix}.$$

Exercise 7.5 Only the linear transformation TS is defined, with $A_{TS} = A_T A_S$,

$$A_{TS} = \begin{pmatrix} 1 & 0 & -1 \\ 0 & 1 & -1 \end{pmatrix} \frac{1}{2} \begin{pmatrix} 3 & 5 & -1 \\ 1 & 5 & -1 \\ -1 & -3 & 1 \end{pmatrix} = \begin{pmatrix} 2 & 4 & -1 \\ 1 & 4 & -1 \end{pmatrix}.$$

Exercise 7.6 The linear transformation T is given by $T(\mathbf{x}) = A\mathbf{x}$, where A is a 3×4 matrix. The simplest way to answer the questions is to construct the matrix whose columns are the images of the standard basis vectors, $T(\mathbf{e}_i)$,

$$A = \begin{pmatrix} 1 & 2 & 5 & x \\ 0 & 1 & 1 & y \\ -1 & 2 & -1 & z \end{pmatrix}.$$

In order to consider the two possibilities in parts (i) and (ii), row reduce this matrix, beginning with $R_3 + R_1$,

$$A \longrightarrow \begin{pmatrix} 1 & 2 & 5 & x \\ 0 & 1 & 1 & y \\ 0 & 4 & 4 & z+x \end{pmatrix} \longrightarrow \begin{pmatrix} 1 & 2 & 5 & x \\ 0 & 1 & 1 & y \\ 0 & 0 & 0 & z+x-4y \end{pmatrix}.$$

(i) By the rank–nullity theorem, $\dim(R(T)) + \dim(N(T)) = \dim V$, and since $T : \mathbb{R}^4 \to \mathbb{R}^3$, $n = 4$. So for the dimensions of $R(T)$ and $N(T)$ to be equal, the subspaces must both have dimension 2. Looking at the reduced form of the matrix, we see that this will happen if

$$x - 4y + z = 0.$$

If the vector $\mathbf{x}$ satisfies this condition, then a basis of $R(T)$ is given by the columns of A corresponding to the leading ones in the row echelon form, which will be the first two columns. So a basis of $R(T)$ is $\{\mathbf{v}_1, \mathbf{v}_2\}$.

You could also approach this question by first deducing from the rank–nullity theorem that $\dim(R(T)) = 2$ as above, so $R(T)$ is a plane in $\mathbb{R}^3$. Therefore, $\{\mathbf{v}_1, \mathbf{v}_2\}$ is a basis, and the Cartesian equation of the plane is given by

$$\begin{vmatrix} 1 & 2 & x \\ 0 & 1 & y \\ -1 & 2 & z \end{vmatrix} = x - 4y + z = 0.$$

The components of the vector $\mathbf{v}_3$ satisfy this equation, and this is the condition that the components of $\mathbf{x}$ must satisfy.

(ii) If the linear transformation has $\dim(N(T)) = 1$, then by the rank–nullity theorem, you know that $\dim(R(T)) = 3$ (and therefore $R(T) = \mathbb{R}^3$), so the echelon form of the matrix A needs to have three leading ones. Therefore, the condition that the components of $\mathbf{x}$ must satisfy is

$$x - 4y + z \neq 0.$$

Now you can continue with row reducing the matrix A to obtain a basis for $N(T)$. The row echelon form of A will have a leading one in the last column (first multiply the last row by $1/(x - 4y + z)$ to get this leading one, then continue to reduced echelon form),

$$A \longrightarrow \cdots \longrightarrow \begin{pmatrix} 1 & 2 & 5 & x \\ 0 & 1 & 1 & y \\ 0 & 0 & 0 & 1 \end{pmatrix} \longrightarrow \begin{pmatrix} 1 & 2 & 5 & 0 \\ 0 & 1 & 1 & 0 \\ 0 & 0 & 0 & 1 \end{pmatrix}$$

$$\longrightarrow \begin{pmatrix} 1 & 0 & 3 & 0 \\ 0 & 1 & 1 & 0 \\ 0 & 0 & 0 & 1 \end{pmatrix}.$$

So a basis of $\ker(T)$ is given by the vector $\mathbf{w} = (-3, -1, 1, 0)^T$.

Exercise 7.7 The easiest method to determine the required values of λ is to evaluate the determinant of the matrix whose columns are these vectors, and then find for what values of λ the determinant is zero:

$$\begin{vmatrix} 1 & 1 & 2 \\ 3 & -1 & 0 \\ -5 & 1 & \lambda \end{vmatrix} = -4 - 4\lambda = 0.$$

So you can conclude that the set of vectors is a basis for all values of λ except $\lambda = -1$.

Therefore, each of the sets B and S is a basis of $\mathbb{R}^3$. There are two methods you can use to find the transition matrix P from S coordinates to B coordinates. One way is to write down the transition matrix P_B from B coordinates to standard, and the transition matrix P_S from S coordinates to standard, and then calculate $P = P_B^{-1} P_S$. Alternatively, you can use the fact that the columns of P are the B coordinates of the basis S vectors. Since the first two vectors of each basis are the same, we will do the latter. We have,

$$[\mathbf{v}_1]_B = \begin{bmatrix} 1 \\ 0 \\ 0 \end{bmatrix}_B \quad \text{and} \quad [\mathbf{v}_2]_B = \begin{bmatrix} 0 \\ 1 \\ 0 \end{bmatrix}_B,$$

so it only remains to find $[\mathbf{s}]_B$. This can be done by Gaussian elimination. To find constants a, b, c, such that $\mathbf{s} = a\mathbf{v}_1 + b\mathbf{v}_2 + c\mathbf{b} = A\mathbf{x}$, we reduce the augmented matrix:

$$\begin{pmatrix} 1 & 1 & 2 & 2 \\ 3 & -1 & 0 & 0 \\ -5 & 1 & 1 & 3 \end{pmatrix} \longrightarrow \cdots \longrightarrow \begin{pmatrix} 1 & 0 & 0 & -\frac{1}{2} \\ 0 & 1 & 0 & -\frac{3}{2} \\ 0 & 0 & 1 & 2 \end{pmatrix}.$$

It follows that

$$[\mathbf{s}]_B = \begin{bmatrix} -\frac{1}{2} \\ -\frac{3}{2} \\ 2 \end{bmatrix}.$$

(You should carry out this row reduction and all the omitted calculations.) Therefore, the transition matrix P from S coordinates to B coordinates is

$$P = ([\mathbf{v}_1]_B, [\mathbf{v}_2]_B, [\mathbf{s}]_B) = \begin{pmatrix} 1 & 0 & -\frac{1}{2} \\ 0 & 1 & -\frac{3}{2} \\ 0 & 0 & 2 \end{pmatrix}.$$

Then, if $[\mathbf{w}]_S = \begin{bmatrix} 1 \\ 2 \\ 2 \end{bmatrix}_S$,

$$[\mathbf{w}]_B = P[\mathbf{w}]_S = \begin{pmatrix} 1 & 0 & -\frac{1}{2} \\ 0 & 1 & -\frac{3}{2} \\ 0 & 0 & 2 \end{pmatrix} \begin{bmatrix} 1 \\ 2 \\ 2 \end{bmatrix}_S = \begin{bmatrix} 0 \\ -1 \\ 4 \end{bmatrix}_B.$$

You can check this result by finding the standard coordinates of $\mathbf{w}$ from each of these. (You will find $\mathbf{w} = (7, 1, 3)^T$.)

Exercise 7.8 To show that each of the vectors in the sets S and B are in W, you just need to substitute the components of each vector into the equation of the plane and show that the equation is satisfied. For example, $x - 2y + 3z = (2) - 2(1) + 3(0) = 0$. Each set contains two linearly independent vectors (neither is a scalar multiple of the other), and you know that a plane is a two-dimensional subspace of $\mathbb{R}^3$. Two linearly independent vectors in a vector space of dimension 2 are a basis, so each of the sets S and B is a basis of W.

Since $x - 2y + 3z = (5) - 2(7) + 3(3) = 0$, $\mathbf{v} \in W$. Its coordinates in the basis S are easily found (because of the zeros and ones in the basis vectors),

$$\begin{pmatrix} 5 \\ 7 \\ 3 \end{pmatrix} = a \begin{pmatrix} 2 \\ 1 \\ 0 \end{pmatrix} + b \begin{pmatrix} -3 \\ 0 \\ 1 \end{pmatrix} \implies a = 7, \; b = 3,$$

$$\implies [\mathbf{v}]_S = \begin{bmatrix} 7 \\ 3 \end{bmatrix}_S.$$

To find the transition matrix M from B to S, you can use the fact that $M = ([\mathbf{b}_1]_S, [\mathbf{b}_2]_S)$, where $B = \{\mathbf{b}_1, \mathbf{b}_2\}$. As we just saw for the vector $\mathbf{v}$, because of the zeros and ones in the basis S, we must have,

$$[\mathbf{b}_1]_S = \begin{bmatrix} 1 \\ 1 \end{bmatrix}_S \quad \text{and} \quad [\mathbf{b}_2]_S = \begin{bmatrix} 2 \\ 1 \end{bmatrix}_S.$$

Then the required transition matrix M is

$$M = \begin{pmatrix} 1 & 2 \\ 1 & 1 \end{pmatrix},$$

so $[\mathbf{x}]_S = M[\mathbf{x}]_B$ for all $\mathbf{x} \in W$. The matrix M is much easier to calculate directly than its inverse matrix, which is why the question was posed this way. However, you need to use the matrix M^{-1} to change from S coordinates to B coordinates.

$$M^{-1} = \begin{pmatrix} -1 & 2 \\ 1 & -1 \end{pmatrix}, \quad \text{so} \quad [\mathbf{v}]_B = \begin{pmatrix} -1 & 2 \\ 1 & -1 \end{pmatrix} \begin{bmatrix} 7 \\ 3 \end{bmatrix}_S = \begin{bmatrix} -1 \\ 4 \end{bmatrix}_B,$$

which is easily checked (by calculating $\mathbf{v} = -\mathbf{b}_1 + 4\mathbf{b}_2$).

Exercise 7.9 The answer, using the notation in Section 7.4, is $A_{[B,B']} = P_{B'}^{-1} A P_B$, where $T(\mathbf{x}) = A\mathbf{x}$ for all $\mathbf{x} \in \mathbb{R}^3$. Now,

$$A = \begin{pmatrix} 0 & 1 \\ -5 & 13 \\ -7 & 16 \end{pmatrix}, \quad P_B = \begin{pmatrix} 3 & 5 \\ 1 & 2 \end{pmatrix}$$

and

$$P_{B'}^{-1} = \begin{pmatrix} 1 & -1 & 0 \\ 0 & 2 & 1 \\ -1 & 2 & 2 \end{pmatrix}^{-1} = \begin{pmatrix} \frac{2}{3} & \frac{2}{3} & -\frac{1}{3} \\ -\frac{1}{3} & \frac{2}{3} & -\frac{1}{3} \\ \frac{2}{3} & -\frac{1}{3} & \frac{2}{3} \end{pmatrix},$$

where we have omitted the details of the calculation of this inverse. It then follows that

$$A_{[B,B']} = P_{B'}^{-1} A P_B = \begin{pmatrix} 1 & 3 \\ 0 & 1 \\ -2 & -1 \end{pmatrix}.$$

Exercise 7.10 $C^\infty(\mathbb{R})$ is not empty; for example, the zero function is in this set, and so is e^x. By standard results of calculus, the sum of two differentiable functions is again differentiable; and a scalar multiple of a differentiable function is differentiable. Thus, if $f, g \in C^\infty(\mathbb{R})$, then their sum $f + g$ is also differentiable arbitrarily often, and the same is true of αf for any $\alpha \in \mathbb{R}$. So $C^\infty(\mathbb{R})$ is closed under scalar multiplication.

The function D is a mapping $D : C^\infty(\mathbb{R}) \to C^\infty(\mathbb{R})$ since $D(f) = f'$ where $f' : x \mapsto f'(x)$ and f' can also be differentiated arbitrarily often. To show D is a linear operator, you only need to show that it is a linear function. We have:

$$D(f + g) = (f + g)' = f' + g' = D(f) + D(g)$$

and

$$D(\alpha f) = (\alpha f)' = \alpha f' = \alpha D(f).$$

These are just the rules of differentiation which you encounter in calculus; that is, the derivative of the sum of two functions is the sum of the derivatives, and the derivative of a scalar multiple of a function is the scalar multiple of the derivative.

Chapter 8 exercises

Exercise 8.1 To diagonalise the matrix A, first find the characteristic equation and solve for the eigenvalues.

$$|A - \lambda I| = \begin{vmatrix} 4 - \lambda & 5 \\ -1 & -2 - \lambda \end{vmatrix} = \lambda^2 - 2\lambda - 3 = (\lambda - 3)(\lambda + 1) = 0.$$

The eigenvalues are $\lambda = 3$ and $\lambda = -1$. Next find a corresponding eigenvector for each eigenvalue:

$$\lambda_1 = -1 : \quad A + I = \begin{pmatrix} 5 & 5 \\ -1 & -1 \end{pmatrix} \to \begin{pmatrix} 1 & 1 \\ 0 & 0 \end{pmatrix} \implies \mathbf{v}_1 = \begin{pmatrix} -1 \\ 1 \end{pmatrix}.$$

$$\lambda_2 = 3 : \quad A - 3I = \begin{pmatrix} 1 & 5 \\ -1 & -5 \end{pmatrix} \to \begin{pmatrix} 1 & 5 \\ 0 & 0 \end{pmatrix} \implies \mathbf{v}_2 = \begin{pmatrix} -5 \\ 1 \end{pmatrix}.$$

Then you can choose

$$P = \begin{pmatrix} -1 & -5 \\ 1 & 1 \end{pmatrix}, \quad \text{then} \quad D = \begin{pmatrix} -1 & 0 \\ 0 & 3 \end{pmatrix}, \quad \text{and} \quad P^{-1}AP = D.$$

Check that your eigenvectors are correct by calculating AP,

$$AP = \begin{pmatrix} 4 & 5 \\ -1 & -2 \end{pmatrix} \begin{pmatrix} -1 & -5 \\ 1 & 1 \end{pmatrix} = \begin{pmatrix} 1 & -15 \\ -1 & 3 \end{pmatrix} = (-1\mathbf{v}_1 \quad 3\mathbf{v}_2) = PD.$$

This checks that $A\mathbf{v}_1 = (-1)\mathbf{v}_1$ and $A\mathbf{v}_2 = 3\mathbf{v}_1$. You can also check your answer by finding P^{-1} and calculating $P^{-1}AP$:

$$P^{-1}AP = \frac{1}{4} \begin{pmatrix} 1 & 5 \\ -1 & -1 \end{pmatrix} \begin{pmatrix} 1 & -15 \\ -1 & 3 \end{pmatrix} = \begin{pmatrix} -1 & 0 \\ 0 & 3 \end{pmatrix} = D.$$

Exercise 8.2 We'll provide an answer to this question, but leave the calculations to you. If you have carried out the steps carefully and checked that $AP = PD$, you should have a correct answer.

The characteristic polynomial is $-\lambda^3 + 14\lambda^2 - 48\lambda$, which is easily factorised as $-\lambda(\lambda - 6)(\lambda - 8)$. So the eigenvalues are $0, 6, 8$. Corresponding eigenvectors, respectively, are calculated to be non-zero scalar multiples of

$$\begin{pmatrix} 1 \\ -1 \\ 2 \end{pmatrix}, \quad \begin{pmatrix} 1 \\ 2 \\ 2 \end{pmatrix}, \quad \begin{pmatrix} 1 \\ 4 \\ 0 \end{pmatrix}.$$

We may therefore take

$$P = \begin{pmatrix} 1 & 1 & 1 \\ -1 & 2 & 4 \\ 2 & 2 & 0 \end{pmatrix} \qquad D = \mathrm{diag}(0, 6, 8),$$

and then $P^{-1}AP = D$. Your answer is correct as long as $AP = PD$ and your eigenvectors are scalar multiples of the ones given (taken in any order as the columns of P, as long as D matches).

Exercise 8.3 The matrix A has only one eigenvalue, $\lambda = 1$. The corresponding eigenvectors are all the non-zero scalar multiples of $(1, 0)^{\mathrm{T}}$, so there cannot be two linearly independent eigenvectors, and hence the matrix is not diagonalisable.

The eigenvalues of the matrix B are 0 and 2. Since this matrix has distinct eigenvalues, it can be diagonalised.

Exercise 8.4 If M is an $n \times n$ matrix and λ is a real number such that $M\mathbf{v} = \lambda\mathbf{v}$ for some non-zero vector $\mathbf{v}$, then λ is an eigenvalue of M with corresponding eigenvector $\mathbf{v}$.

Exercise 8.5 We will give this solution in some detail. We have

$$A\mathbf{v} = \begin{pmatrix} 6 & 13 & -8 \\ 2 & 5 & -2 \\ 7 & 17 & -9 \end{pmatrix} \begin{pmatrix} 1 \\ 0 \\ 1 \end{pmatrix} = \begin{pmatrix} -2 \\ 0 \\ -2 \end{pmatrix} = -2 \begin{pmatrix} 1 \\ 0 \\ 1 \end{pmatrix},$$

which shows that $\mathbf{v}$ is an eigenvector of A corresponding to the eigenvalue $\lambda_1 = -2$.

The fact that $T(\mathbf{x}) = A\mathbf{x} = \mathbf{x}$ for some non-zero vector $\mathbf{x}$, tells us that $\mathbf{x}$ is an eigenvector of A corresponding to the eigenvalue $\lambda_2 = 1$.

To find the eigenvector **x**, we solve $(A - I)\mathbf{x} = 0$,

$$\begin{pmatrix} 5 & 13 & -8 \\ 2 & 4 & -2 \\ 7 & 17 & -10 \end{pmatrix} \longrightarrow \begin{pmatrix} 1 & 2 & -1 \\ 5 & 13 & -8 \\ 7 & 17 & -10 \end{pmatrix} \longrightarrow \begin{pmatrix} 1 & 2 & -1 \\ 0 & 3 & -3 \\ 0 & 3 & -3 \end{pmatrix}$$

$$\longrightarrow \begin{pmatrix} 1 & 2 & -1 \\ 0 & 1 & -1 \\ 0 & 0 & 0 \end{pmatrix} \longrightarrow \begin{pmatrix} 1 & 0 & 1 \\ 0 & 1 & -1 \\ 0 & 0 & 0 \end{pmatrix}.$$

So a suitable eigenvector is

$$\mathbf{v}_2 = \begin{pmatrix} -1 \\ 1 \\ 1 \end{pmatrix}.$$

To check this,

$$A\mathbf{v} = \begin{pmatrix} 6 & 13 & -8 \\ 2 & 5 & -2 \\ 7 & 17 & -9 \end{pmatrix} \begin{pmatrix} -1 \\ 1 \\ 1 \end{pmatrix} = \begin{pmatrix} -1 \\ 1 \\ 1 \end{pmatrix}.$$

To diagonalise A, we need to find the remaining eigenvalue and eigenvector. You can do this by finding the characteristic equation $|A - \lambda I| = 0$ and solving for λ. But there is an easier way using $|A|$. Evaluating the determinant – say, by using the cofactor expansion by row 1 – you should find that $|A| = -6$. Since the determinant is the product of the eigenvalues, and since we already have $\lambda_1 = -2$ and $\lambda_2 = 1$, we can deduce that the third eigenvalue is $\lambda_3 = 3$. Then a corresponding eigenvector is obtained from solving $(A - 3I)\mathbf{v} = \mathbf{0}$,

$$A - 3I = \begin{pmatrix} 3 & 13 & -8 \\ 2 & 2 & -2 \\ 7 & 17 & -12 \end{pmatrix} \longrightarrow \cdots \longrightarrow \begin{pmatrix} 1 & 0 & -\frac{1}{2} \\ 0 & 1 & -\frac{1}{2} \\ 0 & 0 & 0 \end{pmatrix}.$$

So we can let

$$\mathbf{v}_3 = \begin{pmatrix} 1 \\ 1 \\ 2 \end{pmatrix}.$$

Then take P to be the matrix whose columns are these three eigenvectors (in any order) and take D to be the diagonal matrix with the corresponding eigenvalues in corresponding columns. For example, if you let

$$P = \begin{pmatrix} 1 & -1 & 1 \\ 0 & 1 & 1 \\ 1 & 1 & 2 \end{pmatrix}, \quad \text{then} \quad D = \begin{pmatrix} -2 & 0 & 0 \\ 0 & 1 & 0 \\ 0 & 0 & 3 \end{pmatrix}.$$

We have already checked the first two eigenvectors, so we should now check the last one, either by calculating AP, or just multiplying,

$$A\mathbf{v}_3 = \begin{pmatrix} 6 & 13 & -8 \\ 2 & 5 & -2 \\ 7 & 17 & -9 \end{pmatrix} \begin{pmatrix} 1 \\ 1 \\ 2 \end{pmatrix} = \begin{pmatrix} 3 \\ 3 \\ 6 \end{pmatrix} = 3\mathbf{v}_3.$$

Then we know that $P^{-1}AP = D$.

The linear transformation $T(\mathbf{x}) = A\mathbf{x}$ can be described as: T is a stretch by a factor of three along the line $\mathbf{x} = t\mathbf{v}_3$, $t \in \mathbb{R}$; a stretch by a factor of two and reversal of direction for any vector on the line $\mathbf{x} = t\mathbf{v}_1$; and T fixes every vector on the line $\mathbf{x} = t\mathbf{v}_2$, $t \in \mathbb{R}$.

Exercise 8.6 We have

$$A\mathbf{x} = \begin{pmatrix} -1 & 1 & 2 \\ -6 & 2 & 6 \\ 0 & 1 & 1 \end{pmatrix} \begin{pmatrix} 1 \\ 1 \\ 1 \end{pmatrix} = \begin{pmatrix} 2 \\ 2 \\ 2 \end{pmatrix} = 2 \begin{pmatrix} 1 \\ 1 \\ 1 \end{pmatrix} = 2\mathbf{x},$$

so $\mathbf{x}$ is an eigenvector with corresponding eigenvalue $\lambda = 2$. The characteristic polynomial of A is $p(\lambda) = -\lambda^3 + 2\lambda^2 + \lambda - 2$. Since $\lambda = 2$ is a root, we know that $(\lambda - 2)$ is a factor. Factorising, we obtain

$$p(\lambda) = (\lambda - 2)(-\lambda^2 + 1) = -(\lambda - 2)(\lambda - 1)(\lambda + 1),$$

so the other eigenvalues are $\lambda = 1, -1$. Corresponding eigenvectors are, respectively, $(1, 0, 1)^T$ and $(0, -2, 1)^T$. We may therefore take

$$P = \begin{pmatrix} 1 & 1 & 0 \\ 0 & 1 & -2 \\ 1 & 1 & 1 \end{pmatrix}, \qquad D = \mathrm{diag}(1, 2, -1).$$

Check that $AP = ((1)\mathbf{v}_1 \ \ 2\mathbf{v}_2 \ \ (-1)\mathbf{v}_3) = PD$.

Exercise 8.7 Expanding the characteristic equation

$$\begin{vmatrix} -\lambda & 0 & -2 \\ 1 & 2 - \lambda & 1 \\ 1 & 0 & 3 - \lambda \end{vmatrix} = 0$$

by the first row, you should find that $(\lambda - 2)$ is a common factor in the two terms, so we can keep things simple, factor this out and not have to grapple with a cubic polynomial. The matrix A does not have three distinct eigenvalues. The eigenvalues turn out to be $\lambda_1 = 2$, with multiplicity 2, and $\lambda_3 = 1$. So we first check that we can find two linearly independent eigenvectors for $\lambda_1 = 2$. We solve $(A - 2I)\mathbf{v} = \mathbf{0}$.

$$A - 2I = \begin{pmatrix} -2 & 0 & -2 \\ 1 & 0 & 1 \\ 1 & 0 & 1 \end{pmatrix} \longrightarrow \begin{pmatrix} 1 & 0 & 1 \\ 0 & 0 & 0 \\ 0 & 0 & 0 \end{pmatrix}.$$

Since the matrix $(A - 2I)$ clearly has rank 1, it is also clear that $\dim(N(A - 2I)) = 2$; the eigenspace has dimension 2. The solutions are

$$\mathbf{x} = s \begin{pmatrix} 0 \\ 1 \\ 0 \end{pmatrix} + t \begin{pmatrix} -1 \\ 0 \\ 1 \end{pmatrix}, \quad s, t \in \mathbb{R}.$$

Let

$$\mathbf{v}_1 = \begin{pmatrix} 0 \\ 1 \\ 0 \end{pmatrix}, \quad \mathbf{v}_2 = \begin{pmatrix} -1 \\ 0 \\ 1 \end{pmatrix}.$$

An eigenvector for $\lambda_3 = 1$ is $\mathbf{v}_3 = (-2, 1, 1)^\mathsf{T}$. These three vectors form a linearly independent set. Therefore, we may take

$$P = \begin{pmatrix} -2 & -1 & 0 \\ 1 & 0 & 1 \\ 1 & 1 & 0 \end{pmatrix}, \quad D = \mathrm{diag}(1, 2, 2).$$

You should check your result by calculating AP.

The eigenspace for $\lambda = 2$ is two-dimensional and has a basis consisting of $(-1, 0, 1)^\mathsf{T}$ and $(0, 1, 0)^\mathsf{T}$. It is a plane in $\mathbb{R}^3$. The eigenspace for $\lambda_3 = 1$ is a line in $\mathbb{R}^3$ with basis $(-2, 1, 1)^\mathsf{T}$.

Exercise 8.8 If 0 is an eigenvalue of A, then, by definition, there is an eigenvector $\mathbf{x}$ corresponding to eigenvalue 0. That means $\mathbf{x} \neq \mathbf{0}$ and that $A\mathbf{x} = 0\mathbf{x} = \mathbf{0}$. So $A\mathbf{x} = \mathbf{0}$ has the non-trivial solution $\mathbf{x}$. Conversely, if there is some non-trivial solution to $A\mathbf{x} = \mathbf{0}$, then we have a non-zero $\mathbf{x}$ with $A\mathbf{x} = 0\mathbf{x}$, which means that 0 is an eigenvalue (and $\mathbf{x}$ a corresponding eigenvector).

Exercise 8.9 Let $\mathbf{v}_1$, $\mathbf{v}_2$ be two eigenvectors of a matrix A corresponding to eigenvalues λ_1 and λ_2 respectively, with $\lambda_1 \neq \lambda_2$. Then $\mathbf{v}_1$, $\mathbf{v}_2$ are linearly independent. To show this, let

$$a_1\mathbf{v}_1 + a_2\mathbf{v}_2 = \mathbf{0}$$

be a linear combination which is equal to the zero vector. If we can show that this equation only has the trivial solution, $a_1 = a_2 = 0$, then the vectors are linearly independent. Multiply this equation through, first by λ_2 and then by A. Since $A\mathbf{v}_1 = \lambda_1\mathbf{v}_1$ and $A\mathbf{v}_2 = \lambda_2\mathbf{v}_2$, we obtain,

$$a_1\lambda_2\mathbf{v}_1 + a_2\lambda_2\mathbf{v}_2 = \mathbf{0}$$

and

$$a_1 A\mathbf{v}_1 + a_2 A\mathbf{v}_2 = a_1\lambda_1\mathbf{v}_1 + a_2\lambda_2\mathbf{v}_2 = \mathbf{0}.$$

Now, subtracting the first equation from the last equation in the line above, we have

$$a_1(\lambda_1 - \lambda_2)\mathbf{v}_1 = \mathbf{0}.$$

But $\mathbf{v}_1 \neq \mathbf{0}$ since it is an eigenvector, and $(\lambda_1 - \lambda_2) \neq 0$ since $\lambda_1 \neq \lambda_2$. Therefore, $a_1 = 0$. Returning to the equation $a_1\mathbf{v}_1 + a_2\mathbf{v}_2 = \mathbf{0}$, we conclude that $a_2\mathbf{v}_2 = \mathbf{0}$. But again, $\mathbf{v}_2 \neq \mathbf{0}$ since it is an eigenvector, so $a_2 = 0$ and we are done.

To use an inductive argument, we now assume that the statement is true for $n - 1$ eigenvectors and show that this implies it is true for n eigenvectors. In this way, $n = 2 \implies n = 3 \implies n = 4 \implies \cdots$ and so on.

So assume the statement that 'eigenvectors corresponding to $n - 1$ different eigenvalues are linearly independent' is true, and assume we have n eigenvectors, $\mathbf{v}_i$, corresponding to n different eigenvalues, λ_i. Let

$$a_1\mathbf{v}_1 + a_2\mathbf{v}_2 + \cdots + a_{n-1}\mathbf{v}_{n-1} + a_n\mathbf{v}_n = \mathbf{0}$$

be a linear combination which is equal to the zero vector. Multiply this equation through, first by λ_n and then by A. Since $A\mathbf{v}_i = \lambda_i\mathbf{v}_i$ for each i, we have

$$a_1\lambda_n\mathbf{v}_1 + a_2\lambda_n\mathbf{v}_2 + \cdots + a_{n-1}\lambda_n\mathbf{v}_{n-1} + a_n\lambda_n\mathbf{v}_n = \mathbf{0}$$

and

$$a_1 A\mathbf{v}_1 + a_2 A\mathbf{v}_2 + \cdots + a_{n-1} A\mathbf{v}_{n-1} + a_n A\mathbf{v}_n$$
$$= a_1\lambda_1\mathbf{v}_1 + a_2\lambda_2\mathbf{v}_2 + \cdots + a_{n-1}\lambda_{n-1}\mathbf{v}_{n-1} + a_n\lambda_n\mathbf{v}_n = \mathbf{0}.$$

Subtracting the the first equation from the last equation in the lines above, we have

$$a_1(\lambda_1 - \lambda_n)\mathbf{v}_1 + a_2(\lambda_2 - \lambda_n)\mathbf{v}_2 + \cdots + a_{n-1}(\lambda_{n-1} - \lambda_n)\mathbf{v}_{n-1} = \mathbf{0}.$$

But we have assumed that $n - 1$ eigenvectors corresponding to distinct eigenvalues are linearly independent, so all the coefficients are zero. Since, also, $(\lambda_i - \lambda_n) \neq 0$ for $i = 1, \ldots, n - 1$, we can conclude that $a_1 = a_2 = \cdots = a_{n-1} = 0$. This leaves us with $a_n\mathbf{v}_n = 0$, from which we conclude that $a_n = 0$. Therefore, the n eigenvectors are linearly independent.

Exercise 8.10 Since A can be diagonalised, we have $P^{-1}AP = D$ for some P, where $D = \text{diag}(\lambda_1, \ldots, \lambda_n)$, these entries being the eigenvalues of A. It is given that all $\lambda_i \geq 0$. We have $A = PDP^{-1}$.

If $B^2 = A$, we must have

$$D = P^{-1}AP = P^{-1}B^2P = P^{-1}BPP^{-1}BP = (P^{-1}BP)^2.$$

Therefore, let

$$B = P\text{diag}(\sqrt{\lambda_1}, \sqrt{\lambda_2}, \ldots, \sqrt{\lambda_n})P^{-1}.$$

Then reversing the above steps,

$$B^2 = P\,\text{diag}(\sqrt{\lambda_1}, \sqrt{\lambda_2}, \ldots, \sqrt{\lambda_n})P^{-1}P\text{diag}(\sqrt{\lambda_1}, \sqrt{\lambda_2}, \ldots, \sqrt{\lambda_n})P^{-1}$$
$$= P\,\text{diag}(\sqrt{\lambda_1}^2, \sqrt{\lambda_2}^2, \ldots, \sqrt{\lambda_n}^2)P^{-1}$$
$$= PDP^{-1} = A,$$

and we are done.

Chapter 9 exercises

Exercise 9.1 You diagonalised this matrix in Exercise 8.1. If

$$P = \begin{pmatrix} -1 & -5 \\ 1 & 1 \end{pmatrix}, \quad \text{and} \quad D = \begin{pmatrix} -1 & 0 \\ 0 & 3 \end{pmatrix},$$

then $P^{-1}AP = D$. Therefore, the matrix A^n is given by

$$A^n = PD^nP^{-1} = \begin{pmatrix} -1 & -5 \\ 1 & 1 \end{pmatrix}\begin{pmatrix} (-1)^n & 0 \\ 0 & 3^n \end{pmatrix}\frac{1}{4}\begin{pmatrix} 1 & 5 \\ -1 & -1 \end{pmatrix}$$
$$= \frac{1}{4}\begin{pmatrix} -(-1)^n + 5(3^n) & -5(-1)^n + 5(3^n) \\ (-1)^n - 3^n & 5(-1)^n - 3^n \end{pmatrix}.$$

Incidentally, since the matrix A contains only integer entries, any power of A will also contain only integer entries. Therefore, each of the entries in the expression for A^n is an integer; in particular,

$$\frac{(-1)^n - 3^n}{4}$$

is an integer. (Try this for some number n, say $n = 5$.)

Exercise 9.2 We solve this using matrix powers. We could, of course, use a change of variable instead. Notice that the system can be written as

$$\mathbf{x}_{t+1} = \begin{pmatrix} 1 & 4 \\ \frac{1}{2} & 0 \end{pmatrix} \mathbf{x}_t, \quad \text{where } \mathbf{x}_t = \begin{pmatrix} x_t \\ y_t \end{pmatrix}.$$

This is $\mathbf{x}_{t+1} = A\mathbf{x}_t$, where A is the matrix whose powers we calculated in Example 9.4. The solution (using the result from Example 9.4) is

$$\mathbf{x}_t = A^t \mathbf{x}_0$$

$$= \frac{1}{6} \begin{pmatrix} 2(-1)^t + 4(2^t) & -8(-1)^t + 8(2^t) \\ -(-1)^t + 2^t & 4(-1)^t + 2(2^t) \end{pmatrix} \begin{pmatrix} 1000 \\ 1000 \end{pmatrix}$$

$$= \begin{pmatrix} -1000(-1)^t + 2000(2^t) \\ 500(-1)^t + 500(2^t) \end{pmatrix}.$$

That is,

$$x_t = -1000(-1)^t + 2000(2^t), \quad y_t = 500(-1)^t + 500(2^t).$$

Exercise 9.3 The system of difference equations can be expressed as $\mathbf{x}_{t+1} = A\mathbf{x}_t$, where

$$A = \begin{pmatrix} 7 & 0 & -3 \\ 1 & 6 & 5 \\ 5 & 0 & -1 \end{pmatrix}, \quad \mathbf{x}_t = \begin{pmatrix} x_t \\ y_t \\ z_t \end{pmatrix}.$$

You need to diagonalise A. Expanding the determinant by column two,

$$|A - \lambda I| = \begin{vmatrix} 7 - \lambda & 0 & -3 \\ 1 & 6 - \lambda & 5 \\ 5 & 0 & -1 - \lambda \end{vmatrix} = (6 - \lambda)(\lambda^2 - 6\lambda + 8),$$

so the eigenvalues are $\lambda = 6, 4, 2$. Next find an eigenvector for each eigenvalue. You should find that if, for example, you set

$$P = \begin{pmatrix} 1 & 0 & 3 \\ -3 & 1 & -7 \\ 1 & 0 & 5 \end{pmatrix} \text{ and } D = \begin{pmatrix} 4 & 0 & 0 \\ 0 & 6 & 0 \\ 0 & 0 & 2 \end{pmatrix}, \text{ then } P^{-1}AP = D.$$

Then the solution to the system of difference equations is given by $\mathbf{x}_t = PD^t P^{-1} \mathbf{x}_0$, so it only remains to find P^{-1} and then multiply the matrices. But before that, you should check that $AP = PD$ so that you know that your eigenvalues and eigenvectors are correct!

$$AP = \begin{pmatrix} 7 & 0 & -3 \\ 1 & 6 & 5 \\ 5 & 0 & -1 \end{pmatrix} \begin{pmatrix} 1 & 0 & 3 \\ -3 & 1 & -7 \\ 1 & 0 & 5 \end{pmatrix}$$

$$= \begin{pmatrix} 4 & 0 & 6 \\ -12 & 6 & -14 \\ 4 & 0 & 10 \end{pmatrix} = (4\mathbf{v}_1 \quad 6\mathbf{v}_2 \quad 2\mathbf{v}_3).$$

Since the eigenvalues are distinct, you know that your eigenvectors are linearly independent (as long as you have chosen one eigenvector for each eigenvalue). Then P will be invertible, and using either row operations or the cofactor method,

$$P^{-1} = \frac{1}{2} \begin{pmatrix} 5 & 0 & -3 \\ 8 & 2 & -2 \\ -1 & 0 & 1 \end{pmatrix},$$

so $P^{-1}\mathbf{x}_0 = \frac{1}{2} \begin{pmatrix} 5 & 0 & -3 \\ 8 & 2 & -2 \\ -1 & 0 & 1 \end{pmatrix} \begin{pmatrix} -1 \\ 2 \\ 1 \end{pmatrix} = \begin{pmatrix} -4 \\ -3 \\ 1 \end{pmatrix}.$

Then $\mathbf{x}_t = PD^t P^{-1}\mathbf{x}_0$,

$$\begin{pmatrix} x_t \\ y_t \\ z_t \end{pmatrix} = \begin{pmatrix} 1 & 0 & 3 \\ -3 & 1 & -7 \\ 1 & 0 & 5 \end{pmatrix} \begin{pmatrix} 4^t & 0 & 0 \\ 0 & 6^t & 0 \\ 0 & 0 & 2^t \end{pmatrix} \begin{pmatrix} -4 \\ -3 \\ 1 \end{pmatrix}$$

$$= \begin{pmatrix} 1 & 0 & 3 \\ -3 & 1 & -7 \\ 1 & 0 & 5 \end{pmatrix} \begin{pmatrix} -4(4^t) \\ -3(6^t) \\ 1(2^t) \end{pmatrix}.$$

Therefore, the required sequences are

$$x_t = -4(4^t) + 3(2^t)$$
$$y_t = 12(4^t) - 3(6^t) - 7(2^t) \qquad \text{for } t \in \mathbb{Z}, \ t \geq 0.$$
$$z_t = -4(4^t) + 5(2^t)$$

Notice that you will get the same answer even if you used a different (correct) matrix P and corresponding matrix D.

To check your result, first see that the initial conditions are satisfied. Substituting $t = 0$ into the equations, we get $x_0 = -1, y_0 = 2, z_0 = 1$ as required. Next look at $\mathbf{x}_1$. Using the original difference equations,

$$\mathbf{x}_1 = A\mathbf{x}_0 = \begin{pmatrix} 7 & 0 & -3 \\ 1 & 6 & 5 \\ 5 & 0 & -1 \end{pmatrix} \begin{pmatrix} -1 \\ 2 \\ 1 \end{pmatrix} = \begin{pmatrix} -10 \\ 16 \\ -6 \end{pmatrix}.$$

From the solutions,

$$x_1 = -4(4) + 3(2) = -10, \quad y_1 = 12(4) - 3(6) - 7(2) = 16,$$
$$z_1 = -4(4) + 5(2) = -6.$$

Exercise 9.4 Eigenvectors are given, so there is no need to determine the characteristic polynomial to find the eigenvalues. Simply multiply A times the given eigenvectors in turn (or you can do all three at once by multiplying AP). For example,

$$A\mathbf{v}_1 = A \begin{pmatrix} 1 \\ -1 \\ 1 \end{pmatrix} = \begin{pmatrix} -3 \\ 3 \\ -3 \end{pmatrix} = -3 \begin{pmatrix} 1 \\ -1 \\ 1 \end{pmatrix},$$

so -3 is an eigenvalue and $\mathbf{v}_1$ is a corresponding eigenvector. The other two are eigenvectors for eigenvalue 3. Since $\mathbf{v}_2$ and $\mathbf{v}_3$ are clearly linearly independent (neither being a scalar multiple of the other), if

$$P = \begin{pmatrix} 1 & -3 & -1 \\ -1 & 0 & 1 \\ 1 & 1 & 0 \end{pmatrix},$$

then

$$P^{-1}AP = \text{diag}(-3, 3, 3) = D.$$

The system of difference equations is $\mathbf{x}_{t+1} = A\mathbf{x}_t$. Let $\mathbf{u}_t = (u_t, v_t, w_t)^{\mathrm{T}}$ be given by $\mathbf{u}_t = P^{-1}\mathbf{x}_t$. Then the system is equivalent to $\mathbf{u}_{t+1} = D\mathbf{u}_t$, which is

$$u_{t+1} = -3u_t, \quad v_{t+1} = 3v_t, \quad w_{t+1} = 3w_t.$$

This has solutions

$$u_t = (-3)^t u_0, \quad v_t = 3^t v_0, \quad w_t = 3^t w_0, .$$

We have to find u_0, v_0, w_0. Now $\mathbf{u}_0 = (u_0, v_0, w_0)^{\mathrm{T}} = P^{-1}\mathbf{x}_0$, and (as can be determined by the usual methods),

$$P^{-1} = \begin{pmatrix} 1/3 & 1/3 & 1 \\ -1/3 & -1/3 & 0 \\ 1/3 & 4/3 & 1 \end{pmatrix},$$

so

$$\mathbf{z}_0 = P^{-1}\mathbf{x}_0 = \begin{pmatrix} 1/3 & 1/3 & 1 \\ -1/3 & -1/3 & 0 \\ 1/3 & 4/3 & 1 \end{pmatrix} \begin{pmatrix} 1 \\ 1 \\ 0 \end{pmatrix} = \begin{pmatrix} 2/3 \\ -2/3 \\ 5/3 \end{pmatrix}.$$

The solution $\mathbf{x}_t$ is therefore

$$\mathbf{x}_t = P\mathbf{z}_t = \begin{pmatrix} 1 & -3 & -1 \\ -1 & 0 & 1 \\ 1 & 1 & 0 \end{pmatrix} \begin{pmatrix} (2/3)(-3)^t \\ (-2/3)3^t \\ (5/3)3^t \end{pmatrix}$$

$$= \begin{pmatrix} (2/3)(-3)^t + (1/3)3^t \\ -(2/3)(-3)^t + (5/3)3^t \\ (2/3)(-3)^t - (2/3)3^t \end{pmatrix}.$$

The term $x_5 = (2/3)(-3)^5 + (1/3)3^5 = 2(-81) + (81) = -81$.

Exercise 9.5 This is a Markov process, as it consists of a total population distributed into two states, and the matrix A satisfies the criteria to be a transition matrix: (1) the entries are positive and (2) the sum of the entries in each column is 1.

Interpreting the system, each year 40% of those living by the sea move to the oasis (60% remain) and 20% of those living in the oasis move to the sea.

To solve the system, we need to diagonalise the matrix A. First find the eigenvalues:

$$|A - \lambda I| = \begin{vmatrix} 0.6 - \lambda & 0.2 \\ 0.4 & 0.8 - \lambda \end{vmatrix}$$
$$= 0.48 - 1.4\lambda + \lambda^2 - 0.08$$
$$= \lambda^2 - 1.4\lambda + 0.4$$
$$= (\lambda - 1)(\lambda - 0.4) = 0,$$

so $\lambda = 1$ and $\lambda = 0.4$ are the eigenvalues.

We find corresponding eigenvectors by solving $(A - \lambda I)\mathbf{v} = \mathbf{0}$:

$$\lambda_1 = 1: \quad A - I = \begin{pmatrix} -0.4 & 0.2 \\ 0.4 & -0.2 \end{pmatrix} \rightarrow \begin{pmatrix} 1 & -\frac{1}{2} \\ 0 & 0 \end{pmatrix} \Longrightarrow \mathbf{v}_1 = \begin{pmatrix} 1 \\ 2 \end{pmatrix}$$

$$\lambda_2 = 0.4: \quad A - 0.4I = \begin{pmatrix} 0.2 & 0.2 \\ 0.4 & 0.4 \end{pmatrix} \rightarrow \begin{pmatrix} 1 & 1 \\ 0 & 0 \end{pmatrix} \Longrightarrow \mathbf{v}_2 = \begin{pmatrix} -1 \\ 1 \end{pmatrix}.$$

Then $\mathbf{x}_t = PD^t P^{-1}\mathbf{x}_0$. The initial distribution is $\mathbf{x}_0 = (0.5, 0.5)^{\mathrm{T}}$,

$$\begin{pmatrix} x_t \\ y_t \end{pmatrix} = \begin{pmatrix} 1 & -1 \\ 2 & 1 \end{pmatrix} \begin{pmatrix} 1^t & 0 \\ 0 & (0.4)^t \end{pmatrix} \frac{1}{3} \begin{pmatrix} 1 & 1 \\ -2 & 1 \end{pmatrix} \begin{pmatrix} 0.5 \\ 0.5 \end{pmatrix}$$

$$= \begin{pmatrix} 1 & -1 \\ 2 & 1 \end{pmatrix} \begin{pmatrix} 1^t & 0 \\ 0 & (0.4)^t \end{pmatrix} \begin{pmatrix} \frac{1}{3} \\ -\frac{1}{6} \end{pmatrix}$$

$$= \frac{1}{3} \begin{pmatrix} 1 \\ 2 \end{pmatrix} - \frac{1}{6}(0.4)^t \begin{pmatrix} -1 \\ 1 \end{pmatrix}.$$

The expressions for x_t and y_t are

$$x_t = \frac{1}{3} + \frac{1}{6}(0.4)^t, \quad y_t = \frac{2}{3} - \frac{1}{6}(0.4)^t.$$

As $t \to \infty$, $\mathbf{x}_t \to (1/3, 2/3)^{\mathrm{T}}$. In terms of the original total population of 210 inhabitants, we multiply $\mathbf{x}_t$ by 210, so the long-term population distribution is 70 inhabitants living by the sea and 140 inhabitants living in the oasis.

Exercise 9.6 (a) The matrix B is a scalar multiple of A, $B = 10\,A$.

Let λ be an eigenvalue of B with corresponding eigenvector $\mathbf{v}$, so that $B\mathbf{v} = \lambda\mathbf{v}$. Then substituting $10A$ for B, we have $10A\mathbf{v} = \lambda\mathbf{v}$ and so

$$Av = \frac{\lambda}{10}v.$$

Therefore, A and B have the same eigenvectors, v, and $\lambda/10$ is the corresponding eigenvalue of A.

The matrix A is the transition matrix of a Markov chain because:

1. All the entries are non-negative ($a_{ij} \geq 0$).
2. The sum of the entries in each column is 1.

Since $\lambda = 1$ is an eigenvalue of a Markov chain, we can deduce that $10\lambda = 10$ is an eigenvalue of B.

(b) To find an eigenvector for $\lambda = 10$, solve $(B - 10I)x = 0$ by reducing $B - 10I$:

$$\begin{pmatrix} -3 & 2 & 2 \\ 0 & -8 & 4 \\ 3 & 6 & -6 \end{pmatrix} \longrightarrow \begin{pmatrix} 1 & 2 & -2 \\ -3 & 2 & 2 \\ 0 & -2 & 1 \end{pmatrix} \longrightarrow \cdots \longrightarrow \begin{pmatrix} 1 & 0 & -1 \\ 0 & 1 & -\frac{1}{2} \\ 0 & 0 & 0 \end{pmatrix}.$$

So an eigenvector for $\lambda = 10$ is $v_1 = \begin{pmatrix} 2 \\ 1 \\ 2 \end{pmatrix}$.

To find the other eigenvalues, we find the characteristic equation. Expanding the determinant by the first column,

$$|B - \lambda I| = \begin{vmatrix} 7 - \lambda & 2 & 2 \\ 0 & 2 - \lambda & 4 \\ 3 & 6 & 4 - \lambda \end{vmatrix}$$

$$= (7 - \lambda)(\lambda^2 - 6\lambda - 16) + 3(2\lambda + 4) = 0.$$

Factoring the quadratic, there is a common factor of $\lambda + 2$ in the two terms, which can be factored out, avoiding a cubic equation. We have

$$|B - \lambda I| = -(\lambda + 2)[(\lambda - 7)(\lambda - 8) - 6]$$
$$= -(\lambda + 2)(\lambda^2 - 15\lambda + 50)$$
$$= -(\lambda + 2)(\lambda - 10)(\lambda - 5).$$

So the eigenvalues are $\lambda = 10, \ 5, \ -2$.

We then find the corresponding eigenvectors. Solving $(B - 5I)v = 0$, by reducing $B - 5I$,

$$\begin{pmatrix} 2 & 2 & 2 \\ 0 & -3 & 4 \\ 3 & 6 & -1 \end{pmatrix} \longrightarrow \cdots \longrightarrow \begin{pmatrix} 1 & 1 & 1 \\ 0 & 3 & -4 \\ 0 & 0 & 0 \end{pmatrix}$$

$$\longrightarrow \cdots \longrightarrow \begin{pmatrix} 1 & 0 & \frac{7}{3} \\ 0 & 1 & -\frac{4}{3} \\ 0 & 0 & 0 \end{pmatrix}.$$

So an eigenvector for $\lambda = 5$ is $\mathbf{v}_2 = \begin{pmatrix} -7 \\ 4 \\ 3 \end{pmatrix}$.

For $\lambda = -2$, we have

$$(B + 2I) = \begin{pmatrix} 9 & 2 & 2 \\ 0 & 4 & 4 \\ 3 & 6 & 6 \end{pmatrix} \longrightarrow \cdots \longrightarrow \begin{pmatrix} 1 & 2 & 2 \\ 0 & 1 & 1 \\ 0 & 0 & 0 \end{pmatrix} \longrightarrow \begin{pmatrix} 1 & 0 & 0 \\ 0 & 1 & 1 \\ 0 & 0 & 0 \end{pmatrix}.$$

So an eigenvector for $\lambda = -2$ is $\mathbf{v}_3 = \begin{pmatrix} 0 \\ -1 \\ 1 \end{pmatrix}$.

If

$$P = \begin{pmatrix} 2 & -7 & 0 \\ 1 & 4 & -1 \\ 2 & 3 & 1 \end{pmatrix} \quad \text{and} \quad D = \begin{pmatrix} 10 & 0 & 0 \\ 0 & 5 & 0 \\ 0 & 0 & -2 \end{pmatrix},$$

then $P^{-1}BP = D$. The eigenvectors and eigenvalues must be listed in corresponding columns.

To check,

$$BP = \begin{pmatrix} 7 & 2 & 2 \\ 0 & 2 & 4 \\ 3 & 6 & 4 \end{pmatrix} \begin{pmatrix} 2 & -7 & 0 \\ 1 & 4 & -1 \\ 2 & 3 & 1 \end{pmatrix} = \begin{pmatrix} 20 & -35 & 0 \\ 10 & 20 & 2 \\ 20 & 15 & -2 \end{pmatrix} = PD.$$

Why are you being asked to *check*? So that you know you do have the correct eigenvalues and eigenvectors. This gives you an opportunity to look for and correct any minor mistakes you may have made.

By part (a), the eigenvalues and corresponding eigenvectors of A are $\lambda = 1$ with eigenvector $\mathbf{v}_1$, $\lambda = 0.5$ with corresponding eigenvector $\mathbf{v}_2$ and $\lambda = -0.2$ with corresponding eigenvector $\mathbf{v}_3$.

(c) The long-term distribution of a Markov chain is given by the eigenvector for $\lambda = 1$. Therefore, the distribution is proportional to the entries of the vector $\mathbf{v}_1$:

$$\frac{1}{5} \begin{pmatrix} 2 \\ 1 \\ 2 \end{pmatrix} 1000 = \begin{pmatrix} 400 \\ 200 \\ 400 \end{pmatrix}.$$

That is, 400 will be employed full-time, 200 will be employed part-time and 400 will remain unemployed. So a total of 600 will be employed.

Notice that you did not need to use the initial conditions and you did not need to find the solution to $\mathbf{x}_t = A\mathbf{x}_{t-1}$ to answer this question. This would have been a perfectly acceptable method, but one which would take much more time. You only needed to know that since $(0.5)^t \to 0$ and $(-0.2)^t \to 0$ as $t \to \infty$, the eigenvector corresponding to $\lambda = 1$ will give the long-term distribution. It must be a distribution vector;

that is, the components of the column vector must sum to 1, which is why we needed the factor $1/5$. When multiplied by the total population of $1,000$, this gives the distribution of workers.

Exercise 9.7 We can express the system of differential equations in matrix form as $y' = Ay$, where

$$A = \begin{pmatrix} 4 & 5 \\ -1 & -2 \end{pmatrix}, \qquad y = \begin{pmatrix} y_1(t) \\ y_2(t) \end{pmatrix}.$$

This is the matrix you diagonalised in Exercise 8.1 (and used in Exercise 9.1). If

$$P = \begin{pmatrix} -1 & -5 \\ 1 & 1 \end{pmatrix} \quad \text{and} \quad D = \begin{pmatrix} -1 & 0 \\ 0 & 3 \end{pmatrix}, \quad \text{then} \quad P^{-1}AP = D.$$

To solve $y' = Ay$, you set $y = Pz$ to define new functions $z = (z_1(t), z_2(t))^T$. Then $y' = (Pz)' = Pz'$ and $Ay = A(Pz) = APz$, so that

$$y' = Ay \iff Pz' = APz \iff z' = P^{-1}APz = Dz.$$

The system $z' = Dz$ is uncoupled; the equations are

$$z_1' = -z_1. \qquad z_2' = 3z_2$$

with solutions

$$z_1(t) = z_1(0)e^{-t}, \qquad z_2(t) = z_2(0)e^{3t}.$$

To find $z_1(0)$, $z_2(0)$, we use $z = P^{-1}y$. Since $y_1(0) = 2$, $y_2(0) = 6$, we have

$$\begin{pmatrix} z_1(0) \\ z_2(0) \end{pmatrix} = \frac{1}{4} \begin{pmatrix} 1 & 5 \\ -1 & -1 \end{pmatrix} \begin{pmatrix} 2 \\ 6 \end{pmatrix} = \begin{pmatrix} 8 \\ -2 \end{pmatrix}.$$

So the solution of the original system is

$$y = Pz = \begin{pmatrix} -1 & -5 \\ 1 & 1 \end{pmatrix} \begin{pmatrix} 8e^{-t} \\ -2e^{3t} \end{pmatrix};$$

that is,

$$y_1(t) = -8e^{-t} + 10e^{3t}$$
$$y_2(t) = 8e^{-t} - 2e^{3t}.$$

To check the solution, we first note that the initial conditions are satisfied by substituting $t = 0$ into the equations,

$$y_1(0) = -8(1) + 10(1) = 2 \quad \text{and} \quad y_2(0) = 8(1) - 2(1) = 6.$$

Next check that you obtain the same values for $\mathbf{y}'(0)$ from both the original equations and the derivatives of the solutions. From the original system,

$$\begin{pmatrix} y_1'(0) \\ y_2'(0) \end{pmatrix} = \begin{pmatrix} 4 & 5 \\ -1 & -2 \end{pmatrix} \begin{pmatrix} 2 \\ 6 \end{pmatrix} = \begin{pmatrix} 38 \\ -14 \end{pmatrix}.$$

Differentiating the solution functions:

$$\begin{aligned} y_1'(t) &= 8e^{-t} + 30e^{3t} \\ y_2'(t) &= -8e^{-t} - 6e^{3t} \end{aligned} \implies \begin{cases} y_1'(0) = 38 \\ y_2'(0) = -14 \end{cases}.$$

Exercise 9.8 This system of differential equations can be expressed as $\mathbf{y}' = A\mathbf{y}$, where

$$A = \begin{pmatrix} -1 & 1 & 2 \\ -6 & 2 & 6 \\ 0 & 1 & 1 \end{pmatrix}, \qquad \mathbf{y} = \begin{pmatrix} y_1 \\ y_2 \\ y_3 \end{pmatrix}.$$

You diagonalised this matrix in Exercise 8.6. Using this result, if

$$P = \begin{pmatrix} 1 & 1 & 0 \\ 0 & 1 & -2 \\ 1 & 1 & 1 \end{pmatrix} \quad \text{and} \quad D = \begin{pmatrix} 1 & 0 & 0 \\ 0 & 2 & 0 \\ 0 & 0 & -1 \end{pmatrix},$$

then $P^{-1}AP = D$. To find the general solution of the system of differential equations, we define new functions $\mathbf{z} = (z_1(t), z_2(t), z_3(t))^{\mathrm{T}}$ by setting $\mathbf{y} = P\mathbf{z}$. Then substituting into $\mathbf{y}' = A\mathbf{y}$, we have

$$\mathbf{y}' = (P\mathbf{z})' = P\mathbf{z}' = A\mathbf{y} = A(P\mathbf{z}) = AP\mathbf{z}$$

and hence

$$\mathbf{z}' = P^{-1}AP\mathbf{z} = D\mathbf{z}.$$

The general solution of $\mathbf{z}' = D\mathbf{z}$ is

$$\begin{pmatrix} z_1 \\ z_2 \\ z_3 \end{pmatrix} = \begin{pmatrix} \alpha e^t \\ \beta e^{2t} \\ \gamma e^{-t} \end{pmatrix},$$

so the general solution of the original system is given by

$$\mathbf{y} = \begin{pmatrix} y_1 \\ y_2 \\ y_3 \end{pmatrix} = P\mathbf{z} = \begin{pmatrix} 1 & 1 & 0 \\ 0 & 1 & -2 \\ 1 & 1 & 1 \end{pmatrix} \begin{pmatrix} \alpha e^t \\ \beta e^{2t} \\ \gamma e^{-t} \end{pmatrix};$$

that is,

$$y_1(t) = \alpha e^t + \beta e^{2t}$$
$$y_2(t) = \beta e^{2t} - 2\gamma e^{-t}$$
$$y_3(t) = \alpha e^t + \beta e^{2t} + \gamma e^{-t}$$

for arbitrary constants $\alpha, \beta, \gamma \in \mathbb{R}$.

Exercise 9.9 The system of differential equations is $\mathbf{y}' = A\mathbf{y}$, where A is the matrix

$$A = \begin{pmatrix} 4 & 0 & 4 \\ 0 & 4 & 4 \\ 4 & 4 & 8 \end{pmatrix}.$$

This is the matrix we used in the example of Section 9.2.5 (and which we diagonalised in Example 8.23), so we have already done most of the work for this solution. Using the same matrices P and D as we did there, we set $\mathbf{y} = P\mathbf{z}$ to define new functions $\mathbf{z} = (z_1(t), z_2(t), z_3(t))^{\mathrm{T}}$, and then find the solutions to $\mathbf{z}' = D\mathbf{z}$. We have

$$\begin{pmatrix} z_1' \\ z_2' \\ z_3' \end{pmatrix} = \begin{pmatrix} 4 & 0 & 0 \\ 0 & 0 & 0 \\ 0 & 0 & 12 \end{pmatrix} \begin{pmatrix} z_1 \\ z_2 \\ z_3 \end{pmatrix}$$

with solutions:

$$z_1(t) = z_1(0)e^{4t}, \quad z_2(t) = z_2(0)e^{0t} = z_2(0), \quad z_3(t) = z_3(0)e^{12t}.$$

Since the initial conditions are essentially the same, $\mathbf{z}(0) = P^{-1}\mathbf{y}(0)$, so

$$\begin{pmatrix} z_1(0) \\ z_2(0) \\ z_3(0) \end{pmatrix} = \frac{1}{6} \begin{pmatrix} -3 & 3 & 0 \\ -2 & -2 & 2 \\ 1 & 1 & 2 \end{pmatrix} \begin{pmatrix} 6 \\ 12 \\ 12 \end{pmatrix} = \begin{pmatrix} 3 \\ -2 \\ 7 \end{pmatrix}$$

and the solutions are given by

$$\mathbf{y} = \begin{pmatrix} y_1 \\ y_2 \\ y_3 \end{pmatrix} = P\mathbf{z} = \begin{pmatrix} -1 & -1 & 1 \\ 1 & -1 & 1 \\ 0 & 1 & 2 \end{pmatrix} \begin{pmatrix} 3e^{4t} \\ -2 \\ 7e^{12t} \end{pmatrix}, \qquad t \in \mathbb{R};$$

that is,

$$y_1(t) = -3e^{4t} + 2 + 7e^{12t}$$
$$y_2(t) = 3e^{4t} + 2 + 7e^{12t}$$
$$y_3(t) = -2 + 14e^{12t}.$$

Notice that the eigenvalue $\lambda = 0$ causes no problems here. You can check the result using $\mathbf{y}(0)$ and $\mathbf{y}'(0)$.

Exercise 9.10 To answer the first question, put A into reduced row echelon form,

$$A = \begin{pmatrix} 5 & -8 & -4 \\ 3 & -5 & -3 \\ -1 & 2 & 2 \end{pmatrix} \longrightarrow \cdots \longrightarrow \begin{pmatrix} 1 & 0 & 4 \\ 0 & 1 & 3 \\ 0 & 0 & 0 \end{pmatrix}.$$

Setting the non-leading variable z to be equal to t, a general solution of the system of equations $A\mathbf{x} = \mathbf{0}$ is

$$\mathbf{x} = \begin{pmatrix} -4t \\ -3t \\ t \end{pmatrix} = t \begin{pmatrix} -4 \\ -3 \\ 1 \end{pmatrix} = t\mathbf{v}_3, \qquad t \in \mathbb{R}.$$

So a basis of $N(A)$ is $\{\mathbf{v}_3\}$.

To show that $\mathbf{v}_1$ is an eigenvector of A,

$$A\mathbf{v}_1 = \begin{pmatrix} 5 & -8 & -4 \\ 3 & -5 & -3 \\ -1 & 2 & 2 \end{pmatrix} \begin{pmatrix} 2 \\ 1 \\ 0 \end{pmatrix} = \begin{pmatrix} 2 \\ 1 \\ 0 \end{pmatrix} = \mathbf{v}_1,$$

so $\mathbf{v}_1$ is an eigenvector with corresponding eigenvalue $\lambda = 1$.

Next find all the eigenvectors of A which correspond to $\lambda = 1$ by solving $(A - I)\mathbf{v} = \mathbf{0}$:

$$A - I = \begin{pmatrix} 4 & -8 & -4 \\ 3 & -6 & -3 \\ -1 & 2 & 1 \end{pmatrix} \longrightarrow \begin{pmatrix} 1 & -2 & -1 \\ 0 & 0 & 0 \\ 0 & 0 & 0 \end{pmatrix},$$

with solution

$$\begin{pmatrix} x \\ y \\ z \end{pmatrix} = \begin{pmatrix} 2s + t \\ s \\ t \end{pmatrix} = s \begin{pmatrix} 2 \\ 1 \\ 0 \end{pmatrix} + t \begin{pmatrix} 1 \\ 0 \\ 1 \end{pmatrix} = s\mathbf{v}_1 + t\mathbf{v}_2, \qquad s, t \in \mathbb{R}.$$

Therefore, the matrix A has an eigenvalue $\lambda = 1$ of multiplicity 2 with two linearly independent eigenvectors. Since $A\mathbf{x} = \mathbf{0}$ has a non-trivial solution, we know that $\lambda = 0$ is the third eigenvalue of A, with corresponding eigenvector $\mathbf{v}_3$. Then an invertible matrix P and a diagonal matrix D such that $P^{-1}AP = D$ are given by

$$P = \begin{pmatrix} 2 & 1 & -4 \\ 1 & 0 & -3 \\ 0 & 1 & 1 \end{pmatrix}, \qquad D = \begin{pmatrix} 1 & 0 & 0 \\ 0 & 1 & 0 \\ 0 & 0 & 0 \end{pmatrix}.$$

You should now check that the eigenvectors are correct by showing $AP = PD$.

Using this diagonalisation, $A = PDP^{-1}$, so that $A^n = PD^nP^{-1}$. But $D^n = D$ for $n \geq 1$, since the entries on the diagonal are either 1 or 0.

Therefore,

$$A^n = PD^nP^{-1} = PDP^{-1} = A.$$

A matrix with the property that $A^n = A$ for all $n \geq 1$ is said to be *idempotent* (meaning the same for all powers). We shall see more about these in Chapter 12.

The solution to the system of difference equations given by $x_{t+1} = Ax_t$ is $x_t = A^t x_0$. So for all $t \geq 1$, this is just $x_t = Ax_0$. Given the initial conditions, we have

$$\mathbf{x}_t = A\mathbf{x}_0 = \begin{pmatrix} 5 & -8 & -4 \\ 3 & -5 & -3 \\ -1 & 2 & 2 \end{pmatrix} \begin{pmatrix} 1 \\ 1 \\ 1 \end{pmatrix} = \begin{pmatrix} -7 \\ -5 \\ 3 \end{pmatrix}, \qquad t \geq 1.$$

Therefore, the sequences are:

$$x_t : \{1, -7, -7, -7, \ldots\}, \qquad y_t : \{1, -5, -5, -5, \ldots\},$$
$$z_t : \{1, 3, 3, 3, \ldots\}.$$

Chapter 10 exercises

Exercise 10.1 Property (iii) of the definition of inner product follows from the fact that

$$\langle A, A \rangle = \sum_{i=1}^{m} \sum_{j=1}^{n} a_{ij}^2 \geq 0$$

is the sum of positive numbers and this sum equals 0 if and only if for every i and every j, $a_{ij} = 0$, which means that A is the zero matrix, which in this vector space is the zero vector. Property (i) is easy to verify, as also is (ii).

Exercise 10.2 We have:

$$\|\mathbf{x} + \mathbf{y}\|^2 + \|\mathbf{x} - \mathbf{y}\|^2 = \langle \mathbf{x} + \mathbf{y}, \mathbf{x} + \mathbf{y} \rangle + \langle \mathbf{x} - \mathbf{y}, \mathbf{x} - \mathbf{y} \rangle$$
$$= \langle \mathbf{x}, \mathbf{x} \rangle + 2\langle \mathbf{x}, \mathbf{y} \rangle + \langle \mathbf{y}, \mathbf{y} \rangle + \langle \mathbf{x}, \mathbf{x} \rangle - 2\langle \mathbf{x}, \mathbf{y} \rangle + \langle \mathbf{y}, \mathbf{y} \rangle$$
$$= 2\langle \mathbf{x}, \mathbf{x} \rangle + 2\langle \mathbf{y}, \mathbf{y} \rangle$$
$$= 2\|\mathbf{x}\|^2 + 2\|\mathbf{y}\|^2.$$

Exercise 10.3 The set $W \neq \emptyset$ because $\mathbf{0} \in W$ since $\langle \mathbf{0}, \mathbf{v} \rangle = 0$. Suppose $\mathbf{x}, \mathbf{y} \in W$ and $\alpha, \beta \in \mathbb{R}$. Because $\mathbf{x} \perp \mathbf{v}$ and $\mathbf{y} \perp \mathbf{v}$, we have (by definition) $\langle \mathbf{x}, \mathbf{v} \rangle = \langle \mathbf{y}, \mathbf{v} \rangle = 0$. Therefore,

$$\langle \alpha\mathbf{x} + \beta\mathbf{y}, \mathbf{v} \rangle = \alpha\langle \mathbf{x}, \mathbf{v} \rangle + \beta\langle \mathbf{y}, \mathbf{v} \rangle = \alpha(0) + \beta(0) = 0,$$

and hence $\alpha\mathbf{x} + \beta\mathbf{y} \perp \mathbf{v}$; that is, $\alpha\mathbf{x} + \beta\mathbf{y} \in W$. Therefore, W is a subspace. In fact, W is the set $\{\mathbf{x} \mid \langle \mathbf{x}, \mathbf{v} \rangle = 0\}$, which is the hyperplane through the origin with normal vector $\mathbf{v}$.

The proof that $S^{\perp}$ is a subspace is similar. The vector $\mathbf{0}$ is in $S^{\perp}$ since $\langle \mathbf{0}, \mathbf{v} \rangle = 0$ for all $\mathbf{v} \in S$, so $S^{\perp}$ is non-empty. If $\mathbf{x}, \mathbf{y} \in S^{\perp}$ and $\alpha, \beta \in \mathbb{R}$, then $\mathbf{x}$ and $\mathbf{y}$ are each orthogonal to all the vectors in S; that is, if $\mathbf{v} \in S$, then $\langle \mathbf{x}, \mathbf{v} \rangle = \langle \mathbf{y}, \mathbf{v} \rangle = 0$. Therefore,

$$\langle \alpha\mathbf{x} + \beta\mathbf{y}, \mathbf{v} \rangle = \alpha \langle \mathbf{x}, \mathbf{v} \rangle + \beta \langle \mathbf{y}, \mathbf{v} \rangle = \alpha(0) + \beta(0) = 0 \quad \text{for all } \mathbf{v} \in S.$$

Therefore, $\alpha\mathbf{x} + \beta\mathbf{y} \in S^{\perp}$, so $S^{\perp}$ is a subspace of $\mathbb{R}^n$. The subspace $S^{\perp}$ is known as the orthogonal complement of S.

Exercise 10.4 If P is an orthogonal matrix, then $P^{\mathrm{T}}P = I$. Using the fact that the product of two $n \times n$ determinants is the determinant of the product, we have

$$\det(P^{\mathrm{T}})\det(P) = \det(P^{\mathrm{T}}P) = \det(I) = 1.$$

But $\det(P^{\mathrm{T}}) = \det(P)$, so this becomes $(\det(P))^2 = 1$, which implies that $\det(P) = \pm 1$.

Exercise 10.5 To show this is an inner product, we need to show it satisfies each of the three properties in the definition.
(i) You can show that $\langle \mathbf{x}, \mathbf{y} \rangle = \langle \mathbf{y}, \mathbf{x} \rangle$ for all $\mathbf{x}, \mathbf{y} \in \mathbb{R}^2$ by letting $\mathbf{x} = (x_1, x_2)^{\mathrm{T}}$, $\mathbf{y} = (y_1, y_2)^{\mathrm{T}}$, multiplying out the matrix product $\mathbf{x}^{\mathrm{T}}A\mathbf{y}$ and using properties of real numbers. But there is an easier way. The given matrix A is symmetric, so $A^{\mathrm{T}} = A$. Since $\mathbf{x}^{\mathrm{T}}A\mathbf{y}$ is a 1×1 matrix, it is also symmetric. Therefore,

$$\langle \mathbf{x}, \mathbf{y} \rangle = \mathbf{x}^{\mathrm{T}}A\mathbf{y} = (\mathbf{x}^{\mathrm{T}}A\mathbf{y})^{\mathrm{T}} = \mathbf{y}^{\mathrm{T}}A^{\mathrm{T}}\mathbf{x} = \mathbf{y}^{\mathrm{T}}A\mathbf{x} = \langle \mathbf{y}, \mathbf{x} \rangle.$$

(ii) We next show that $\langle \alpha\mathbf{x} + \beta\mathbf{y}, \mathbf{z} \rangle = \alpha \langle \mathbf{x}, \mathbf{z} \rangle + \beta \langle \mathbf{y}, \mathbf{z} \rangle$ for all $\mathbf{x}, \mathbf{y}, \mathbf{z} \in \mathbb{R}^2$ and all $\alpha, \beta \in \mathbb{R}$. This follows from the rules of matrix algebra:

$$\langle \alpha\mathbf{x} + \beta\mathbf{y}, \mathbf{z} \rangle = (\alpha\mathbf{x} + \beta\mathbf{y})^{\mathrm{T}}A\mathbf{z} = (\alpha\mathbf{x}^{\mathrm{T}} + \beta\mathbf{y}^{\mathrm{T}})A\mathbf{z}$$
$$= \alpha\mathbf{x}^{\mathrm{T}}A\mathbf{z} + \beta\mathbf{y}^{\mathrm{T}}A\mathbf{z} = \alpha \langle \mathbf{x}, \mathbf{z} \rangle + \beta \langle \mathbf{y}, \mathbf{z} \rangle.$$

(iii) Finally, we need to show that $\langle \mathbf{x}, \mathbf{x} \rangle \geq 0$ for all $\mathbf{x} \in \mathbb{R}^2$, and $\langle \mathbf{x}, \mathbf{x} \rangle = 0$ if and only if $\mathbf{x} = \mathbf{0}$. If $\mathbf{x} = \mathbf{0}$, then $\langle \mathbf{x}, \mathbf{x} \rangle = \mathbf{0}^{\mathrm{T}}A\mathbf{0} = 0$, so it just remains to show that $\langle \mathbf{x}, \mathbf{x} \rangle > 0$ if $\mathbf{x} \neq \mathbf{0}$.

Assume $\mathbf{x} = (x_1, x_2)$ is any non-zero vector in $\mathbb{R}^2$. If $x_2 = 0$, then $x_1 \neq 0$ (otherwise, we would have the zero vector), and

$$\langle \mathbf{x}, \mathbf{x} \rangle = (x_1 \quad 0) \begin{pmatrix} 5 & 2 \\ 2 & 1 \end{pmatrix} \begin{pmatrix} x_1 \\ 0 \end{pmatrix} = 5x_1^2 > 0.$$

If $x_2 \neq 0$, then

$$\langle \mathbf{x}, \mathbf{x} \rangle = (x_1 \quad x_2) \begin{pmatrix} 5 & 2 \\ 2 & 1 \end{pmatrix} \begin{pmatrix} x_1 \\ x_2 \end{pmatrix} = 5x_1^2 + 4x_1 x_2 + x_2^2$$

$$= x_2^2 \left(5\frac{x_1^2}{x_2^2} + 4\frac{x_1}{x_2} + 1 \right) = x_2^2 (5t^2 + 4t + 1), \quad \text{for } t \in \mathbb{R}.$$

Now $f(t) = 5t^2 + 4t + 1$ is a quadratic function whose graph is a parabola. To see if it crosses the x axis, we look for the solutions of $5t^2 + 4t + 1 = 0$, which are given by the quadratic formula: $t = (-4 \pm \sqrt{16 - 4(5)(1)})/10$. So there are no real solutions, therefore $5t^2 + 4t + 1$ is either always strictly positive or strictly negative, and if $t = 1$, for example, it is positive, so we can conclude that

$$\langle \mathbf{x}, \mathbf{x} \rangle = x_2^2 (5t^2 + 4t + 1) > 0 \quad \text{for all } \mathbf{x} \neq \mathbf{0} \in \mathbb{R}^2.$$

Therefore, this is an inner product on $\mathbb{R}^2$.

(a) Using this inner product, for the given vectors $\mathbf{v}$ and $\mathbf{w}$,

$$\langle \mathbf{v}, \mathbf{w} \rangle = (1 \quad 1) \begin{pmatrix} 5 & 2 \\ 2 & 1 \end{pmatrix} \begin{pmatrix} -1 \\ 2 \end{pmatrix} = (7 \quad 3) \begin{pmatrix} -1 \\ 2 \end{pmatrix} = -1.$$

(b) The norm of $\mathbf{v}$ satisfies

$$\|\mathbf{v}\|^2 = \langle \mathbf{v}, \mathbf{v} \rangle = (1 \quad 1) \begin{pmatrix} 5 & 2 \\ 2 & 1 \end{pmatrix} \begin{pmatrix} 1 \\ 1 \end{pmatrix} = (7 \quad 3) \begin{pmatrix} 1 \\ 1 \end{pmatrix} = 10,$$

so $\|\mathbf{v}\| = \sqrt{10}$.

(c) You need to find the set of all vectors $\mathbf{x} = (x, y)^T$ for which $\langle \mathbf{v}, \mathbf{x} \rangle = 0$; that is,

$$\langle \mathbf{v}, \mathbf{x} \rangle = (1 \quad 1) \begin{pmatrix} 5 & 2 \\ 2 & 1 \end{pmatrix} \begin{pmatrix} x \\ y \end{pmatrix} = (7 \quad 3) \begin{pmatrix} x \\ y \end{pmatrix} = 7x + 3y = 0.$$

Therefore,

$$S^\perp = \left\{ \begin{pmatrix} x \\ y \end{pmatrix} \,\middle|\, 7x + 3y = 0 \right\}.$$

A basis of $S^\perp$ is

$$\left\{ \begin{pmatrix} -3 \\ 7 \end{pmatrix} \right\}.$$

(d) A basis of $S = \text{Lin}(\mathbf{v})$ is the vector $\mathbf{v}$. Therefore, you need to express $\mathbf{w}$ as a linear combination of $\mathbf{v}$ and the basis vector $\mathbf{n} = (-3, 7)^T$ of $S^\perp$.

That is, find a, b such that

$$\begin{pmatrix} -1 \\ 2 \end{pmatrix} = a \begin{pmatrix} 1 \\ 1 \end{pmatrix} + b \begin{pmatrix} -3 \\ 7 \end{pmatrix}.$$

You can solve this directly by writing out the equations and solving them simultaneously, or by using an inverse matrix,

$$\begin{pmatrix} a \\ b \end{pmatrix} = \frac{1}{10} \begin{pmatrix} 7 & 3 \\ -1 & 1 \end{pmatrix} \begin{pmatrix} -1 \\ 2 \end{pmatrix} = \frac{1}{10} \begin{pmatrix} -1 \\ 3 \end{pmatrix}.$$

Check your answer:

$$\begin{pmatrix} -1 \\ 2 \end{pmatrix} = -\frac{1}{10} \begin{pmatrix} 1 \\ 1 \end{pmatrix} + \frac{3}{10} \begin{pmatrix} -3 \\ 7 \end{pmatrix}.$$

(e) The linearly independent vectors $\mathbf{v}$ and $\mathbf{n}$ are orthogonal under this inner product, so all you need to do is to normalise them. You already know the length of $\mathbf{v}$; the length of $\mathbf{n}$ is given by

$$\|\mathbf{n}\|^2 = \langle \mathbf{n}, \mathbf{n} \rangle = (-3 \quad 7) \begin{pmatrix} 5 & 2 \\ 2 & 1 \end{pmatrix} \begin{pmatrix} -3 \\ 7 \end{pmatrix} = (-1 \quad -1) \begin{pmatrix} -3 \\ 7 \end{pmatrix} = 10,$$

so $\|\mathbf{n}\| = \sqrt{10}$. Therefore,

$$\mathbf{u}_1 = \frac{1}{\sqrt{10}} \begin{pmatrix} 1 \\ 1 \end{pmatrix}, \quad \mathbf{u}_2 = \frac{1}{\sqrt{10}} \begin{pmatrix} -3 \\ 7 \end{pmatrix}$$

form an orthonormal basis of $\mathbb{R}^2$ with the given inner product.

Exercise 10.6 To start with,

$$\mathbf{u}_1 = \mathbf{v}_1/\|\mathbf{v}_1\| = (1/\sqrt{2})(1, 0, 1, 0)^\mathrm{T}.$$

Then we let

$$\mathbf{w}_2 = \mathbf{v}_2 - \langle \mathbf{v}_2, \mathbf{u}_1 \rangle \mathbf{u}_1 = \begin{pmatrix} 1 \\ 2 \\ 1 \\ 1 \end{pmatrix} - \frac{2}{\sqrt{2}} \frac{1}{\sqrt{2}} \begin{pmatrix} 1 \\ 0 \\ 1 \\ 0 \end{pmatrix} = \begin{pmatrix} 0 \\ 2 \\ 0 \\ 1 \end{pmatrix}.$$

Now check that $\mathbf{w}_2 \perp \mathbf{u}_1$ by calculating $\langle \mathbf{w}_2, \mathbf{u}_1 \rangle = 0$. Then

$$\mathbf{u}_2 = \frac{\mathbf{w}_2}{\|\mathbf{w}_2\|} = \frac{1}{\sqrt{5}} \begin{pmatrix} 0 \\ 2 \\ 0 \\ 1 \end{pmatrix}.$$

Next (and you should fill in the missing steps),

$$\mathbf{w}_3 = \mathbf{v}_3 - \langle \mathbf{v}_3, \mathbf{u}_2 \rangle \mathbf{u}_2 - \langle \mathbf{v}_3, \mathbf{u}_1 \rangle \mathbf{u}_1 = \cdots = \begin{pmatrix} -1 \\ -1/5 \\ 1 \\ 2/5 \end{pmatrix}.$$

Now let $\mathbf{w}_3' = (-5, -1, 5, 2)^T$ (for ease of calculation) and check that $\mathbf{w}_3'$ is perpendicular to both $\mathbf{u}_1$ and $\mathbf{u}_2$.

Normalising $\mathbf{w}_3$, we obtain

$$\mathbf{u}_3 = \frac{1}{\sqrt{55}}(-5, -1, 5, 2)^T.$$

The required basis is $\{\mathbf{u}_1, \mathbf{u}_2, \mathbf{u}_3\}$.

Exercise 10.7 Choose two linearly independent vectors in the plane as a basis. (This is most easily done by choosing one of the components to be 0, another to be equal to 1, say, and then solving the equation for the third. By choosing the zeros differently, you obtain linearly independent vectors.) For example, let

$$\mathbf{v}_1 = \begin{pmatrix} 2 \\ 1 \\ 0 \end{pmatrix}, \quad \mathbf{v}_2 = \begin{pmatrix} 3 \\ 0 \\ -1 \end{pmatrix}.$$

Then $\{\mathbf{v}_1, \mathbf{v}_2\}$ is a basis of W. Now use Gram–Schmidt orthonormalisation. Set $\mathbf{u}_1 = (2/\sqrt{5}, 1/\sqrt{5}, 0)^T$ and

$$\mathbf{w} = \begin{pmatrix} 3 \\ 0 \\ -1 \end{pmatrix} - \left\langle \begin{pmatrix} 3 \\ 0 \\ -1 \end{pmatrix}, \begin{pmatrix} \frac{2}{\sqrt{5}} \\ \frac{1}{\sqrt{5}} \\ 0 \end{pmatrix} \right\rangle \begin{pmatrix} \frac{2}{\sqrt{5}} \\ \frac{1}{\sqrt{5}} \\ 0 \end{pmatrix}$$

$$= \begin{pmatrix} 3 \\ 0 \\ -1 \end{pmatrix} - \begin{pmatrix} \frac{12}{5} \\ \frac{6}{5} \\ 0 \end{pmatrix} = \begin{pmatrix} \frac{3}{5} \\ -\frac{6}{5} \\ -1 \end{pmatrix}.$$

The vector $\mathbf{w}_2 = (3, -6, -5)^T$ is parallel to $\mathbf{w}$. (This is a good time to check that $\mathbf{w}_2 \perp \mathbf{u}_1$ and also that $\mathbf{w}_2 \in W$.) Now set $\mathbf{u}_2 = (3/\sqrt{70}, -6/\sqrt{70}, -5/\sqrt{70})^T$. The set $\{\mathbf{u}_1, \mathbf{u}_2\}$ is an orthonormal basis of W.

To extend this to an orthonormal basis of $\mathbb{R}^3$, note that W is a plane with normal vector $\mathbf{n} = (1, -2, 3)^T$, so $\mathbf{n}$ is perpendicular to every vector in W. If you set $\mathbf{u}_3 = (1/\sqrt{14}, -2/\sqrt{14}, 3/\sqrt{14})^T$, then $\{\mathbf{u}_1, \mathbf{u}_2, \mathbf{u}_3\}$ is an orthonormal basis of $\mathbb{R}^3$ as required.

Chapter 11 exercises

Exercise 11.1 A can be orthogonally diagonalised because it is symmetric. The characteristic polynomial of A is

$$|A - \lambda I| = \begin{vmatrix} 7 - \lambda & 0 & 9 \\ 0 & 2 - \lambda & 0 \\ 9 & 0 & 7 - \lambda \end{vmatrix}$$

$$= (2 - \lambda)[(7 - \lambda)(7 - \lambda) - 81]$$
$$= (2 - \lambda)(\lambda^2 - 14\lambda - 32)$$
$$= (2 - \lambda)(\lambda - 16)(\lambda + 2),$$

where we have expanded the determinant using the middle row. So the eigenvalues are 2, 16, -2. An eigenvector for $\lambda = 2$ is given by reducing the matrix $A - 2I$:

$$(A - 2I) = \begin{pmatrix} 5 & 0 & 9 \\ 0 & 0 & 0 \\ 9 & 0 & 5 \end{pmatrix} \cdots \longrightarrow \cdots \begin{pmatrix} 1 & 0 & 0 \\ 0 & 0 & 1 \\ 0 & 0 & 0 \end{pmatrix}.$$

This means $x = z = 0$. So we may take $(0, 1, 0)^{\text{T}}$. This already has length 1 so there is no need to normalise it. (Recall that we need three eigenvectors which are of length 1.) For $\lambda = -2$, we find that an eigenvector is $(-1, 0, 1)^{\text{T}}$ (or some multiple of this). To normalise (that is, to make of length 1), we divide by its length, which is $\sqrt{2}$, obtaining $(1/\sqrt{2})(-1, 0, 1)^{\text{T}}$. For $\lambda = 16$, we find a normalised eigenvector is $(1/\sqrt{2})(1, 0, 1)$. It follows that if we let

$$P = \begin{pmatrix} 0 & -1/\sqrt{2} & 1/\sqrt{2} \\ 1 & 0 & 0 \\ 0 & 1/\sqrt{2} & 1/\sqrt{2} \end{pmatrix},$$

then P is orthogonal and $P^{\text{T}} A P = D = \text{diag}(2, -2, 16)$. Check this!

Exercise 11.2 To show that $\mathbf{v}_1$ is an eigenvector of A, find $A\mathbf{v}_1$.

$$A\mathbf{v}_1 = \begin{pmatrix} 2 & 1 & -2 \\ 1 & 2 & 2 \\ -2 & 2 & -1 \end{pmatrix} \begin{pmatrix} 1 \\ 1 \\ 0 \end{pmatrix} = \begin{pmatrix} 3 \\ 3 \\ 0 \end{pmatrix} = 3\mathbf{v}_1,$$

so $\mathbf{v}_1$ is an eigenvector corresponding to $\lambda_1 = 3$.

For the eigenvectors,

$$A - 3I = \begin{pmatrix} -1 & 1 & -2 \\ 1 & -1 & 2 \\ -2 & 2 & -4 \end{pmatrix} \longrightarrow \begin{pmatrix} 1 & -1 & 2 \\ 0 & 0 & 0 \\ 0 & 0 & 0 \end{pmatrix},$$

with solutions

$$\mathbf{x} = \begin{pmatrix} s - 2t \\ s \\ t \end{pmatrix} = s \begin{pmatrix} 1 \\ 1 \\ 0 \end{pmatrix} + t \begin{pmatrix} -2 \\ 0 \\ 1 \end{pmatrix} = s\mathbf{v}_1 + t\mathbf{v}_2, \qquad s, t \in \mathbb{R}.$$

Therefore, a basis of the eigenspace is $\{\mathbf{v}_1, \mathbf{v}_2\}$.

To orthogonally diagonalise this matrix, you need to make this basis into an orthonormal basis of the eigenspace, and you need to find another eigenvalue and corresponding eigenvalue. You have choices available to you as to how to do each one.

You can find the remaining eigenvalue by finding the characteristic equation, $|A - \lambda I| = 0$, and then find the corresponding eigenvector. Alternatively, you know that the eigenspace of $\lambda_1 = 3$ is a plane in $\mathbb{R}^3$, and you can deduce the normal to this plane from the reduced row echelon form of $A - 3I$ to be the vector $\mathbf{v}_3 = (1, -1, 2)^\mathrm{T}$, so this must be the third eigenvector. Then you can find the corresponding eigenvalue by

$$A\mathbf{v}_3 = \begin{pmatrix} 2 & 1 & -2 \\ 1 & 2 & 2 \\ -2 & 2 & -1 \end{pmatrix} \begin{pmatrix} 1 \\ -1 \\ 2 \end{pmatrix} = \begin{pmatrix} -3 \\ 3 \\ -6 \end{pmatrix} = -3\mathbf{v}_3.$$

So $\lambda_3 = -3$. You still need to obtain an orthonormal basis for the eigenspace of $\lambda_1 = 3$. Using Gram–Schmidt, set $\mathbf{u}_1 = (\frac{1}{\sqrt{2}}, \frac{1}{\sqrt{2}}, 0)^\mathrm{T}$, then

$$\mathbf{w} = \begin{pmatrix} -2 \\ 0 \\ 1 \end{pmatrix} - \left\langle \begin{pmatrix} -2 \\ 0 \\ 1 \end{pmatrix}, \begin{pmatrix} \frac{1}{\sqrt{2}} \\ \frac{1}{\sqrt{2}} \\ 0 \end{pmatrix} \right\rangle \begin{pmatrix} \frac{1}{\sqrt{2}} \\ \frac{1}{\sqrt{2}} \\ 0 \end{pmatrix}$$

$$= \begin{pmatrix} -2 \\ 0 \\ 1 \end{pmatrix} + \frac{2}{\sqrt{2}} \begin{pmatrix} \frac{1}{\sqrt{2}} \\ \frac{1}{\sqrt{2}} \\ 0 \end{pmatrix} = \begin{pmatrix} -1 \\ 1 \\ 1 \end{pmatrix}.$$

Now check that $\mathbf{w}$ is indeed orthogonal to the other two eigenvectors.

Taking the unit eigenvectors, you can set

$$P = \begin{pmatrix} \frac{1}{\sqrt{2}} & -\frac{1}{\sqrt{3}} & \frac{1}{\sqrt{6}} \\ \frac{1}{\sqrt{2}} & \frac{1}{\sqrt{3}} & -\frac{1}{\sqrt{6}} \\ 0 & \frac{1}{\sqrt{3}} & \frac{2}{\sqrt{6}} \end{pmatrix} \quad \text{and} \quad D = \begin{pmatrix} 3 & 0 & 0 \\ 0 & 3 & 0 \\ 0 & 0 & -3 \end{pmatrix}.$$

Then $P^\mathrm{T} = P^{-1}$ (P is an orthogonal matrix) and

$$P^\mathrm{T} A P = P^{-1} A P = D.$$

Exercise 11.3 The matrix representing the quadratic form $q(x, y, z)$ is

$$A = \begin{pmatrix} -4 & 3 & 1 \\ 3 & -2 & -2 \\ 1 & -2 & -4 \end{pmatrix}.$$

The first two principal minors are

$$a_{11} = -4 \quad \text{and} \quad \begin{vmatrix} -4 & 3 \\ 3 & -2 \end{vmatrix} = -1.$$

The first is negative and the second negative. If the matrix (and the quadratic form) were positive definite, both should be positive. If it were negative definite, the first should be negative and the second positive. So it is neither. Since $|A| = 10 \neq 0$, the quadratic form is indefinite.

For the second quadratic form, $f(x, y, z) = \mathbf{x}^\mathrm{T} B\mathbf{x}$, the matrix is

$$B = \begin{pmatrix} -4 & 1 & 3 \\ 1 & -2 & -2 \\ 3 & -2 & -4 \end{pmatrix}.$$

For this matrix, the principal minors are

$$b_{11} = -4, \qquad \begin{vmatrix} -4 & 1 \\ 1 & -2 \end{vmatrix} = 7, \qquad |B| = -6.$$

Therefore, this quadratic form is negative definite.

Exercise 11.4 We found an orthogonal matrix P for which $D = \mathrm{diag}(2, -2, 16)$. Changing the order of the columns to satisfy the condition $\lambda_1 \geq \lambda_2 \geq \lambda_3$, let $Q = (\mathbf{u}_1, \mathbf{u}_2, \mathbf{u}_3)$ be the matrix

$$Q = \begin{pmatrix} 1/\sqrt{2} & 0 & -1/\sqrt{2} \\ 0 & 1 & 0 \\ 1/\sqrt{2} & 0 & 1/\sqrt{2} \end{pmatrix}, \quad \text{so that } D = \begin{pmatrix} 16 & 0 & 0 \\ 0 & 2 & 0 \\ 0 & 0 & -1 \end{pmatrix}.$$

Then, if $\mathbf{x} = Q\mathbf{z}$, with $\mathbf{x} = (x, y, z)^\mathrm{T}$ and $\mathbf{z} = (X, Y, Z)^\mathrm{T}$, we have

$$f(x, y, z) = \mathbf{x}^\mathrm{T} A\mathbf{x} = \mathbf{z}^\mathrm{T} D\mathbf{z} = 16X^2 + 2Y^2 - 2Z^2.$$

Since A has both positive and negative eigenvalues, the quadratic form is indefinite.

To find $\mathbf{a} = (a, b, c)$ such that $f(a, b, c) = -8$, look at the expression for $f(x, y, z)$ in the coordinates with respect to the basis B of eigenvectors of A, which are the columns of Q. The unit eigenvector $\mathbf{u}_1 = (-1/\sqrt{2}, 0, 1/\sqrt{2})^\mathrm{T}$ has B coordinates, $[0, 0, 1]_B$ and will therefore give the value $f(x, y, z) = -2Z^2 = -2$. So to obtain the value $f(x, y, z) = -8$, we can take the vector $[0, 0, 2]_B$, which in standard coordinates is $2\mathbf{u}_3$. You can check by substituting these values into f that, indeed, $f(-2/\sqrt{2}, 0, 2/\sqrt{2}) = -8$.

Exercise 11.5 If $\mathbf{e}_1, \mathbf{e}_2, \ldots, \mathbf{e}_n$ are the standard basis vectors in $\mathbb{R}^n$, then $\mathbf{e}_i$ has 1 as its ith component and 0 elsewhere. Then, if A is positive definite,

$$\mathbf{e}_i^\mathrm{T} A\mathbf{e}_i = a_{ii} > 0,$$

since A is positive definite.

The converse of this statement, however, is far from true. There are many matrices with positive numbers on the main diagonal which are *not* positive definite. For example, the matrix

$$A = \begin{pmatrix} 1 & 2 \\ 2 & 1 \end{pmatrix}$$

has eigenvalues $\lambda = 3, -1$, so it is indefinite.

Exercise 11.6　The matrix $B^T B$ is a $k \times k$ symmetric matrix since $(B^T B)^T = B^T (B^T)^T = B^T B$. To show it is a positive definite matrix, we need to show that $\mathbf{x}^T B^T B \mathbf{x} \geq 0$ for any vector $\mathbf{x} \in \mathbb{R}^n$, and $\mathbf{x}^T B^T B \mathbf{x} = 0$ if and only if $\mathbf{x} = \mathbf{0}$. We have, for all $\mathbf{x} \in \mathbb{R}^n$,

$$\mathbf{x}^T B^T B \mathbf{x} = (B\mathbf{x})^T (B\mathbf{x}) = \langle B\mathbf{x}, B\mathbf{x} \rangle = ||B\mathbf{x}||^2,$$

which is positive for all $B\mathbf{x} \neq \mathbf{0}$; and $||B\mathbf{x}||^2 = 0$ if and only if $B\mathbf{x} = \mathbf{0}$.

Since rank$(B) = k$, the reduced row echelon form of B will have a leading one in every column, so the only solution of $B\mathbf{x} = \mathbf{0}$ is $\mathbf{x} = \mathbf{0}$. Therefore, $\mathbf{x}^T (B^T B) \mathbf{x} > 0$ for all $\mathbf{x} \neq \mathbf{0}$, and $\mathbf{x}^T (B^T B) \mathbf{x} = 0$ if and only if $\mathbf{x} = \mathbf{0}$. Hence the matrix $B^T B$ is positive definite.

If an $n \times n$ symmetric A is positive definite, then all of its eigenvalues are positive, so 0 is not an eigenvalue of A. Therefore, the system of equations $A\mathbf{x} = \mathbf{0}$ has no non-trivial solution, and so A is invertible.

Exercise 11.7　The matrix A is

$$A = \begin{pmatrix} 1 & -2 & -1 \\ -2 & 5 & 3 \\ -1 & 3 & 2 \end{pmatrix}.$$

To determine if A is positive definite, negative definite or indefinite, we consider the principal minors:

$$(a_{11}) = 1 > 0, \quad \begin{vmatrix} 1 & -2 \\ -2 & 5 \end{vmatrix} = 1 > 0$$
$$|A| = 1(10 - 9) + 2(-4 + 3) - 1(-6 + 5) = 0.$$

Since $a_{11} = 1 > 0$, the matrix A is not negative definite. Since $|A| = 0$, one of the eigenvalues of A is 0, so A is not positive definite.

To determine if A is indefinite, we need to find the eigenvalues. Expanding the characteristic equation,

$$|A - \lambda I| = \begin{vmatrix} 1 - \lambda & -2 & -1 \\ -2 & 5 - \lambda & 3 \\ -1 & 3 & 2 - \lambda \end{vmatrix}$$
$$= -\lambda^3 + 8\lambda^2 - 3\lambda = -\lambda(\lambda^2 - 8\lambda + 3).$$

The roots are $\lambda = 0$ and, using the quadratic formula,

$$\lambda = \frac{8 \pm \sqrt{64 - 12}}{2}.$$

So both roots of the quadratic equation are positive. Therefore, the matrix is not indefinite, it is positive semi-definite. Therefore, there is no point (a, b, c) for which $f(a, b, c) < 0$.

Exercise 11.8 We have $q(x, y) = \mathbf{x}^T A \mathbf{x} = 5x^2 - 6xy + 5y^2 = 2$. This time we orthogonally diagonalise the matrix

$$A = \begin{pmatrix} 5 & -3 \\ -3 & 5 \end{pmatrix}$$

using $Q = (\mathbf{w}_1, \mathbf{w}_2)$, where

$$Q = \begin{pmatrix} -\frac{1}{\sqrt{2}} & \frac{1}{\sqrt{2}} \\ \frac{1}{\sqrt{2}} & \frac{1}{\sqrt{2}} \end{pmatrix}.$$

Then

$$Q^T A Q = D = \begin{pmatrix} 8 & 0 \\ 0 & 2 \end{pmatrix},$$

since $A\mathbf{w}_1 = 8\mathbf{w}_1$ and $A\mathbf{w}_2 = 2\mathbf{w}_2$. Then, if $\mathbf{x} = Q\mathbf{z}$, in the new coordinates, $\mathbf{z} = (X, Y)^T$, the equation becomes

$$\mathbf{x}^T A \mathbf{x} = \mathbf{z}^T D \mathbf{z} = 8X^2 + 2Y^2 = 2,$$

or $4X^2 + Y^2 = 1$. To sketch this, we first need to find the positions of the new X and Y axes. This time we must rely only on the eigenvectors. The new X axis is in the direction of the vector $\mathbf{w}_1$, and the new Y axis is in the direction of $\mathbf{w}_2$. (So this linear transformation is actually a reflection about a line through the origin.) We first draw these new axes, and then sketch the ellipse in the usual way. This time it intersects the X axis in $X = \pm(1/2)$ and the Y axis in $Y = \pm 1$, as shown below. Notice that this is exactly the same ellipse as in Example 11.41 (as it should be!); we merely used a different change of basis to sketch it.

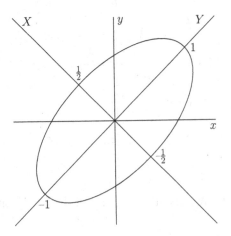

Exercise 11.9 The matrix A is

$$A = \begin{pmatrix} 9 & 2 \\ 2 & 6 \end{pmatrix}.$$

Its eigenvalues are 5 and 10. Since these are both positive, the quadratic form is positive definite. This also indicates that there is an orthogonal matrix P such that $P^T A P = D = \mathrm{diag}(5, 10)$, and so that $\mathbf{x}^T A \mathbf{x} = 10$ is an ellipse.

The vectors $\mathbf{v}_1 = (-1, 2)^T$ and $\mathbf{v}_2 = (2, 1)^T$ are eigenvectors of A corresponding to the eigenvalues 5 and 10, respectively. If

$$P = \begin{pmatrix} \frac{2}{\sqrt{5}} & -\frac{1}{\sqrt{5}} \\ \frac{1}{\sqrt{5}} & \frac{2}{\sqrt{5}} \end{pmatrix} \quad \text{and} \quad D = \begin{pmatrix} 10 & 0 \\ 0 & 5 \end{pmatrix},$$

then $P^T A P = D$. Set $\mathbf{x} = P\mathbf{z}$ with $\mathbf{x} = (x, y)^T$, $\mathbf{z} = (X, Y)^T$. Then the curve $\mathbf{x}^T A \mathbf{x} = 10$ becomes

$$\mathbf{x}^T A \mathbf{x} = \mathbf{z}^T D \mathbf{x} = 10 X^2 + 5 Y^2 = 10,$$

which is an ellipse. The linear transformation defined by P is a rotation, but not by an angle which we recognise. Therefore, the images of the x and y axes under P are found by looking at the images of the standard basis vectors under the linear transformation defined by P. Thus, the direction of the positive X axis is given by $\mathbf{v}_2 = (2, 1)^T$ and the direction of the positive Y axis is given by the vector $\mathbf{v}_1 = (-1, 2)^T$. We now sketch the ellipse in standard position on the X and Y axes.

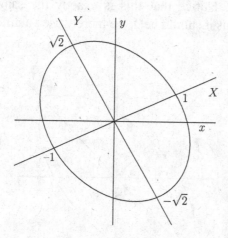

It intersects the X axis at $X = \pm 1$ and the Y axis at $Y = \pm\sqrt{2}$.

Exercise 11.10 The vectors $\mathbf{v}_1 = (1, 1)^T$ and $\mathbf{v}_2 = (-1, 1)^T$ are eigen-vectors of A since

$$A\mathbf{v}_1 = \begin{pmatrix} a & b \\ b & a \end{pmatrix} \begin{pmatrix} 1 \\ 1 \end{pmatrix} = \begin{pmatrix} a+b \\ b+a \end{pmatrix} = (a+b)\mathbf{v}_1$$

and

$$A\mathbf{v}_2 = \begin{pmatrix} a & b \\ b & a \end{pmatrix} \begin{pmatrix} -1 \\ 1 \end{pmatrix} = \begin{pmatrix} -a+b \\ -b+a \end{pmatrix} = (a-b)\mathbf{v}_2.$$

So the eigenvalues are $\lambda_1 = a + b$ and $\lambda_2 = a - b$, respectively.

Exercise 11.11 To sketch the curve $x^2 + y^2 + 6xy = 4$ in the xy-plane, we write

$$x^2 + y^2 + 6xy = \mathbf{x}^T A \mathbf{x} \qquad \text{with} \quad A = \begin{pmatrix} 1 & 3 \\ 3 & 1 \end{pmatrix},$$

and orthogonally diagonalise A. The eigenvalues of A are $\lambda_1 = 4$ and $\lambda_2 = -2$, with corresponding eigenvectors $\mathbf{v}_1 = (1, 1)^T$ and $\mathbf{v}_2 = (-1, 1)^T$. Therefore, we can take

$$P = \begin{pmatrix} \frac{1}{\sqrt{2}} & -\frac{1}{\sqrt{2}} \\ \frac{1}{\sqrt{2}} & \frac{1}{\sqrt{2}} \end{pmatrix}, \qquad D = \begin{pmatrix} 4 & 0 \\ 0 & -2 \end{pmatrix},$$

so that $P^{-1}AP = P^T AP = D$ and P defines a rotation anticlockwise by $\pi/4$ radians. Then setting $\mathbf{x} = P\mathbf{z}$, with $\mathbf{z} = (X, Y)^T$,

$$x^2 + y^2 + 6xy = \mathbf{x}^T A \mathbf{x} = \mathbf{z}^T D \mathbf{z} = 4X^2 - 2Y^2 = 4.$$

So we need to sketch the hyperbola $2X^2 - Y^2 = 2$ on the new X, Y axes. This intersects only the X axis at $X = \pm 1$. However, sketching the hyperbola is more difficult than sketching an ellipse: we also need to sketch the asymptotes so we know its shape. The asymptotes are $Y = \pm\sqrt{2}X$. You can find the equations of these asymptotes in the standard x, y coordinates using $\mathbf{z} = P^T\mathbf{x}$ to substitute for X and Y in the two equations, $Y = \sqrt{2}X$ and $Y = -\sqrt{2}X$, and then sketch the two lines in x, y coordinates. But we are only required to do a sketch, so you can reason as follows. The line $Y = \sqrt{2}X$ has a steeper slope than the line $Y = X$, and because we have rotated by $\pi/4$ radians anticlockwise, the line $Y = X$ is just the (old) y axis. Using this idea, you can sketch the asymptotes. and then the hyperbola.

Knowing the points of intersection with the old axes helps here. We have the points of intersection with the new X axis are $X = \pm 1$. The points of intersection with the old x and y axes are given by setting, respectively, $y = 0$ and $x = 0$ in the original equation, $x^2 + y^2 + 6xy = 4$. These are $x = \pm 2$ and $y = \pm 2$. Here is the sketch.

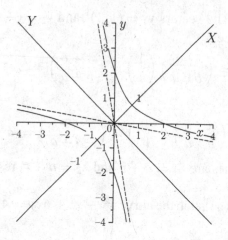

Chapter 12 exercises

Exercise 12.1 Suppose $\dim(S) = r$. (Let us assume that $0 < r < n$, the cases $r = 0$ and $r = n$ being easy to deal with separately.) The proof of Theorem 12.13 shows us that there is an orthonormal basis of $\mathbb{R}^n$ of the form $\{\mathbf{e}_1, \mathbf{e}_2, \ldots, \mathbf{e}_r, \mathbf{e}_{r+1}, \ldots, \mathbf{e}_n\}$, where $\{\mathbf{e}_1, \ldots, \mathbf{e}_r\}$ is a basis of S and $\{\mathbf{e}_{r+1}, \ldots, \mathbf{e}_n\}$ is a basis of $S^\perp$. Make sure you understand why. So this means that $\dim(S^\perp) = n - r = n - \dim(S)$.

Exercise 12.2 If $\mathbf{z} \in S^\perp$, then, for all $\mathbf{s} \in S$, $\langle \mathbf{z}, \mathbf{s} \rangle = 0$. Now, let $\mathbf{s} \in S$. Then, for any $\mathbf{z} \in S^\perp$, we have $\langle \mathbf{z}, \mathbf{s} \rangle = 0$. So, for all $\mathbf{z} \in S^\perp$, we have $\langle \mathbf{s}, \mathbf{z} \rangle = 0$. But this shows that $\mathbf{s}$ is orthogonal to every member of $S^\perp$. That is, $\mathbf{s} \in (S^\perp)^\perp$. Hence, $S \subseteq (S^\perp)^\perp$. Now, by the previous exercise,

$$\dim((S^\perp)^\perp) = n - \dim(S^\perp) = n - (n - \dim(S)) = \dim(S).$$

So $S \subseteq (S^\perp)^\perp$, and both are subspaces of the same dimension. Hence $S = (S^\perp)^\perp$.

Exercise 12.3 The orthogonal projection of $\mathbb{R}^4$ onto the subspace spanned by the vectors $(1, 0, 1, 0)^\mathrm{T}$ and $(1, 2, 1, 2)^\mathrm{T}$ is the same as the orthogonal projection onto $R(A)$, where A is the rank 2 matrix

$$A = \begin{pmatrix} 1 & 1 \\ 0 & 2 \\ 1 & 1 \\ 0 & 2 \end{pmatrix}.$$

The projection is therefore represented by the matrix

$$P = A(A^{\mathrm{T}}A)^{-1}A^{\mathrm{T}} = \begin{pmatrix} \frac{1}{2} & 0 & \frac{1}{2} & 0 \\ 0 & \frac{1}{2} & 0 & \frac{1}{2} \\ \frac{1}{2} & 0 & \frac{1}{2} & 0 \\ 0 & \frac{1}{2} & 0 & \frac{1}{2} \end{pmatrix}.$$

(Check this!) That is, for $\mathbf{x} = (x, y, z, w)^{\mathrm{T}} \in \mathbb{R}^4$, the orthogonal projection of $\mathbf{x}$ onto the subspace is given by

$$\mathbf{x} \mapsto \begin{pmatrix} \frac{1}{2}x + \frac{1}{2}z \\ \frac{1}{2}y + \frac{1}{2}w \\ \frac{1}{2}x + \frac{1}{2}z \\ \frac{1}{2}y + \frac{1}{2}w \end{pmatrix}.$$

Exercise 12.4 You have already shown in Activity 12.24 that the only eigenvalues of an idempotent matrix are $\lambda = 1$ and $\lambda = 0$. Since A can be diagonalised, there is an invertible matrix P and a diagonal matrix $D = \mathrm{diag}(1, \dots, 1, 0, \dots, 0)$ such that $P^{-1}AP = D$. Let $\mathbf{v}_1, \mathbf{v}_2, \dots, \mathbf{v}_n$ denote the column vectors of P with $\mathbf{v}_1, \dots, \mathbf{v}_i$ being eigenvectors for $\lambda = 1$ and $\mathbf{v}_{i+1}, \dots, \mathbf{v}_n$ eigenvectors for $\lambda = 0$. Then $B = \{\mathbf{v}_1, \mathbf{v}_2, \dots, \mathbf{v}_n\}$ is a basis of $\mathbb{R}^n$ and $\{\mathbf{v}_1, \dots, \mathbf{v}_i\}$ is a basis of the eigenspace, $E(1)$, for eigenvalue $\lambda = 1$.

If $\mathbf{y} \in R(A)$, then $\mathbf{y} = A\mathbf{x}$ for some $\mathbf{x} \in \mathbb{R}^n$. Now, $\mathbf{x}$ can be written as a unique linear combination $\mathbf{x} = a_1\mathbf{v}_1 + a_2\mathbf{v}_2 + \cdots + a_n\mathbf{v}_n$. We therefore have

$$\begin{aligned} \mathbf{y} = A\mathbf{x} &= A(a_1\mathbf{v}_1 + a_2\mathbf{v}_2 + \cdots + a_n\mathbf{v}_n) \\ &= a_1 A\mathbf{v}_1 + a_2 A\mathbf{v}_2 + \cdots + a_n A\mathbf{v}_n \\ &= a_1\mathbf{v}_1 + a_2\mathbf{v}_2 + \cdots + a_i\mathbf{v}_i \end{aligned}$$

since $A\mathbf{v}_j = \mathbf{v}_j$ for $j = 1, \cdots, i$ and $A\mathbf{v}_j = \mathbf{0}$ for $j = i+1, \cdots, n$. Therefore, $\mathbf{y} \in E(1)$, so $R(A) \subseteq E(1)$. Now let $\mathbf{y} \in E(1)$. Then $\mathbf{y} \in \mathbb{R}^n$ and $A\mathbf{y} = \mathbf{y}$, so $\mathbf{y} \in R(A)$. This completes the proof.

Exercise 12.5 You can show directly that $A^2 = A$ by multiplying the matrices. Alternatively, you can use the diagonalisation. An invertible matrix P and a diagonal matrix D such that $P^{-1}AP = D$ are

$$P = \begin{pmatrix} 2 & 1 & -4 \\ 1 & 0 & -3 \\ 0 & 1 & 1 \end{pmatrix}, \qquad D = \begin{pmatrix} 1 & 0 & 0 \\ 0 & 1 & 0 \\ 0 & 0 & 0 \end{pmatrix}.$$

Then $D^2 = D$, and since $A = PDP^{-1}$,

$$A^2 = PDP^{-1}PDP^{-1} = PD^2P^{-1} = PDP^{-1} = A,$$

so A is idempotent.

Because A is idempotent, the linear transformation given by $T(\mathbf{x}) = A\mathbf{x}$ is idempotent, and is therefore a projection from $\mathbb{R}^3$ onto the subspace $U = R(T)$ parallel to the subspace $W = N(T)$; that is, $U = R(A)$ and $W = N(A)$. It is not an orthogonal projection because A is not symmetric.

The null space of A, namely $N(A)$, is the same as the eigenspace corresponding to the eigenvalue $\lambda = 0$. Since A is idempotent, by Exercise 12.4, $R(A)$ is the eigenspace corresponding to the eigenvalue $\lambda = 1$.

Exercise 12.6 In matrix form, we want the least squares solution to $A\mathbf{z} = \mathbf{b}$, where

$$A = \begin{pmatrix} 1 & -1 \\ 1 & 0 \\ 1 & 1 \\ 1 & 2 \end{pmatrix}, \quad \mathbf{z} = \begin{pmatrix} a \\ b \end{pmatrix}, \quad \mathbf{b} = \begin{pmatrix} 0 \\ 1 \\ 3 \\ 9 \end{pmatrix}.$$

So a least squares solution is

$$\mathbf{z} = (A^T A)^{-1} A^T \mathbf{b} = \begin{pmatrix} 1.8 \\ 2.9 \end{pmatrix}.$$

(We've omitted the calculations, but you can check this.) So a best-fit linear relationship is

$$Y = 1.8 + 2.9X.$$

Exercise 12.7 In matrix form, this is $A\mathbf{z} = \mathbf{b}$, where

$$A = \begin{pmatrix} 1 & 0 & 0 \\ 1 & 1 & 1 \\ 1 & 2 & 4 \\ 1 & 3 & 9 \end{pmatrix}, \quad \mathbf{z} = \begin{pmatrix} a \\ b \\ c \end{pmatrix}, \quad \mathbf{b} = \begin{pmatrix} 3 \\ 2 \\ 4 \\ 4 \end{pmatrix}.$$

Here, A has rank 3, and the theory above tells us that a least squares solution will be

$$\mathbf{z} = (A^{\mathrm{T}}A)^{-1}A^{\mathrm{T}}\mathbf{b} = \begin{pmatrix} 2.75 \\ -0.25 \\ 0.25 \end{pmatrix}.$$

(Details of the calculation are omitted.) So the best-fit model of this type is $Y = 2.75 - 0.25X + 0.25X^2$.

Chapter 13 exercises

Exercise 13.1 To solve $z^4 = -4$, write $z = re^{i\theta}$ and

$$-4 = 4e^{i\pi} = 4e^{i(\pi+2n\pi)}.$$

Then $z^4 = (re^{i\theta})^4 = 4e^{i(\pi+2n\pi)}$, so that $r^4 = 4$ and $4\theta = \pi + 2n\pi$. Therefore, $r = \sqrt{2}$ and $\theta = \frac{\pi}{4}, \frac{3\pi}{4}, \frac{5\pi}{4}, \frac{7\pi}{4}$ will give the four complex roots. These are:

$$z_1 = \sqrt{2}e^{i\pi/4} = \sqrt{2}\frac{1}{\sqrt{2}} + i\sqrt{2}\frac{1}{\sqrt{2}} = 1 + i,$$

$$z_2 = \sqrt{2}e^{i3\pi/4} = -\sqrt{2}\frac{1}{\sqrt{2}} + i\sqrt{2}\frac{1}{\sqrt{2}} = -1 + i,$$

$$z_3 = \sqrt{2}e^{i5\pi/4} = -1 - i = \bar{z}_2,$$

$$z_4 = \sqrt{2}e^{i7\pi/4} = 1 - i = \bar{z}_1.$$

To factorise $z^4 + 4$ into a product of quadratic factors with real coefficients, write the polynomial as a product of linear factors in conjugate pairs. For the first conjugate pair,

$$(z - z_1)(z - \bar{z}_1) = z^2 - 2\mathrm{Re}(z_1)z + z_1\bar{z}_1 = z^2 - 2z + 2.$$

In the same way, $(z - z_2)(z - \bar{z}_2) = z^2 + 2z + 2$, so that

$$z^4 + 4 = (z^2 - 2z + 2)(z^2 + 2z + 2).$$

Exercise 13.2 The determinant of a complex matrix is calculated in the same way as for a real matrix. You should find that

$$|A| = \begin{vmatrix} 1 & i \\ 1+i & -1 \end{vmatrix} = -1 - (i)(i + 1) = -1 - i^2 - i = -i.$$

You can solve the system using row reduction, or by finding A^{-1} (exactly as you do for a 2×2 real matrix). Then, $\mathbf{x} = A^{-1}\mathbf{b}$. We have

$$A^{-1} = -\frac{1}{i}\begin{pmatrix} -1 & -i \\ -1-i & 1 \end{pmatrix} = \frac{1}{i}\begin{pmatrix} 1 & i \\ 1+i & -1 \end{pmatrix}$$

so that

$$\mathbf{x} = \frac{1}{i}\begin{pmatrix} 1 & i \\ 1+i & -1 \end{pmatrix}\begin{pmatrix} 1 \\ i \end{pmatrix} = \frac{1}{i}\begin{pmatrix} 0 \\ 1 \end{pmatrix} = \begin{pmatrix} 0 \\ -i \end{pmatrix}.$$

Exercise 13.3 The set W_1 is not a subspace of $M_2(\mathbb{R})$ since

$$\begin{pmatrix} 1 & 0 \\ 0 & 1 \end{pmatrix} \in W_1 \quad \text{but} \quad (-1)\begin{pmatrix} 1 & 0 \\ 0 & 1 \end{pmatrix} \notin W_1.$$

But W_2 is a subspace of $M_2(\mathbb{C})$. It is closed under addition and under scalar multiplication since any complex number w can be written as $w = z^2$ for some complex number z.

Lin(S) denotes a subspace of $M_2(\mathbb{R})$ consisting of all real linear combinations of the matrices, and Lin(S) denotes a subspace of $M_2(\mathbb{C})$ consisting of the much larger set of all complex linear combinations of the matrices. These are very different sets; however, the basis will be the same in each.

To find a basis, we need to eliminate any 'vectors' which are linear combinations of the other vectors in the spanning set. Note that

$$\begin{pmatrix} 2 & 1 \\ 0 & 2 \end{pmatrix} + \begin{pmatrix} 1 & 2 \\ 0 & 1 \end{pmatrix} = \begin{pmatrix} 3 & 3 \\ 0 & 3 \end{pmatrix} \quad \text{and} \quad \begin{pmatrix} -1 & -1 \\ 0 & -1 \end{pmatrix} = -\frac{1}{3}\begin{pmatrix} 3 & 3 \\ 0 & 3 \end{pmatrix},$$

so that a basis (a linearly independent spanning set) of Lin(S) consists of the first two matrices; that is,

$$\left\{ \begin{pmatrix} 2 & 1 \\ 0 & 2 \end{pmatrix}, \begin{pmatrix} 1 & 2 \\ 0 & 1 \end{pmatrix} \right\}.$$

Hence Lin(S) is two-dimensional, considered either as a subspace of $M_2(\mathbb{R})$ or of $M_2(\mathbb{C})$.

Exercise 13.4 The eigenvalues of a complex matrix are calculated in the same way as for a real matrix. The characteristic equation is

$$|A - \lambda I| = \begin{vmatrix} 1-\lambda & i \\ 1+i & -1-\lambda \end{vmatrix} = \lambda^2 - 1 - i - i^2 = \lambda^2 - i = 0.$$

So the eigenvalues are the solutions of the equation $\lambda^2 = i$. To solve this, write $\lambda = re^{i\theta}$. Then one solution is obtained by setting

$$\lambda^2 = (re^{i\theta})^2 = r^2 e^{i2\theta} = i = e^{i\pi/2}.$$

Equating the moduli and arguments, we obtain $r = 1$ and $\theta = \pi/4$, so that

$$\lambda_1 = e^{i\pi/4} = \cos\left(\frac{\pi}{4}\right) + i\sin\left(\frac{\pi}{4}\right) = \frac{1}{\sqrt{2}} + i\frac{1}{\sqrt{2}}.$$

The other eigenvalue can be obtained by realising that $\lambda_2 = -\lambda_1$ (or using another expression for i, such as $\lambda^2 = (re^{i\theta})^2 = i = e^{5\pi/2}$). The other eigenvalue is $\lambda_2 = -\lambda_1 = -\frac{1}{\sqrt{2}} - i\frac{1}{\sqrt{2}}.$

Exercise 13.5 The eigenvalues of A are $2 \pm i$. If

$$P = \begin{pmatrix} 1+i & 1-i \\ 2 & 2 \end{pmatrix} \quad \text{and} \quad D = \begin{pmatrix} 2+i & 0 \\ 0 & 2-i \end{pmatrix},$$

then $P^{-1}AP = D$.

Exercise 13.6 (a) To show that $\mathbf{v}$ is an eigenvector, calculate $A\mathbf{v}$,

$$A\mathbf{v} = \begin{pmatrix} 5 & 5 & -5 \\ 3 & 3 & -5 \\ 4 & 0 & -2 \end{pmatrix} \begin{pmatrix} 0 \\ 1 \\ 1 \end{pmatrix} = \begin{pmatrix} 0 \\ -2 \\ -2 \end{pmatrix}.$$

Hence $\mathbf{v}$ is an eigenvector with corresponding eigenvalue $\lambda = -2$.
(b) Solve $(A - \lambda I)\mathbf{v} = \mathbf{0}$ for $\lambda = 4 + 2i$ to see if it is an eigenvalue and to find a corresponding eigenvector.

$$(A - \lambda I) = \begin{pmatrix} 1-2i & 5 & -5 \\ 3 & -1-2i & -5 \\ 4 & 0 & -6-2i \end{pmatrix}$$

$$\to \cdots \to \begin{pmatrix} 1 & 0 & -\frac{3}{2} - \frac{1}{2}i \\ 0 & 5 & -\frac{5}{2} - \frac{5}{2}i \\ 0 & 0 & 0 \end{pmatrix}.$$

Hence $\mathbf{x} = (3+i, \ 1+i, \ 2)^{\mathrm{T}}$ is an eigenvector corresponding to the eigenvalue $\lambda = 4 + 2i$.
(c) Since the matrix A is real, complex eigenvalues appear in conjugate pairs, so that $4 - 2i$ is also an eigenvalue with corresponding eigenvector $\bar{\mathbf{x}} = (3-i, \ 1-i, \ 2)^{\mathrm{T}}$. If

$$P = \begin{pmatrix} 3+i & 3-i & 0 \\ 1+i & 1-i & 1 \\ 2 & 2 & 1 \end{pmatrix} \quad \text{and} \quad D = \begin{pmatrix} 4+2i & 0 & 0 \\ 0 & 4-2i & 0 \\ 0 & 0 & -2 \end{pmatrix},$$

then $P^{-1}AP = D$.

A cannot be unitarily diagonalised. To show this, either show that A is not normal (show $A^*A \neq AA^*$), or show that it is not possible to form an orthonormal basis of eigenvectors of A. For example, $\langle \mathbf{x}, \mathbf{v} \rangle = \mathbf{v}^* \mathbf{x} = 3 + i \neq 0$.

Exercise 13.7 Calculate $\langle \mathbf{v}_1, \mathbf{v}_2 \rangle$ to show that the vectors are orthogonal. Then check that each vector has unit length.

Extend this to an orthonormal basis of $\mathbb{C}^3$ by taking any vector not in $\text{Lin}(S)$ and using Gram–Schmidt. For example, use $(1, 0, 0)^T$ and set

$$\mathbf{w} = \begin{pmatrix} 1 \\ 0 \\ 0 \end{pmatrix} - \left\langle \begin{pmatrix} 1 \\ 0 \\ 0 \end{pmatrix}, \begin{pmatrix} 1 \\ i \\ 0 \end{pmatrix} \right\rangle \frac{1}{2} \begin{pmatrix} 1 \\ i \\ 0 \end{pmatrix}$$

$$- \left\langle \begin{pmatrix} 1 \\ 0 \\ 0 \end{pmatrix}, \begin{pmatrix} 1+2i \\ 2-i \\ 2+2i \end{pmatrix} \right\rangle \frac{1}{18} \begin{pmatrix} 1+2i \\ 2-i \\ 2+2i \end{pmatrix}$$

to obtain

$$\mathbf{w} = \frac{1}{9} \begin{pmatrix} 2 \\ -2i \\ -3+i \end{pmatrix}.$$

Then, if

$$\mathbf{v}_3 = \frac{1}{3\sqrt{2}} \begin{pmatrix} 2 \\ -2i \\ -3+i \end{pmatrix},$$

the vectors $\{\mathbf{v}_1, \mathbf{v}_2, \mathbf{v}_3\}$ are an orthonormal basis of $\mathbb{C}^3$. (You should check that $\mathbf{v}_3$ is orthogonal to the other two vectors.)

Exercise 13.8 The answers to these questions are contained in the text.

Exercise 13.9 Let λ be an eigenvalue of A and let $\mathbf{x}$ be an eigenvector corresponding to λ, so that $A\mathbf{x} = \lambda\mathbf{x}$. As A is unitary, $AA^* = A^*A = I$, so that $\mathbf{x}^* A^* A \mathbf{x} = \mathbf{x}^* I \mathbf{x} = \mathbf{x}^* \mathbf{x}$. Also,

$$\mathbf{x}^* A^* A \mathbf{x} = (A\mathbf{x})^*(A\mathbf{x}) = (\lambda\mathbf{x})^*(\lambda\mathbf{x}) = \lambda^* \lambda \mathbf{x}^* \mathbf{x} = |\lambda|^2 \mathbf{x}^* \mathbf{x}.$$

Equating the two expressions for $\mathbf{x}^* A^* A \mathbf{x}$, we have $\mathbf{x}^* \mathbf{x} = |\lambda|^2 \mathbf{x}^* \mathbf{x}$, from which we obtain,

$$(1 - |\lambda|^2)\mathbf{x}^* \mathbf{x} = 0.$$

As $\mathbf{x}$ is an eigenvector, $\mathbf{x}^* \mathbf{x} = \|\mathbf{x}\|^2 \neq 0$, so we can conclude that $|\lambda|^2 = 1$ and so $|\lambda| = 1$ (since the modulus of a complex number is a real non-negative number).

Exercise 13.10 The eigenvalues of the real symmetric matrix A are $2, -2, 16$. You diagonalised this matrix in Exercise 11.1. An orthogonal matrix P and a diagonal matrix D which orthogonally diagonalise A are

$$P = \frac{1}{\sqrt{2}} \begin{pmatrix} 0 & -1 & 1 \\ \sqrt{2} & 0 & 0 \\ 0 & 1 & 1 \end{pmatrix}, \qquad D = \begin{pmatrix} 2 & 0 & 0 \\ 0 & -2 & 0 \\ 0 & 0 & 16 \end{pmatrix},$$

so that $P^{\mathrm{T}} A P = D$.

Then $A = 2E_1 - 2E_2 + 16E_3$, where

$$E_1 = \begin{pmatrix} 0 & 0 & 0 \\ 0 & 1 & 0 \\ 0 & 0 & 0 \end{pmatrix}, \qquad E_2 = \frac{1}{2} \begin{pmatrix} 1 & 0 & -1 \\ 0 & 0 & 0 \\ -1 & 0 & 1 \end{pmatrix},$$

$$E_3 = \frac{1}{2} \begin{pmatrix} 1 & 0 & 1 \\ 0 & 0 & 0 \\ 1 & 0 & 1 \end{pmatrix}.$$

A quick calculation should convince you that these matrices have the required properties.

Since the matrices E_1, E_2, E_3 have these properties, then for any real numbers $\alpha_1, \alpha_2, \alpha_3$, and any positive integer n, we can conclude that $(\alpha_1 E_1 + \alpha_2 E_2 + \alpha_3 E_3)^n = \alpha_1^n E_1 + \alpha_2^n E_2 + \alpha_3^n E_3$. In particular, if $B = \alpha_1 E_1 + \alpha_2 E_2 + \alpha_3 E_3$ and $B^3 = A$, then to find a matrix B as required, we use

$$B = 2^{1/3} E_1 + (-2)^{1/3} E_2 + 16^{1/3} E_3,$$

which (after simplification) gives us

$$B = 2^{-2/3} \begin{pmatrix} 1 & 0 & 3 \\ 0 & 2 & 0 \\ 3 & 0 & 1 \end{pmatrix}.$$

You should check all this.

Exercise 13.11 (a) The proof that $A^* A$ is Hermitian is straightforward: $(A^* A)^* = A^* (A^*)^* = A^* A$, so $A^* A$ is Hermitian. Every Hermitian matrix is normal, so $A^* A$ is normal. (This can also be proved directly.)

(b) We have

$$\mathbf{v}^* (A^* A) \mathbf{v} = (A\mathbf{v})^* (A\mathbf{v}) = \langle A\mathbf{v}, A\mathbf{v} \rangle \geq 0$$

and $\mathbf{v}^* (A^* A) \mathbf{v} = 0$ only if $A\mathbf{v} = \mathbf{0}$ by properties of the inner product. Since A has full column rank, $A\mathbf{v} = \mathbf{0}$ has only the trivial solution

$\mathbf{v} = 0$. So $\mathbf{v}^*(A^*A)\mathbf{v} > 0$ for all $\mathbf{v} \in \mathbb{C}^n$, $\mathbf{v} \neq \mathbf{0}$, which shows that A^*A is positive definite.

Similarly, if $A^*A\mathbf{v} = \mathbf{0}$, then multiplying both sides of this equality on the left by $\mathbf{v}^*$ we can conclude that $A\mathbf{v} = \mathbf{0}$, and hence that $\mathbf{v} = \mathbf{0}$. That is, $A^*A\mathbf{v} = \mathbf{0}$ has only the trivial solution $\mathbf{v} = \mathbf{0}$, which implies that A^*A is invertible.

Exercise 13.12 The matrix B is not Hermitian, but you can easily check that it is normal and can therefore be unitarily diagonalised. To diagonalise, we see immediately from the matrix B that $\lambda = 1$ is an eigenvalue with corresponding eigenvector $\mathbf{x}_1 = (0,\ 0,\ 1)^\mathrm{T}$. Solving the characteristic equation, you should find that the other two eigenvalues are $\lambda_2 = 1 + i$ and $\lambda_3 = 1 - i$ with corresponding eigenvectors $\mathbf{x}_2 = (1,\ 1,\ 0)^\mathrm{T}$ and $\mathbf{x}_3 = (1,\ -1,\ 0)^\mathrm{T}$, respectively. Note that these eigenvectors are mutually orthogonal. Normalising, we have,

$$P = \frac{1}{\sqrt{2}} \begin{pmatrix} 1 & 1 & 0 \\ 1 & -1 & 0 \\ 0 & 0 & \sqrt{2} \end{pmatrix} = (\mathbf{u}_1 \mathbf{u}_2 \mathbf{u}_3)$$

and

$$D = \begin{pmatrix} 1+i & 0 & 0 \\ 0 & 1-i & 0 \\ 0 & 0 & 1 \end{pmatrix}.$$

Then $B = (1 + i)E_1 + (1 - i)E_2 + E_3$, where $E_i = \mathbf{u}_i \mathbf{u}_i^*$, $i = 1, 2, 3$. We find

$$E_1 = \frac{1}{2} \begin{pmatrix} 1 & 1 & 0 \\ 1 & 1 & 0 \\ 0 & 0 & 0 \end{pmatrix}, \qquad E_2 = \frac{1}{2} \begin{pmatrix} 1 & -1 & 0 \\ -1 & 1 & 0 \\ 0 & 0 & 0 \end{pmatrix},$$

$$E_3 = \begin{pmatrix} 0 & 0 & 0 \\ 0 & 0 & 0 \\ 0 & 0 & 1 \end{pmatrix}.$$

Therefore,

$$B = \frac{1+i}{2} \begin{pmatrix} 1 & 1 & 0 \\ 1 & 1 & 0 \\ 0 & 0 & 0 \end{pmatrix} + \frac{1-i}{2} \begin{pmatrix} 1 & -1 & 0 \\ -1 & 1 & 0 \\ 0 & 0 & 0 \end{pmatrix} + \begin{pmatrix} 0 & 0 & 0 \\ 0 & 0 & 0 \\ 0 & 0 & 1 \end{pmatrix}.$$

Index

Printed in the United States
by Baker & Taylor Publisher Services